Membrane-Based Separations in Metallurgy

Membrane-Based Separations in Metallurgy

Principles and Applications

Edited by

Lan Ying Jiang

School of Metallurgy and Environment
Central South University, Changsha, Hunan Province, China

Na Li

School of Chemical Engineering and Technology
Xi'an Jiaotong University, Xi'an, Shanxi Province, China

ELSEVIER

Elsevier
Radarweg 29, PO Box 211, 1000 AE Amsterdam, Netherlands
The Boulevard, Langford Lane, Kidlington, Oxford OX5 1GB, United Kingdom
50 Hampshire Street, 5th Floor, Cambridge, MA 02139, United States

Library of Congress Cataloging-in-Publication Data
A catalog record for this book is available from the Library of Congress

British Library Cataloguing-in-Publication Data
A catalogue record for this book is available from the British Library

ISBN: 978-0-12-803410-1

Publisher: John Fedor
Acquisition Editor: Kostas Marinakis
Editorial Project Manager: Sarah Jane Watson
Production Project Manager: Vijayaraj Purushothaman
Designer: Victoria Pearson

Typeset by Thomson Digital

Contents

List of Contributors

Yi-Ming Cao Dalian Institute of Chemical Physics, Chinese Academy of Science, Dalian, Liaoning Province, China

Yue-Hui He School of Powder Metallurgy, Central South University, Changsha, Hunan Province, China

Lan Ying Jiang School of Metallurgy and Environment, Central South University, Changsha, Hunan Province, China

Xiao-Long Lu State Key Laboratory of Separation Membranes and Membrane Processes, Tianjin Polytechnic University, Tianjin, China

Chong-Jian Tang School of Metallurgy and Environment, Central South University, Changsha, Hunan Province, China

Hui Wang Beijing Liquid Cube Technology Ltd., China

Liang Wu School of Chemistry and Materials Science, University of Science and Technology of China, Hefei, Anhui Province, China

Lian-Sheng Xiao School of Metallurgy and Environment, Central South University, Changsha, Hunan Province, China

Wei-Hong Xing National Engineering Research Center for Special Separation Membranes, Nanjing Tech University, Nanjing, Jiangsu Province, China

Tong-Wen Xu School of Chemistry and Materials Science, University of Science and Technology of China, Hefei, Anhui Province, China

Zhen-Liang Xu State Key Laboratory of Chemical Engineering, Membrane Science and Engineering R&D Lab, Chemical Engineering Research Center, East China University of Science and Technology, Shanghai, China

Foreword

Nowadays metallic materials hold very critical positions in social economy worldwide. Besides the conventional industries, metals and metal alloys are applied in a wide range of high-tech industries, such as high-speed transport, new energy, advanced manufacturing, and health care. The relationship between man and man, man and nature, and man and space beyond the earth are being modified continuously with the development of these fields.

Absolutely indispensable at present stage and in the foreseeable future to the production of metals and metal alloys is metallurgical industry. With great advancement in metallurgical science and engineering in the past decades, however, some problems persist. The most serious ones, particularly for developing countries, are the dissipation of natural resources and pollution to environment. They are among the elements that doom the global sustainability a Utopian and must be addressed seriously and pragmatically.

In response, national and international policies and regulations are established or amended to facilitate the economic and societal development in a manner without trading off the welfare of future generations. Meanwhile science, technology, and process engineering undoubtedly are the major supports for actualizing the targets at the level of concrete operation and practice.

Due to the effort of scientists and engineers worldwide, membrane-based separation has been playing an important role in constructing "green metallurgy." For example, the ion-exchange membrane is utilized in uranium metallurgy, which greatly simplifies the process flow. Iron and steel industry is the masterpiece in making use of membrane for water treatment; a recommendable example is the desalination of integrated wastewater. Application of metal membranes has successfully helped to purify and recycle gaseous emissions. Briefly, the technology has already permeated into the domain of all the metal categories, and new applications are being continually identified, even for synthetic metals. The major modes for membrane to be involved in metallurgical industry are summarized as follows: replacing or supplementing original separation technologies to improve overall process efficiency; helping achieve comprehensive utilization of resources and reduce waste.

Conventional membrane separation, such as microfiltration, ultrafiltration, nanofiltration, reverse osmosis, electrodialysis draws support from the physiochemical properties of the solid membrane "pore" for discrimination among particles. "New concept membranes," for example, membrane contactors, membrane reactors, coupling membrane media with biological, chemical, or other processes in specific manners, shows great promise for tasks otherwise unrealistic. To combine membrane-based separation efficiently with metallurgical process, two aspects are very critical. For most of the membrane categories, first, high-separation properties with sound stabilities under in-service conditions must be ensured. Additionally, a series of phenomena occurring during separation, for instance, fouling, wetting, polarization, which impact negatively the mass and heat transfer efficiency, should be controlled with great care.

EU countries, USA, and Japan were the leading forces in the early years of developing and utilizing membrane technologies. The fast development of China and many other countries in past one or two decades has delivered many important achievements, based on incremental or totally independent innovation. Ceramic membranes and ion-exchange membranes, and their applications in China are excellent models.

In pace with the social development, metallurgical industry per se is experiencing continuous progress and some new situations are appearing, like the upgraded scale in metal production and demand, the exploitation of unconventional resources, and the adoption of novel extractive technologies or processes. The potential problems associated with these situations regarding resource conservation and environmental protection will require further attention of scientists from both academia and industry.

Considering all these facts, a book is needed to comprehensively and systematically summarize the contribution of membrane technologies in changing the old "smoke stack" image of metallurgy in terms of scientific and technological principals. More critically, the pros and cons of the membrane technologies should be evaluated and compared, so as to provide clear guidance for our future effort.

Appreciations go to Dr. Lan Ying Jiang, Dr. Na Li, and contributors of each chapter for their great effort in writing this book. The information displayed will present good references to those who are potential users or innovators of membrane technology and metallurgy. And I believe its publication and distribution will greatly facilitate the development of applying membranes in metallurgical industry.

Ding-Fan Qiu

PART 1

Introduction

CHAPTER 1

Metallurgy: Importance, Processes, and Development Status

L.Y. Jiang

School of Metallurgy and Environment, Central South University, Changsha, Hunan Province, China

1.1 Introduction

Broadly speaking, metallurgy covers the entire chain concerning metallic materials production, from mineral exploration, mining, concentration, smelting, refining, to material engineering. In a narrow sense, metallurgy is about extraction of metals from ores, concentrates, or other resources.

Among 98 naturally occurring elements on earth, 18 are nonmetals, 7 are metalloids, and the remaining are all metals. Over the past 10,000 years, metals and alloys have been playing the most important role in the technological progress of human kind [1]. In China, the earliest sites that have yielded metal objects date back to the late fourth and third millennia before the Christian Era (BCE) [2]. Quite early metal-using communities in China are identified in Gansu, Xinjiang, Shandong, Liaoning, Inner Mongolia, and the Central Plain. Table 1.1 displays the estimated periods of earliest metal-relevant activities in different regions of the world [2]. Copper was generally regarded as the metal that has the longest history of being used by mankind. The use of copper in ancient China goes back to 3000 BCE. The influence of metals is so profound throughout history that some of them have become synonymous with different eras of human development (eg, the Copper Age, the Bronze Age, the Iron Age, the Nuclear Age and the Titanium Age).

The division into ferrous and nonferrous metals is the foremost classification. Ferrous metals only encompass iron, chromium, manganese, and their alloys. The latter two have been intensively and extensively used for fabricating steel alloys (eg, manganese steel and chromium steel by using wolframite, pyrolusite, and chromite), hence categorized as ferrous metals. The resources for ferrous metals are abundant. The notable features of ferrous metals are the excellent comprehensive mechanical properties, stable quality, and low price.

A nonferrous metal is a metal, including alloys, that does not contain iron (eg, ferrite) in appreciable amounts. Generally being more expensive than ferrous metals, nonferrous metals are used mainly for some of their desirable properties, such as low density (eg, aluminum), high conductivity (eg, copper), nonmagnetic identity, and resistance to corrosion (eg, zinc).

Membrane-Based Separations in Metallurgy
http://dx.doi.org/10.1016/B978-0-12-803410-1.00001-3

Table 1.1: The periods* of earliest metal-using activities at different regions worldwide

Archaeological Age	The Chalcolithic Age		The Forepart of Bronze	Bronzes of the Middle and Late Bronzes			
Metal Age	**Native Copper**	**Smelting Copper**	**"The Forepart of Bronze"**		**Early Iron-mongeny**	**Roman Ironmongeny**	**Later Stage Ironmongeny**
Anatolia		5000	3000	2000	1000	30	
Egypt							
Palestine		3500	2900	1900	1000	100	
Mesopotamia	7000	4000	3500	2800	1200	AD 100	
Aegean Sea		3300	2500	2000	1000	150	
Italy		3000	2000	1200	800	250	
Iberia		4500	1500	1000	700	200	
The southeast Europe	5000	2200	3000	1500	700	AD 100	
The northwest Europe		2200	1800	1200	500	50	AD 400
Eastern Europe and Russia			1500	700	400	—	
the Far East			1500	1300	700	—	
West African					500		
East Africa					200	AD 1	
South America		AD 700	AD 1000	AD 1400	—	—	
North America	4000						

*BC for all the periods except otherwise stated.

Table 1.2 summarizes the classification for all the nonferrous metals [3]. The table has been updated by incorporating the synthetic metals. Important basic nonferrous metals that are in great need in modern industries include aluminum, copper, lead, nickel, tin, and zinc, and their alloys such as brass. Precious or noble metals include gold, silver, and metals of platinum group. The concept of "rare metal" is a generic term for metals that have very small total reserve, low level of abundance (generally lower than 100 ppm), and scattered occurrence. In light of this point, rare earth metal, noble metal, and radioactive metal are all essentially rare metal.

In the early 1970s, the post-World War II prosperity in world metal consumption came to a halt. However, a renewed boom is now emerging in the metals sector, with metallic materials holding very critical positions in social economy worldwide. The applications of metals have been permeating into almost all the aspects of human society [4–6]. As for ferrous metals [4,5], the most important areas include high-rise buildings, deep underground and marine facilities, large-span and heavy-load bridges, light-weight energy-saving vehicles, high-speed watercraft, oil extraction and long-distance transmission pipeline, large-scale storage containers, high-speed railway, energy facilities, construction machinery, precision instrument, aerospace industry, etc. The applications of nonferrous metals are also extensive [6]. Power generation, power grid, and construction require huge amount of copper and aluminum; the

Table 1.2: Classification of nonferrous metals

Classification		Metal Elements	Descriptions
Light nonferrous metals		Aluminum (Al); magnesium (Mg); sodium (Na); potassium (K); calcium (Ca); strontium (Sr); barium (Ba)	Nonferrous metals with density lower than 4.5 g cm^{-3}
Heavy nonferrous metals		Copper (Cu); lead (Pb); zinc (Zn); nickel (Ni); cobalt (Co); tin (Sn); cadmium (Cd); bismuth (Bi); antimony (Sb); mercury (Hg)	Nonferrous metals with density higher than 4.5 g cm^{-3}
Rare metals	Light metal	Lithium (Li); rubidium (Rb); cesium (Cs); beryllium (Be)	They have very low densities (0.53–1.87 g cm^{-3}) and very active chemically.
	Refractory metal	Titanium (Ti); zirconium (Zr); hafnium (Hf); vanadium (V); niobium (Nb); tantalum (Ta); molybdenum (Mo); tungsten (W)	They commonly have high melting point, generally higher than 1650°C, strong corrosion resistance, and can form very hard and refractory compounds with some nonmetals.
	Scattered metal	Gallium (Ga); indium (In); thallium (Tl); germanium (Ge); selenium (Se); tellurium (Te); rhenium (Re)	Their distribution in the lithosphere is highly scattered, and they generally have no independent minerals.
	Rare earth metals	Lanthanum (La), cerium (Ce), praseodymium (Pr), neodymium (Nd), promitnium (Pm), samarium (Sm), europium (Eu), gadolinium (Gd), terbium (Tb), dysprosium (Dy), holmium (Ho), erbium (Er), denebium (Tm), ytterbium (Yb), lutetium (Lu), scandium (Sc), yttrium (Yt)	Their nuclear structures and physicochemical properties are quite similar. They are concomitant inside the ores.
	Radioactive metals	Naturally occuring elements: Francium (Fr); polonium (Po), radium (Ra), actinium (Ac), thorium (Th), protacnium (Pa), uranium (U); neptunium (Np); plutonium (Pu). Synthetic elements: technetium (Tc), neptunium (Np), plutonium (Pu), americium (Am), curkelium (Cm), berkelium (Bk), californium (Cf), einsteinium (Es), fermim (Fm), mendelevium (Md), nobelium (No), lawrencium (Lw), rutherfordium (Rf), dubnium (Db), seaborgium (Sg), bohriium (Bh), hassium (Hs), meitnerium (Mt), damstadium (Ds), roentgenium (Rg), copernicium (Cn), ununtium (Uut), flerovium (Fl), ununpentium (Uup), livermorium (Lv), ununseptium (Uus), ununoctium (Uuo)	/
Noble metals		Gold (Au), silver (Ag), platinum (Pt), iridium (Ir), osmium (Os), ruthenium (Ru), palladium (Pd), rhodium (Rh)	They are difficult to extract and have excellent chemical stability.

Table 1.3: The domestic and global demands on ferrous and nonferrous metals

Metal Category		Copper		Aluminum		Nonferrous metals (total)		Ferrous metal (crude steel, total)	
	Year	2003	2011	2003	2011	2003	2011	2009	2012
Global	Production (10^4 tons)	1,500	1,900	3,000	4,350	6,500	9,084	122,000	159,400
	Demand (10^4 tons)	1,510	1,850	2,996	4,239	6,310.6	8,542	122,300	152,700
	Production (10^4 tons)	178	520	650	1,550	1,200	3,250	56,800	73,400
	Demand (10^4 tons)	300	770	588	1,806	1,310.2	3,479.2	57,100	66,200

demand of aluminum in transportation equipment manufacturing is always high; machinery and electronics need copper, aluminum, nickel, tin, zinc, tungsten, tantalum, etc.; titanium, chromium, rare earth metals, and other strategic metals are the basic raw materials for aerospace industry and military hardware manufacturing. Put simply, a modern world without metal is inconceivable. Table 1.3 displays the demands on metal products in China and at abroad [7,8]. Obviously, the demands take on an increasing trend in the past decade for both ferrous and nonferrous metals. Taking copper as an example, its global demand grows from 15 million tons in 2003 to 19 million tons in 2011.

1.2 Extractive Metallurgy

As aforementioned, metallurgy sensu lato actually covers a quite extensive field. This part is solely contributed to extractive metallurgy. The raw materials or the ores contain not only the target metals, but also other metals and gangues. Extractive metallurgical process consists of the removal and elimination of the gangue, decomposition of the mineral and collecting and sometimes refining of the metal. Due to the varieties in nature and composition of the ores, energy availability, and sometimes consideration in environmental protection, the methods used in extractive metallurgy are diverse. In terms of the key conversion—essentially of a chemical nature—involved for metal extraction from minerals, pyro-, hydro-, electro-, and bio-metallurgy are generally classified. In effect, none of the technologies may in itself suffice for the treatment of a particular material, and in most cases a combination of two or more approaches is adopted.

1.2.1 Pyrometallurgy

Pyrometallurgy is a technology conducted at high temperature [9,10]. The minerals in the ore or concentrate will be partially or completely converted into another form of chemical compound or elementary substances by a series of high-temperature physico-chemical changes, and enriched in gas, liquid, or solid products. After being formed into big particles by sintering, the raw material or the concentrate is subject to oxidizing roasting. The subsequent procedure is smelting, generating slag consisting of gangue, flux and fuel ash, and matte (nonferrous sulfide and iron sulfide eutectic) or a liquid metal containing a small quantity of impurities. For smelting three technologies have been developed; namely smelting reduction, oxidation blowing, and matte smelting. With these operations, the separation of target metals and impurities is achieved.

The thermal energy for the pyrometallurgical process is in most cases supplied by fuel burning, and sometimes by exothermic reaction inherent of the chemical reactions. For the latter, an example is oxidizing roasting and smelting of sulfide ore. In addition, thermal reduction of metal is a self-heating process. Refining is used to remove the small amount of impurities. For requirements on highly purified metals, zone melting is used.

Pyrometallurgy has been intensively used for production of copper, zinc, nickel, etc. The technologies for these metals are relatively mature. The flow diagram for copper pyrometallurgy is shown in Fig. 1.1. The principal procedures include copper matte generation by smelting, copper blister formation by blowing, fire refining, and electrolytic refining. Blister copper oxidation is refined in reverberatory furnace, followed by cast of copper anode for electrolytic refining.

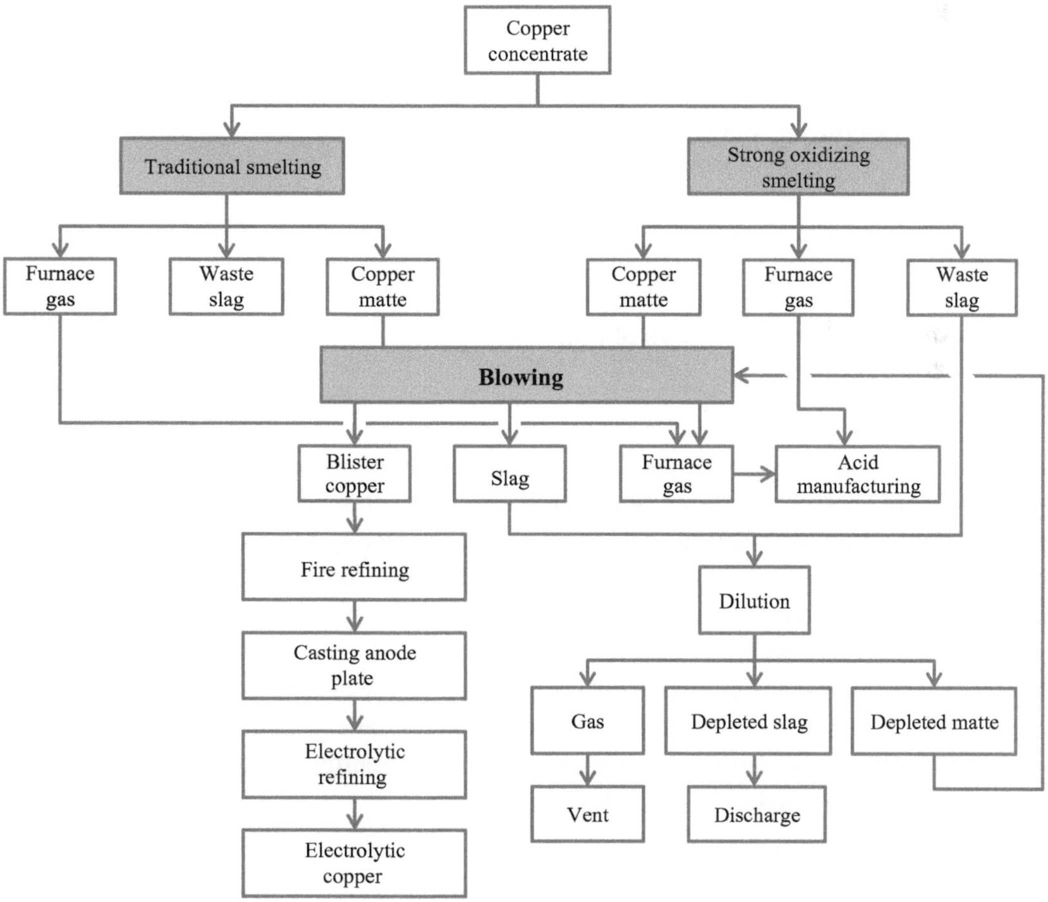

Figure 1.1: The flow chart of copper pyrometallurgy.

Special metallurgy is a branch of pyrometallurgy mainly pointing at the needs from high-tech, defense areas, etc., such as high-quality special steel, refractory alloys (eg, tungsten, molybdenum, niobium, tantalum), active metals (eg, titanium, zirconium), highly purified metals (eg, zero-inclusion steel). The technologies belonging to this branch comprise chiefly plasma metallurgy, vacuum induction degassing and pouring, cold crucible levitation melting, electron-beam smelting, electro-slag remelting [10,11].

Plasma metallurgy: Plasma metallurgy utilizes heat generated by plasma flow as the thermal resource [12]. Several gases (eg, argon (Ar), hydrogen (H_2)) and their mixtures can be used as media in plasma gun for achieving different metallurgical targets. For example, inert gas plasma is used for smelting of metals with high melting point and active metal, as well as purification of some metals and metal alloys. Application of H_2 or H_2 containing gas is feasible for recovering several metals (eg, iron, aluminum, cobalt, niobium, tantalum, zirconium, tungsten) from their oxides. In addition, plasma flow is capable of decomposing nickel carbonyl ($Ni(CO)_4$), generating ultrafine Ni powders.

The potential advantages of using plasma arc as a heat source in metallurgical process are summarized as follows [11]: high temperature (5000–30,000K); rapid temperature rise due to concentrated energy and fast plasma flow; favorable conditions for gas–liquid–solid reaction kinetics; stable electric arc and small noise.

Since its advent in 1962, plasma metallurgy has been regarded as a very promising technology. Applications of plasma jet flow in secondary remelting of steel and producing Cr steel have been industrialized. The hindrance for its penetration into other fields at industrial level is mainly due to problems associated with the life span of high-power water-cooling plasma gun and plasma thermal efficiency.

1.2.2 Hydrometallurgy

Traditional hydrometallurgy relies on liquid phase chemistry to obtain metals or their compounds from ores, concentrates, and recycled or residual materials [13]. The chemical reactions refer to mainly oxidation, reduction, neutralization, hydrolysis, and complexation. The temperatures employed in the process range from lower than 100°C to around 300°C. They are much lower than those used in pyrometallurgy. The primary procedures for hydrometallurgical process are leaching, separation, enrichment, and extraction [14]. The technologies used across the entire extractive process include dissolution using acid, base, cyanide and organic solvent, chemical precipitation, membrane separation, crystallization, ion-exchange, solvent extraction, stripping, etc. Hydrometallurgy is dominantly used in the production of zinc, alumina (Al_2O_3), and some rare metals.

Fig. 1.2 illustrates the flow diagram of a traditional zinc hydrometallurgy process. The key steps are the neutral leaching and acid leaching that transfer zinc from solid concentrate and slag, respectively, into aqueous solution. Thereafter, electrowinning (EW) is used to obtain

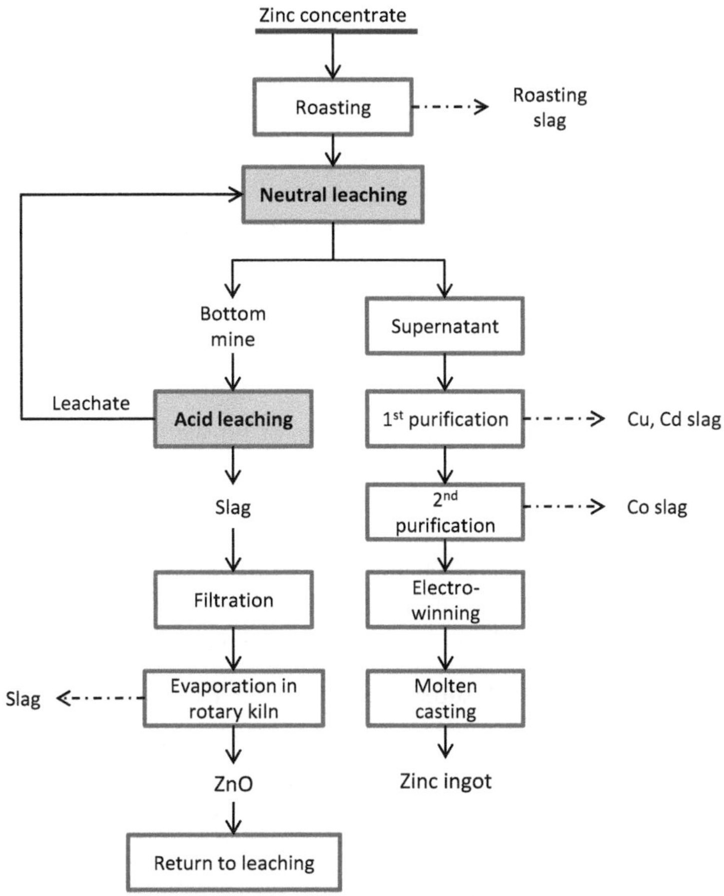

Figure 1.2: The flow chart of traditional zinc hydrometallurgy.

elementary zinc which is then cast into ingot. Meanwhile, the Al_2O_3 recovered from bottom mine by acid leaching is recycled back to neutral leaching.

Hydrometallurgy is most suitable for treating low-grade ores, but with relatively high cost [15]. With the development of science and technology, more and more innovations are developed and applied for improving the efficiency of hydrometallurgy [16–18]. Leaching and extraction are two principal steps. As for leaching, in addition to conventional acid/base approaches, researches have revealed high pressure leaching, fluidized leaching, and in situ leaching. The methods for extraction are more abundant; microwave extraction, ultrasonic extraction, predispersed extraction, magnetic field-assisted solvent extraction, liquid membrane extraction, nonequilibrium solvent extraction, etc. have been proposed based on the conventional solvent extraction. Some of the technologies are introduced as follows.

High pressure leaching: In leaching process involving gas (eg, oxygen (O_2)) as a chemically active reagent, raising the gaseous pressure is conducive to the shift of reaction equilibrium toward product side; hence the leaching efficiency or yield is improved. In addition, it brings about abatement in operation cost and phasedown of environmental pollution [17]. High pressure leaching has been penetrating in ammonia (NH_3) leaching, alkali leaching, and acid leaching. The first pressurized heavy metal refinery plant was installed in China in 1993.

Fluidized leaching: Due to the effect of flowing liquid, the mineral or concentrates particles are suspended in the leaching progress. This method exhibits several superiorities over the conventional leaching: complete replacement of mechanical stirring by hydraulic operation; low liquid–solid ratio; highly concentrated leachate even with low-grade ores; continuous operation; formation of concentration gradient inside single unit; compact equipment and small footprint.

In situ leaching: It is also known as chemical mining, or no-well mining. The conventional excavation is abandoned; the leachant is applied in situ to dissolve the valuable metal by selective chemical reaction.

Mechanical activation leaching: Under the action of mechanical force, the mineral crystals generate various internal or lattice defects, shifting the structure to a state characterized of higher energy level. Consequently, the chemical reactivity is enhanced.

Supercritical fluid solvent extraction: A supercritical fluid (SCF) is any fluid at a temperature and pressure above its critical point, where distinct liquid and gas phases do not exist. SCF synergizes the characteristics of low gas phase, such as viscosity and high diffusivity, and liquid phase such as high density and good dissolving ability. The solubility of various components in the SCF can be adjusted by varying temperature and pressure. Employment of SCF in metallurgy has brought about following benefits: it consumes less energy relative to traditional solvent extraction; more efficient management of element distribution factor can be achieved by optimizing temperature and pressure, leading to higher separation factor between different metals. SCFs now commonly used are CO_2, NH_3, ethylene (C_2H_4), propylene, water, etc.

Reverse micelles solvent extraction (RMSE): In nature, RMSE is liquid–liquid solvent extraction. The surfactant added into the liquid system develops by self-assembly into reversed micelles dispersed inside the organic phase. They have internal hydrophilic microenvironment. During extraction, bio-molecules that have low solubility in organic phase can exist inside hydrophilic interior of reversed micelles. This characteristic helps prevent de-naturation of bio-molecules that happens in conventional extraction. Research is initiated to transfer the approach to metal extraction.

Electrophoresis extraction: It is a separation technology coupling electrophoresis and solvent extraction. On one hand, it enhances mass transfer and extraction yield by utilizing electrical potential; on the other hand, the liquid–liquid (ie, solvent-water) interface presents resistance to prevent convective diffusion. Electrophoresis extraction shows good performance in treating substances that are difficult to tackle with by conventional solvent extraction.

1.2.3 Electrometallurgy

In electric metallurgy, one of the key transformations is electrodeposition or electroplating of metal at the cathode via metal ion reduction under the action of electric current [10,19]. This mechanism is used in electrolytic metallurgy and electrolytic refining. Depending on the state of electrolyte, electrolytic metallurgy can be divided into solution electrolysis and molten salt electrolysis. For electro-refining, the metal anode with a certain level of impurity dissolves into solution to form metal ions. The ions subsequently deposit on the cathode. In the process, the inclusion from the anode is left in the solution. In molten salt electrolysis, electricity not only participates in the redox reaction, but also provides thermal energy for melting the metal salt.

Electrometallurgy was first introduced in the extraction of Cu from low-grade Cu ores. The extraction rate is up to 80–90% with a relatively low production cost. At present, it has been used in the extraction of cadmium, antimony, cobalt, chromium, iron, gallium, manganese, silver, and so on. The technological process of kryocide–Al_2O_3 molten salt electrolytic metallurgy for producing aluminum is displayed in Fig. 1.3. Al_2O_3 after pretreatment is heated inside the electrolytic cell to form molten salt. Subject to the strong electric current, it releases elementary Al in its molten state that is subsequently cast into ingot.

Slurry electrolysis: It is a technology that integrates mineral leaching, partial purification of leachate, and EW in one single system [20]. The novelty of the design is that the anode reaction in EW, which is previously ignored, is now harnessed for mineral oxidization leaching. The metal ions generated then transport through the ion-exchange membrane to enter the cathode chamber. Thereafter, the ions precipitate on the cathode via reduction reaction. At the same time, the mineral slurry in anode is delivered to solid/liquid separation, with the liquid product being refluxed to electrolytic tank and the slag for other treatment. This design enables the simplification of flow diagram and extremely high recovery rate. It has been industrialized for bismuth metallurgy.

1.2.4 Biometallurgy

Biometallurgy, also known as biological metallurgy or bioleaching [16,21], is a technique that combines the biological function of bacteria with conventional hydrometallurgy. The microorganisms or bacteria used are generally chemoautotrophic. They feed on minerals, obtaining energy for their metabolism by bio-oxidation. Simultaneously, the target component in the mineral becomes soluble in water. Subsequently the metal can be extracted directly from the aqueous solution. It is a relatively new technology, facing the challenges of slow reaction rate, poor adaptability of the bacteria to the application environment, and so on. Genetic engineering has been suggested by many scientists as the principal solution.

Currently, investigation of biometallurgy application has been conducted on extraction of gallium, selenium, germanium, scandium, fine manganese, etc. Compared with conventional

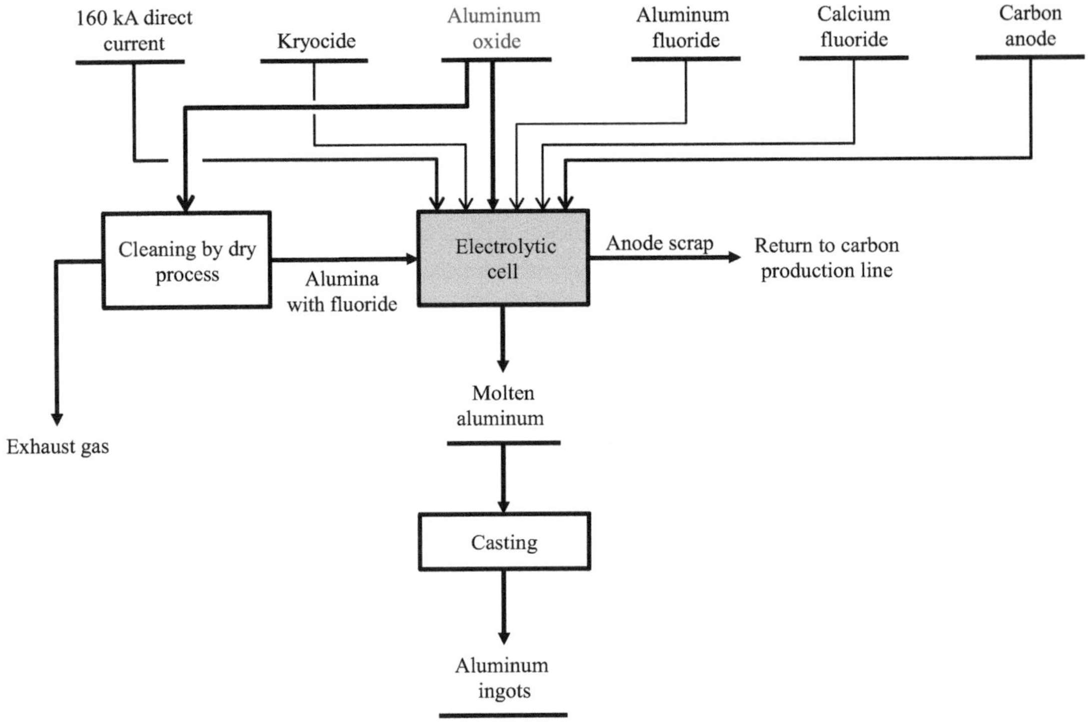

Figure 1.3: The flow chart of aluminum electrometallurgy.

oxidation, bio-oxidation is superior in low cost and no pollution. It has been identified as a promising alternative for exploiting low-grade and recalcitrant sulfide ores. The potential applications that may be industrialized in the foreseeable future include the following:

1. tank leaching for some metal sulfide concentrates from mineral flotation;
2. oxidation pretreatment for heap leaching of refractory gold ore;
3. bioleaching of oxide ore;
4. metal extraction from aqueous solutions.

1.3 Current Status of Metallurgical Industry

1.3.1 Progress in China

Due to infrastructure construction in developing western region and grand development of real estate, the demand on steel and in response the output has taken on a rapid increasing trend in past 10 years in China. In 2006, the steel output in China exceeded 0.42 billion tons; in 2010, the crude steel production reached 0.62 billion tons, contributing to 44% of world's total. In 2013, the steel output in China was 0.78 billion tons, accounting for more than 50% of the world's total. So to speak, China has already established its position in world iron and steel industry. Nevertheless, there exist several challenges for China to improve its competitiveness and maintain sustainability of this specific sector in the future [22]. Briefly speaking, they are imbalanced development, low level of internationalization, and constraint by resource and environment.

The first problem is the low level of technical intervention in the products. There exists a large gap between China and foreign countries in high-quality steel production. The international frontier has been moved to the development of advanced properties (ie, high cleanliness, high homogeneity, ultrafine microstructure, and high precision), diversification in variety and specification, and advanced manufacturing technologies for advanced materials (ie, composite materials, functionalized materials, environmentally benign material) [23]. In China, however, some outmoded products are still manufactured on a large scale whereas the advanced products, such as special steel, particularly high-end and critical categories, share a low portion in domestic market [23,24]. The aforementioned comparison also implies another important aspect of the first problem, the unreasonable product breed structure. Therefore, the situation of heavy dependence on importing high performance materials persists [25].

The further growth in the scale for the industry has been constrained by the following factors; namely, lack of high quality feedstock, water resource scarcity, energy intensiveness, and environmental pollution.

The ores of iron, manganese, and chromium in China are characterized of being low grade and complex, leading to difficulty in mining and mineral dressing. The iron ore reserve identified in China is 57.6 billion tons, but the economic base reserve is only 16.7 billion tons. The import rate was 71% for the iron ores in 2009. In effect, there exists a substantial part of low grade complex symbiotic ore (eg, high phosphorus iron ore, vanadium titanium iron ore [26]. If utilization of this part could be actualized by using advanced technologies, the raw material supply will be doubled for iron and steel metallurgy in China, greatly relaxing the bottleneck effect of iron ore resources due to reliance on import and price fluctuation of high-grade ores.

In 2002, water consumption per ton of steel for industrial use is 159 m^3 in China, and the total is 25.4 billion m^3 [27]. Referring to the status of water resource in China indicates that water scarcity is a serious problem, which constrains iron and steel entrepreneurs. The degree of the constraint is heterogeneous, with the problem being especially serious in Northern China. In the late 1970s and early 1980s, new intake of water per ton of steel in developed countries has already been reduced to around 6 m^3. To solve the problem in China necessitates the consideration of following factor: (1) the old one-through water supply system is routinely used in many plants to save investment and operation cost; (2) the water recycling system makes no discrimination between clean and turbid water and takes no measures on water quality stabilization, causing increase in water consumption and discharge; (3) the amount of water used in production process is large; (4) there lack efficient technologies for some wastewaters' treatment.

Coal is still the major primary energy resource for metallurgical industry. The coal reserve is affluent in China but with only 25–27% being suitable for coking. Hereinto, only 19% is high quality coking coal and 13% rich coal. In addition, the energy consumption per ton of steel in manufacturing process is several times of that in developed countries. Therefore the energy supply shortage has long been there and become intensified in past decade. Another problem is regarding the recovery of waste heat. The technologies for the sensible heat released from various steps, such as blast furnace slag and gas, steel slag, sintering and coking gas etc. have not been utilized efficiently [28,29].

Yet another challenge is environmental pollution. The most serious problem is green house gas (eg, CO_2) emission. For producing every ton of steel, 2.6 tons of CO_2 will be released, along with substantial amount of SO_2, NO_x and dust; the average level is <2 tons in foreign developed countries [30]. Reduction of total feculence or wastes is undoubtedly a critical target for ferrous metallurgy in the future.

Regarding nonferrous metallurgy in China, reasonable industrial structure and relatively complete system have been constructed with the development in the past decades of years. In the new century, the investment on nonferrous metallurgy has increased year by year, with progressively improved investment structure. In 2015, the output for refined Cu, electrolytic aluminum, lead, and zinc were 6.5, 24.0, 5.5, and 7.2 million tons, respectively.

In recent years, however, the contradiction between economic downward pressure and excessive production capacity has been intensified for nonferrous metallurgy in China. Particularly for electrolytic aluminum industry, the falling price for nonferrous metals and the soaring price in cost elements, such as electricity, energy, etc., are the major factors leading to the predicament of the overall deficit of aluminum industry in China. In spite of the situation, the production scale is still on the track of expansion. The situation is further compounded by the coexistence of insufficient technology innovation and rising operation cost.

Experiencing tens of years of rapid growth, traditional development pattern for nonferrous metallurgy relying on resource intensiveness, output growth, low labor costs and backward environmental standards is no longer sustainable. The costs for energy, resource, human power, accountancy, environmental protection, etc. are all exhibiting rigid upward trend. One aspect needing some elaboration is environmental cost. For quite a long time, direct discharge of industrial wastes has been a major measure to squeeze operation cost in China. In nonferrous metallurgy, the polluters in gaseous discharge are mainly dust, SO_2 and nitrogen oxides NO_x. Those in wastewater include chemical oxygen demande (COD), ammonia-nitrogen (NH_3-N), petroleum organic matters, six-valence chromium, heavy metals, and a small amount of cyanide. Solid discharge also contains heavy metal residues. Controlling the waste generation and discharge will undoubtedly lower the damage to eco-system and adverse effect on human health.

Insecure supply of metal resources due to geographical concentration as a further problem associated with nonferrous metallurgy is also quite serious. For the two major nonferrous metals, copper and aluminum, the percentages of their external dependence have reached 70 and 50%, respectively. The situation becomes even worse for some other metals, for example, tantalum, niobium and metals from platinum group, with even higher dependence on external resources.

Addressing these critical problems of metallurgy in China, the "*13th Five-year Plan*" proposes the metallurgical development direction focusing on resolving excess production capacity, deploying large-scale structural reorganization, curbing disorderly competition, strengthening product innovation, promoting green development, as well as encouraging technical and product export. In addition, promoting new industries, such as new energy, new energy vehicles, aviation equipment, rail transit, will be a very important lever for pushing demand on new materials, thus boosting the transformation and escalation of metallurgical industry, and solving excess production capability.

1.3.2 European Metallurgical Renaissance [31,32]

Currently, metallurgy is a very important sector of European industrial development. "Metals and alloys can be found in a wide range of products such as airplanes, cars, trains, ships, satellites, propulsion systems, power-generating turbines, batteries, fuel cells, catalytic reactors, wind turbines, magnets, electrical cabling, pipelines, robots, orthopedics, computers and mobile phones, and many, many more." The market for high added-value metal products is huge, totaling 1.3 trillion Euros per annum. The gross domestic product (GDP) associated with metallurgical processes (ie, primary metal production, alloy creation, downstream processing, integrated metal products, and alloy recycling) accounts for 46% of the total in manufacturing industry in European Union (EU).

Metallurgy has also been identified as a potential solution to many problems confronting EU development in the future, such as CO_2 reduction and renewable energy sources, sustainable product development and resource efficiency in general, development of improved health care products for an aging population–to name just a few.

Realizing the importance of metal and metal alloys in tackling aforementioned societal challenges posed by energy, renewable, climate change, health, etc., in 2012, the Materials Science and Engineering Expert Committee (MatSEEC) at European Science Foundation presented the document entitled *Metallurgy Europe–A Renaissance Programme for 2012—2022*. On this basis, a 7-year project of *Metallurgy Europe* was initiated in Sep. 2014. In order to facilitate the implementation of follow-up activities, a report named *Metallurgy made in and for Europe: The Perspective of Producers and End—Users Roadmap was announced*, making explicit analysis and setting appropriate bound for the many problems encountered by European metallurgical industries, along with the development objectives and themes.

According to the program, for EU securing leading position in metallurgy in the coming decade, it needs to intensify and integrate its efforts from industrial, academic, and research sectors to support the development of the entire materials value chain—from material discovery, alloy design, processing, optimization, to scale-up and in-service deployment.

As an application-orientated research program, requirements on advanced metals, metal alloys, and metal-containing composites are suggested addressing both the entrenched domain of manufacturing industries and the potential fields in tackling the newly emerging societal problems. Surrounding these 17 key requirements, 50 cross-sectoral metallurgical R&D topics have been identified as having high strategic and technical values in dealing with the challenges.

The 50 topics are classed into three directions: material discovery; novel design, metal processing, and optimization; and fundamental understanding of metallurgy. Within these topics, promotion of novel materials identification is the major part due to its directly targeting the new requirements from products and applications, as well as economic and societal development. Improvement of product characteristics and properties via a series of designs and processing is also emphasized; the processes such as severe plastic deformation, incremental sheet forming, additive manufacturing, etc. have been incorporated. Fundamental researches that make predictive revelation and facilitate understanding of various behaviors of metal products are also given significant attention. Included in this program are also the recycling, refinement, and purification processes.

1.4 Conclusions

In this chapter, some basic concepts about metallic materials, metallurgical processes in broad and narrow senses, and development status of metallurgical industry in China and EU have been introduced. Metals and their alloys are of great importance in many fields of societal activities in modern world. In the new century, its potentials in tackling the challenges facing future development of human being have been identified. Novel metal materials and optimized processes are necessitated in providing products satisfying the requirements in terms of advanced properties, environmental protection, and resource conservation. Environmental pollution is severe problem, particularly in regions where some backward technologies and outmoded management pattern are still in function. Flowsheet development and technical troubleshooting are therefore important components for this direction. Benefited from the scientific and technological progress, notwithstanding, metallurgical industry has achieved meaningful advances in the past. How to continue this positive relationship to further facilitate global sustainable development will be an important topic. Issues of operations management, worker safety, risk assessment, and finances are also important, but will not be the focuses of the book.

Glossary and Symbols

Al_2O_3	Alumina
Ar	Argon
BCE	Before the christian era
COD	Chemical oxygen demand
CO_2	Carbon dioxide
EU	European union
EW	Electrowinning
GDP	Gross domestic product
H_2	Hydrogen
MatSEEC	Materials science and engineering expert committee
NH_3-N	Ammonia nitrogen
$Ni(CO)_4$	Nickel carbonyl
NO_x	Nitrogen oxides
O_2	Oxygen
RMSE	Reverse micelles solvent extraction
SCF	Supercritical fluid
SO_2	Sulfur dioxide

References

[1] E.C. Bunker, Ancient Bronzes of the Eastern Eurasian Steppes from the Arthur M. Sackler Collections, Arthur M. Sackler Foundation, New York, (1997) pp. 169–171.
[2] K.M. Linduff, J. Mei, Metallurgy in ancient Eastern Asia: retrospect and prospects, J. World Prehist. 22 (3) (2009) 265–281.
[3] J.F. Lancaster, Non-ferrous Metals, Metallurgy of Welding, Springer, Netherlands, (1980) pp. 197–223.
[4] W.F. Li, Analysis on present situation and exploration of development direction for iron and steel industry in our country, J. Econ. Issues 12 (2004) 27–30 (in Chinese).
[5] J.J. Pei, China's iron and steel metallurgy industry situation and development trend analysis, China High-tech Enterprises 7 (2008) 22–122 (in Chinese).
[6] J. Banhart, Manufacture, characterisation and application of cellular metals and metal foams, Prog. Mater. Sci. 46 (6) (2001) 559–632.
[7] Available from: http://www.cnscdc.com/shujulanmu/146810.html.
[8] Available from: http://www.doc88.com/p-97541402561.html.
[9] H.G. Li, Metallurgical Principles, Science Press, Beijing, (2005).
[10] Z.Q. Cui, H.T. Ni, Y. Zheng, Introduction to present situation and the development of nonferrous metallurgy technology, Guangzhou Chem. Ind. 42 (17) (2014) 43–44 (in Chinese).
[11] Z.B. Li, New progress for special metallurgy, China Metall. (1) (2002) 19–21 (in Chinese).
[12] L.S. Wang, New progress for world powder metallurgy, Mech. Eng. China 5 (1991) 24–26 (in Chinese).
[13] H.G. Li, Hydrometallurgy, Central South University Press, Changsha, (2005).
[14] J.B. Li, Domestic copper hydrometallurgical application status, Hydrometall. China 26 (1) (2007) 13–16 (in Chinese).
[15] C.Y. Tian, M.H. Deng, C.H. Lin et al, An environmentally friendly hydrometallurgical technology, Hydrometall. China 33 (1) (2014) 1–3 (in Chinese).
[16] W.P. Liu, D.F. Qiu, H.M. Lu, Recent advances in hydrometallurgy, Min. Metall. Eng. 23 (5) (2003) 39–42 (in Chinese).
[17] D.F. Qiu, Process chemistry and industrial practice of pressure hydrometallurgy, Min. Metall. (4) (1994) 55–67 (in Chinese).
[18] R.J. Ma, New development of hydrometallurgy, Hydrometall. China 26 (1) (2007) 1–9 (in Chinese).

[19] Y.X. Liu, Modern Electrolytic Aluminium, Metallurgical Industry Press, Beijing, (2008).
[20] D.F. Qiu, Clean and efficient extraction metallurgy-Slurry electrolysis, China Eng. Sci. 1 (1) (1999) 67–72.
[21] J.H. Jiang, R.N. Yu, J.Y. Zhu, R. Yi, G.Z. Qiu, Y.H. He, X.H. Liu, Shape-controlled synthesis and properties of manganese sulfide microcrystals via a biomolecule-assisted hydrothermal process, Mater. Chem. Phys. 115 (2009) 502–506.
[22] L.J. Xu, Face challenges to achieve transformation from the steel big country to the iron and steel power, China Steel 5 (2012) 7–10 (in Chinese).
[23] Y.L. Xie, Current Status and Development Trend of Special Steel, Industry in China 86 (2016) 1–6.
[24] K.D. Xu, China's special steel production over the past 60 years, Iron Steel 49 (7) (2014) 2–7.
[25] J.L. Zhang, W.W. Qu, The cause of the rapid growth of China's steel industry in recent years, World Econ. Papers 4 (2014) 1–19.
[26] J.H. Shao, H.Q. Tang, L.Y. Liu. The necessity and research status of technical bottleneck for fluidized iron making technology. Chinese Metal Association Annual Meeting for the Use of Non-blast Furnace Iron-making Technology and Seminar on Comprehensive Utilization of Titanomagnetite, Panzhihua, Sichuan Province, China, 2010 (in Chinese).
[27] T. Sun, T. Wang, F. Shao, Y. Han, Discussion on the current situation and problems of water quota for iron and steel industry in China, China Water Resour. 23 (2015) 52–54 (in Chinese).
[28] H. Li, W. Bao, C. Xiu, Y. Zhang, H. Xu, Energy conservation and circular economy in China's process industries, Energy 35 (2010) 4273–4281.
[29] J.C. Zhou, C.X. Zhang, X.P. Li, W.G. Han, L. Gan, Reasonable recycling methods of waste heat at steel plants based on energy level analysis, Iron Steel 48 (2) (2013) 80–85.
[30] J. Zhang, Y. Zhang, S.S. Zhang, S.L. Liu, C. You, Analysis of factors influencing the carbon dioxide emission in iron and steel industry, Mod. Chem. Ind. 29 (2009) 82–87.
[31] D. Jarvis, D. Raabe, R. Singer, P.V. Houtte, C. Vahlas, N. Alford, K. Kurzydlowski, R. Nieminen, A. Mortensen, H, Gleiter, M. Loretto, Y. Brechet, H.-J. Fecht, Metallurgy Europe—A Renaissance Programme for 2012–2022. European Science Foundation Report, 2011.
[32] R.L. Moss, E. Tzimas, H. Kara, P. Willis, J. Kooroshy, Critical Metals in Strategic Energy Technologies. JRC Scientific and Technical Report, 2012.

CHAPTER 2

Membrane-Based Separation

L.-S. Xiao

School of Metallurgy and Environment, Central South University, Changsha, Hunan Province, China

In this chapter, the basic concepts, materials, and separation principles for the various synthetic membranes that have been investigated for or applied in metallurgical processes are introduced. The conventional membranes that will be covered include microfiltration (MF), ultrafiltration (UF), nanofiltration (NF), reverse osmosis (RO), ion-exchange (IE), and gas separation (GS) membranes [1]. The briefing is also extended to the concepts of forward osmosis (FO) and pressure retarded osmosis (PRO) for which the passion has been rekindled in the past few years. "New concept" or unconventional membrane separations such as membrane gas absorber (MGA), membrane extraction, liquid membrane (LM), membrane distillation (MD), membrane adsorption (MA), and membrane reactor (MR) are receiving intensive attention by academic and industrial sectors and reviewed as well. Fig. 2.1 shows the dendrogram of classification for the synthetic membrane categories.

2.1 Basics About Membrane-Based Separation

2.1.1 Membrane and Membrane Separation

In the most general sense, a membrane is a semipermeable barrier, located between two phases [1]. Its differential resistance to the transport of chemical species is the core for membrane-based separation.

Membrane separation is in effect a process that enriches one or more components from the feed characterized of certain conditions at the other side of the membrane. Process performance is straightforward for user making judgment about membrane application. The most widely used parameters for describing the performance of a membrane process are flux (J) and separation factor (β) as defined in the following:

$$J_t = \frac{M}{At} \tag{2.1}$$

$$\beta_{i/j} = \frac{y_i/y_j}{x_i/x_j} \times 100\% \tag{2.2}$$

http://dx.doi.org/10.1016/B978-0-12-803410-1.00002-5

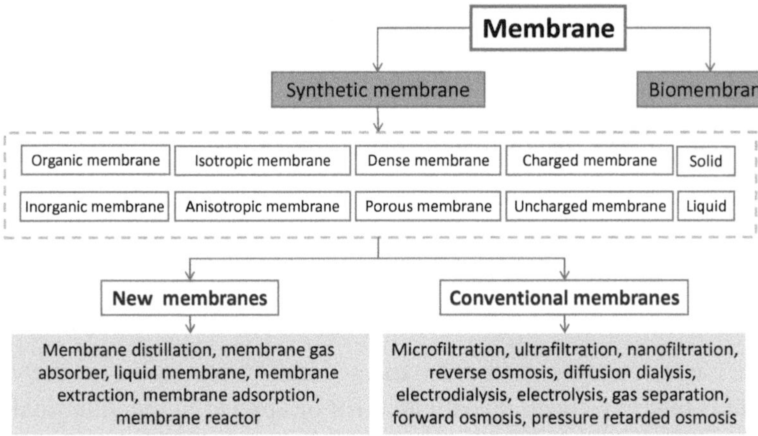

Figure 2.1: Dendrogram correlating different membrane classifications.

where M is the total mass transferred over time t through a membrane, A is the membrane area, x and y are the concentrations of components in the feed and permeate, respectively, and *subscripts* i and j indicate the components to be separated. Flux is obtained from the amount of permeant collected from a laboratory setup at a certain time interval divided by membrane area, as defined in Eq. 2.1, while separation factor is defined as the concentration ratio of two components in a binary system, as defined in Eq. 2.2.

In membrane separation, in principle, the chemical potential gradient across the membrane acts as a driving force on the individual component for their passage through the membrane. The rate for the transport of penetrants in the separation process is determined by the driving force in conjunction with the mobility and concentration of the species within the membrane [2]. The flux, J_i, of a component, i, is given by the following equation:

$$J_i = L_i \frac{d\mu_i}{dx} \tag{2.3}$$

where $d\mu_i/dx$ is the chemical potential gradient and L_i is a coefficient of proportionality (not necessarily constant) correlating chemical potential with flux.

Depending on the characteristics of separation target system, chemical potential is generally externalized in one of the series physical parameters, including electric field, concentration, pressure, or temperature [3].

a. A hydrostatic pressure difference between two phases across a membrane leads to a volume flux and a separation of chemical species with different hydrodynamic permeability through the membrane.
b. A concentration difference between two phases separated by a membrane can lead to a transport of matter and a separation of various chemical species when the diffusivity

and the concentration of the various chemical species in the membrane are different for different components.

c. A difference in the electrical potential between two phases separated by a membrane can lead to a transport of matter and a separation of various chemical species when the different charged particles show different mobility in the membrane.
d. A temperature difference per se between the two phases cannot lead to mass transport through the membrane. However, it can arouse other factors (e.g., vapor pressure difference) that can be utilized to drive the orientated movement of the particles.

The phenomenological relationships between the mass flows through the membrane to the magnitude of specific driving force are summarized in Table 2.1 [3]. Fick's law, for example, correlates the flow of substance with a concentration gradient; the constant of proportionality is the diffusion coefficient. Ohm's law describes the relation between an electrical current and an electrical potential gradient, whereas Fourier's law describes the relation between heat transport and a temperature gradient. In broad sense, the proportionality represents the resistance level of a membrane toward the transport of the penentrant.

In most situations, one driving force is independent of the influence of another. Nevertheless, there exist some processes where the driving forces show interdependences. A concentration gradient across a membrane when driving the mass flow causes the build-up of a hydrostatic pressure difference under certain conditions, and the phenomenon is called osmosis. Similarly, a gradient in hydrostatic pressure may lead to a volume flow and meanwhile the formation of a concentration accumulation, which is called RO. Another example is a temperature gradient across a membrane which may cause a flow of heat accompanied by mass transfer. This process is thermo-diffusion named with Soret effect. The reversal of the process involves a mass flow twinned with an energy flux. This is known as the Dufour effect. In addition, fluxes as well as driving forces are coupled and the flow of

Table 2.1: Phenomenological relationships between various fluxes and the corresponding driving forces [3]

Relationship	Flux	Driving Force	Constant of Proportionality
Fick's law	Mass	Concentration difference	Diffusion coefficient
$J = -D\Delta C$	J	ΔC	D
Ohm's law	Electricity	Electrical potential difference	Electrical resistance
$I = \Delta U/R$	I	ΔU	R
Fourier's law	Heat	Temperature difference	Heat conductivity
$Q = k\Delta T$	Q	ΔT	K
Hagen–Poiseuille's law $V = h_d\Delta P$	Volume	Pressure difference	Hydrodynamic permeability h_d
	V	ΔP	

one component is accompanied by that of another. One example of the coupled fluxes is the transport of bound water with an ion which is driven across a membrane by an electrical potential gradient. In GS and pervaporation, coupling effect of two components has also been reported [4]. There are generally two types of coupled transport, that is, the thermodynamic and kinetic coupling.

At a given driving force the decisive factors for the magnitude of a flux are the penetrants' mobility and concentration. As for the mobility, its magnitude is more closely associated with the molecular size of the penetrants and the membrane structural details from molecular to macroscopic scales (e.g., the aggregation state, the rigidity, the charge state, etc.). The latter constitute the so-called intrinsic membrane properties, which is basic and critical for obtaining separation process with high efficiency. Generally, the concentration of the penetrant within the membrane varies with interaction between the penetrant species and the membrane active material [5]. In some cases (e.g., GS), the role of the penetrant condensability is also important. A very high selectivity can be achieved for the most condensable component (e.g., carbon dioxide (CO_2), hydrocarbons) [6].

In summary, the mass transfer through a membrane is rather specified. The term "membrane," therefore, "involves a great variety of materials and structures and a membrane can often be better described in terms of what it does rather than what it is" [3].

2.1.2 Membrane Structure and Configuration

Generally, the conventional membranes refer to MF, UF, NF, RO, IE, GS membranes, etc. For this group of membranes, their framework is completely composed of solid material (e.g., polymer, metal, ceramics, and their derivatives). More critically, the separation mechanism and efficacy are not related to any media other than the entire hierarchy of structures within the solid framework [5]. The "structure" here mentioned of a membrane relates to its internal components and their arrangements. Subatomic structure involves electrons within the individual atoms and interactions with their nuclei. On an atomic level, structure encompasses the organization of atoms or small molecules relative to one another. Actually, the first two levels to a greater extent are recognized as something intrinsic of membrane material. The next larger structural realm, which contains large groups of atoms that are normally agglomerated together, is termed "subnano to microscopic structure." Finally, the structural elements that may be viewed with the naked eye are termed "macroscopic structure." The latter two levels are usually adjusted by engineering of membrane formation.

Within the scope of the "new concept" category, the membranes also contain solid frameworks, except for LM that is composed solely of liquid. Theoretically, nevertheless, the separation efficiency that finally can be achieved is not determined by the solid framework, of which the major function is largely reduced to mechanical support. Instead, for some membrane processes, factors dominating the differentiation are the interaction of molecules to be separated with the material, sometimes mixture, percolating into or locating at the other side of the porous solid framework, which are unanimously in liquid state. The membranes that tally with the aforementioned descriptions include membrane extraction and MGA. For supported liquid membrane (SLM), as will be introduced in the forthcoming section, the fluid phase (stripping phase) at the other side of the membrane is undoubtedly very critical to the separation. MD is a process that achieves the separation of (more) volatile and less volatile or nonvolatile components inside liquid solution. As its name implies, vapor pressure difference is the principal factor for MD discriminating efficiency. The role of the solid framework is not limited to providing mechanical strength. Its pore size, porosity, and thickness interfere with how large the flux could be reached. Another aspect is that the stability of most membranes in the second category relies on the properties (e.g., pore size, oleophilicity) of the solid framework to a certain degree.

Solid framework is indispensible for actualizing the separation function of both types of membranes. Synthetic solid membranes are generally classified into isotropic or anisotropic [2]. Fig. 2.2 displays the schematics for dense isotropic and various anisotropic membranes. Isotropic membranes are uniform in chemical composition and physical characteristics through the entire membrane. Anisotropic membranes are nonuniform over the membrane cross-section; simply stated they consist of two layers which vary in polymer chain arrangement, and sometimes chemical composition. In nature, in most cases, its cross-section is featured with graded transition from the relatively compact skin layer to the truly open region. As shown in Fig. 2.2, anisotropic structure may be integrally skinned or composite.

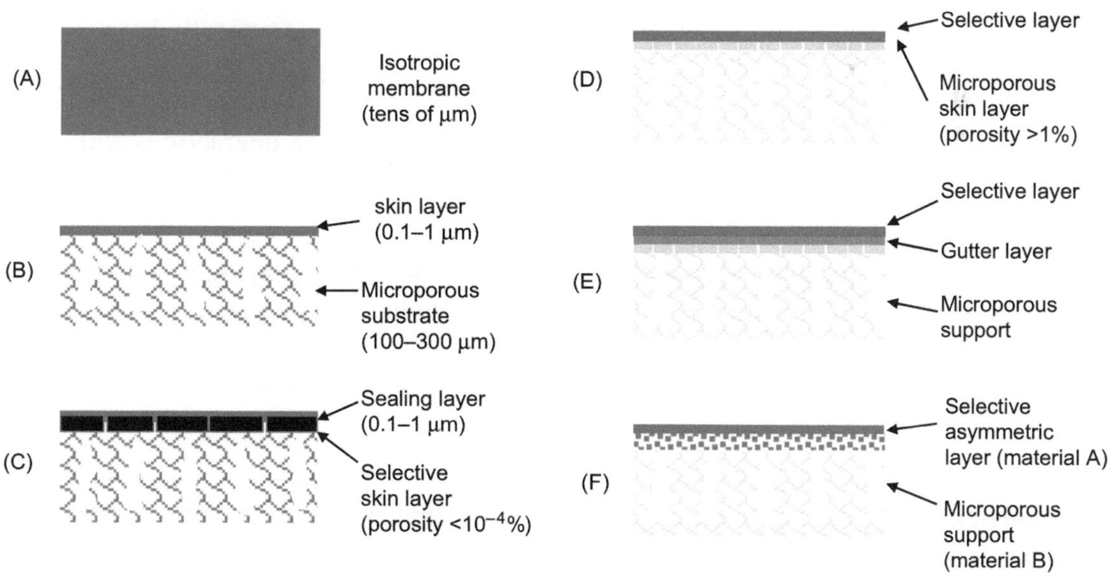

Figure 2.2: Schematics of dense isotropic membrane and some typical anisotropic membranes. (A) Isotropic; (B) integrally skinned asymmetric; (C) multicomponent ("caulked"); (D) single-layer thin film composite; (E) multilayer thin film composite; (F) asymmetric composite.

For anisotropic membranes, using polymeric ones as an example, the structural factors that are crucial to membrane separation performance are associated principally with the layer that presents the highest resistance toward mass transport, which is usually located at membrane skin. These factors include [5]: chemically specified segmental composition of the polymer comprising the skin; spatial steric relationships in the segmental repeat units of the selective polymer skin; nano- to microscale arrangement of polymer chain at the top skin. Concerning the first factor, it is the nature of the membrane material. As for the latter two, their features are dependent on both material intrinsic properties and membrane fabrication conditions.

Membranes used for commercial GS, RO, NF applications are usually asymmetric or anisotropic [7]. In an ideal asymmetric membrane the support is essentially "invisible" in the transport process, so the thickness of the denser skin is of utmost importance in determining the flux through the overall membrane. This is the principal motivation for using an anisotropic structure. In practice, however, the relationship between the skin and support in the anisotropic structure plays an important role in overall separation properties of anisotropic membrane.

Both isotropic and anisotropic membranes can be further divided into subcategories of microporous and dense. For anisotropic membranes, the characteristics of skin are the indicator for judging its classification. Dense membranes lack pores or only contain pores that are beyond any microscopic observation at present. Among conventional group of membranes, dense membranes are used in RO, FO, GS, electrodialysis (ED), and diffusion dialysis (DD). IE membranes (IEMs) for DD and ED typically have little gradation in structure between the two faces; however, they may be supported on a fabric for maintaining mechanical stability [8]. In addition, the two faces of the active membrane may have different functional groups to optimize interactions with the respective external phase. Among the dense membranes, the channels of selective skin differ slightly, depending on the penetrant size.

For the new types of membranes, the solid framework, though not playing an active role in the separation, needs structural tuning to meet the specific requirements in respective separation. In SLM, for example, the minimum transmembrane pressure required to push the impregnating phase out of the largest pores follows the calculation by Laplace equation [9].

Membranes can be fabricated into flat sheet, hollow fiber, and tubular configurations. The modules with all these configurations have been commercialized. Common hollow fibers used in industry today are anisotropic with a relatively denser outer or inner layer connected to a porous support. One advantage of hollow fiber membranes is that they have higher packing density or more surface area per unit volume than flat sheet membranes [10].

2.1.3 Membrane Materials

A wide range of materials have been investigated or applied for membrane fabrication. As for conventional membranes, the materials are broadly classified into polymers, ceramics, metals, and composites. For some new types of membranes or membrane contactors, particularly the LM and SLM, the liquid for facilitating the transport and dominating the separation is very critical. Besides satisfying the requirements from separation efficiency, membrane materials are supposed to contribute to resistance toward failure of multinary structures due to decomposition by chemical reagents, microorganisms or high temperature, deformation by various mechanical stresses, fouling, etc.

2.1.3.1 Polymers

Polymers are still the most popular materials for membrane fabrication and the vast majority of membranes used commercially are polymer-based. Many of the polymers are organic compounds that are chemically based on carbon, hydrogen, and other nonmetallic elements; furthermore, they have very large molecular structures. Polymers include the familiar plastic and rubber materials. The architecture of polymer chain could be linear or branched. The numbers of polymers are really tremendous, owing to the many different ways for carbon to connect with other atoms. The categories that are available for membrane formation and have relatively longer history of investigation and application include polysulfone (PSF), polyethersulfone (PES), cellulose acetate (CA), polyimide (PI), polyvinyl alcohol (PVA), polyvinyl fluoride (PVDF), and polydimethylsiloxane (PDMS). Polymers typically have low densities, are thermal and electrical insulators, do not have high strengths, and may be extremely flexible. Novel synthesis, chemical modification, and blending are the three major approaches for generating new polymeric materials meeting special separation requirements, such as charging and chemical stability.

2.1.3.2 Metals

Metals dominate the periodic table and a substantial part of metals are synthetic. Metallic materials are normally combinations of metallic elements. They have large numbers of nonlocalized electrons; that is, these electrons are not bound to particular atoms. Many properties of metals are directly attributable to these electrons. Metals are extremely good conductors of electricity and heat and not transparent to visible light. In addition, metals are quite strong, yet deformable, which accounts for their extensive use in structural applications.

The metals that have been used for porous membrane preparation principally are stainless steel, nickel (Ni), titanium (Ti), etc. and their alloys. Dense membranes based on palladium (Pd) and its alloys have been used for many years for the selective transport of hydrogen (H_2). These membranes have practically infinite selectivity for H_2. However, their large-scale industrial applications are limited due to the high price [11].

2.1.3.3 Ceramics [12,13]

Ceramics are compounds between metallic and nonmetallic elements. The conventional materials that fall within this classification comprises clay minerals, cement, and glass. Advanced ceramics are not necessarily clay-based. Instead, they are either based on oxides, nonoxides, or combinations of the two; the most typical oxides are alumina (Al_2O_3) and zirconia (ZrO_2). Nonoxides are often carbides, borides, nitrides, and silicides, for example, boron carbide (B_4C), silicon carbide (SiC), and molybdenum disilicide ($MoSi_2$).

This group of materials is typically insulative to the passage of electricity and heat, and is more resistant to high temperatures and harsh environments than metals and polymers. With regard to mechanical behavior, ceramics are hard but very brittle. Currently, both porous and dense ceramic membranes have been developed, and are being hotspots for several separations.

2.1.3.4 Composites [7,14]

A number of composite materials have been engineered that consist of more than one material type. A composite is designed to consolidate the best characteristics or advantages of each of the component material. For example, fiber glass acquires strength from the glass and flexibility from the polymer. Many of the recent material developments have involved composite materials. As for membrane separation, composite composed of polymer and inorganic filler are extensively investigated for a series of separations. The organic/inorganic composite has another name of mixed matrix material (MMM) in some situations [7]. As for GS and pervaporation, for example, carbon molecular sieve (CMS), zeolite, metal-organic framework etc. loaded MMM have been intensively followed till present.

2.1.3.5 Liquid [15]

In several membrane contactors (e.g. MGA, MD, membrane extraction), the most important material is the liquid phase determining the selection. For MGA, usually aqueous solutions, sometimes containing chemical agents (e.g., amine, alkaline) to facilitate chemical absorption, are used. In LM or membrane extraction, organic solutions are the major choice. It comprises extractant, diluent, matrix modifier, flocculant, brightening agent, surfactant, etc. The most critical chemical is extranctant, of which the molecular design or selection is based on their selectivity toward the components to be separated in the feed phases. The other chemicals inside LM are also organic compounds. In supported liquid membrane (SLM), the liquid organic phase is embedded inside the solid support and acts as an immobilized phase. Regarding metal ions or compounds universally in metallurgical process fluids, the common extractants used are chelating agents such as LIX 84-I, LIX 64N, LIX 62N, LIX 860, and so forth, or acidic such as di-(2-ethylhexyl)phosphoric acid (D2EHPA), PC88A, di-(2-ethylhexyl) phosphoric acid (Cyanex272) (phosphoric acid derivatives), and so forth [15]. Ionic liquid

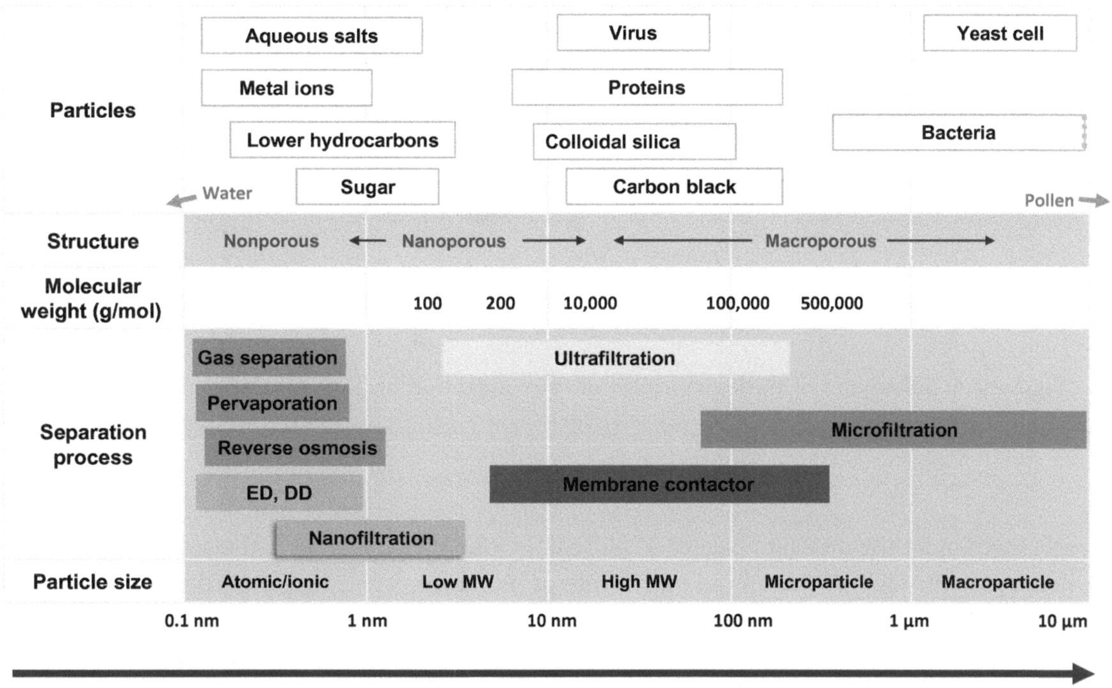

Figure 2.3: The spectrum of various separation processes useful for different particles.

(IL) is an emerging category of liquid material receiving increased attention in membrane separation [16,17]. They can be used for both gas absorption and metal ions extraction.

2.2 Fundamentals of Some Technically Relevant Separations

In Fig. 2.3, the conventional membrane separation processes with their range of applications in terms of particle size are shown. In general, more than one type of membrane technology is available to solve a certain practical mass separation task. The selection of the proper membrane and the integrated or hybrid system depends on several factors, such as the nature of the constituents in a mixture, the volume of the solution to be handled, the degree of separation required and, particularly in large scale industrial processes, the cost of the process.

2.2.1 Hydraulic and Osmotic Pressure–Driven Processes

2.2.1.1 Microfiltration and ultrafiltration [18]

MF and UF are pressure-driven processes, with the selectivity being determined by a simple sieving effect. The operating principle of UF and MF is shown schematically in Fig. 2.4. The feed solution to be treated is brought by convection to the surface of a microporous

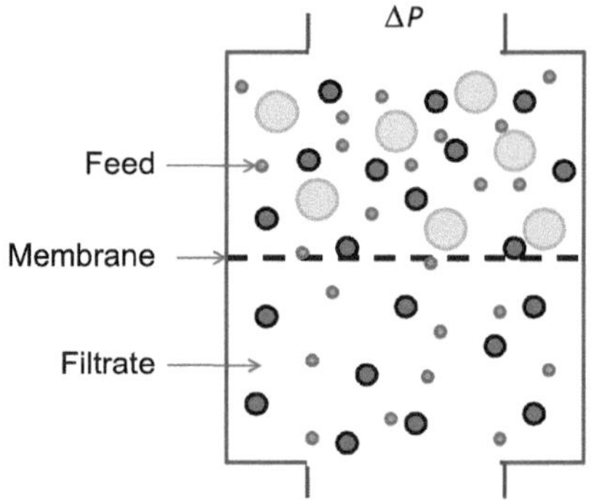

Figure 2.4: Schematic of basic principles of microfiltration and ultrafiltration separations.
Adapted from Reference [3].

membrane. Under the driving force of a hydrostatic pressure difference, the solvent and some particles pass through the membrane as filtrate or permeate, while larger particles are rejected by the membrane in the retentate. Transport in these membranes appears to be best described by a pore-flow model. The membrane flux, indicating the filtration rate through the membrane is described by the following relation:

$$J_v = \Delta p / R_h \tag{2.4}$$

where J_v is the filtration rate or flux, Δp is the hydrostatic pressure difference across the membrane, and R_h is the hydrodynamic resistance of the membrane.

The term MF is applied when particles with diameters in the range of 0.1–10 μm are separated from solvent and other components with low molecular weight (MW). The separation process is called UF when the components to be selectively removed are true single molecules or particles no larger than 0.3 μm in diameter. In UF, the filtrate or the retentate, or both can be of interest for further application. In MF, generally only the filtrate is of interest and the particles trapped in the microporous structure of the membrane are discarded with the used membrane.

2.2.1.2 Nanofiltration

The pore size of NF membranes is typically in the range of 0.5–2 nm, corresponding to a molecular weight cut-off (MWCO) of 300–500 Da. Also driven by hydraulic pressure difference, however, NF mechanism is actually complex, as compared with MF and UF.

Transport through the NF membrane principally occurs by a combination of convection (permanent pore flow) and solution–diffusion (transient pore flow) [19,20]. The latter is a function of solvation and partition. The transport of neutral solutes is via the steric mechanism (size-based exclusion). Inside charged membrane, molecules with net valence are influenced by electrical potential as well [21,22]. The equilibrium between the charged membrane and the bulk solution is characterized by an electric potential, the Donnan potential, which tends to retain ionic species. This Donnan exclusion allows the removal of ions with a size below the pore size of the membrane. Briefly, the separation performance of NF membranes will be determined by the sieving (steric hindrance) effect and/or Donnan (electrostatic) effect [20]. There are three types of charges on NF membrane surface [23]: (1) the ionizable surface charging groups, (2) the weak ion exchange capacity, and (3) in some cases ions adsorbed from the contacting solution to the membrane surface. Electrostatic repulsion or attraction varies with the ion valence and the fixed charge of the membrane.

A number of models have been suggested for describing the mass transfer of noncharged and charged compounds in NF. One of the models was proposed by Kedem and Katchalsky based on the nonequilibrium thermodynamics [24], using the following equations to describe the solvent and solute permeating process through membrane:

$$J_{\mathrm{v}} = L_{\mathrm{p}}(\Delta p - \sigma\Delta\pi) \tag{2.5}$$

$$J_{\mathrm{s}} = -(P\Delta x)\frac{dc}{dx} + (1-\sigma)\bar{C}J_{\mathrm{v}}c \tag{2.6}$$

where J_{v} and J_{s} are the solvent and solute flux, respectively. L_{p}, σ, and P refer to the pure water permeability (hydraulic permeation coefficient), reflection coefficient, and solute permeability, individually. σ is approximating zero with dilute solution [25]. The hydraulic pressure is indicated by Δp, and the osmotic pressure by $\Delta\pi$. $\bar{C}$ is the average concentration over the thickness of membrane selective skin. c is the solute concentration in solution and Δx is the membrane thickness.

Integration over the membrane effective thickness delivers Spiegler–Kedem equations:

$$R = 1 - \frac{c_{\mathrm{p}}}{c_{\mathrm{f}}} = \frac{\sigma(1-F)}{1-\sigma F} \tag{2.7}$$

$$F = \exp\left\{-\frac{(1-\sigma)J_{\mathrm{v}}}{P}\right\} \tag{2.8}$$

Here R is the real rejection and F is a parameter. c_{p} and c_{f} represent the permeate and feed concentrations, respectively.

When using low pressure in NF separation, the water flux is low and the flux of solute molecules permeating by convection is correspondingly low. At high pressure, both the water flux and the convective flux of solutes are high. Diffusion, nevertheless, is almost independent of pressure for dissolved molecules. As a result, rejection increases with pressure.

Yet another phenomenon that might affect the ion species transport is dielectric exclusion. There are two principal hypotheses about the nature of this interaction, namely, the "image forces" and the "solvation energy barrier" [26–28]. Both mechanisms are associated with spatial confinement and nanolength scales that are present in NF membrane and are effectively charge based exclusion [28].

Compared with other hydraulic pressure–driven membrane separations, the most distinctive characteristics of NF are the low rejection of monovalent ions and high rejection of divalent or multivalent ions. This property makes NF very promising for metallurgical process involving separation of various metal ions. Based on the pore size, NF is also suitable for removing relatively small organics.

2.2.1.3 Forward osmosis, reverse osmosis, and pressure-retarded osmosis

FO, or direct osmosis, refers to water transport across a selectively permeable membrane from a phase of higher water chemical potential to the other side with lower water chemical potential [29]. The membrane allows passage of water, but rejects solute molecules (e.g., organics, ions). Typically, FO process results in concentration of the feed stream and dilution of a highly concentrated stream which is referred to as draw solution (DS). Diluted DS is then treated to generate purified water as final product and recover its concentration for recycling in FO.

The membranes in FO are effectively nonporous, and the transport of penetrant through membrane obeys the solution–diffusion mechanism. Water molecules sorb into the membrane surface at feed side, then diffuse down the chemical potential gradient across the membrane, and desorb from the downstream face of the membrane. Application of the more general model of mass transport being driven by chemical potential gradients rather than concentration gradients leads to the transport equation for solution–diffusion of FO, which is in the same form of Eq. 2.5. This equation is applicable to not only FO, but also RO and PRO. In FO, specifically, there exists no external pressure, namely, ΔP is zero.

Within the context of solution–diffusion mechanism for dense membrane, L_p is represented by following equation [21]:

$$L_p = \frac{DSV}{RTl} \tag{2.9}$$

where D is the water diffusivity in the membrane, S is the water solubility in the membrane, V is the molar volume of water, R is the ideal gas constant, T is the absolute temperature, and l is the membrane selective layer thickness.

Osmotic pressure is estimated using following equation [30]:

$$\pi = i\phi cRT \tag{2.10}$$

where i is the number of ions produced during dissociation of solute (or dissociation factor); ϕ, the osmotic coefficient, C, the concentration of all solutes in solution.

Unlike water flux, which is driven by both applied transmembrane pressure and osmotic pressure, the salt flux is only a function of salt concentration [31]:

$$J_s = B(c_f - c_p) \quad (2.11)$$

where B is the salt permeability constant describing the physical characteristics of the membrane. Analogous to L_p of water in the solution–diffusion equation, B is given by

$$B = \frac{D_s K_s}{l} \quad (2.12)$$

where D_s is the salt diffusivity and K_s is the salt partition coefficient, in the membrane.

As FO is an osmotically driven process that requires very low hydraulic pressure, the effects of fouling are depressed, and membrane compaction is of less concern. Commercial utilization of FO, though limited, has expanded to wastewater treatment and water purification, seawater desalination, food processing, and pharmaceutical applications.

If substantial hydraulic pressure ΔP is applied, the process is no longer FO. Depending on the relative magnitude of ΔP and $\Delta \pi$ (σ is 1 by default), two other types of processes, similarly based on solution–diffusion mechanism, are resulted: PRO, for which $\Delta P < \Delta \pi$, and RO with $\Delta P > \Delta \pi$. The schematics comparing FO, RO, and PRO are illustrated in Fig. 2.5A [29]. The FO point, PRO zone, and RO zone, along with the flux reversal point, are illustrated in Fig. 2.5B [32].

RO is most widely used among the all the osmosis processes for water treatment. As the name indicates, RO relies on hydraulic pressure to oppose, and exceed the osmotic pressure of an aqueous feed solution to produce purified water. In RO, salt transport across a membrane is as important as water transport. Instead of reporting salt flux values, most membrane performance specifications provide salt rejection values. Salt rejection, R, is defined as follows:

$$R = \left(1 - \frac{c_p}{c_f}\right) \times 100\% \quad (2.13)$$

Furthermore, water flux and salt flux depend on each other. Eq. 2.14 relates the water flux, J_v, and the salt flux, J_s:

$$\frac{c_w}{c_p} = \frac{J_v}{J_s} \quad (2.14)$$

where c_w is the water concentration in the permeate. By rearranging the aforementioned equations of 2.5, 2.11, and 2.14, the following expression for rejection may be derived:

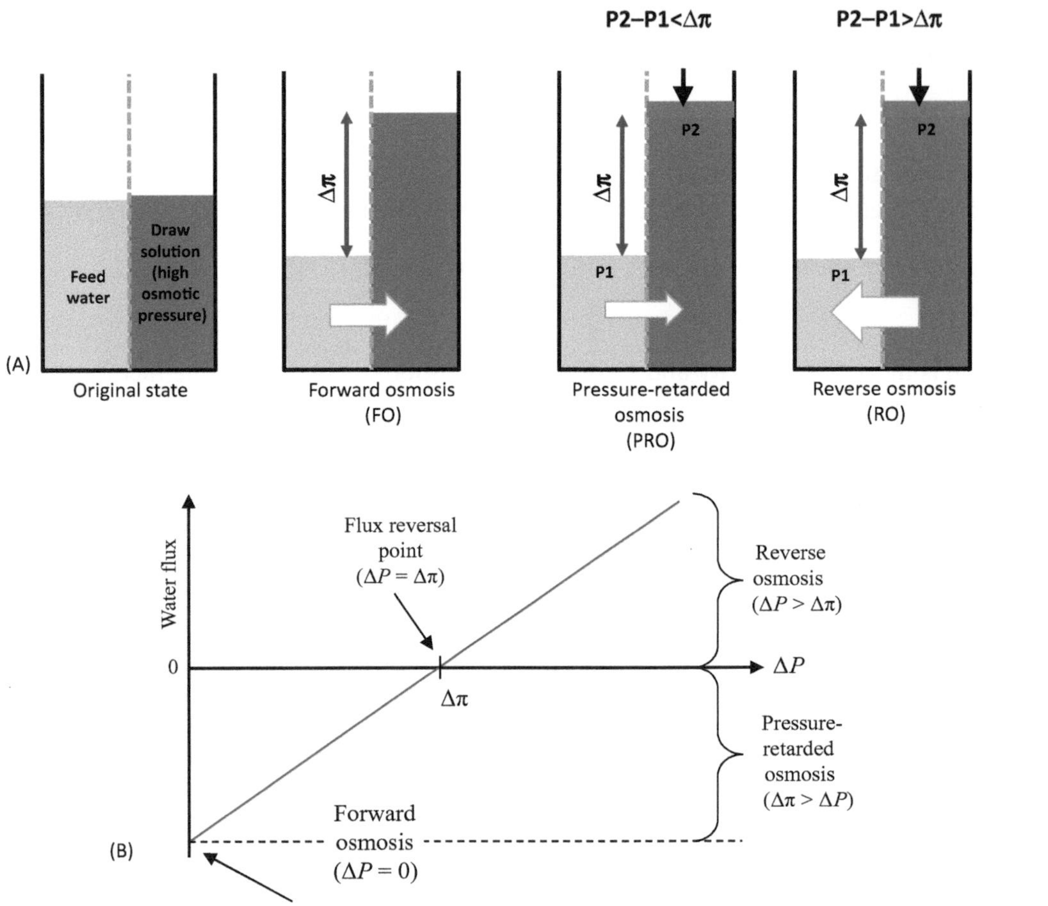

Figure 2.5: (A) Schematics showing the comparison of different membrane osmosis. (B) Direction and magnitude of water flux as a function of applied pressure in FO, PRO, and RO [31].

$$R = \left(\frac{\frac{L_p}{B}(\Delta p - \Delta \pi)}{1 + \frac{L_p}{B}(\Delta p - \Delta \pi)} \right) \times 100\% \tag{2.15}$$

This equation relates salt rejection to the physical properties of the membrane (which influence L_p and B), the applied transmembrane pressure difference, and the osmotic pressure difference between the permeate and the feed. Eq. 2.15 allows one to predict the salt rejection of the membrane based on the experimental conditions and the membrane properties.

PRO can be viewed as an intermediate process between FO and RO, where hydraulic pressure is applied in the opposite direction of the osmotic pressure gradient (similar to RO). However, the net water flux is still in the direction of the concentrated draw solution (similar to FO). A very special application of PRO, as in contrast with other membranes, is energy production [33]. In PRO, water from a solution with low salinity permeates through a membrane into a pressurized solution with high salinity. Power is obtained upon depressurizing the permeate using a hydro-turbine. In PRO, the power density equals the product of the water flux and the hydraulic pressure difference across the membrane:

$$W = J_w \Delta p = A(\Delta\pi - \Delta p)\Delta p \tag{2.16}$$

By differentiating Eq. 2.16 with respect to Δp, it can be shown that W reaches a maximum (W_{max}) when $\Delta p = \Delta\pi/2$. Substituting this value for Δp in Eq. (2.17) yields

$$W_{max} = A\frac{(\Delta\pi)^2}{4} \tag{2.17}$$

2.2.2 Processes Using Ion-Exchange Membranes

Similar to IE resins, IEM contains functional groups that can be dissociated or ionized. A membrane with fixed positively charged ions is referred to as an anion-exchange membrane (AEM), and it is capable to bind anions in the surrounding solution in contact, whereas, a membrane containing fixed negatively charged ions is called a cation-exchange membrane (CEM) [34]. A bipolar membrane (BPM), another kind of charged membrane, consists of anion exchange resin layer (AL), cation exchange resin layer (CL), and an intermediate catalytic layer packed together [34]. The charged membranes can be dense or microporous, but are most commonly very finely microporous in wet state, with the pore walls carrying fixed positively or negatively charged ions.

The separation of ions by AEM or CEM is achieved mainly by exclusion of ions carrying the same charge as the fixed ions of the membrane structure, and to a much lesser extent, by the pore size. The mechanisms are schematized in Fig. 2.6. The separation is affected by the charge and concentration of the ions in solution. For example, monovalent ions are

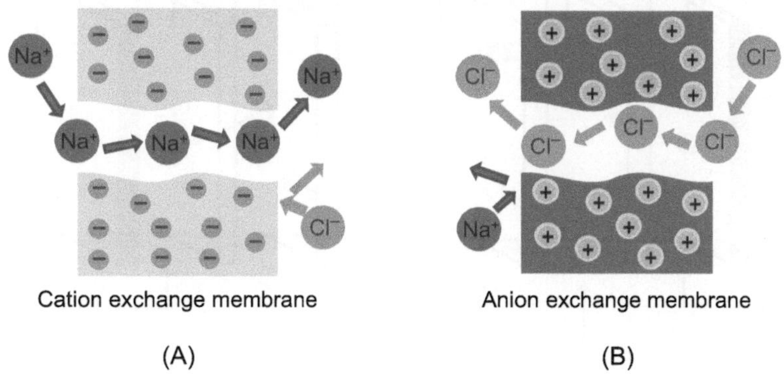

Figure 2.6: Separation mechanism of ion-exchange membrane.

excluded less effectively than divalent ions and, in solutions of high ionic strength, selectivity decreases. The selective permeation (*P*) of the membrane is given as follows:

$$P(\%) = \frac{\Delta t_r}{\Delta t_o} \times 100 = \frac{\bar{t}_m - t_s}{1 - t_s} \times 100 \tag{2.18}$$

$\bar{t}_m$ is the transfer number of counter-ion inside the membrane and t_s is the transfer number of counter-ion in the solution.

Generally, the mass transfer through the IEM comprises convective flow, diffusional flow, and conductive flow. Nernst–Planck equation, which only considers the one-dimensional transfer perpendicular to the membrane surface is used as the basic model:

$$\bar{J}_i = -\bar{D}_i \frac{d\bar{c}_i}{dx} + \frac{Z_i F \bar{c}_i \bar{D}_i}{RT} \frac{d\varphi}{dx} + c_i \bar{V}_i \tag{2.19}$$

where $\bar{J}_i$ is the mass transfer rate of ion *i* through membrane (mol cm^{-2} s^{-1}); $\bar{D}_i$ is the diffusion coefficient of ion *i* inside membrane (cm^2 s^{-1}); $\bar{c}_i$ is the concentration of ion *i* inside membrane (mol m^{-3}); $\bar{V}_i$ is the liquid transfer rate inside membrane (cm s^{-1}); φ is the electrical potential (*V*); Z_i is the valence of ion *i*; and *F* is the Faraday constant.

2.2.2.1 Electrodialysis

In practical application, ED processes generally employ multicompartment system or electrodialyzer to achieve desalination. A typical ED system design and the working principle are shown in Fig. 2.7 [35]. With their selective permeability, AEMs only allow anions to pass through, whereas CEMs are just the opposite, respectively. Under the action of electric

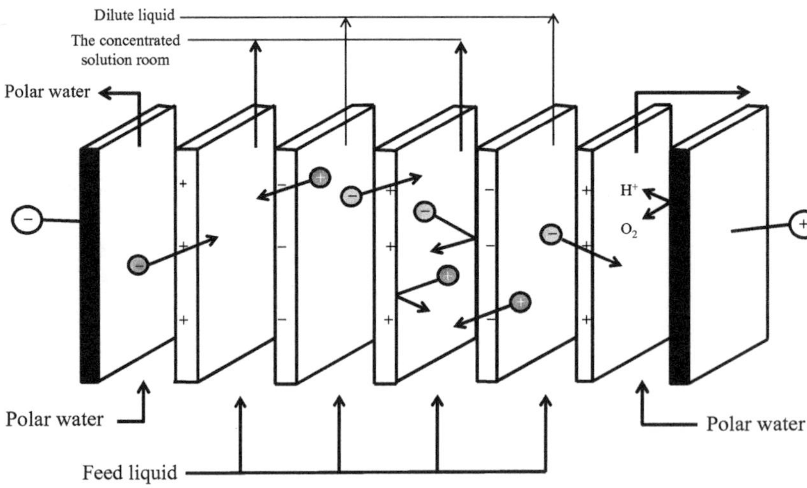

Figure 2.7: Working principle of electrodialysis using AEM and CEM [35].

field, ions in water would make orientated migration along the corresponding electrical potential gradient. The cations and anions from the dilute solution compartment would move to the adjacent concentrate solution compartment. Because of the alternative arrangement of AEM and CEM, compounded with the electrical potential gradient, ions accumulated in the concentrating compartment will not move away, so as to realize the enrichment of inorganic salts and generation of desalinized water.

In addition to seawater desalination, salt production from seawater, electronics, medicine, petrochemical, food, beer, and other fields, ED can be applied to deal with waste water, recovery of precious metals.

Some phenomena occurring in ED process are very critical to the overall process efficiency. They are concentration polarization and related current polarization, membrane fouling, and membrane poisoning. In addition, there exist several adjoint mass transfers or phenomena that are detrimental to ED efficiency, such as common ion transfer, water ionization into H^+, and OH^-.

2.2.2.2 Diffusion dialysis

In DD, the driving force is concentration difference. The typical applications of DD include separation of acid or alkaline from inorganic salts [36]. The DD device is composed of IEM, liquid distribution plate, enhanced plate, liquid flow frame, and some other components. Each unit is divided into dialysis chamber and diffusion chamber by a piece of homogeneous AEM or CEM. The waste acid (waste alkali) and accepting liquid (tap water) are filled into each side of the membrane. The concentrations of acid (alkali) and accompanying salt inside the dialysis chamber are far higher than those of accepting water. Driven by this concentration gradient, acid (alkali) and the salts have a tendency to transfer to the other side. Attributing to the selective permeability of the membrane, anions (cations) of acid (alkali) are attracted by the membrane and pass through the membrane to enter the diffusion chamber. Based on the electric neutrality principle, H^+ (OH^-) has small hydration radius and lower valence will also penetrate the membrane. In this respect, acid (alkali) is separated from the salts. The mechanism using acid recovery is illustrated schematically in Fig. 2.8. The most critical disadvantage of DD is the low concentration of the product.

2.2.2.3 Membrane electrolysis

Membrane electrolysis is also a process using IEM. It not only possesses the characteristics of ED, but also uses a direct current (DC) to cause an otherwise nonspontaneous chemical reaction, in principle, redox reaction [37]. Compared with conventional electrolysis, combination with membrane efficiently separates the products evolved at two electrodes.

ED can be combined with several types of electrolysis. They are production of acid and base, cathodic metal electrowinning (EW), anodic dissolution, and valence control. In some ED

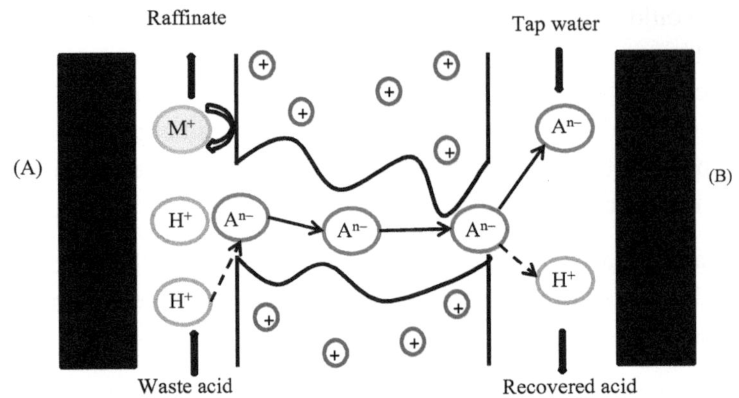

Figure 2.8: Mechanism of diffusion dialysis in recycling waste acid.

processes, BPMs are used to replace AEMs or CEMs, or being inserted between them inside a traditional electrodialyzer. Inside the direct current electric field, as shown in Fig. 2.9 [38] illustrating simply the system configuration and working principle, water contained between the two layers of BPM is ionized into H^+ and OH^- in catalytic layer which then pass through CL and AL layers, respectively, and become the source of H^+ and OH^- for the overall process. The migration of water and ions inside BPM comprises three steps: (1) water molecules

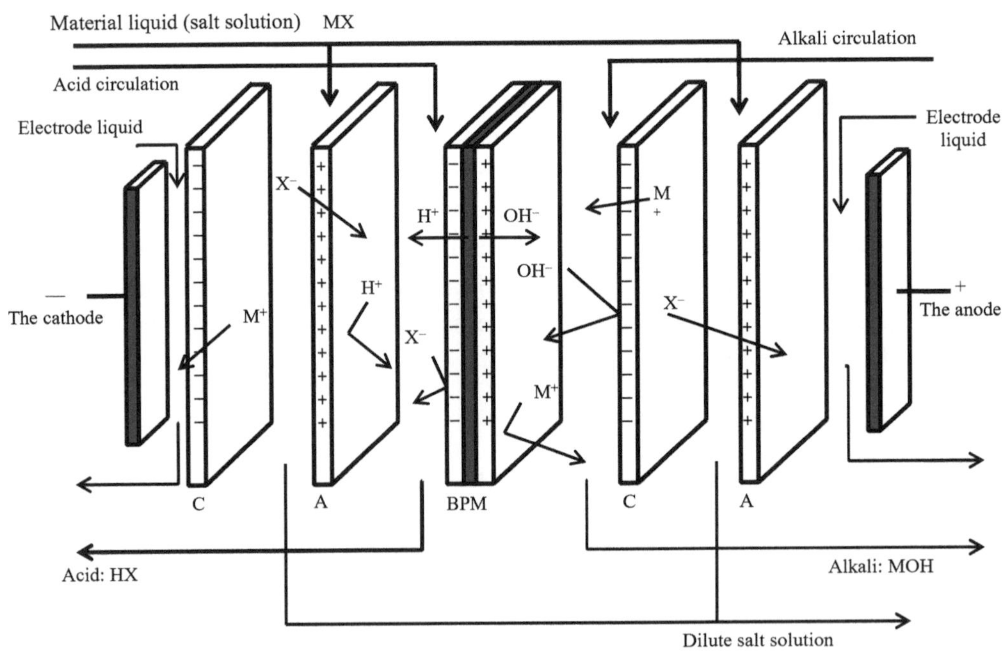

Figure 2.9: The mechanism schematic for electrodialysis using bipolar membrane.

dissolve on membrane surface, (2) water molecules move to AL and CL interfaces, (3) water is hydrolyzed into H^+ and OH^-, (4) H^+ and OH^- transfer to CL and AL surfaces, respectively, under the action of electric field.

2.2.3 Gas Separation

In molecular-level GS, or fraction of different gas molecules, the membrane permits the preferential transport of one or more component(s) of a feed mixture. As a result, the bulk phase on one side of the membrane will be enriched of certain components, while the other side almost depleted of them. This transport of gas species across the membrane is caused by chemical potential gradient due to concentration or pressure difference or both across the membrane.

Various mechanisms for gas transport through the membrane have been proposed, depending on the penetrant and membrane properties, as well as their relationship (e.g., interaction). These mainly include Knudsen diffusion, molecular sieving, and solution–diffusion (Fig. 2.10) [2,6]. These models have been found to be applicable only in limited gas/material systems.

The permeation of gas through porous membrane abides by Knudsen diffusion, Poiseuille flow, or in some cases both. The proportion of Knudsen diffusion to Poiseuille flow is governed by the ratio of the pore radius to the mean free path of gas molecules. In Knudsen diffusion, the ratio is much less than 1. In order to function as a molecular sieve, membranes must have pore diameter that are in between those of the gas molecules to be separated. The pore with size smaller than 0.5 nm will generate separation factor larger than 10. GS can

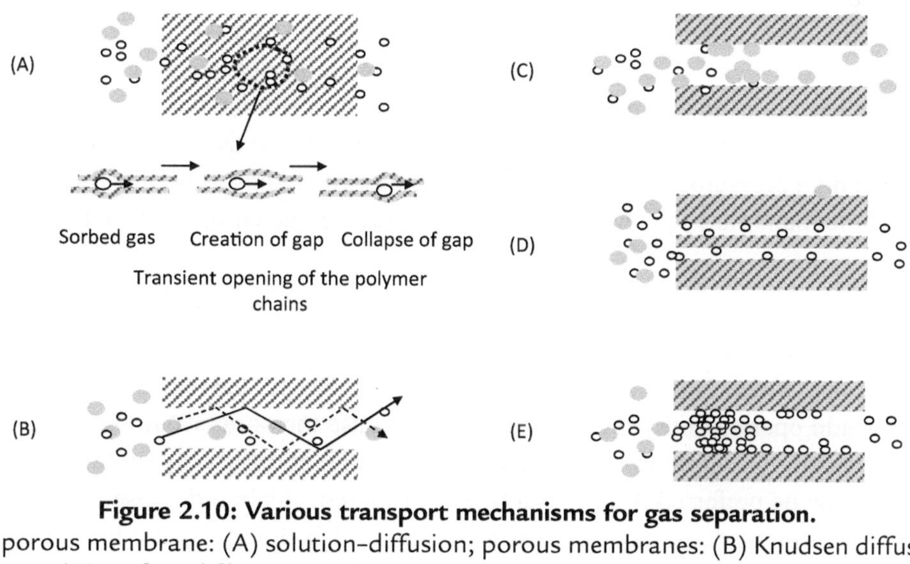

Figure 2.10: Various transport mechanisms for gas separation.
Nonporous membrane: (A) solution–diffusion; porous membranes: (B) Knudsen diffusion; (C) surface diffusion; (D) molecular sieving; (E) capillary condensation.

also be affected by partial condensation. Selective adsorption of the more strongly adsorbed components of a gas mixture onto the pore surface followed by the surface diffusion of the adsorbed molecule across the pore can also facilitate the separation. The solution–diffusion mechanism for dense membrane consists of three steps: (1) adsorption at the upstream boundary; (2) activated diffusion (solubility) through the membrane, and (3) desorption or evaporation on the other side.

2.2.4 Membrane Contactor

A membrane contactor is a process that achieves gas/liquid, or liquid/liquid mass transfer without dispersion of one phase within another [39]. By using a suitable membrane configuration such as a hollow fiber or a flat sheet, fluids to be interacted flow on two sides of the membrane and the interface between two contacting phase forms at the mouth of each membrane pore.

In membrane contactor, the membrane generally imparts trivial or no selectivity to the separation [40,41]. In principle, all traditional stripping, scrubbing, absorption, evaporation, distillation, crystallization, emulsification, liquid/liquid extraction and mass transfer catalysis processes or equilibrium-limited chemical reactions can be conducted in this manner. Currently, the most widely investigated membrane contactor designs include MGA, membrane extraction, emulsion liquid membrane (ELM), SLM, MD, etc.

2.2.4.1 Membrane gas absorption

MGA is a combination of membrane separation and gas absorption [42]. The gas to be treated and the liquid absorbant are not in direct contact with each other, but flow on each sides of membrane. Microporous membrane is mainly used in this technology. In the situation with porous structure, the membrane itself is not selective, but a barrier isolating absorbant from gaseous feed. The pores on the microporous membrane are presumed to be big enough to theoretically allow gas molecules to go through the membrane to the other side. The process does not need high pressure, and mainly relies on the selective absorption of liquid on the other side of the membrane to achieve the aim of removing one component in the mixed gases.

Two modes of operation are possible in gas/liquid contactors: wetted mode and dry (or nonwetted) mode. Wetted mode occurs when the pores are filled with liquid, for example, if the liquid phase is aqueous and a hydrophilic membrane is used. Conversely, a hydrophobic membrane would operate in dry mode in this case because the pores would be filled with gas. Dry mode is usually preferred to take advantage of the higher diffusivity in the gas; however, wetted mode may be preferred if there is a fast or instantaneous liquid phase reaction and as a result the gas phase resistance controls [43]. Fig. 2.11 is the schematic for dry mode membrane absorption [39].

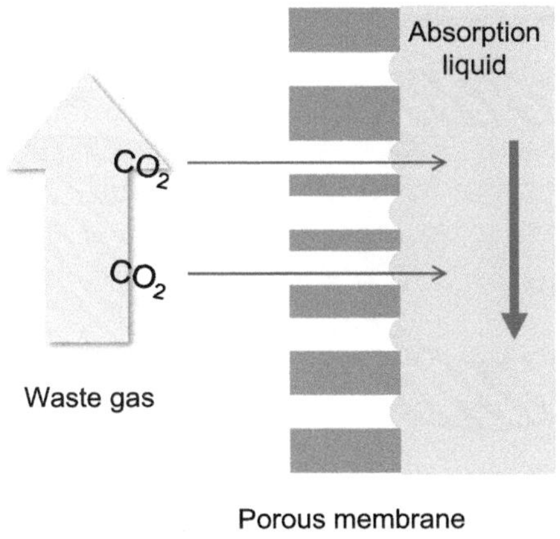

Figure 2.11: Schematics of dry mode membrane gas absorption.
Adapted from [23].

2.2.4.2 Membrane extraction and its derivatives

In principle, membrane extraction is a combination of membrane with solvent extraction [44–46]. The extraction mechanisms used in membrane extraction are classified as (1) compound formation, (2) ion-pair formation, and (3) preferential solvation. The transfer or reaction for selective permeation of solutes from one phase to the other takes place at the interface. The membrane processes that have been categorized as ME include typical ME, ELM, and SLM.

Shown in Fig. 2.12 are a typical membrane extraction and the corresponding resistance. The bulk phases are separated by a solid porous membrane. The inner structure of the membrane is usually wetted by the organic phase. With this operation mode, the only liquid/liquid interface is located at the aqueous side of the solid membrane.

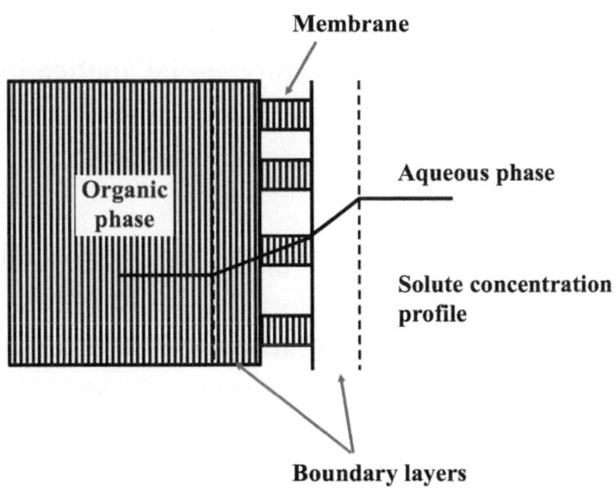

Figure 2.12: Typical membrane pertraction and the resistance model.

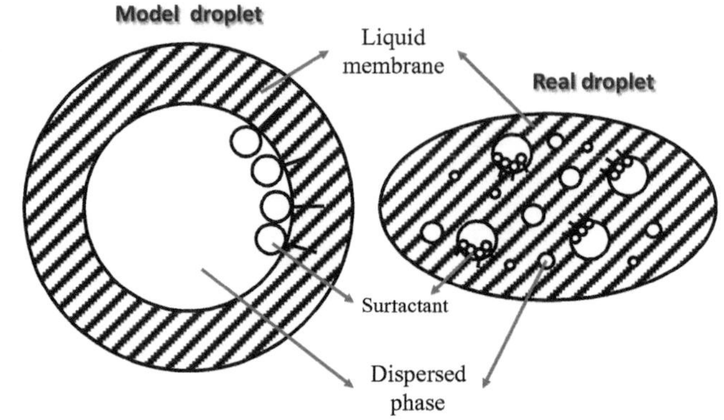

Figure 2.13: Schematic representation of a W/O/W emulsion droplet ([10]).

The solute of interest will encounter three resistances in series upon transfer from the aqueous to the organic phase; namely, the aqueous phase boundary layer, the membrane, and the organic phase boundary layer [39].

The derivatives for membrane extraction include ELM, and SLM. ELM is also known as liquid pertraction or, simply, pertraction [47,48]. The schematic of ELM structure and mass transfer is represented in Fig. 2.13 [49]. Therefore pertraction is a combination of extraction, permeation, and stripping of the desired species and can be treated as a continuous multistage extraction–stripping process. The membranes usually consist of a water-immiscible organic membrane (M) layer separating an aqueous source phase, or feed (s.p. or F), consisting of a mixture of metal ions or other particles of interest, from a receiving aqueous phase (r.p. or R), in which the particles of interest are concentrated preferentially. In other words, ELMs are double water-in-oil-in-water emulsions (W/O/W). They are stabilized by the employment of suitable surfactant substances. Demulsification is needed to collect the final product from the inner R phase. Sometimes the relative position of the phases may be reversed, resulting in "oil-in-water" emulsions dispersed in a continuous organic phase. There already exist commercial applications of ELM processes on an industrial scale, for example zinc (Zn) removal at a textile plant in Austria, cyanide removal from waste liquors at a gold(Au) processing plant in China, and phenol removal from wastewater [50].

SLM is another branch or refinement of LM. It is often formed by immobilizing a suitable extractant in the pores of solid membrane support. The SLM thus formed is interposed between the feed and the strip phase. In a similar manner to ELM, different steps inclusive extraction, and stripping are combined to a single step in SLM [51]. The schematic for basic principle of SLM is exhibited in Fig. 2.14. The mass transfer process in the SLM process consists of several important segments: (1) diffusion of target particles from the bulk of the feed to membrane surface; (2) diffusion of carrier from stripping side interface

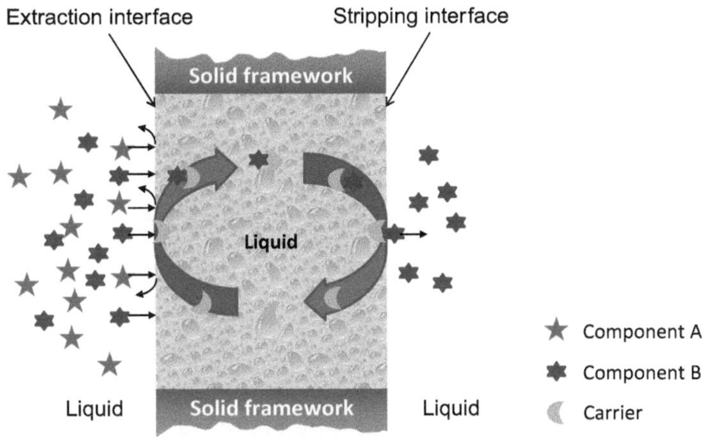

Figure 2.14: The schematic for basic principle of supported liquid membrane.
Adapted from [11].

toward feed side interface; (3) combination or reaction of particles in feed with carrier in membrane at corresponding interface; (4) diffusion of the complex from feed side interface to stripping side interface; (5) upon reaching the membrane-strip phase interface, the strip reaction takes place, regenerating the carrier and liberating the targeted particles; the regenerated carrier diffuses back to the feed side membrane surface, as in segment (2); (6) diffusion of the liberated target particles from the surface of the membrane to the bulk strip phase. In real application of SLM, the segments and the chemical reactions can be very complicated.

For LMs, there exist three modes for the mass transfer of targeted particles: passive transfer by preferential dissolution and diffusion, facilitated transfer by introducing chemical reaction in stripping phase, and facilitated transfer by chemical reaction inside membrane.

Relative to traditional solvent extraction–stripping process, SLM eliminates the equilibrium limitation inherent to solvent extraction by continuously circulating stripping solution. In the SLM process, therefore, the target species can be transported across the membrane against their concentration gradient. This makes it economically feasible to use tailor-made expensive extractants.

In the past two decades, membrane extraction has been developed to avoid some problems intrinsic of conventional liquid–liquid extraction. The problems are solvent loss, emulsion formation, phase separation, and the shortcomings due to flooding and loading in column contactors (e.g., packed towers) and mixer-settlers [39]. Although membrane extraction processes have a slower rate of mass transfer compared to traditional operation if membrane resistance is significant [39], the property offered by higher surface area/volume associated with microporous hollow-fiber modules can partially counteract this negative impact.

2.2.4.3 Membrane distillation

There are several reviews about MD in recent years [52]. In MD process, a microporous hydrophobic membrane is in contact with an aqueous solution on one side (feed or retentate). The hydrophobic nature of the membrane prevents a mass transfer in liquid phase through the membrane and instead creates a vapor–liquid interface at the pore entrance. Driven by vapor pressure difference, volatile compounds evaporate, diffuse, and/or flow by convection across the membrane pores, and are condensed and/or removed on the opposite side (permeate or distillate) of the system.

The specific methods used to carry away distillate also activate the vapour pressure gradient across the membrane, which are the basis for classifying MD configurations. In direct contact membrane distillation DCMD, a flowing liquid (often pure water) with temperature lower than that of feed flow through the permeate channel of membrane module and directly in contact with the membrane surface. In air gap membrane distillation (AGMD), alternatively, the component in vapor state is recovered on a cold condensing surface separated from the membrane by an air gap. For vacuum membrane distillation (VMD), vacuum is applied at the distillate side to lower the vapor pressure. Another category is sweep gas membrane distillation (SGMD), in which a cold inert gas passes through the permeate channel to take away the vapor molecules; the condensation takes place outside the membrane module. In osmotic membrane distillation (OMD), the difference in vapor pressures of the target component(s) in the solutions at two sides of the membrane is generated by using concentrated saline or hypertonic solution in distillate side. The above configurations are schematized in Fig. 2.15 [53].

The transmembrane flux in MD is often expressed as a linear function of the vapor pressure difference Δp across the membrane:

$$J = C\Delta p \tag{2.20}$$

where J is the molar flux. The coefficient C is a function of the membrane structural properties (eg. pore size, thickness, porosity, and tortuosity), physical and chemical properties of the vapor transporting across the membrane (i.e. MW and diffusivity), and operative conditions. The relationship between the transmembrane flux and the different membrane characteristic parameters is given as:

$$J \infty \frac{r^a \varepsilon}{\delta_m \tau} \tag{2.21}$$

where r is the mean pore size of the membrane pores, a factor whose value equals 1 or 2 for Knudsen diffusion and viscous fluxes, respectively, δ_m the membrane thickness, ε the membrane porosity, and τ the membrane tortuosity. Some new categories of membrane

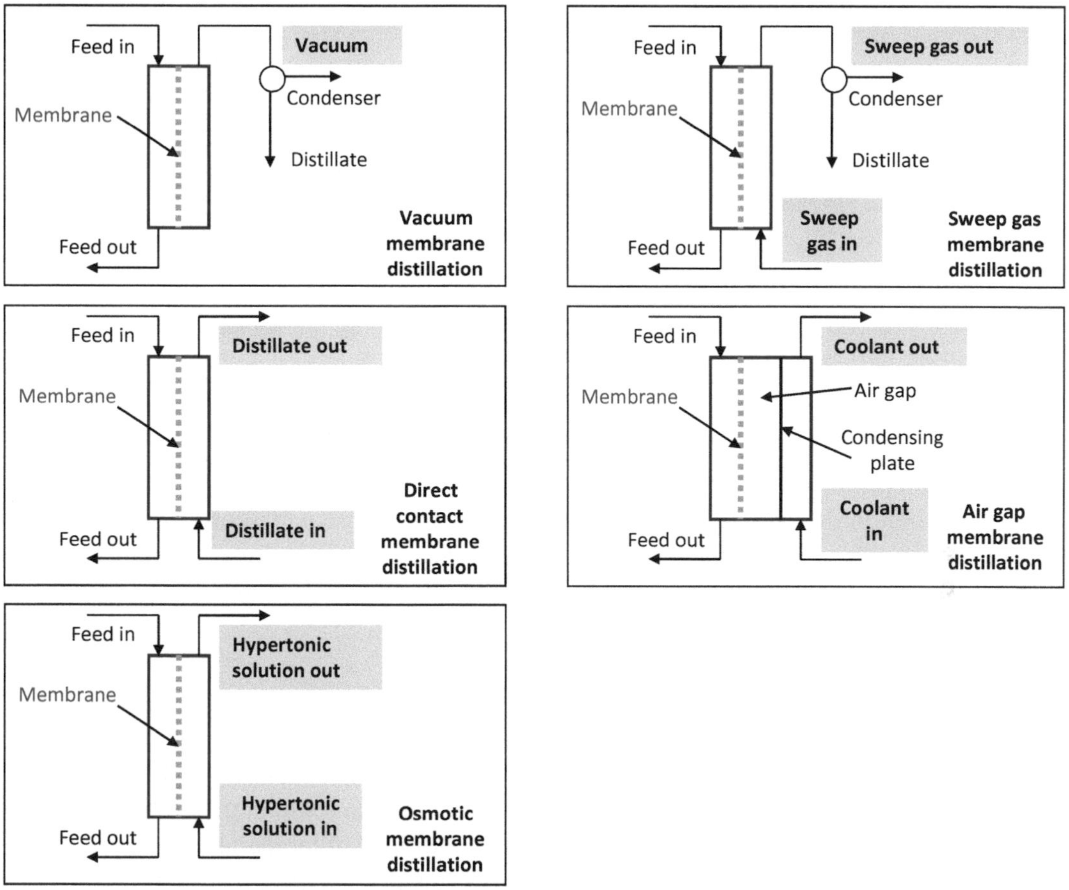

Figure 2.15: Schematics of membrane distillation in different configurations.

contactor, for example, electrophoretic membrane contactor have also been developed, but will not be dwelt on here [54].

2.2.5 Membrane Adsorption

Affinity membrane, also known as MA, synergizing membrane separation and adsorption has been explored for metal removal from wastewater and recovery of noble metals [55]. It was initially developed as an alternative to traditional packed bed for improved mass transfer for bioseparation. The schematic of one full cycle of MA inclusive loading, washing, and elution is displayed in Fig. 2.16. The loading procedure terminates at saturation point when the effluent starts to present almost the same concentration of target penetrant as feed. The key to the separation efficiency of MA lies with the interaction of

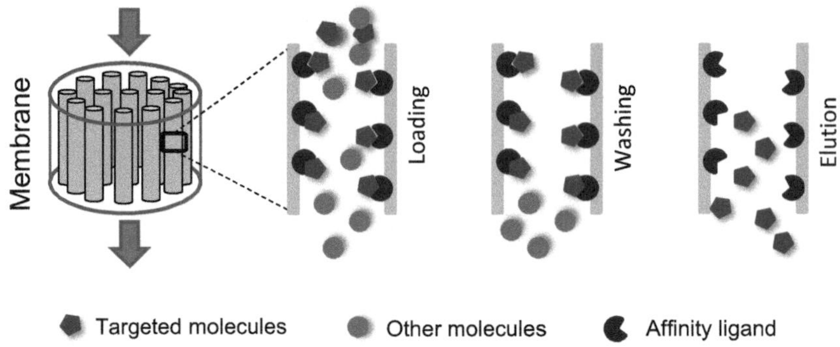

Figure 2.16: The operation of one purification cycle in membrane absorption.

ligands with the penetrants. The membrane used in MA is porous, and transport of solute of interest to the binding sites of ligands occurs primarily by convection. The overall mass transfer rate is elevated in MA which allows shorter residence time for proper adsorption. Unfortunately, the conventional membranes have relatively low adsorption capacity, probably due to their low surface area.

2.2.6 Membrane Reactor

The use of membranes in chemical reactors is motivated principally by reaction equilibrium shift via membrane separation, leading to a higher conversion in a single pass [56,57]. The shift also allows obtaining a given conversion less severe conditions of temperature and pressure. A schematic layout for MRs with different membrane functions is presented in Fig. 2.17. In light of the role of membranes, the functions of membrane may be classified as: (a) extractor, (b) distributor, and (c) contactor [57]. Concerning the role of "extractor," the membrane allows selective permeation of some reaction products from the reaction system; this may improve the yield of a desired component in a multiple reaction system. With a membrane performing as a "distributor," introduction of a reactant in a controlled manner through a permeating membrane may allow the regulation of the reaction, leading to better yield and improved control. As for the role of "contactor," it improves the contact between reactants and catalyst; the membrane itself may act as a catalyst or catalyst may be impregnated on the membrane. Yet another MR mode is call "reaction coupling" in which two reactions taking place on either side of the membrane complement each other. In the first situation, the product of the reaction on one side is the reactant for the reaction on the other side. Another case is that the endothermicity of one reaction is compensated by the exothermicity of the other. In addition to enhancing yield, MR also allows hot separation of products and eliminates the need for cooling a reaction to prevent reaction reversal. This task requires special membranes that can sustain extreme temperature. The most widely applied MR in wastewater treatment is membrane bioreactor (MBR) which synergizes the membrane filtration with biodegradation. In recent years, the domain of MR has

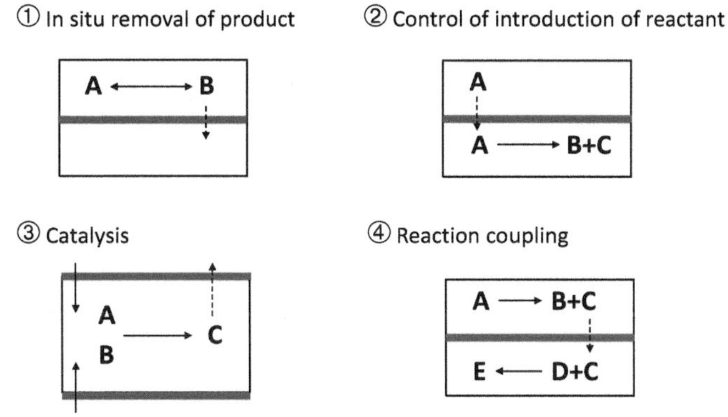

Figure 2.17: Graphs showing the different roles of a membrane in membrane reactors.

been expanded by combination with a series of novel processes, which lead to the proposition of the name “MXR”.

2.3 Brief History of Development

Table 2.2 summarizes the major events in the history of membrane science and technology. Studies of membrane phenomena are traced to the work of the 18th century philosopher scientists. For example, spontaneous passage of water through animal bladder was discovered by Abbe Nollet in 1748 for describing the permeation of water through a diaphragm. Thereafter, however, no progress was reported for quite a long period of time, 70 years, until 1827, when the word of “osmosis” was introduced by Dutrochet. From then on, a phase of prosperity began in the scientific and technological development for membrane separation. The time span between two important findings or inventions in succession is 10 years by average, as shown.

ELM was invented by N. N. Li in 1968, which created a new era in membrane history [49]. From their initial application for the fractionation of hydrocarbons, ELMs have diversified into the recovery of heavy metal ions, the removal of contaminants from wastewater, heterogeneous reactions replacing conventional catalytic processes, treatment of disorders in the blood stream, and the extraction (followed by reaction in some cases) of products of fermentations. Another important technology also associated with liquid/liquid interfacial mass transfer, membrane extraction, was forwarded by Kiani et al. in 1984.

The progress in the industrial or commercial uses of membrane separation is not synchronous with the theoretical development. As shown in Table 2.3, which displays the history of membrane industry, throughout the 19th and during the early 20th centuries, membranes had no practical applications. Instead, they were used mainly in laboratory for developing physical/chemical theories. The seminal discovery that transformed membrane separation from a laboratory to an industrial process was the development, in the early

Table 2.2: History of membrane science and technology development

Year	Scientist(s)	Events
1748	Nollet	Water permeation through pig bladder, which is osmosis phenomenon.
1827	Dutrochet	Introduction of term "osmosis".
1831	Mitchell	Development of gas permeating rubber membrane.
1855	Fick	Diffusion law; preparation of synthetic semipermeable membrane.
1861–1966	Graham	Discovery of different permeation rate of gases; discovery of dialysis phenomenon
1860–1977	Van't Hoff, Tranbe, Preffer	Osmotic pressure law.
1906	Kahlenberg	Observation of hydrocarbon/ethanol solution selective permeation through rubbery film.
1917	Kober	Introduction of term "pervaporation".
1911	Donnan	Donnan distribution law; investigation about formation of molecule charging, charge distribution, Donnan electrodialysis and balance phenomenon in coupled transport.
1922	Zsigmondy, Bachman, Fofirol etc.	Application of microfiltration; preliminary stage of ultrafiltration and reverse osmosis (membrane materials were cellophane and regenerated cellulose).
1920	Mangold, Michaels, Mobain etc.	Observation of reverse osmosis of electrolyte and nonelectrolyte by using celluloid and nitro-cellulose.
1930	Teorell, Meyer, Sievers etc.	Investigation of membrane electro-potential, the basis of electrodialysis and membrane electrode.
1944	William Kolff	The first successful application of synthetic kidney.
1950	Juda, Mcrae	Research about synthetic membrane. Invention of processes such as electrodialysis, microfiltration, and hemodialysis.
1960	Loeb-Sourirajan	Asymmetric reverse osmosis membrane formation by phase inversion.
1968	N. N. Li	Invention of liquid membrane.
1980	Cadotte	Invention of interfacial polymerized composite membrane.

Table 2.3: The history of membrane industrialization or commercialization

Membrane Process	Year	Producer	Current Supplier
MF	1925	Sartorius	Millipore Corp.; Pall Corp.; Asahi Chemical
ED	1960	Ionics Inc.	Ionics Inc.; Tokuyama Soda; Asahi Glass
RO	1965	Haxens Industry General Atomics	Film Tech.; Dow Hydronautics; Nitto Torray; DuPont
DD	1965	Enka (AKZO)	Enka; AKZO; Gambro; Asahi Chemical
UF	1970	Amicon Corp.	Amicon Corp.; Koch Eng. Inc.; Nitto Denko
Controlled release	1975	Alza Corp.	Alza Corp.; Ciba, SA
GS	1980	Permea (DOW)	Permea; Air Prod.; Ube Ind.; Hoechst; Celanese
Pervaporation	1990	GFT GmbH	Sulzer; BP; MTR

1960s, of the Loeb–Sourirajan process for making defect-free, high-flux, anisotropic RO membranes [58].

Extraction of uranium (U) was the metallurgical process that first employed membrane separation technology at industrialized scale [54]. The enrichment of uranium-235 (U235) as a uranium hexafluoride (UF_6) was conducted using porous, metallic membranes. Since the MW ratio of UF_6 isotopomers is about 1.008, and the gas diffusion enrichment factor was low (1.0043), the single stage separation effect is not high enough to obtain qualified product. A large number of stages have to be employed to achieve higher enrichment of the desired product. The cascade design is frequently used in nuclear industry, especially for isotope separation.

Laboratory breakthrough regarding MF membrane development was realized in 1922 by Zsigmondy and coworkers. Soon after this event, industrial level production of MF membrane was launched by Sartorius Company in 1925. Around 1980s, application of dead-end filtration by MF for removing heavy metals (e.g., Zn^{2+}) was initiated for wastewater discharged from galvanized steel sheet production. In addition, adsorption of heavy metals using MF membrane combined with multifunctional groups possessing adsorption properties was first ventured during that period of that time [59,60]. In 1988, Bhave and Fleming from Alcoa Company investigated the performance of MF in oil/water separation [61].

The pioneering research work in China using MF for metallurgical separation was conducted by Song et al. [62]. Flat sheet membranes with a mean pore size of 0.2 μm was employed to treat oily wastewater. The findings about rejection and flux indicated that the oil content in the permeate liquid could be kept lower than 30 mg L^{-1}. What was concluded was that MF was suitable for situations of which the requirements on wastewater discharge were not strict or recycled water not demanding.

The large-scale application of UF in industry was not actualized until 1980s [63]. A major field using UF is treatment of oil waste in iron and steel metallurgy. Raymond made a detailed description about the mechanism and technology for using UF in treating wasted emulsified lubrication oil, and reported the experimental results [64]. William published a patent in the late 1980s about using UF to treat wastewater containing oil as well [65]. A high shear and rotated UF system was reported by Reed et al. to process oily wastewater [66]. The turbulant flow generated in the system reduced the concentration polarization and prevented the flux attenuation. In addition, the oil concentration in the retentate increased from 5% to 65%. In the mid 1980s, Shanghai Baoshan Iron and Steel Plant introduced from ABCOR Company of the United States a tubular MF system for treating wastewater containing emulsified oil in lubrication and cooling. The inner diameter and length of a single tube are 25.4 and 3048 mm, respectively. The material for the membrane is PVDF and the MWCO of the membrane is 20,000 g mol^{-1}. A two-stage recycling system was employed with the oil concentration being raised from 2% to 10% in the first stage and 10% to 50% in the second stage. An average flux of 33.4 kg m^{-2} h^{-1} was obtained at an inlet pressure of 4.5 MPa and a feed temperature of 55°C. The system has been installed for many years with stable performance.

The industrial application of RO technology started from 1960s for water treatment. In effect, the first recorded synthetic membrane was prepared in 1867 by Moritz Traube [67]. The membrane was a film prepared by precipitation of copper ferrocyanide ($Cu_2Fe(CN)_6$), for studying osmosis. RO commercial debut was made when Gulf General Atomics and Aerojet General employed the Loeb-Sourirajan CA membranes in spiral wound modules to purify water. Slightly lagging behind, RO application for metallurgy was initiated around 1970s. According to Koge and Ushikoshi's report [68], a closed RO separation cycle using polyamide (PA) membrane modules was adopted to recover aluminum (Al) from the electrophosresis paint. The recovery has led to obvious economic benefit. The polishing process in electrolytic Al production delivers waste phosphoric acid (H_3PO_4). Application of DDS HR298 RO membrane could recover 96–98% acid, as found in Thoren's report [69]. Slater et al. utilized ultrathin RO membrane to remove cadmium (Cd) from the metal processing water [70]. The initial concentration of Cd in wastewater is 165 mg L^{-1} and its removal rate obtained was higher than 99.9%; the removal rates for other metals were higher than 98%. As a result, the water recovered has a high quality. Mok et al. concentrated and purified the large quantity of lithium hydroxide (LiOH) solution generated in the electroplating process [71].

In Nicod et al.'s work, complexing reinforced NF has successfully separated chromium (Cr) and strontium (Sr) [72]. Nowadays, the typical applications of NF are associated with copper (Cu) metallurgy, which will be introduced in Chapter 4. One is the concentration of Cu in heap leaching solution for follow-up solvent extraction and EW. The other is recycling of barren liquid in Cu EW.

IEM related membrane separation is another process that has already been identified as a very important performance enhancer for metallurgy. One application of significance is producing acids and bases from the corresponding salts by catalyzed water dissociation using BPM. Another is concerning tungston (W) metallurgy. Utilizing membrane electrolysis in the production of sodium metatungstate ($Na_6H_2W_{12}O_{40}$) reduces the generation of sodium sulfate (Na_2SO_4) that impacts environment negatively if discharged. In addition, its benefits include economizing acid and base used in the production process, recovery of base, and obtaining H_2 and oxygen (O_2). This technology has already been commercialized. One installation was previously in Hungaria, and is now operated by General Electric. The development history of IEM and related processes is drawn in Fig. 2.18 and introduced by Paidar et al. [37]. The first industrial application of IEM occurred in 1940s. Nowadays novel IEM processes, membrane of hybrid materials, etc. dominate the R&D activities.

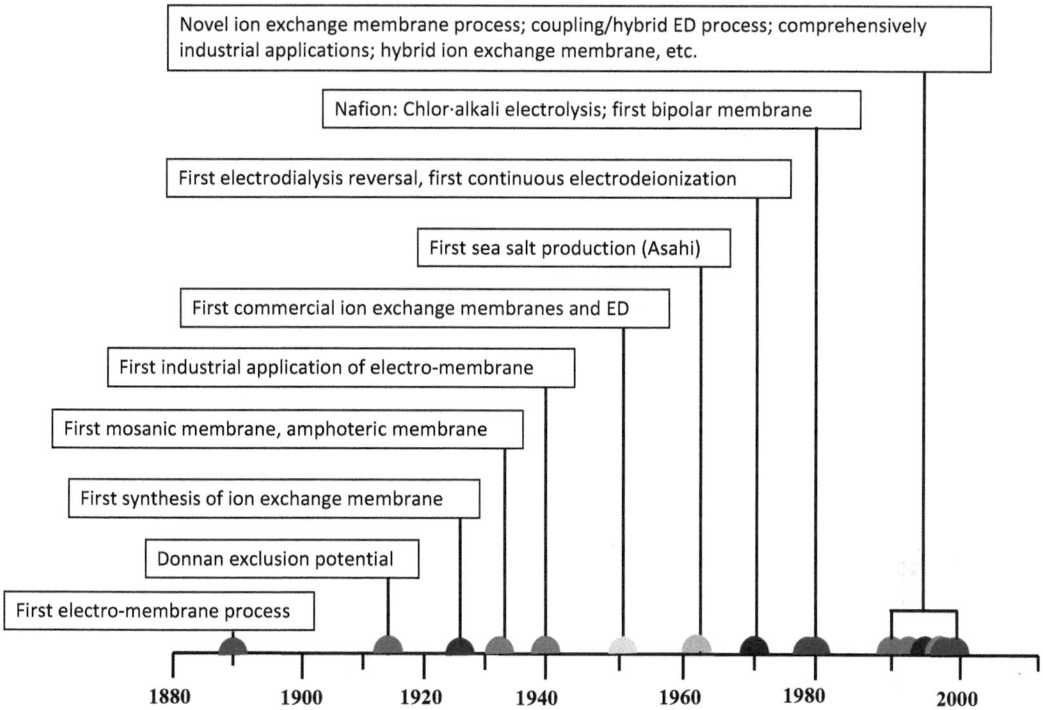

Figure 2.18: Time line visualization of ion exchange membrane development and their related processes.

With more than 30 years development, membrane-based separations have successfully penetrated into many fields and the market size keeps increasing. More detailed information regarding development of China in terms of manufacturer, market share, application status (particularly in water treatment) could be referred to book series entitled "Report for sustainable development strategy of China water treatment industry" [73]. Membrane separation plays important role throughout the entire metallurgical process from mining to final material engineering and processing. In terms of metal categories, metallurgy of ferrous metal, heavy nonferrous, light nonferrous, rare earth, noble metal, radioactive metal, etc. all receive contributions from membrane technology, but to different degree. Encouragingly, efforts are not stopping, but are now being extended to new fields.

2.4 Conclusions

This chapter takes the readers to basic knowledge about membrane-based separation. This includes the concepts of membrane and membrane process, driving force classification, multilevel membrane structures, membrane materials, and some mass transfer mechanisms. Both the conventional and nonconventional membranes are covered. A briefing about the development history of membrane separation, particularly in metallurgical industry, is also given. The chapter is put here as a preparation for the forthcoming sections detailing the membrane applications in metallurgy of various metals and development of special membranes.

Glossary and Symbols

AEM	Anion-exchange membrane
AGMD	Air gap membrane distillation AGMG
AL	Anion exchange resin layer
Al	Aluminum
Al_2O_3	Alumina
B_4C	Boron carbide
CA	Cellulose acetate
Cd	Cadmium
CEM	Cation-exchange membrane
CL	Cation exchange resin layer
CO_2	Carbon dioxide
Cr	Chromium
Cu	Copper
$Cu_2Fe(CN)_6$	Copper ferrocyanide
Cyanex 272	Di-isooctylphosphinic acid
DC	Direct current
DCMD	Direct contact membrane distillation
DD	Diffusion dialysis
DS	Draw solution
D2EHPA	Di-(2-ethylhexyl)phosphoric acid
ED	Electrodialysis
ELM	Emulsion liquid membrane
EW	Electrowinning
FO	Forward osmosis
GS	Gas separation
HTU	Height of a transfer unit
H_2	Hydrogen
H_3PO_4	Phosphoric acid
IE	Ion-exchange
IEM	IE membranes
IL	Ionic liquid
LiOH	Lithium hydroxide
LM	Liquid membrane
MA	Membrane adsorption
MD	Membrane distillation
MF	Microfiltration
MGA	Membrane gas absorber
MR	Membrane reactor
MMM	Mixed matrix material
$MoSi_2$	Molybdenum disilicide
MW	Molecular weight
MWCO	Molecular weight cut-off
$Na_6H_2W_{12}O_{40}$	Sodium metatungstate

Na_2SO_4	Sodium sulfate
NF	Nanofiltration
Ni	Nickel
OMD	Osmotic membrane distillation
O_2	Oxygen
PA	Polyamide
PDMS	Polydimehtylsiloxane
PES	Polyethersulfone
PI	Polyimide
PRO	Pressure retarded osmosis
PSF	Polysulfone
PVA	Polyvinyl alcohol
PVDF	Polyvinyl fluoride
RO	Reverse osmosis
SGMD	Sweep gas membrane distillation
SiC	Silicon carbide
SLM	Supported liquid membrane
Sr	Strontium
UF	Ultrafiltration
Ti	Titanium
VMD	Vacuum membrane distillation
W/O/W	Water-in-oil-in-water emulsions
ZrO_2	Zirconia
Zn	Zinc

Nomenclature

a	A factor whose value equals 1 or 2 for Knudsen diffusion and viscous fluxes
A	Membrane area
B	Salt permeability constant describing the physical characteristics of membrane
c	Concentration in solution
c_f	Salt concentration in the feed solution
c_p	Salt concentration in the permeate solution
c_w	Water concentration in permeate
$\overline{c}_i$	Concentration of ion *i* inside membrane
C	Concentration of all solutes
C	Coefficient in membrane distillation as a function of membrane structural properties, physical and chemical properties of the vapor transported across membrane, and operative conditions
$\overline{C}$	Average concentration over the thickness of membrane selective skin
D	Water diffusivity in the membrane
$\overline{D}_i$	Diffusion coefficient of ion *i* inside membrane
D_s	Salt diffusivity in membrane
F	Faraday constant
i	Number of ions produced during dissociation of solute (or dissociation factor)
J	Water vapor flux in membrane distillation
$\overline{J}_i$	Mass transfer rate of ion *i* through membrane
J_v	Solvent (water) flux in hydraulic pressure driven separation
J_s	Solute flux
K_s	Salt partition coefficient in the membrane
l	Membrane selective layer thickness
L_p	Pure water permeability (hydraulic permeation coefficient)

L_i	A coefficient of proportionality
M	Total mass transferred through membrane
MW	Molecular weight
P	Solute permeability
Δp	(hydrostatic, vapor) Pressure difference across membrane
r	Mean pore size of membrane pores
R	Real rejection in nanofiltration
R	Ideal gas constant
R_h	Hydrodynamic resistance of membrane
S	Water solubility in membrane
t	Time
$\bar{t}_m$	Transfer number of counter-ion inside membrane
t_s	Transfer number of counter-ion in solution
T	Absolute temperature
V	Molar volume of water
$\bar{V}_i$	Liquid transfer rate inside membrane
W	Power density in pressure-retarded osmosis
x and y	Concentrations of components in the feed and permeate, respectively
Δx	Membrane selective layer thickness
Z_i	Valence of ions

Greek Symbols

β	Separation factor
δm	Membrane selective layer thickness
ε	Membrane porosity
ϕ	Osmotic coefficient
μi	Chemical potential
π	Osmotic pressure
σ	Reflection coefficient
τ	Membrane tortuosity
φ	Electrical potential

Subscripts

f	Feed
i	Component i
j	Component j
m	Membrane
p	Permeate
s	Solute
v	Solvent
w	Water

References

[1] H. Strathman, Membrane separation processes: current relevance and future opportunities, AICHE J. 47 (2001) 1078–1087.
[2] R.W. Baker, Membrane Technology and Applications, John Wiley & Sons, Ltd, Hoboken, NJ, (2004).
[3] H. Strathman, Membrane separation processes, J. Membr. Sci. 9 (1981) 121–189.

[4] P. Shao, R.Y.M. Huang, Review-polymeric membrane pervaporation, J. Membr. Sci. 287 (2007) 162–179.
[5] W.J. Koros, G.K. Flemming, S.M. Jordan, T.H. Kim, H.H. Hoehn, Polymeric membrane materials for solution-diffusion based permeation separation, Prog. Polym. Sci. 13 (1988) 339–401.
[6] P. Pandey, R.S. Chauhan, Membranes for gas separation, Prog. Polym. Sci. 26 (2001) 853–893.
[7] T.S. Chung, L.Y. Jiang, Y. Li, S. Kulprathipanja, Mixed matrix membranes (MMMs) comprising organic polymers with dispersed inorganic fillers for gas separation, Prog. Polym. Sci. 32 (4) (2007) 483–507.
[8] T. Yoshinaga, Ion Exchange Membranes: Fundamentals and Applications, second ed., Elsevier Science, Amsterdam, the Netherlands, (2010).
[9] Z.W. Song, L.Y. Jiang, Optimization of morphology and performance of PVDF hollow fiber for direct contact membrane distillation using experimental design, Chem. Eng. Sci. 101 (2013) 130–143.
[10] T.D. Waite, A.G. Fane, A.I. Schater, Nanofiltration: Principles and Applications, Elsevier Advanced Technology, Amsterdam, the Netherlands, (2005).
[11] A.L. Athayde, R.W. Baker, P. Nguyen, Metal composite membranes for hydrogen separation, J. Membr. Sci. 94 (1994) 299–311.
[12] H. Jiang, L. Memg, R. Chen, W. Jin, W. Xing, N. Xu, Progress on porous ceramic membranes reactors for heterogeneous catalysis over ultrafine and nano-sized catalysts, Chin. J. Chem. Eng. 21 (2013) 205–215.
[13] Y. Wei, W. Yang, J. Caro, H. Wang, Dense ceramic oxygen permeable membranes and catalytic membrane reactors, Chem. Eng. J. 220 (2013) 185–203.
[14] M. Rezakazemi, A.E. Amooghin, M. Mehdi, M.M Montazer-Rahmati, A.F. Ismail, T. Matsuura, State-of-the-art membrane based CO_2 Separation using mixed matrix membranes (MMMs): An overview on current status and future directions, Prog. Polym. Sci. 39 (2014) 817–861.
[15] P.K. Parhi, Supported liquid membrane principle and its practices: a short review, J. Chem. (2013) 1–11.
[16] J. Wang, J. Luo, S. Feng, H. Li, Y. Wan, X. Zhang, Review-recent development of ionic liquid membranes, Green Energy Environ. 1 (2016) 1–19.
[17] R. Leyma, S. Platzer, F. Jirsaa, W. Kandioller, R. Krachler, B.K. Keppler, Novel thiosalicylate-based ionic liquids for heavy metal extractions, J. Hazard. Mater. 314 (2016) 164–171.
[18] L.J. Zeman, A.L. Zydney, Microfiltration and Ultrafiltration: Principles and Applications, Marcel Dekker, New York, (1996).
[19] A. Lhassani, M. Rumeau, D. Benjelloun, M. Pontie, Selective demineralisation of water by nanofiltration application to the defluorination of brackish water, Water Res. 35 (2001) 3260–3264.
[20] N. Hilal, H. A1-Zoubi, N.A. Darwish, A.W. Mohammad, M. Abu Arabid, A comprehensive review of nanofiltration membranes: treatment, pretreatment, modelling, and atomic force microscopy, Desalination 170 (2004) 281–308.
[21] J.G. Wijmans, R.W. Baker, The solution-diffusion model: a review, J. Membr. Sci. 107 (1995) 1–21.
[22] Van. B. der Bruggen, C. Vandecasteele, T. Van Gestel, W. Doyenb, R. Leysen, A review of pressure-driven membrane processes in wastewater treatment and drinking water production, Environ. Prog. 22 (2003) 46–56.
[23] M.D. Afonso, G. Hagmeyer, R. Gimbel, Streaming potential measurements to assess the variation of nanofiltration membranes surface charge with the concentration of salt solutions, Sep. Purif. Technol. 22–23 (2001) 529–541.
[24] X. J. Duan, Study on Charged Nanofiltration Membrane Performance and Rejection Mechanism to Salt Solution, Master's thesis, Chinese Marine University, 2010.
[25] C. Tu, W. Shang, X. Wang, Reflection coefficient of nanofiltration membranes in single electrolyte solution, J. Chem. Ind. Eng. (Chinese) 58 (2007) 1507–1513.
[26] A.E. Yaroshchuk, Rejection mechanisms of NF membranes, Serono. Sym. (1998) 9–12.
[27] W.R. Bowen, J.S. Welfoot, Modelling the performance of membrane nanofiltration-critical assessment and model development, Chem. Eng. Sci. 57 (2002) 1121–1137.
[28] D.L. Oatley, L. Llenas, R. Pérez, P.M. Williams, X. Martínez-Lladó, M. Rovira, Review of the dielectric properties of nanofiltration membranes and verification of the single oriented layer approximation, Adv. Colloid Interface 173 (2012) 1–11.

[29] T.Y. Cath, E. Amy, M. Childress, Elimelech, Forward osmosis: principles, applications, and recent developments, J. Membr. Sci. 281 (2006) 70–87.
[30] Q. Ge, M. Ling, T.-S. Chung, Draw solutions for forward osmosis processes: developments, challenges, and prospects for the future, J. Membr. Sci. 442 (2013) 225–237.
[31] M. Mulder, Basic Principles of Membrane Technology, Kluwer Academic Publishers, Dordrecht, The Netherlands, (1991).
[32] K.L. Lee, R.W. Baker, H.K. Lonsdale, Membranes for power generation by pressure-retarded osmosis, J. Membr. Sci. 8 (1981) 141–171.
[33] A. Achilli, A.E Childress, Pressure retarded osmosis: from the vision of Sidney Loeb to the first experimental installation—Review, Desalination 261 (2010) 205–211.
[34] T.W. Xu, Review-ion exchange membranes: state of their development and perspective, J. Membr. Sci. 263 (2005) 1–29.
[35] L. Alvarado, A. Chen, Electrodeionization: principles, strategies and applications, Electrochim. Acta 132 (2014) 583–597.
[36] J. Luo, C. Wu, T. Xu, Y. Wu, Diffusion dialysis-concept, principle and applications, J. Membr. Sci. 366 (2011) 1–16.
[37] M. Paidar, V. Fateev, K. Bouzek, Membrane electrolysis—history, current status and perspective, Electrochem. Acta (2016) Accepted.
[38] A.C. Gupta, Preparation of ion exchange membranes by grafting acrylic acid into pre-irradiated polymer films. 1. Grafting into polyethylene, Eur. Polym. J. 25 (1989) 1137.
[39] A. Gabelmana, S.-T. Hwang, Hollow fiber membrane contactors, J. Membr. Sci. 159 (1999) 61–106.
[40] A. Kiani, R.R. Bhave, K.K. Sirkar, Solvent extraction with immobilized interfaces in a microporous hydrophobic membrane, J. Membr. Sci. 20 (1984) 125–145.
[41] P.H.M. Feron, A.E. Jansen, CO_2 separation with polyolefin membrane contactors and dedicated absorption liquids: performances and prospects, Sep. Purif. Technol. 27 (2002) 231–242.
[42] L. Yuan, Z. Wang, S.C. Wang, Membrane absorption technology and its application in removing acid gas research, Membr. Sci. Technol. 22 (4) (2002) 55–59 (in Chinese).
[43] S. Karoor, K.K. Sirkar, Gas absorption studies in microporous hollow fiber membrane modules, Ind. Eng. Chem. Res. 32 (1993) 674–684.
[44] N.A. Delia, L. Dahuron, E.L. Cussler, Liquid–liquid extractions with microporous hollow fibers, J. Membr. Sci. 29 (1986) 309–319.
[45] P.R. Alexander, R.W. Callahan, Liquid–liquid extraction and stripping of gold with microporous hollow fibers, J. Membr. Sci. 35 (1987) 57–71.
[46] R. Prasad, K.K. Sirkar, Microporous membrane solvent extraction, Sep. Sci. Technol 22 (1987) 619–640.
[47] J. De Gyves, E.R. De San Miguel, Metal ion separations by supported liquid membranes, Ind. Eng. Chem. Res. 38 (1999) 2182–2202.
[48] F. Valenzuela, C. Fonseca, C. Basualto, O. Correa, C. Tapia, J. Sapag, Removal of copper ions from a waste mine water by a liquid emulsion membrane method, Miner. Eng. 18 (2005) 33–40.
[49] P.R. Patnaik, Liquid emulsion membranes: principles, problems and applications in fermentation processes, Biotechnol. Adv. 13 (1995) 105–208.
[50] P.S. Kulkarni, S. Mukhopadhyay, S.K. Ghosh, Liquid membrane process for the selective recovery of uranium from industrial leach solutions, Ind. Eng. Chem. Res. 48 (2009) 3118–3125.
[51] N.M. Kocherginsky, Q. Yang, L. Seelam, Recent advances in supported liquid membrane technology, Sep. Purif. Technol. 53 (2007) 171–177.
[52] A. Alkhudhiri, N. Darwish, N. Hilal, Membrane distillation: a comprehensive review, Desalination 287 (2012) 2–18.
[53] E. Curcio, E. Drioli, Membrane distillation and related operations—a review, Sep. Purif. Rev. 34 (2005) 35–86.
[54] A. Kumar Pabby, A. Maria Sastre, State-of-the-art review on hollow fibre contactor technology and membrane-based extraction processes, J. Membr. Sci. 430 (2013) 263–230.

[55] H. Li, X. Wang, L. Cao, X. Zhang, C. Yang, Gold-recovery PVDF membrane functionalized with thiosemicarbazide, Chem. Eng. J. 280 (2015) 399–408.
[56] J. Zaman, A. Chakma, Review-inorganic membrane reactors, J. Membr. Sci. 92 (1994) 1–28.
[57] A. Iulianelli, P. Ribeirinha, A. Mendes, A. Basile, Methanol steam reforming for hydrogen generation via conventional and membrane reactors: a review, Renew. Sustain. Energy Rev. 29 (2014) 355–368.
[58] S. Loeb and S. Sourirajan, Sea water demineralization by means of an osmotic membrane, in: Saline Water Conversion–II, Advances in Chemistry Series Number 28, American Chemical Society, Washington, DC, 1963, pp. 117–132.
[59] D. Bhattacharyya, J.A. Hestekin, S.M.C. Ritchie, L.G. Bachas, Functionalised membranes remove and recover dissolved heavy metals, Membr. Technol. 110 (1999) 441.
[60] S.M.C. Ritchie, K.E. Kissick, L.G. Bachas, S.K. Sikdar, C. Parikh, D. Bhattacharyya, Polycysteine and other polyamino acid functionalized microfiltration membranes for heavy metal capture, Environ. Sci. Technol. 35 (2001) 3252–3258.
[61] C.M. Wang, Investigation on treating oil emulsion wastewater. Master's Thesis, Nanjing University of Chemical Engineering, 1999.
[62] H. Song, R. Field, T. Arnot, Treatment of emulsified oily wastewater by microfiltration, Water Treat. Technol. (in Chinese) 202 (1994) 95–98.
[63] F.L. Lou, X.F. Lu, G.M. Liang, Treatment of metallurgical oil emulsion wastewater using ultrafiltration. The First National (China) Symposium on Application of Membrane-based Separation in Metallurgical Industry, 1999, pp. 193–197.
[64] M.D. Raymond, Ultrafiltration for oily wastewater treatment, J. Am. Soc. Lubric. 38 (4) (1982) 219–222.
[65] W. W. William. System and process for processing used emulsion coolant, US Patent 4738781, 1988.
[66] B.E. Reed, Treatment of oily wastes using high-shear rotary ultrafiltration, J. Environ. Eng. 123 (12) (1997) 1234–1242.
[67] E.A. Mason, From pig bladders and cracked jars to polysulfones: an historical perspective on membrane transport, J. Membr. Sci. 60 (1991) 125–145.
[68] K. S. Koga, Ushikoshi, Desalination 23 (1977) 105.
[69] T. Thorsen, Recovery of phosphoric acid with RO, Desalination 53 (1985) 217.
[70] C. Slater, A. Ferrari, P. Wisniewski, J. Environ. Sci. Health A22 (1987) 707.
[71] K.W. Mok, P.J. Pickering, J.E. V. Broome, US Patent 9859385, 1998.
[72] L. Nicod, F. Chitry, E. Gaubert, M. Lemaire, H. Barnier, J. Incl. Phenom. Macrocycl. Chem. 34 (1999) 145.
[73] X. Zheng, Chapter one: overview of global membrane market, in: X. Zheng, W.Y. Song, Z.W. Wang, Z.X. Zhang (Eds.), Report for Sustainable Development Strategy of China Water Treatment Industry, China Renmin University Press, Bejing, 2016.

PART 2

Membrane-Based Separation for Metallurgical Process Improvement and Wastewater Treatment

CHAPTER 3

Ferrous Metallurgy

Y.-M. Cao*, Y.-H. He**

**Dalian Institute of Chemical Physics, Chinese Academy of Science, Dalian, Liaoning Province, China*
***School of Powder Metallurgy, Central South University, Changsha, Hunan Province, China*

3.1 Iron and Steel

3.1.1 Introduction

Iron and steel industry is very critical for a country's economy. In the past decade, the fast development in China with the expansion of industrialization, urbanization, etc. has been raising the consumption demand of iron and steel. In 2011, the steel output in China was 782 million tons, accounting for more than 50% of the world total [1]. The process involved in iron and steel industry is very complicated. An integrated iron and steel process is illustrated in Fig. 3.1, comprising principally three parts, which are ironmaking, steelmaking, and steel rolling. A brief introduction of the three sectors is given as follows [2].

Ironmaking: Blast furnace is the core equipment of the ironmaking. The major raw materials including ironstone (e.g. ferric oxide (Fe_2O_3)), air, coke, limestone, etc. are added to the furnace. With the physical and chemical reactions at high temperature, iron oxides are reduced to liquid metal iron. Thereafter, the molten slag covering the surface of the molten iron is removed from the taphole, cooled by water, whereas the hot metal is delivered into hot metal ladle via passing through taphole. The iron is then shipped to the mixer, for homogenizing the composition and the temperature, which is the preparation step for steelmaking.

Steelmaking: Molten iron is delivered in hot state to the steelmaking workshop, hoisted to converter platform by metal ladle crane, and subsequently transferred into the furnace. Clean steel scraps or cast iron pieces is added according to the amount of the load, temperature, and composition, in order to improve the steel production. After loading is completed, the slag making materials (e.g. lime, dolomite), and steel scraps added to the furnace, and oxygen is blown into it. During the physical and chemical reactions in the furnace, carbon (C) in the hot metal is oxidized into carbon monoxide (CO) and carbon dioxide (CO_2), which is then eliminated in the form of air bubbles. All of the silicon (Si) and part of the Fe are oxidized, releasing a large amount of heat at the same time. When the temperature in the oven reaches about 1400°C, slag making materials begin to melt and react; the harmful elements in the molten iron such as sulfur (S) and phosphorus (P) are oxidized and transferred to the liquid steel slag, thus purifying and meanwhile

Membrane-Based Separations in Metallurgy
http://dx.doi.org/10.1016/B978-0-12-803410-1.00003-7

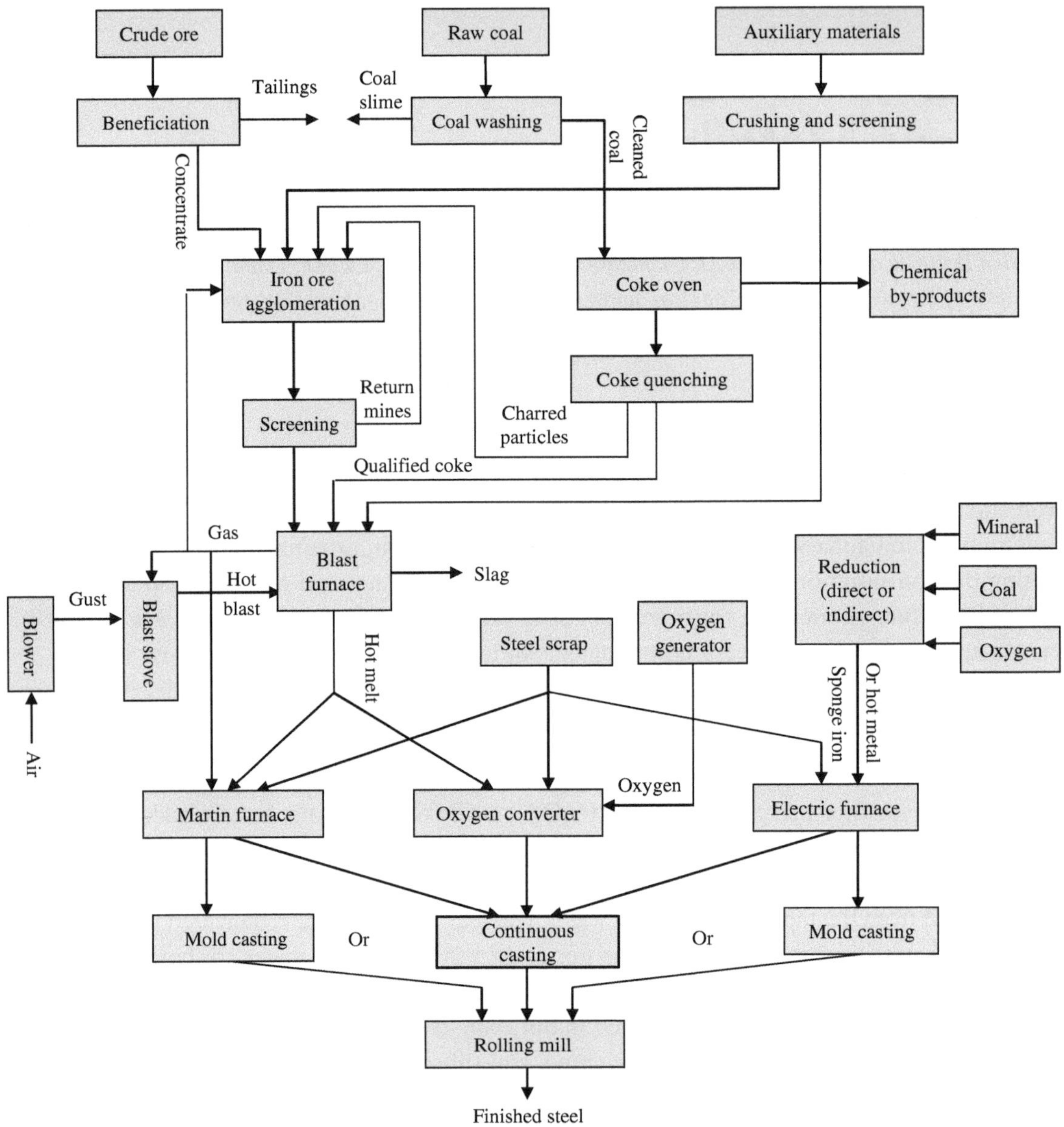

Figure 3.1: Production process in integrated iron and steel enterprise (not including system for supplying steam, water, electricity, etc.) [1,2].

controlling the composition of molten steel. On condition that the composition and temperature satisfy the requirements, tapping will be conducted by adding alloy pieces, such as ferrosilicon (Fe-Si), ferromanganese (Fe-Mn), and silicon–manganese (Si-Mn) to further adjust the composition of the molten steel. The following step is degassing. At the same time, a suitable amount of aluminum (Al) wire is added to the ladle, which is also useful for managing desired properties of the steel product. Finally, the steel ladle is delivered by crane to the continuous casting platform.

Steel rolling: Advanced iron and steel enterprises generally adopt continuous casting and rolling process. The continuous casting billet is cut directly inside the furnace. After adjusting the temperature of the billet to suit the requirements of rolling, it is sent to rolling mill, where the end product of steel is obtained after roughing, intermediate rolling, and finish rolling.

In iron and steel manufacturing system, the material and energy consumption are intensive, conjugated with discharge of huge amount of waste containing myriads of polluters. The wastewater discharged from primary iron and steel plants reached 0.63 billion m^3 in 2011 in China [3]. Therefore, nowadays the iron and steel industry, particularly in China, is greatly challenged against the outline of more stringent regulations on environmental protection. In the following section, the applications of membrane-based separation in iron and steel manufacturing will be summarized, with emphasis on waste treatment.

3.1.2 Treatment of Acidic Wastewater

During the hot rolling or heat treatment of various kinds of steel, O_2 from the air reacts with the Fe on the surface of the steel to form a crust that is made-up of a mixture of iron oxides. The presence of oxides (or scale) on the surface of the steel is undesired when the steel is to be subsequently shaped, cold rolled, coated, or treated by, for example, galvanizing, electroplating, porcelain enamel.

The major role of pickling is to get high quality metal surface by chemical treatment to remove the adherent scale composed mainly of rust and impurities. Waste pickling solutions from stainless steel industries generally contain metal complexes, acids, and free anions [4]. Fig. 3.2 shows the scheme of a pickling line with the respective waste streams.

The pickling step prior to the galvanization comprises treating the metal surface with acid in order to remove the adherent scale composed mainly of rust and impurities. The series of effluents generated in this process generally contain high concentrations of zinc chloride ($ZnCl_2$), ferrous chloride ($FeCl_2$), and hydrochloric acid (HCl), and some minor components including organic compounds, such as hydrogen (H_2) evolution reaction (HER) inhibitors, other heavy metals (eg, Pb, Cr) and metal oxide particles [5]. A Fe rich effluent will be obtained when new metal pieces are dealt with, whereas effluent having more Zn is produced when galvanized pieces are treated. In some situation, both kinds of pieces are treated, obtaining a mixture effluent.

When pickling is employed for follow-up steel rolling, different solutions—mixture of hydrofluoric (HF) and nitric acid (HNO_3) or sulfuric acid (H_2SO_4) and its mixture with HF or hydrogen peroxide (H_2O_2), Fe, and other metal ions (eg, Cr, Ni)—are resulted [6]. Table 3.1 summarizes the compositions of different pickling solutions [4]. Sometimes H_2SO_4 is used instead of HNO_3 in pickling baths for stainless steel treatment to make the bath less

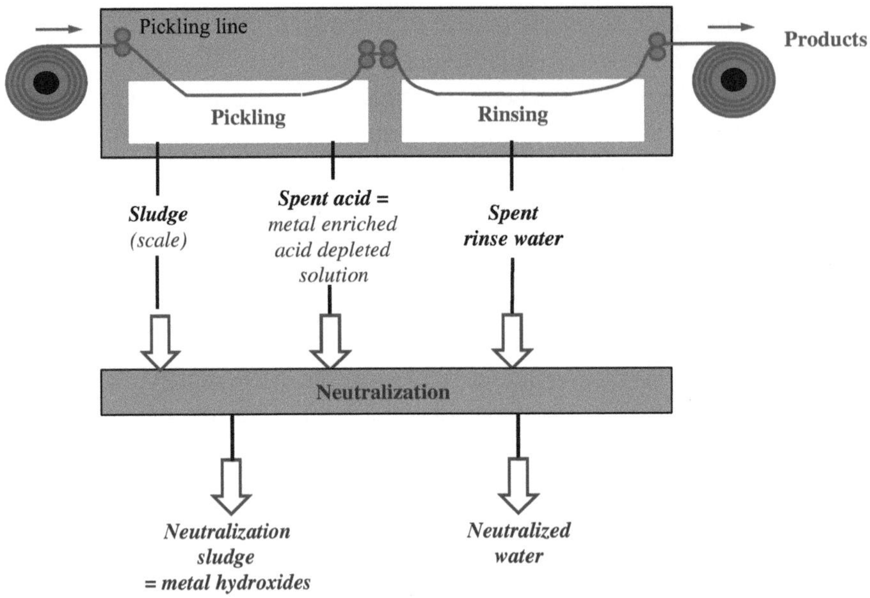

Figure 3.2: Waste streams generated in pickling lines [4].

aggressive. In the past, H_2SO_4 was very often applied for pickling; however nowadays in most cases it is replaced by HCl because the latter ensures better surface quality and a superior finish effect, faster pickling resulting in increase in steel throughputs and more economical acid regeneration [7]. When pickling with a HNO_3/HF mixture is carried out, HNO_3 oxidizes Fe, Cr, and nickel (Ni), while fluorides form stable complexes with these metals. Fluoride complexes will precipitate out, if the concentration of metals in the pickling bath exceeds 5%. Thereafter, the spent bath is renewed and fresh acid is added [8].

During pickling, the acid in solution is consumed, while the content of dissolved metals increases, leading to additional depression in pickling efficiency with each cycle of reuse. Pickling solutions are considered spent or useless when the acid concentration decreases by 75–85%, coupled with metal content increase even up to 150–250 g dm^{-3} [4]. These effluents present serious environmental problems due to the hazardous nature. In addition, the amount of spent solution that has to be disposed of is tremendous in metallurgical industry. For example, the steelwork plants in European Union (EU) deliver 300,000 m^3 of spent pickling solutions each year. The other acidic solutions mainly refer to rinsing and electroplating wastewaters.

The commonly applied precipitation–filtration technique prior to wastewater discharge comprises three steps: conversion of Fe^{2+} to Fe^{3+} by H_2O_2 oxidization, precipitation of the heavy metals as hydroxides by using lime or caustic soda to tune the pH up to 7–8, and finally the solution settlement and filtration of the generated sludge. With this approach, nevertheless, valuable metals are irrecoverably lost. Complete or partial recovery or

Table 3.1: Compositions of selected real spent pickling solutions [4]

Acid type	Acid conc. (g dm^{-3})	Zn (g dm^{-3})	Fe (g dm^{-3})	Other
H_2SO_4	150	–	70	–
	92	–	78	–
HCl	237	~80	90-96 Fe(II), or 84 Fe(II), 8.3 Fe(III)	6.4 M Cl^-, less than 400 ppm of Mn, Pb, Al, Cr, Ni, Cd, Cu, Co
	150	–	50	–
	101	–	8.9 Fe(III)	–
	90	80	30 Fe(II)	6 M Cl^-
	85	25	160 (158 Fe(II), 2 Fe(III))	Traces of Cd
	80	4.65	88.5 (incl. 1.6 Fe(II))	–
	~70	34	204	0.04 g dm^{-3} Cr
	~40	110	88.5	7.7 M Cl^-
	33	12	106Fe(II)	–
	20-60	–	80—110 Fe(II)	–
	18	26	140	0.09 g dm^{-3} Pb, 0.08 g dm^{-3} Ni, 0.03 g dm^{-3} Cu, 0.09 g dm^{-3} Al
	~10	70	92	–
	30	20	120	–
Mixed acids HNO_3 HF	180-200		28—30 (max. 35)	10-15 g dm^{-3} Cr(III), 5-10 g dm^{-3} Ni(II)
	40-45			
	120-150	–	30-45 Fe(III)	5-10 g dm^{-3}Cr(III), 3-5 g dm^{-3} Ni(II)
	15-30			–
	125	–	40	
	30			

regeneration of the spent acids and metals is state-of-the-art, for achieving cost effective and sustainable stainless steel production. Many different technical approaches for pickling solution regeneration are proposed and investigated [6], such as ion exchange, liquid–liquid extraction, pyrometallurgical method, and solvent extraction. Regel-Rosocka [4] has given a comprehensive review on the pros and cons of the different methods in regenerating spent pickling solutions. The membrane techniques that have been reported for treating these solutions include diffusion dialysis (DD), membrane distillation (MD), electrodialysis (ED) with bipolar membranes (BPM), cation-exchange membranes (CEMs), Neosepta membranes and anion-exchange membranes (AEMs), membrane electrolysis, and/or nanofiltration (NF).

DD was first used by companies from United States in recovering acid from iron and steel pickling solution [9]. In China, Shanghai Silicon Steel Plant adopted the technology for

Table 3.2: Concentrations of metal ions in the picking solution of steels [9]

Metal ions	Fe	Ni	Cr	Mn
Content (g L^{-1})	4.138	0.452	0.789	0.073

recovery of H_2SO_4 as early as 1970s; the AEM used was S-203, a kind of noncrosslinked quaternary ammonium polysulfone (PSF) membrane. DD has also been in use in Sweden for treatment of stainless steel pickling baths (HF/HNO_3) [4].

Wang et al. investigated the efficiency of DD in recovery of HNO_3 from stainless steel pickling solution [9]. The concentration of the acid in waste is 1.38 mol L^{-1} and the major metal ions with their corresponding concentrations are shown in Table 3.2. Two types of AEMs, DSV and APS supplied by Japan Asahi Co. Ltd., were investigated. The experimental results shown in Fig. 3.3 indicate that the acid recovery, η, kept decreasing as the feed flow rate increased from 100 to 240 mL h^{-1}. This is because higher feed flow rate shortens the retention time that is critical for adsorption and diffusion of ions inside membrane matrix. In addition, the impact of feed flow rate is greater on the retention time reduction than the abatement of boundary layer thickness. As a result, the mass transfer of HNO_3 inside membrane becomes lower. Due to the same mechanism, the rejection efficiency of metal ions increased at higher feed flow rate. Both membranes exhibit rejection coefficients $> 95\%$ for all the metal ions, with those of APS being slightly better. Contents of metal ions in permeate with different feed flow rates are listed in Table 3.3. It is also found that for both APS and DSV membranes, the rejection is improved with higher concentration of HNO_3. This is due to the balance between the two effects caused by higher level of the original

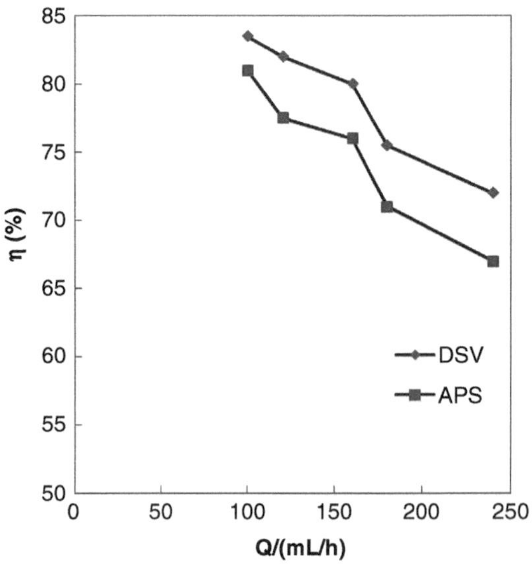

Figure 3.3: Relationship between recovery rate and nitric acid feed flow rate [9].

Table 3.3: Composition of metal irons in the recovered solution under different feed flow [9]

Volume flow rate (mL h^{-1})	Fe (g L^{-1})		Ni (g L^{-1})		Cr (g L^{-1})	
	DSV	APS	DSV	APS	DSV	APS
100	0.1425	0.0642	0.0094	0.0002	0.0128	0.0023
120	0.1264	0.0456	0.0073	0.0013	0.0090	0.0019
160	0.0832	0.038	0.0072	0.0003	0.0072	0.0009
190	0.0645	0.0232	0.0050	0.0001	0.0055	0.0005
240	0.0628	0.0217	0.0047	0	0.0053	0.0004

acid concentration: (1) the anion concentrations inside membrane become higher, hence stronger mass transfer driving force; (2) ionic strength increases, which reduces the average activity coefficient of electrolyte and diffusion coefficient in solution, and depresses the acid dissociation and hence actual driving force in solution.

DD is regarded as one of the best available technologies [3] because of the low operating and maintenance costs, compact apparatus and space, low energy consumption, considerable reduction in fresh chemicals consumption and wastewater dumping, etc. Its disadvantage is that HCl acid recovered by DD may be contaminated by metal ions with smaller diameter (eg, zinc ions (Zn^{2+}) in recovery of galvanizing pickling solution). These ions transport together with the acid. DD enables effective separation of Fe^{2+} with a separation factor over 100 in relation to Zn^{2+}.

Special industry membrane separation (SIMS) process with NF and/or reverse osmosis (RO) is adopted by HWA Technology (HK) Co. Ltd. in dealing with rinsing and pickling wastewater from steel rolling plant [11]. According to the characteristics of the acidic wastewater and the recovery requirements highlighted in Table 3.4, several-processes are proposed and applied, as shown in Figs. 3.4–3.7. For all the processes, the common feature is the installation of prefilter for retaining suspended solids (SS). In case of the presence of slick oil, additional oil trapping device will be employed.

Table 3.4: Quality data of acidic wastewater of a steel rolling plant [11]

Water quality	Mixed acid rinsing water	Sulfuric acid pickling waste water	Sulfuric acid rinsing water
Water flow Q/(Lh^{-1})	1000	8000	8000
Main ingredients mass concentration (g L^{-1})	HF 4 HNO_3 20 Me* 3	H_2SO_4 300 Me* 65 SS 1	H_2SO_4 20 Me* 5 SS 1
Requirements	Concentrate and recover mixed acid; recovery HNO_3 when its content is greater than 10%.	Purify sulfuric acid, removed metal ions.	Concentrate and recover H2SO4 when its content is higher than 10%.

*Me indicates Metal.

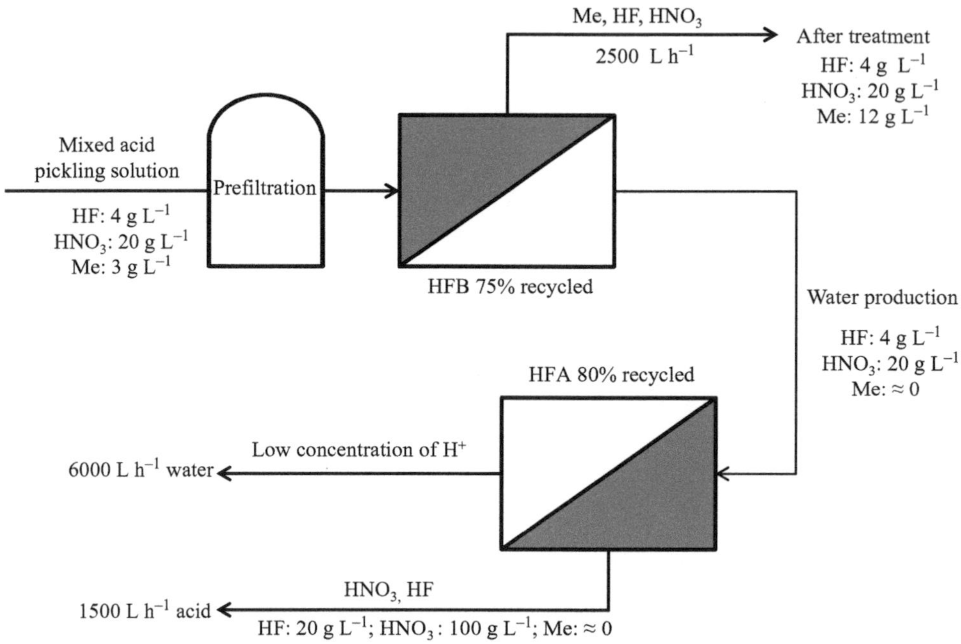

Figure 3.4: Treatment of mixed acid pickling solution by SIMS method.

As for mixed acid rinsing water, the treatment process is given in Fig. 3.4. A SIMS HFB membrane is utilized to remove dissolved organics with molecular weight (MW) higher than 200 g mol^{-1} and the metal ions. The working pressure is 42 bars and the recovery for mixed acids is up to 75%. The concentration of metal ions in the retentate is 4–5 times of the original level, favoring posttreatment or metal recovery. The permeate being almost completely removed of metal ions is delivered to the SIMS HFA membrane system, whereby an increment of four times in concentration for mixed acids is achieved. About 80 vol.% of water passes through the membrane, containing only trivial amount of acids. The operation pressure for the second stage membrane system is 70 bars.

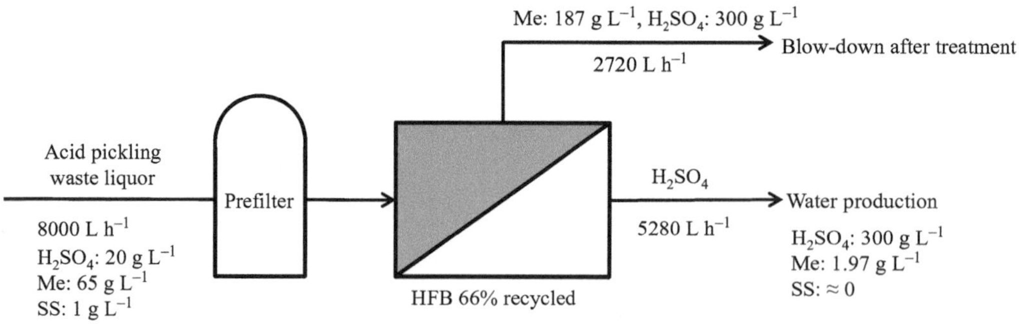

Figure 3.5: Sulfuric acid pickling solution treatment by SIMS method.

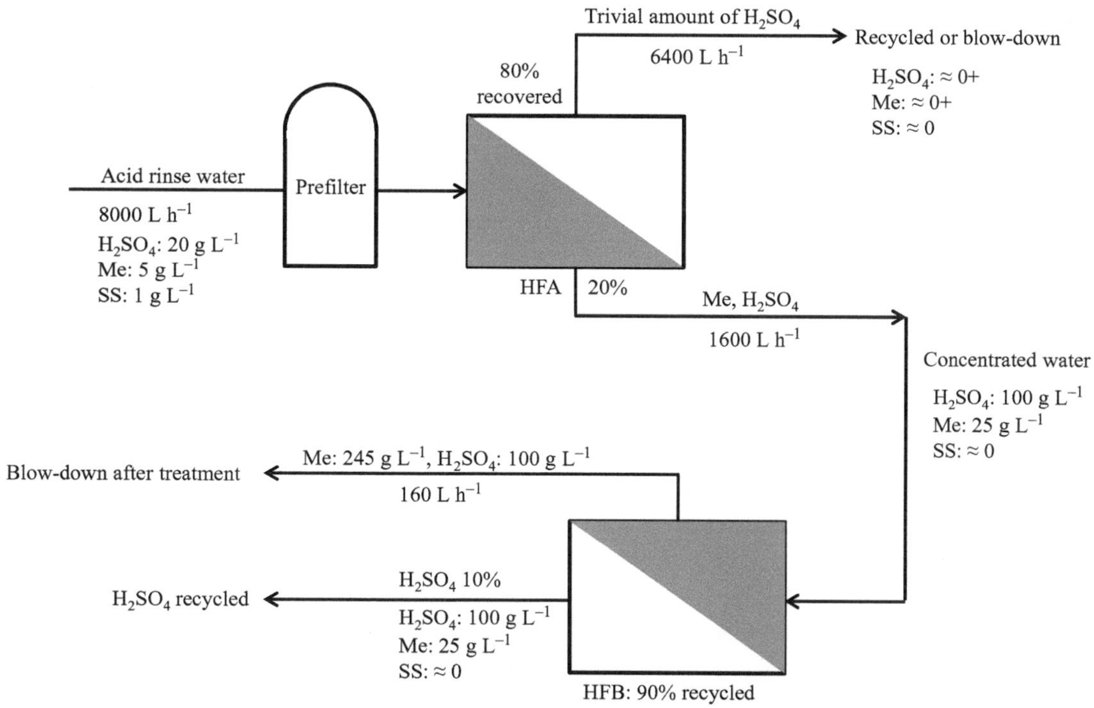

Figure 3.6: Flow diagram for processing sulfuric acid pickling solution using SIMS scheme.

The process proposed for recovering H_2SO_4 pickling solution is shown in Fig. 3.6. The sole SIMS membrane system allows preferential passage of water and H_2SO_4 into permeate, whereas it rejects most of the dissolved organics (MW higher than 200 g mol^{-1}) and metal ions. With this system, 66% of H_2SO_4 is recovered as pure acid and can be reused in pickling step; 98% of metals are concentrated.

As for H_2SO_4 rinsing water, a two-stage integrated membrane process is suggested with its schematic flow diagram demonstrated in Fig. 3.6. In the first membrane stage, the RO system operating at 69 bars intercepts efficiently all the dissolved components in the wastewater; an 80% recovery rate for purified water is resulted. The retentate is then transferred to the second treatment stage using NF by which, metals are enriched in rentent ate with a concentration up to 245 g L^{-1}; meanwhile 90% of H_2SO_4 goes to the permeate side along with 25 g L^{-1} metals.

Considering the similar compositions and treatment requirements of the H_2SO_4 rinsing and pickling solution, a novel scheme integrates the treatment of both streams in one system. As shown in Fig. 3.7, the H_2SO_4 rinsing water concentrated by HFA system is mixed with original H_2SO_4 pickling water. It is necessary to remove the particles carried forward by the pickling solution prior to second stage of membrane process. With HFB system operating

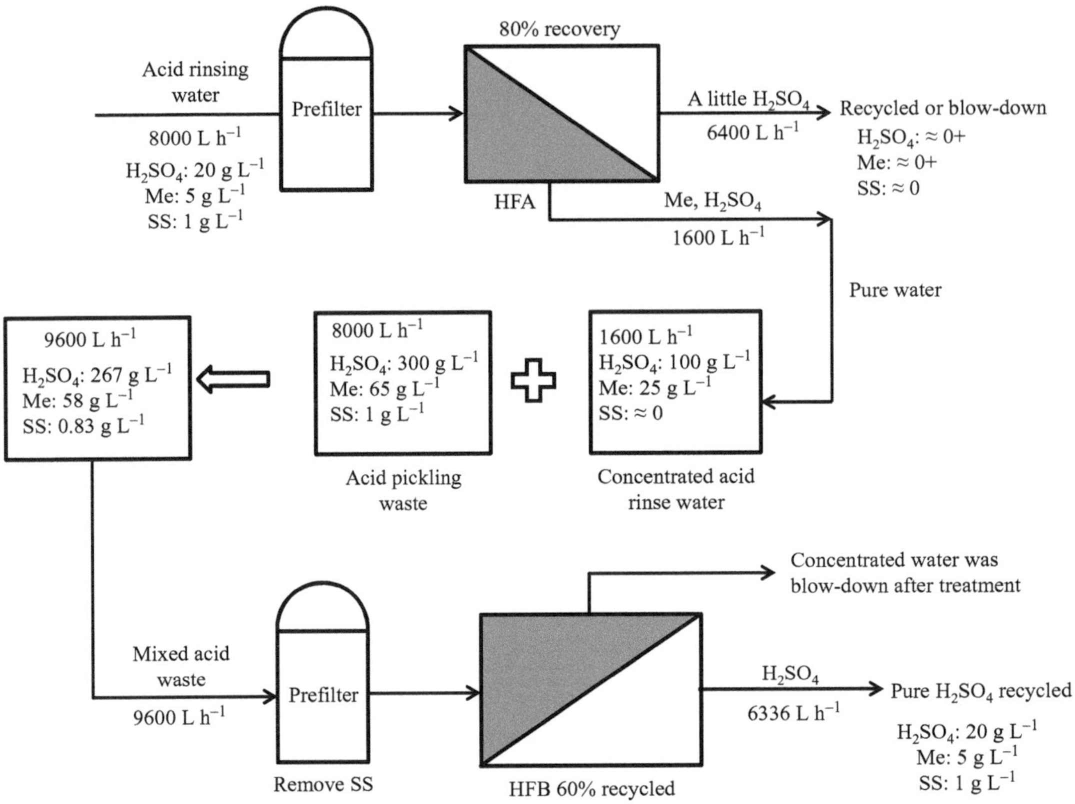

Figure 3.7: Comprehensive processing of waste sulfuric acid and rinse water by SIMS [11].

at 70 bars, 66% of H_2SO_4 in the form of pure acid is recovered for the mixed solution. The overall system takes up 50 m^2 and the investment is around USD 650,000. The operation cost is less than 20 RMB ton^{-1}.

The comparison between the SIMS and the DD for pickling acid treatment is made in Table 3.5. The SIMS schemes introduced previously exhibit the higher level of purification and concentration efficiency, processing capability, and economic benefit. It is well suited for tackling with acidic wastewater with large volume and relatively low concentration.

Combination of NF with crystallization has been investigated for the HF/HNO_3 regeneration in two stainless steel plants in Sweden [12]. Metals are generally crystallized as fluoride salts [mainly hydrated ferric fluoride ($FeF_3{\cdot}3H_2O$)] that can form metal oxides after pyrometallurgical treatment. Fe has been successfully separated from spent pickling solutions by precipitation in the form of β-$FeF_3{\cdot}3H_2O$ in a continuous crystallizer (10 L scale), whereas the solution is concentrated by NF.

MD driven by vapor pressure difference has also been investigated for treating pickling solution. Its major function is the concentration of HCl and inorganic salts [13,14].

Table 3.5: Technological comparison between SIMS and DD [11]

	SIMS	DD
Technical principles	Osmosis driven by external pressure	Natural osmosis and pure water absorption
Concentration of the acid being treated	<30%	>5%
Removal rate of the metal ion	~99%	~90%
Treatment capacity	Any capacity	6–8 ton day^{-1} for a single system
Equipment investment (10 t h^{-1})	<5000,000 CNY	>10,000,000 CNY
Operation cost (10 wt.%)	<20 CNY/DAY	Very low
Preferential applications	<15%, process load is big	>10%, process load is small
Collaborative areas	Reduce significantly the solution volume by concentration using SIMS, followed with acid separation DD.	

The increase of temperature and acid concentration in the feed caused a substantial increase in HCl molar flux through the membrane. The presence of ferric chloride ($FeCl_3$) and the increase of its concentration in the feed also cause an obvious increase in HCl molar flux in comparison with when pure acid is used as a feed. This observation is attributed to the salting out effect. As shown in Fig. 3.8, the addition of $FeCl_3$ to the HCl solution increased the partial pressure of HCl, while lowering that of water vapor (not shown in the figure). The salt was retained in the feed, and the permeate was pure HCl solution.

The experiments were also performed with a real metal pickling solution mainly containing salts of copper (Cu), Fe, Zn, and magnesium (Mg) [14]. The concentration of HCl solution varied from 85.5 to 135.25 g L^{-1}, as shown in Fig. 3.9. The solid residue or inorganic salts

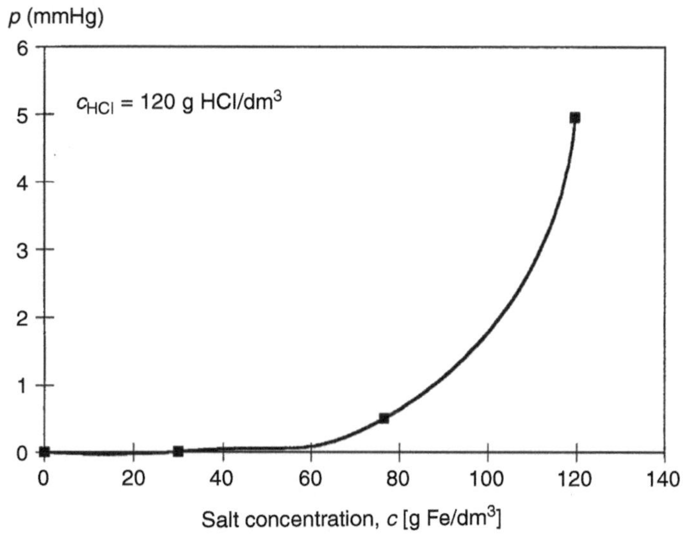

Figure 3.8: Influence of salt concentration in the liquid phase on the partial pressure of HCl [13].

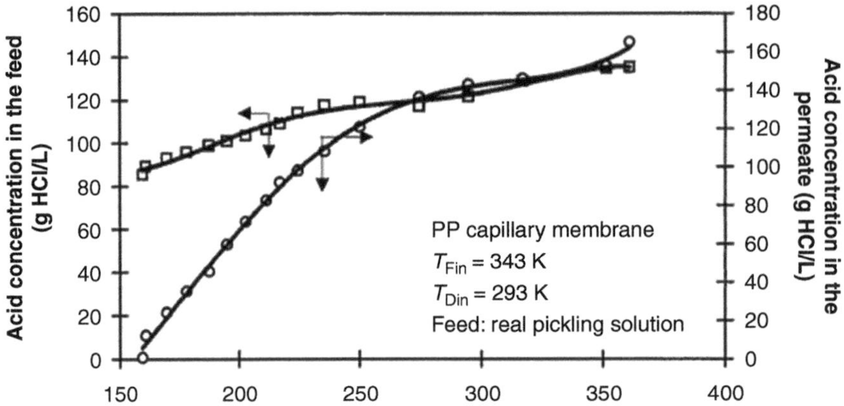

Figure 3.9: Gradual concentration of a raw pickling solution by MD (method I).
The effect of the feed composition on the distillate concentration.

in the raw solution was equal to 259.3 g dm^{-3}. With this amount of salt in the feed, the HCl was efficiently transferred during the first hour of the MD due to the salting out effect, although the acid concentration was rather low. In another research, the feed solutions consisted of a mixture of HCl and H_2SO_4 with various initial concentrations [15]. It was found that during MD process the H_2SO_4 was retained in the feed, attaining progressively increased concentration; whereas HCl was concentrated at the distillate side even at low feed concentrations. A higher H_2SO_4 content (150 g L^{-1}) in the feed facilitated the desorption of HCl and the corresponding HCl flux across the membrane in MD was 20.12 mol m^{-2} d^{-1}, 10 times of that when the H_2SO_4 concentration was only 50 g L^{-1}.

MD is more energy consuming than DD; however the recovery of HCl solution with high purity renders economical benefits, such as decrease in the consumption of alkali used for neutralization of wastewater and simultaneous separation of toxic metals and HCl. Nevertheless, the experiments with the model solutions confirm the necessity of membrane cleaning, especially on the feed side [14]. The formation of salt crystals in the membrane pores may cause membrane wetting.

Yet another membrane technology that has been considered for pickling solution recovery is extraction-stripping. The membrane-based extraction can be operated in two different modes: (1) two modules with the organic phase circulating in one side of each module; (2) one module in which polymeric support is impregnated with the organic phase—supported liquid membrane (SLM).

Membrane extraction is very flexible with the capability of treating effluents with a wide range of metal ions concentration. Nevertheless, proper filtering of the industrial waste is needed to avoid fouling formation and ensure stable operation of the contactor. In addition, careful control of the transmembrane pressure is highly necessitated to prevent breakthrough of one phase into another. Actually at present, none of membrane extraction systems is

Table 3.6: Average spent pickling liquor composition [15]

Component	Concentration (g L^{-1})
HCl	36[a]
Zn	134
Fe(II)	74
Fe(III)	<1
Cl	300[b]
Cr, Cd, Pb, Cu, and Ni	Trace
Grease, bismuth (Bi)	Variable

[a]1.0 mol L^{-1}.
[b]8.5 mol L^{-1}.

Table 3.7: Initial concentrations for laboratory-based HFM experiments [15]

Experiment	Zn (g L^{-1})	Fe (g L^{-1})	HCl (mol L^{-1})	Cl (mol L^{-1})
Industrial waste (extraction)	133	59.8	1.0	7.2
Model solution (extraction)	82	0	1.0	5[a]
Loaded TBP (stripping)	5.5	0	0.025	0.19

[a]Chloride concentration adjusted by addition of NaCl.

working at industrial level. As for SLM, it is characterized of liquid membrane instability and remains a rather laboratory method for preliminary studies.

In Lum et al.'s work [16], the extraction of Zn using tributyl phosphate (TBP) from real galvanizing effluent [from Industrial Galvanizers Australia (Melbourne) Plant] and synthetic solutions using nondispersive hollow fiber membrane extraction and back extraction was investigated. Table 3.6 shows the average spent pickling liquor composition. The initial concentrations for the various solutions in the experiments are displayed in Table 3.7. In the back extraction, water is used for recovering the Zn. It was shown that extraction rate of Zn was higher than that of HCl; however, the stripping experiments showed that HCl is more easily stripped. Therefore, despite the low percentage of acid extraction, the percentage of stripped acid is comparable to that of Zn. In addition, the process was strongly disturbed by the competitive transport of Fe^{3+} if present [16]. In Alguacil et al.'s work, Zn^{2+} transport through supported liquid membrane (SLM) impregnated with 30% Cyanex 923 in Solvesso 100 is controlled by diffusion in the aqueous solution, and needs a long operation time [17].

San Roman et al. proposed an integrated membrane process consisting of a liquid membrane and DD [18]. The flow diagram is illustrated in Fig. 3.10. The first step generates a raffinate highly concentrated in iron chloride and a mixture of HCl and zinc chlorides (i.e. $ZnCl^+$, $ZnCl_2$, $ZnCl_3^-$, $ZnCl_4^{2-}$) in the stripping phase [18]. Subsequently, diffusion or Donnan dialysis (DD) was used to treat the stripping solution from step one in order to further transfer the

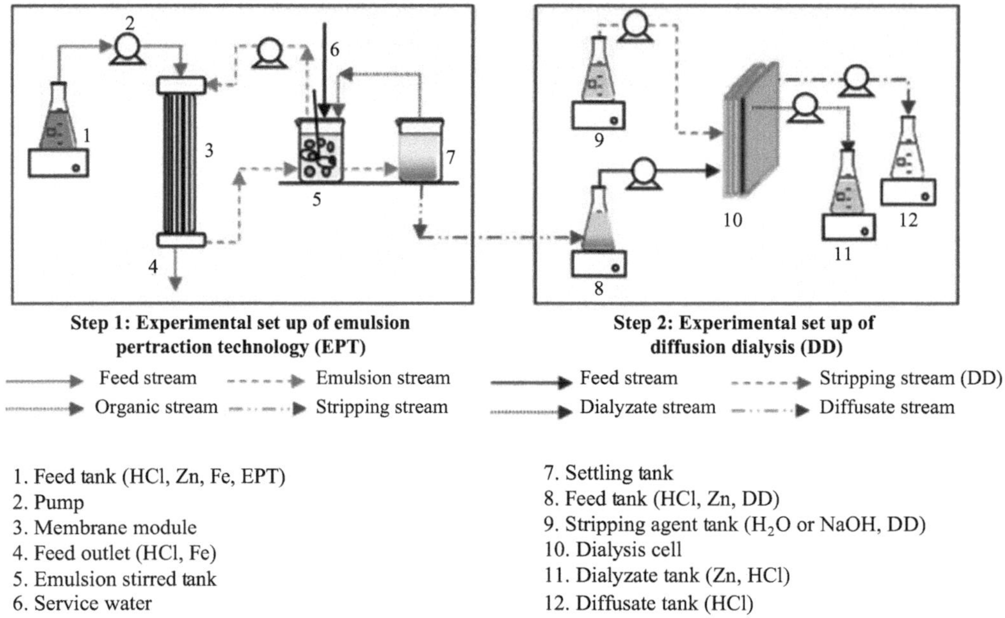

Figure 3.10: Experimental set-up for the integrated membrane process [18].

acid to the diffusate and thus separation from zinc chlorides in dialyzate. In the first step of emulsion pertraction, the spent pickling effluent circulates through the inner side of the hollow fiber, while the emulsion phase flows through the shell channel. The loaded emulsion phase is transferred to the settling tank, where the aqueous stripping product and organic solvent are separated. The aqueous solution, enriched in Zn and HCl is sent to the second step and the emulsion stream is replenished with water. In the second step, the aqueous solution is further treated in the dialysis cell. At the beginning of the DD experiment, the aqueous feed and the stripping agent (water or NaOH solution) entered the bottom part of the compartment 1 of DD system. During the experiments, the dialysis unit separates the feed into two solutions with different concentrations. The dialyzate is mainly made up of zinc chlorocomplexes and the diffusate tank contains hydrochloric acid or sodium chloride. No information regarding the components in emulsion phase is provided.

In DD, NaOH solution as a stripping liquid delivers higher selectivity for Zn/Cl. Driven by the concentration difference across the AEM, the transport of OH^- from the diffusate to the dialyzation happened, coupled with Cl^- moving in opposite direction. Although the $ZnCl_3^-$ and $ZnCl_4^{2-}$, negatively charged, have the tendency to proceed along with Cl^- to the diffusate, their larger size impact negatively on their transport rates.

ED for pickling solution treatment was proposed in 1990s. It can separate acid from the feed, and achieve a high concentration factor that enables the recycling of the acid to the pickling bath. Application of ED as a part of an industrial process to treat solutions containing HCl or H_2SO_4, Fe and Zn ions, has been investigated by Paquay et al. [19]. The large difference between the faradic yields with or without additives can be explained by the complex formation capacity of ions in the solution in the presence of organic compounds. This can improve or weaken the mass transfer of ionic species across the ion-exchange membrane (IEM). It is also concluded by the authors that for the moment, currently available electrodialyzer design is not economically suitable for some "common" industrial applications. It was necessary to develop new electrodialyzers that are more flexible, cheaper, more convenient in operation, and have lower maintenance costs. These objectives can be achieved by considering high flow rate hydrodynamic conditions in the ED equipment. With ED, the problem of undesired by-product formation [eg, chlorine (Cl_2) gas] should be solved; otherwise the membranes would be destroyed as they are sensitive to Cl_2 gas attack.

In spite of the drawbacks in terms of energy intensiveness, high investment costs, undesired sludge, ED unit has been industrialized. One system installed at the Outokumpu plant in Nyby is used to reduce the discharge of nitrates by HNO_3 recycling, and to diminish by 50% fresh water and HNO_3 consumption.

Tokuyama Soda Co. Ltd. of Japan proposed a close circle process as shown in Fig. 3.11 comprising ED and BPM ED to handle waste pickling acid [7,20]. Firstly, the Fe, Cr, and Ni

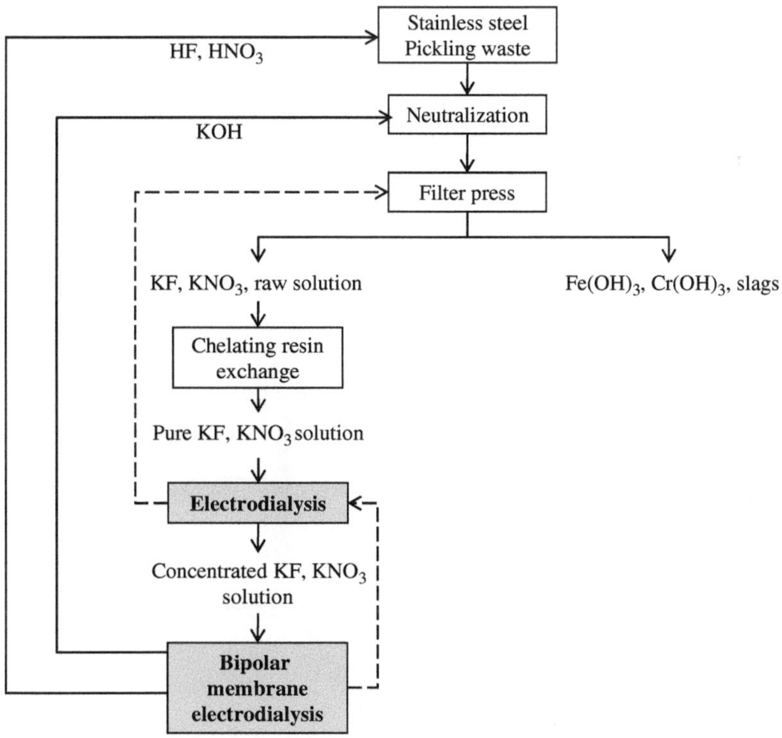

Figure 3.11: Process for treating waste acid from stainless steel pickling [20].

in the pickling solution were converted into hydroxide precipitates by reaction with KOH and separated from the solution by filtration. Thereafter, the potassium fluoride (KF)/potassium nitrate (KNO_3) solution was removed of the trace amount of divalent ions by chelating resin, followed with concentration by ED. The dilute solution produced in ED is recycled to the filtration step and the concentrate was sent to electro-splitting by BPM ED for generating acid (HF and HNO_3) and base (KOH). The two chemicals are then cycled back to pickling and neutralization processes, respectively. The main drawback of the BPM application is fouling by potassium fluorosilicate (K_2SiF_6), when silica is present in spent pickling solution [20].

Fig. 3.12 describes the principle of membrane electrolysis, a kind of membrane reactor (MR) involving the transport of ions through semipermeable membranes and the subsequent electrodeposition of metals on the cathode. Csicsovszki et al. proposed a combination of anion exchange and membrane electrowinning (EW) techniques to recover HCl, and separate Zn from Fe [21]. First, Zn in pickling solution is separated using anion-exchange resin (quaternary ammine type polystyrene-DVB) and eluted with 0.1 M HCl, whereas Fe and acid remain in the effluent. Then Fe is deposited on the Ni cathode at pH1–1.5. Fig. 3.13 reveals the electrolytic cell designed. By analyzing the obtained overall polarization curves, it may be stated that at approximately pH 1–1.5, the reduction of Fe dominates the cathode reaction; however, the evolution of H_2 still exists. The pH of the electrolyte solution was found to be an important factor influencing the process. When pH values were <1, H_2 evolution was found to be excessive and pH values higher than 1.5 did not result in significant improvement in the process of Fe deposition.

In order to recover Zn in only one step and try to prevent Fe codeposition, a CEM, NAFION-117, is used in Carrillo-Abad et al.'s work [5].

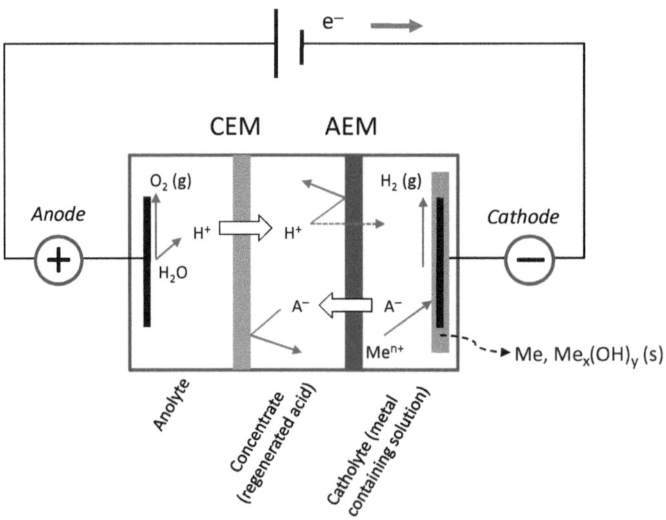

Figure 3.12: Scheme of a membrane electrolysis cell [21].

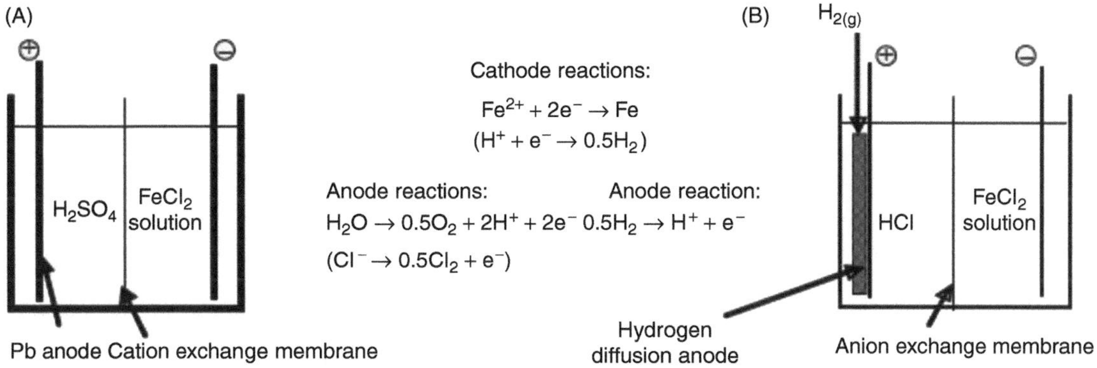

Figure 3.13: Schematic of the designed electrolytic cells.
(A) Cell with inert anode and cation-exchange membrane, (B) cell with hydrogen gas anode and anion-exchange membrane [21].

Previously, electrolysis without chamber partition was attempted in potentiostatic and galvanostatic modes [22] to determine the viability of Zn recovery from spent pickling baths. Zn redissolution was observed at high time values. The reason is that the Fe ions and dissolved Cl_2 gas attack Zn deposits via oxidation [23].

To prevent the Zn redissolution, electrolysis with AEM was initially attempted to avoid Cl_2 presence in the cathodic compartment [5]. This design obtained a Zn conversion closer to 100% and higher current efficiency. However, Fe codeposition was observed when Zn concentration was reduced to a certain level. The deposition in the Fe-Zn system observes the anomalous codeposition [24], in which the less noble metal (i.e. Zn) deposits preferentially, and Fe deposition depends on the Zn-Fe ratio, the applied current, and the pH value.

In research regarding the efficiency of using CEM in electrolysis [5], the anodic compartment contains the spent pickling solution whereas the cathodic one is filled with HCl with or without Zn salt. For AEM mode, the pickling solution is filled in the cathodic chamber. The principles for both modes are illustrated in Fig. 3.14. The average composition of the spent pickling

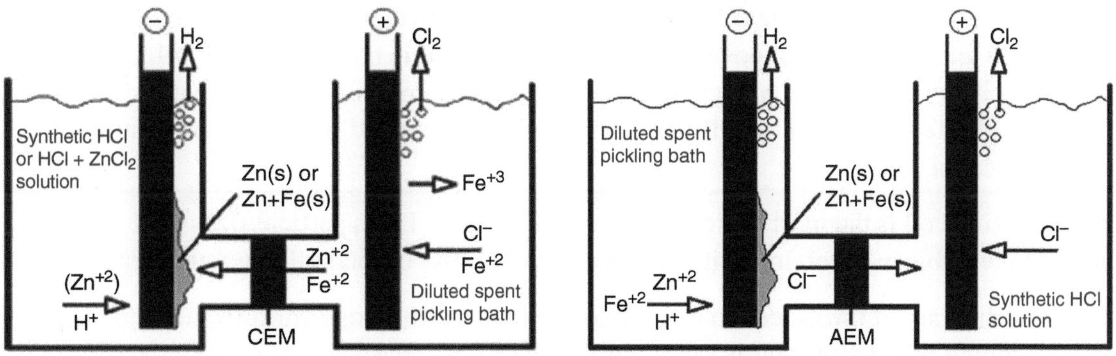

Figure 3.14: Simplified diagram for membrane reactors using CEM and AEM [5].

Table 3.8: Average composition of the spend pickling bath

	Zn	Fe	HCl
M (mol L^{-1})	1.9780	1.0578	2.1380
g L^{-1}	129.3242	58.8129	78.0356

bath used is displayed in Table 3.8. With using membranes, the presence of Cl_2 in cathodic compartment is avoided as CEM is barrier for its transfer. As observed, Zn conversion with AEM was greater than that when CEM is concerned, since the pickling solution was directly put in the cathodic compartment from the beginning of the AEM electrolysis.

The transport of Zn through the CEM is favored as compared with that of Fe. This is due to the fact that ferrous ions (Fe^{2+}) was oxidized to ferric ions (Fe^{3+}) in the anodic compartment and the NAFION-117 can trap trivalent cations more easily than divalent ones. The presence of Zn in the cathodic compartment at a concentration of 0.1 M from the beginning of the electrolysis inhibited Fe deposition and rendered H_2 evolution a secondary reaction.

3.1.3 Oily Wastewater Treatment

A great amount of oily wastewater is produced in iron and steel industries. In cold rolling stainless steel production process, steps such as, annealing, pickling, grinding, polishing, cutting, etc. release wastewater carrying oil and fat. In strip rolling, emulsion cooling and lubrication are utilized to eliminate thermal deformation due to cold rolling, meanwhile generating cold rolling oily wastewater. Alkali oily wastewater is discharged when using alkali solution for degreasing cooling steel strip before open coil annealing. The effluent from hot rolling mill and silicon steel plant is also categorized as oily wastewater [25].

There are four types of existing forms for the oil and fat in wastewater; they are slick oil (>100 μm), dispersed oil (10–100 μm), emulsified oil (0.1–20 μm), and soluble oil (a few nanometers (nm)) [26]. For removal of slick oil, gravity method can lead to high separation efficiency. Soluble oil can be effectively removed by chemical oxidation, adsorption and/or bio-decomposition. [27]. Dispersed oil is not stable, and will change to either bigger oil droplets by aggregation, or turn into emulsification by natural or mechanical effects. Conventionally, dispersed oil it is removed by coarse granulation or coalescence method.

The emulsified oil is the most difficult to deal with [28,29]. It has particle size of 0.1–20 μm and is quite stable in water. In addition, it occupies 20–30% of total oil in wastewater.

The flow diagram of the process using UF as the major technology for oil/water separation of cold rolling oily wastewater is shown in Fig. 3.15 [29]. Wastewaters containing oil and/

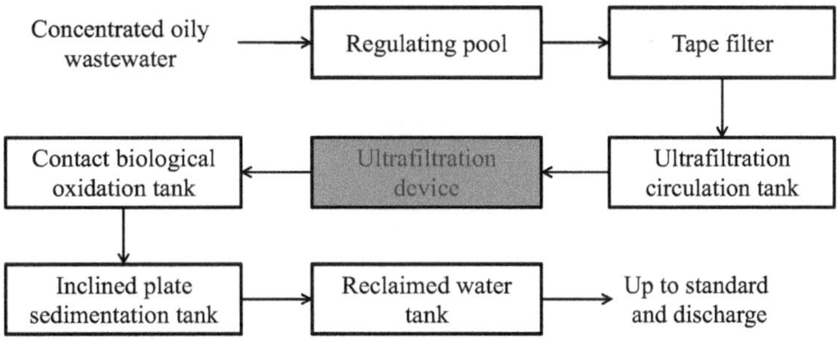

Figure 3.15: Ultrafiltration-based process for treating concentrated oily wastewater [29].

or emulsion from cold rolling production line flow into the wastewater distribution pool and then regulation reservoir. Thereafter, the wastewater is directed into a filtration system as the pretreatment for UF. In this step, the big particles are eliminated to prevent fouling in UF. Effluent from UF is delivered to contact biological oxidation pond for further degradation of chemical oxygen demand (COD), ammonia (NH_3), and other substances. After being treated by inclined plate sedimentation, wastewaters reaching the standard of discharge will be directly drained, whereas those failing to meet the requirements will be recycled to thin oil regulation pool.

Oily wastewater with high degree of the emulsion and using long molecular chain emulsifier tend to cause membrane blockage. Therefore, design of membrane materials with strong hydrophilicity and antifouling properties is imperative. The review about polymeric UF membrane development is given in Chapter 12.

Fang [30] used the inorganic ceramic membrane for dealing with the practical oil emulsion from some cold rolling plant. One month after system's installation and start-up, a ten-day continuous operation was monitored. It was revealed that the content for COD was 1946.7 mg L^{-1}, and for petroleum was 17.48 mg L^{-1}, in the effluent from the system. With every 3–5 days' operation under normal conditions, the flux of the membrane declined, necessitating cleaning. The major components with the membrane foulant include the deposited or adsorbed oil and inorganic ions. The alkaline special detergents and surfactant are effective for taking away these contaminants. The inorganic ceramic membranes exhibited quite good performance in processing the emulsion oil wastewater. The permeate is featured with oil content lower than 20 mg L^{-1}, indicating an oil removal efficiency higher than 99.7%. Besides, the COD removal rate reached 95% and the water quality was very stable, showing no fluctuation with variation of feed water composition. It is also widely recognized that washing and cleaning of ceramic membranes can more easily achieve high recovery of the separation properties.

Table 3.9 summarizes the experimental results of dealing with emulsified oil waste water with ceramic membrane developed by Nanjing Tech University [31]. The mass fraction of oil in

Table 3.9: The treatment effect of inorganic rolling oil emulsion wastewater by ceramic membrane [31]

Items	Inflow	Effluent
Oil/(mg L^{-1})	1000–11000	≤8
COD (mg L^{-1})	2000–50000	200–1000
Oil length (mg L^{-1})	500–1500	≤10
pH	7–14	6–14

the original wastewater is in the range of 1,000–11,000 × 10^{-6} mg L^{-1}. The water generated by ceramic membrane contains only 10 × 10^{-6} mg L^{-1} oil. Economic analysis shows that the ceramic membranes produced by Jiangsu Jiuwu Hi-Tech Co. Ltd. bring about much lower cost (<4 CNY) when applied in oil wastewater processing, as compared with polymeric membranes imported from foreign countries. The breakdown for the cost analysis involves equipment discount, equipment consumption, labor, maintenance, cleaning, and efficiency of oil recovery.

Zhang et al. [32] investigated the efficiency of two types of processes centering on gas energy flocculation (GEF)/electrochemical catalytic oxidation (ECO)/membrane bioreactor (MBR), and UF/ECO/MBR—in tackling cold rolling emulsified wastewater. The wastewater was taken from cold rolling steel plant of Wuhan Iron and Steel Group.

For both processes, ECO is the key step that fundamentally improves the overall removal efficiency of organic polluters. Its contribution to the degrading of the organics is the evolution of active hydroxyl free radical (·OH). With the functioning of electric field, the modifier on the surface of electrode(s) or inside the solution facilitates or restrains the electron transfer on the electrode for the oxidizing agent production.

With GEF step, the water content of scum becomes lower, and simultaneously the oil and suspension in water effluent is low. The effluent is subsequently delivered for electrochemical oxidation step.

The GEF or UF pretreatment has similar effect; both can remove efficiently cold rolling oil emulsion, suspended solids and COD, satisfactorily meeting the requirements of the follow-up processing for improving water quality. Based on the continuous pilot scale operation for 200–3000 L h^{-1} wastewater, analysis has been made about the effect of operating parameters on the removal rate of COD_{Cr}. The experimental results show that the water treated by both processes meets the requirements of recycling to the production process. The GEF/ECO/MBR process delivers more stable water quality, having the value for industrial promotion.

Addressing also the cold rolling oil emulsion wastewater from the same plant mentioned in Zhang et al.'s work [32], Shu et al. explored a new process combining air flotation, UF, and

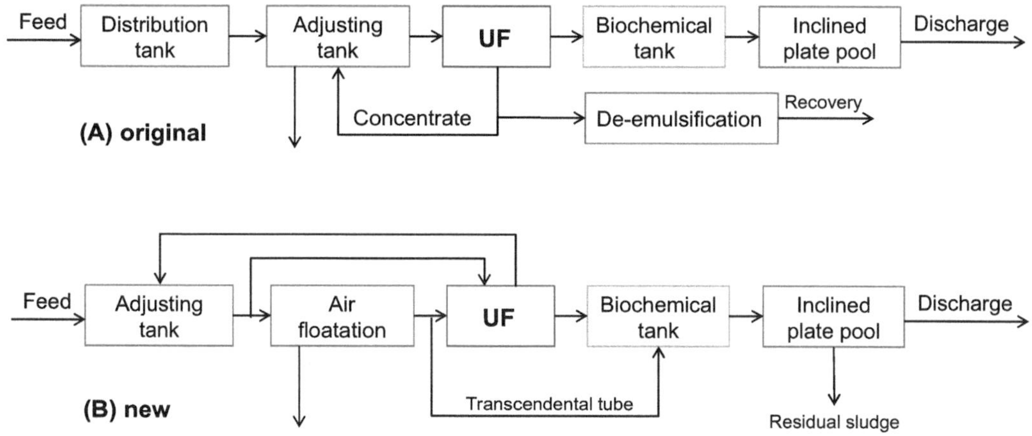

Figure 3.16: (A) The original and (B) the new processes for treating emulsion wastewater [33].

biochemical method [33]. The original process and the new process are schematically shown in Fig. 3.16. The major improvement with the new one is the insertion of air floatation in between the adjustment tank and UF system. It obviously reduces the COD and oil content in the feed to UF, significantly enhancing the effect of UF in removing emulsion. Based on the pilot scale experiment, the combination of flotation, UF, and contact oxidation is technically feasible for dealing with emulsified oil wastewater from cold rolling. With the innovation, the new process is characterized by flexible regulation, good reliability, and low investment and operating costs.

3.1.4 Coking Plant Wastewater Treatment

Coking wastewater comprise a large amount of high MW and refractory organics. The specific sources of wastewater include [34,35]: (1) residual NH_3 water generated in the processes of high temperature destructive distillation of coal and coke-oven gas cooling; (2) water of coal gas end cooling and crude benzene separation generated in gas purification process; (3) wastewater from crude tar processing, fine phenol production, coumarone production, and benzene refining process (4) wastewater from production of fine coke dust and other substances.

At present, the traditional coking wastewater treatment process consists of two stages, with the first stage for the recovery of valuable components from the high concentration wastewater and; the second stage targeting the detoxification by degrading organics such as phenol and cyanogens [36,37].

Currently, the principal process for coking wastewater treatment is the combination of NH_3 distillation and anaerobic/anoxic/oxic (A^2/O) method.

A lot of recent researches have been published regarding application of MBR in coking wastewater treatment. Zhong [38] carried out experimental study on treatment of simulated coking wastewater by aerobic MBR. The experimental results showed that COD and ammonia nitrogen (NH_3-N) removal rate reached more than 95%. It also pointed out that membrane pollution control is very critical for MBR application.

Wang et al. [39] investigated the process involving a special External Composite Membrane Bioreactor (CMBR) in tackling coking wastewater. CMBR is a new technology developed specifically for the treatment of wastewater containing high concentration of NH_3.

The schematic of CMBR is demonstrated in Fig. 3.17, and the process flow diagram with CMBR for the coking wastewater treatment is shown in Fig. 3.18. The quality of effluent corroborates that the process effluent only contains lower than 15 mg L^{-1} NH_3-N. The denitrification by air stripping gets rid of the majority of the NH_3-N. The concentration of the activated sludge inside the biological tank reaches 10–15 g L^{-1}, greatly strengthening the degradation capability. Within the biological tank, the anaerobic, facultative, and aerobic reactions are carried out in turn and through the function of nitrification and denitrification, the residual NH_4^+ and COD have been further removed. At the same time, the organics with higher MW are decomposed by chain scission and removed. Currently the technology has been employed by several coking plants located in Shanxi Province, China.

The anaerobic-anoxic-oxic (A^2O) drainage from coking plant was treated by the Fenton regent pretreatment and microfiltration (MF) [40]. The effects of oxidation on the water quality and membrane fouling were investigated. The experiment results showed that the removal rates of the COD and mineral oil were above 70%, the removal rate by UV254

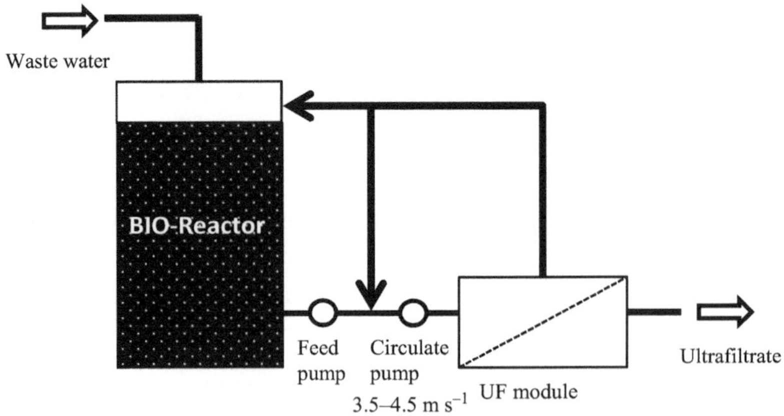

Figure 3.17: System schematic of external membrane bioreactor.

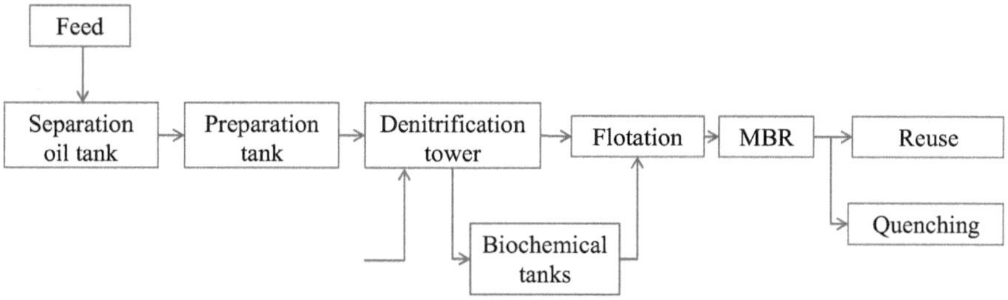

Figure 3.18: Flow chart of coking wastewater treatment process [39].

was 47%, and the removal rates of turbidity and suspended solids were above 80%. The resistance of membrane filtration decreased by 92% and the membrane fouling was controlled effectively.

For the recovery of NH_3 and phenol from the coking plant, hydrophobic microporous hollow fiber membrane absorption has been developed [41]. Experimental results show that NH_3 and phenol could be removed and recovered separately by utilizing their difference in acid–base properties and requirement of pH value. Under optimum operation conditions, the recovery rate of both NH_3 and phenol is over 99.5%. The membrane fouling could be effectively restrained by prefilter and regular rinsing with 2% NaOH solution.

As coking plant effluent is considered recalcitrant with low bioavailability, IEM ED method is used as a pretreatment measure to enhance the removal of COD and NH_3-N [42]. Phenol, the main component of pollutants in coking wastewater, is the research focus and the optimal operating parameters were studied with a bench scale experimental installation. Results using simulated coking wastewater indicated that IEM ED could remove phenol and COD with the removing efficiencies of 84 and 45%, respectively, with a retention time of 2.5 h in the electrolysis cell; and particularly very high rates of NH_3-N removal and recovery at approximately 100% were shown. In general, IEM ED, as a pretreatment method, can improve biodegradable characteristic of coke plant wastewater to facilitate its further biochemical processing.

Membrane process combining ceramic UF, NF, and RO was used in treating desulfurization wastewater from coking process [43]. The sulfur content in the wastewater can be rejected by the ceramic membrane at 100%. NF can separate ammonium thiosulfate ($(NH_4)_2S_2O_3$) from ammonium thiocyanate (NH_4SCN) effectively, with a rejection rate of 95%. The RO is applied to concentrate the NH_4SCN solutions. With four-stage treatment, the overall rejection of NH_4SCN reaches 99%, and the ion conductivity of permeate is below 171 $\mu s\ cm^{-1}$.

Systems using membrane separation have been investigated for desalting wastewater from coking production process in Jinma Coking Co. Ltd. [44]. The water quality of the

Table 3.10: Sewage pollution index (mg L^{-1}) [44]

Project	Ca^{2+}	M^-	Cl^-	pH	Conductivity (μs cm^{-1})	Turbidity (NTU)	COD	NH_3-N	Phenol	Cyanide	Chroma
Standard values				6–9			≤7000	≤300	~800	~20	
Measured values	357.8	389.5	793.5	7.3	6680	150	5800	250	650	15	Brownish red

wastewater is summarized in Table 3.10. One of the processes designed comprise following steps in sequence: biological treatment, flocculation and sedimentation, quartz sand filter, activated carbon filter, decoloration, UF, ion exchange, back to circulating. The efficiency of the process is not stable, with the detain effects and effluent quality deteriorating as the decolorization tower is gradually blocked. In addition, Cl^- in the treated water is relatively high. This is attributed to the membrane blockage by oil sludge and chlorine release by ion exchange. Another design takes biofilm separation as pretreatment for RO, which has not been introduced in detail.

The system using MBR/NF/RO combination was applied for advanced treatment of Baoshan Steel Plant coking wastewater [45]. The full-scale industrial plant was set up on the basis of the pilot-scale test results. The MBR/NF/RO pilot-scale system exhibited continuous and stable treatment performance, and the effluent met internal industrial water quality standards of Baoshan Steel Plant, with the removals of COD and conductivity reaching 98.7% and 95.2%, respectively. The full-scale industrial plant also operated steadily and water production rate was over 60% with the operational cost increase of 4.66 CNY t^{-1}.

The ceramic membrane separation technique is used for filtering the coking substances from the residual ammonia water in the recycling workshop of the coking plant of Anyang Iron & Steel Co. [46]. The problems owing to a too high oil-bearing amount of residual NH_3 liquid, resulting in lowering the wastewater treatment efficiency during the process of coking wastewater treatment, have been solved. Ceramic membrane is oleophobic and exhibits good rejection toward oily component. In addition, it shows high resistance to the strong corrosion of NH_3 solution. The coking wastewater treatment equipment works stably, the treatment effectiveness has been improved, and the treated water quality meets the standard for discharge.

3.1.5 Desalination for Integrated Wastewater

As found in many researches, the so-called "advanced" or "in-depth" treatment almost unanimously refers to desalination and sometimes its combination with one or several pretreatments. This procedure is principally used in industrial sewage wastewater [47]. The industrial sewage water for iron and steel fabrication mainly refers to drainage water of the circulation system.

At present, steel companies in China have generally adopted the integrated wastewater system that combines the process wastewater, industrial sewage water, domestic sewage water, and rain water to form the so-called integrated wastewater. In order to improve the repeating utilization or recycling rate so as to save freshwater consumption, centralized purification of the integrated wastewater is conducted. The treated water will be supplied to the water circulation system for each department [47].

The integrated wastewater contains lots of scrap iron oxide, oil, heavy metals, benzene, phenol, and other organic matters [47]. With conventional treatment, generally, the wastewaters will be removed of most of the turbidity and impurities. In addition, the content of polluters such as COD, oil, etc. will also be reduced to specific level. However, the salinity (ie, ion conductivity, hardness, and concentration of other ions) of the water is still very high. With these characteristics, the water can only be discharged or reused in sintering, iron smelting, steel smelting, rolling, etc. for direct current (DC) slag spraying or sprinkling floor.

A typical desalination process is schematized in Fig. 3.19 and is composed of three parts: (1) preprocessing (from cooling to microflocculation); (2) pretreatment (from activated carbon filtering to ultrafiltration (UF)); (3) RO.

Dual-membrane process using UF and RO desalination is probably the major choice for desalination or advanced treatment of wastewater and its application has been progressively intensified in iron and steel plant. Rizhao Iron and Steel Company employs advanced

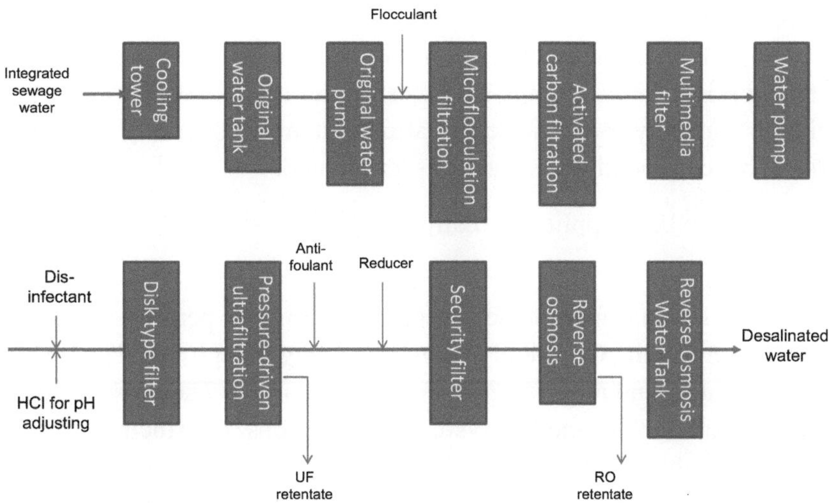

Figure 3.19: Flow diagram for typical process used in integrated wastewater treatment of iron and steel industry [47].

Table 3.11: Major technical parameters and performance of the UF and RO systems in Rizhao Iron & Steel Plant [48]

System	Technical parameters and performance
UF	Operation pressure: 0.25 MPa; system number: 2; system output: 280 m^3/h/system; membrane number: 70/system. Total water yield: 560 m3/h; water recovery: ≥90%.
RO	Operation pressure: 1.6 MPa; system number: 2; system output: 280 m^3/h/system; membrane number: 280/system. Water recovery: ≥75%; desalination rate: ≥96%.

desalination process solely composed of UF, RO, and their respective ancillary equipments to obtain demineralized water that can be reused, consequently reducing the wastewater discharge [48]. The process flow is as follows: original water pool, UF feed pump, self-cleaning filters, UF system, RO feed pump, security filter, and RO system. The technical parameters and performance for UF and RO are summarized in Table 3.11. The comprehensive operating cost based on consumption of electricity, washing, and reagents is 0.487 CNY ton^{-1}. The UF membrane installed in the system is Krystal® by Hyflux Company (Singapore).

RO is selected as the main desalination technology for wastewater treatment in Tangshan Branch of Hebei Iron & Steel Co. Ltd. There are two types of wastewater to be processed: (1) urban reclaimed water that is used as make-up water for several circulation systems after treatment of reducing turbidity and removing hardness; (2) industrial wastewater from circulation drainage, domestic sewage, etc., with a total designed capacity of 3000 m^3.

According to the fact that the salt content of industrial wastewater is high and the composition is complex, the VONTRON RO membranes (FR12-8040, spiral wound module) with good antifouling properties are applied [49]. The pretreatment consolidates the high-density sedimentation tank, V-type filtration tank, multimedia filter, self-cleaning filter, and UF. The system has been working stably, helping save fresh water consumption by 40,000 m^3 day^{-1} and 14.6 million m^3 annually. With the desalination system, outsourcing of freshwater becomes zero and hence ground water intake is terminated.

RO desalination has also been applied in many other iron & steel industries, such as Shougang Group [50].

One of the challenges for dual-membrane technology is the low performance for UF as pretreatment for RO in dealing COD and oil. In processing this group of wastewaters, RO membrane is easily fouled and requires quite frequent cleaning. The presence of oil and COD is the cause for 60–80% of the RO system malfunctions. In principle, the oil content is supposed to be lower than 0.5 mg L^{-1}, and COD lower than 20 mg L^{-1}. Addressing the problem, full flux ceramic membrane is employed by Jin [51]. It is claimed that the special ceramic membranes (brand or producer is not introduced) have good efficiency in intercepting

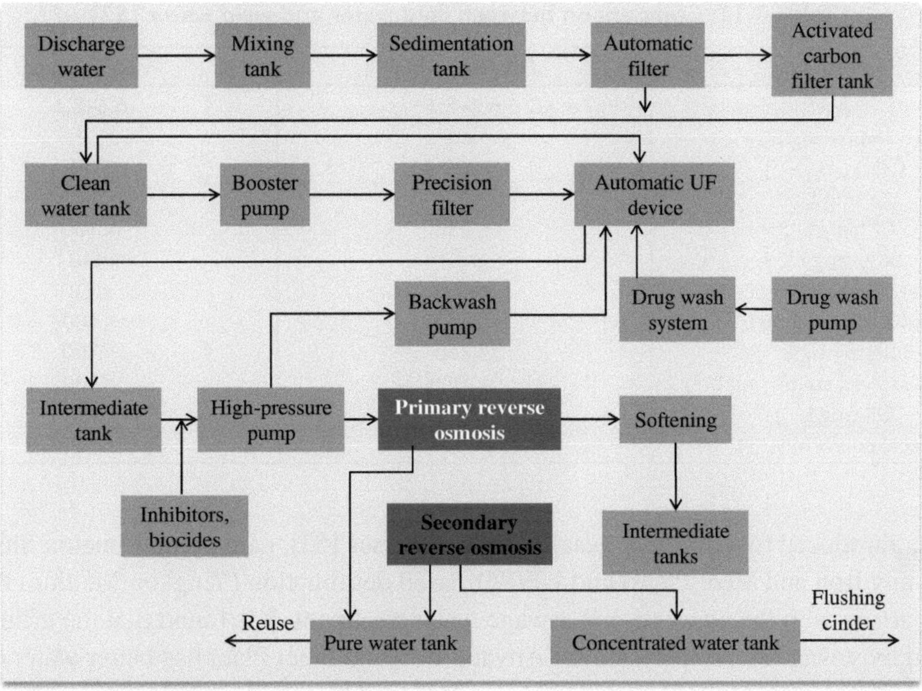

Figure 3.20: Water treatment process with reverse osmosis as the key step [52].

COD and oil substances and higher antifouling capability, attributing to the narrow pore size distribution and unfavorable absorption toward organic components. What's more the flocculants for strengthening UF function with or without reaction can be rejected as well, so that flocculants will not be involved in follow-up RO.

Zhong and Peng used RO as a major technology for dealing with integrated wastewater formed by mixing cooling water, washing drainage, and domestic sewage from plants of Jiu quan Iron & Steel Corporation Organization (JISCO) [52]. The overall process from feeding to water yielding is schematically shown in Fig. 3.20. Finally, a large amount of high quality water (about 90% of feed) is produced, which can be reused in steel production process, while concentrated water from the secondary RO is used for coke quenching. Mechanical filtration was the core technology for RO pretreatment. It comprises filtration, activated carbon filtration, bag filtration, UF etc., assisted by sterilization and sedimentation aiming at removal of suspended solids and microorganism. The pretreatment controls the silting density index (SDI) of effluent below 5 and prevents the presence of a large amount of microorganisms inside the RO system. After tuning-up for a certain period of time, the RO system maintains stability, and the water quality reaches requirements for supplying steel workshop. The quality of product water and feed water of the RO system is shown in Table 3.12.

Table 3.12: Comparison between feed water and yield water [52]

Project	Process water	Product water
pH	6.5–8.5	6.5–8.5
Turbidity (mg L^{-1})	≤5	<1
COD (mg L^{-1})	≤60	<10
Fe (mg L^{-1})	≤0.3	<0.2
Cl^- (mg L^{-1})	≤250	<100
SiO_2 (mg L^{-1})	≤30	<10
Total hardness (mg L^{-1})	≤450	<50
Total alkalinity (mg L^{-1})	≤350	<100
Sulfate (mg L^{-1})	≤250	<100
Total dissolved solids (mg L^{-1})	≤1000	<500
Oil (mg L^{-1})	≤1	<1
Fecal coliform (L^{-1})	≤2000	<200

Fan et al. introduced two different desalination processes [53], namely multimedia filtration/RO (Anyang Iron and Steel Plant) and UF/RO-based desalination (Yingkou Medium Plate). By comparison with the conventional sewage water treatment, it is found that the effluent generated by sewage disposal system of Anyang Iron and Steel Plant has better water quality. Therefore, different desalination processes are deployed by the two plants, considering the dissimilarity. With Yingkou Medium Plate, employment of UF with much smaller filtration diameter can partially offset the low efficiency of conventional sewage water treatment, so as to provide more effective protection for RO. Water yield for each UF device is 235 $m^3\ h^{-1}$. UF has relatively higher investment cost than multimedia filtration. The two systems were designed and provided by Bluestar Environmental Engineering Co. Ltd.

Dual-membrane process was adopted for intensified purification of the sewage water in the desalination water station project of cold rolling plant of Handan Iron & Steel (Group) Company [54]. The process is schematically displayed in Fig. 3.21 and it has already been used in production practice. The silt density index (SDI) for 85% of the water generated by UF is lower than 3 and the overal SDI ≤ 4; the turbidity of the feed water is around 8–17°, whereas that of purified water is ≤ 0.2°. The electroconductivity of influent water for the first stage RO ranges from 2000 to 3500 $\mu s\ cm^{-1}$, and 40–80 $\mu s\ cm^{-1}$ for the first stage desalted water, and 2–4 $\mu s\ cm^{-1}$ for second stage desalted water. The desalination ratio is higher than 97 and 94% for the first and second stages of RO, respectively.

Employment of UF can efficiently remove particles, suspended substance, colloid, bacteria, and viruses. The effluent quality is stable and almost independent of the feed variation. As the pretreatment of RO, it is better than conventional approach in lowering influent SDI value and extending lifespan of the system. The concentrate from second stage RO was sent to plant for flushing cinder. Therefore, there is hardly any discharge of waste acid and caustic liquor. The water from first stage RO was used as make-up water for recycle water system; the effluent from

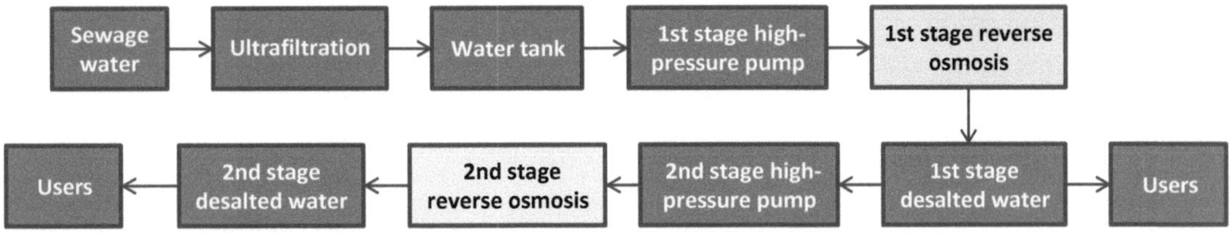

Figure 3.21: The dual-membrane water treatment process adopted in Handan Iron and Steel Company [54].

the second stage RO was very stable in quality and can satisfactorily meet the requirements of being used in systems such as boiler, cold rolling pickling, rinsing, etc. Some problems encountered during the system operation and the solutions are summarized in Table 3.13.

In effect, desalination has also been employed in advanced treatment of process wastewater, such as coking wastewater and cold-rolling wastewater. The researches regarding these applications have already been introduced in the corresponding part in the chapter. Desalination has been widely applied in China and brought about obvious benefits in practical application. However, some problems have also been revealed, for example, the inappropriateness of SDI in evaluating UF effluent, formulation and timing for membrane cleaning, formation of calcium sulfate ($CaSO_4$) particles, and disposal of concentrated saline water. Regarding these problems, the future development direction will be focused on following aspects: more practical water quality parameters for membrane

Table 3.13: Existing problems and corresponding measures

Problems	Reasons	Solutions	Effects
Blockage of security filter cartridge in first stage RO	High concentration of Ca^{2+} and Mg^{2+} in source water; inappropriate scale inhibitor	Use of new type of scale inhibitor and adjustment in concentration and dosage	No blockage during two-month operation
Frequent standby of ultrafiltration	Program for automatic control mismatches actual operation conditions	Adjustment of operation mode and controlling parameters	Decrease the frequency of standby
Relatively high electroconductivity and pH value in second stage RO	High content of NH_4^+ in source water	Addition of acid to neutralize NH^{4+} and change pH value	pH value of purified water produced in first stage RO is reduced to lower than 7; water from second stage is qualified
Organic blockage at initial testing for system operation	Oversized water tank and high temperature; bacteria rapidly reproduce and cannot be sterilized by oxidizing agents efficiently	Addition of nonoxidizing bactericide in the feed to first stage RO according to the air and water temperatures	Organic fouling disappears

separation, optimization of operation conditions, and finally the construction of operation standards for integrated wastewater treatment.

3.1.6 Treatment of Smelting Waste Gas

Ferroalloy is a very important reducing agent for steel making. Its production involves the high temperature reaction of the raw materials including minerals and carbon inside the molten bath of the submerged arc furnace. The production process generates high temperature exhaust containing raw coke oven gas and dust. With more stringent regulations about ferroalloy industry development, targeting more efficient environmental protection, fully enclosed type of furnace is the future direction for the large-scale production of Si-Mn alloy, high carbon ferro-chrome (Fe-Cr), high carbon Fe-Mn, etc. [55]. One of the key problem for operating a ferroalloy industry in the purification of high temperature raw coke oven gas. The coke oven gas contains several components—CO, methane (CH_4), and H_2—that are potential fuels with high combustion value. Nevertheless, the mixed gases are liable to explosion upon contact with air. As mentioned above, dust is another part of the production exhaust. It is rich in rare metal, rare earth metal, and fine carbon powders; these components are valuable resources, but very active and apt to spontaneous ignition in air. In effect due to factors intrinsic of exhaust (e.g. particle size, particle electric resistivity, fluctuation in pressure) and the technologies (medium stability at high temperature, system negative pressure), the conventional technologies such as bag filter, electrostatic precipitator cannot be satisfactorily used in raw coke gas purification.

Zhang et al. investigated the application of metal membranes in treating at high temperature flue gas and dust released from submerged arc furnace [56]. Fig. 3.22 shows some considerations based on aforementioned situation and other factors that motivate the idea of high temperature purification, which makes inorganic membranes a better choice. The pilot scale purification system was installed in some ferroalloy plant in Qinghai Province (China). It is composed of five parts: filtration system, back flushing system, anticondensation system, ash discharging system, and operating system. The operation temperature for gas filtration is 550°C. The dust collection rate is higher than 99.9% and the dust content in effluent gas is lower than 20 mg m^{-3}. The purified gas can be used directly in electricity generation, or production of CH_4 and formic acid.

The metal membrane employed in the system is ferro-aluminum (Fe-Al) intermetallic porous membrane produced by YT Membrane High-tech Co. Ltd. The membrane has good resistance toward high temperature, wearing, oxidation, sulfuration, etc. and several other excellent properties. The YT filtration systems have also been used in many other gas/solid separation processes involving high temperature and harsh environment, for example, the treatment of coal gasification lignite upgrade, treatment of phosphorus exhaust gas. In recent years, application of ceramic membrane in China on gas–solid separation also got rapid

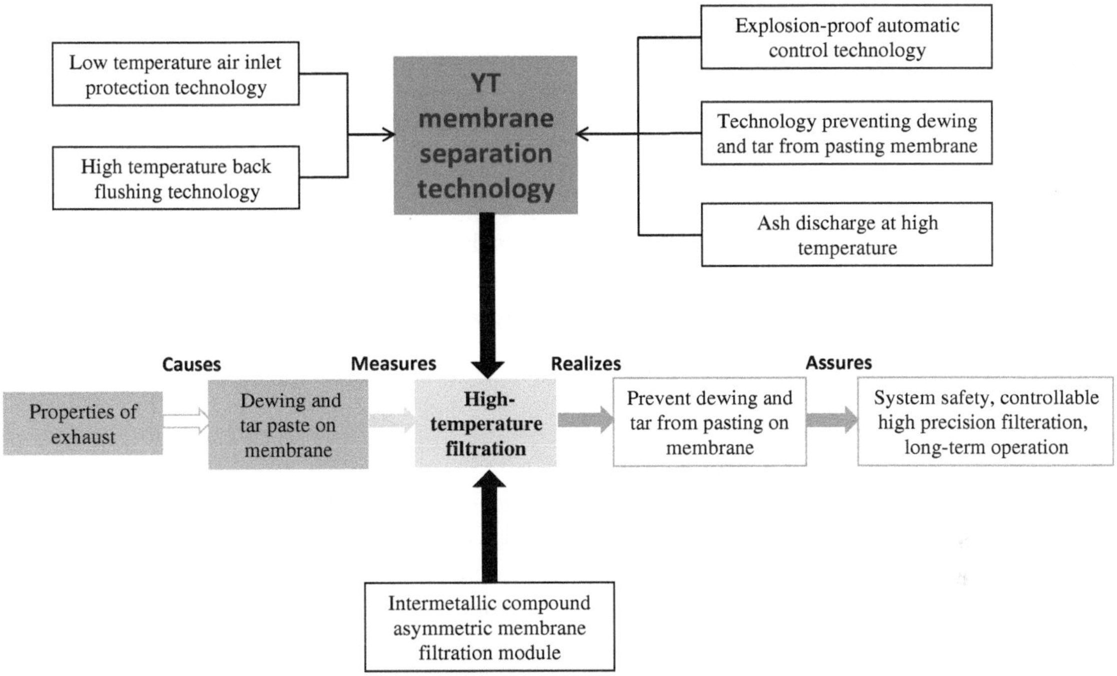

Figure 3.22: Factors necessitating metal membranes in treating coke oven gas and working principle of YT filtration system.

development. Ceramic membrane separation technology in United States, Japan, Germany, and other developed countries has achieved its industrial application, especially in iron and steel industry, coal gasification waste pyrolysis, precious metal recovery, purification of fluidized metal. It has been pointed out by experts that when the temperature rises to 1000°C, ceramic membrane is a more promising technology for high temperature exhaust gas purification.

3.2 Chromium

3.2.1 Introduction

There are about 4.4 billion tons of chromites reserves in the world, with South Africa accounting for about 75–88%. Seventy five percent of Cr extracted is used in metallurgical industry, and about 25% for refractory and chemical industries. Cr is added to the steel as an alloying element, usually in the form of elementary Cr, or Cr alloy (or chrome), to change the characteristics of the steel and improve the toughness and corrosion resistance of steel. Chrome is mainly used for refining Cr superalloy, resistive alloy, precision alloy, casting Fe alloy, Ni-based alloys, electrothermal alloy, drilled alloy, Cu alloys, and others. The derivatives of Cr are widely used in many industries dealing with electroplating, dye, textile, and leather.

The major production methods for Cr are metal thermal reduction and electrolysis. Electrolytic production of Cr accounts for about one-third of total output [57]. At present, chrome ore smelting in China is mainly relying on electric furnace method, followed by vacuum metal thermal reduction, vacuum carbon reduction, and converter. There also exists hydrometallurgical process for Cr production. Cr in many wastewaters is present either as the hexavalent Cr(VI) or the trivalent Cr(III) form. Due to its high toxicity, especially in the hexavalent form, the tolerable or acceptable concentrations for Cr in wastes discharge are controlled by strict norms [58]. The recovery of the Cr is interesting from both environmental and economic points of view. Membrane-based separations regarding Cr treatment are focused in various wastewater treatment and are introduced subsequently.

3.2.2 Application of Nanofiltration

Cr plating is a very important process for improving metal wearing resistance of metal, etc. Additionally in Cr plating industry, chromic acid (H_2CrO_4) is used in pickling and electro-polishing. The discharges from these processes will contain Cr compounds. The traditional treatment processes for above Cr containing wastewater are based on chemical, biological, and electro-chemical methods. For example, one approach reduces Cr(VI) to Cr(III) which is then precipitated at high pH. This method will lead to the waste of Cr.

From 2000, several researches investigated the feasibility of applying NF in tackling various waste streams containing Cr compounds [59–61]. In Hafiane et al.'s work for example [60], it was found that the increase of pH enhanced the chromate (CrO_4^{2-}) removal since it was the predominant species when pH is >8.

In Chen et al.'s work, a novel process, the two-stage nanofiltration (TSNF), was designed for Cr(VI) removal from extremely low pH electroplating wastewater [62]. Fig. 3.23 shows the schematic diagram of the designed TSNF system and the mechanism for treatment. In the first stage, NF is used to retain Cu^{2+} and Ni^{2+} in the concentrate stream and for bichromate ($HCrO_4^-$) to pass through the membrane in the permeate stream. The permeate from the first stage NF system was adjusted in pH by adding 1NNaOH. The final pH is >8 and under this

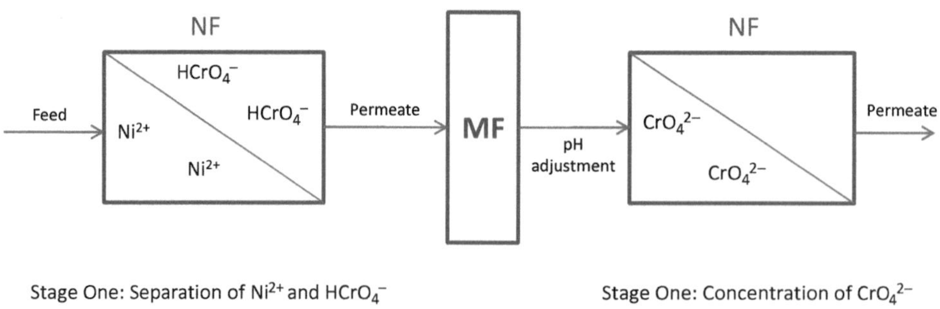

Figure 3.23: Flow chart for the TSNF process.

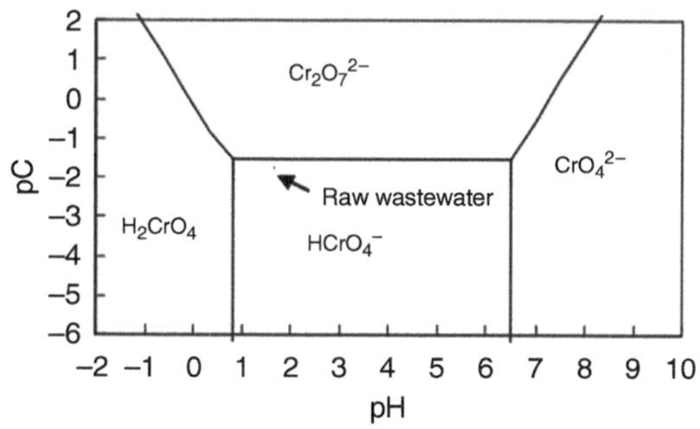

Figure 3.24: Predominance diagram for hexavalent species H_2CrO_4, $HCrO_4^-$, CrO_4^{2-}, $Cr_2O_7^{2-}$ or $HCr_2O_7^-$ based on Eqs. (3.2)–(3.5).

condition, $HCrO_4^-$ was converted to bivalent CrO_4^{2-}. Therefore, in the second stage, rejection of Cr(VI) compounds becomes feasible via NF.

The theoretical basis for designing the aforementioned TSFN is the predominance forms of Cr(VI) compounds as a function of pH value, as shown in Fig. 3.24. Chromium oxide (CrO_3) is the raw material for electroplating. It dissolves in water to form chromic acid (H_2CrO_4) as shown in Eq. (3.1). pH value and Cr(VI) concentration combine to determine the dominant species of Cr(VI) in aqueous solution, based on the reaction equilibriums associated with Eqs. (3.2)–(3.5). They could be H_2CrO_4, $HCrO_4^-$, CrO_4^{2-}, dichromate ($Cr_2O_7^{2-}$) or bi-dichromate ($HCr_2O_7^-$). Eqs. (3.1) to (3.5) are listed subsequently:

$$CrO_3 + H_2O = H_2CrO_4 \tag{3.1}$$

$$H_2CrO_4 \Leftrightarrow HCrO_4^- + H^+ \quad \log K_{a1} = 0.8 \tag{3.2}$$

$$HCrO_4^- \Leftrightarrow CrO_4^{2-} + H^+ \quad \log K_{a2} = -6.5 \tag{3.3}$$

$$2HCrO_4^- \Leftrightarrow Cr_2O_7^{2-} + H_2O \quad \log K_{a3} = 1.52 \tag{3.4}$$

$$Cr_2O_7^{2-} + H^+ \Leftrightarrow HCrO_7^- \quad \log K = 0.07 \tag{3.5}$$

Microfiltration was used between the two NF membranes to prevent possible slight precipitation of nickel oxide (NiO) even if most of the nickel was retained in the first stage.

The result shows that the optimum operation in stage one NF involves 81% system recovery (the ratio of total volume of the permeate stream over the total volume for the feed stream) with 80% mass recovery for both Cr (VI) (in the permeate stream) and Ni (II) (in the concentrate stream); Ni could be concentrated from 94 to 1770 mg L^{-1}. In stage two NF, the concentration of Cr(VI) became 6.6 times of the original (from 905 to 5960 mg L^{-1}) when

Table 3.14: Experimental results of electroplating wastewater by NF90 membrane [63]

Operation conditions		Effluent concentration ($mg\ L^{-1}$)	Rejection (%)	Conductivity ($\mu s\ cm^{-1}$)
Pressure (MPa)	Temperature (°C)			
0.5	20	0.51	99.5	19.8
0.7	20	0.46	99.6	17.6
0.9	20	0.45	99.6	16.1
1.1	20	0.40	99.6	14.3
1.1	25	0.42	99.6	18.8
1.1	30	0.46	99.6	20.3
1.1	35	0.52	99.5	25.1

the system recovery was 90%, and only less than 6% of Cr was lost. The recovered Ni and Cr could be reused in the electroplating to reduce the operating cost and sludge amount.

Zhang et al. treated Cr (VI) wastewater also using NF [63]. The experimental results showed that the removal efficiency of Cr (VI) was higher than 98% and its concentration in permeation was lower than 0.5 mg L^{-1}. The quality of permeate meets the National (China) Discharge Standard (GB8978·1996) or the requirements for being used as rinsing water for electroplated products. As shown in Table 3.14, rejection of Cr (VI) by NF reached 99.5% in a wide range of operation pressure (>0.5 MPa) and temperature (<35°C); the conductivity of permeate was lower than 25.1 μs cm^{-1}. As suggested, the retentate might be further treated for reuse, for example in plating passivation.

3.2.3 Other Membrane Separations

Ji focused on the treatment of wastewater containing Cr (VI) by vacuum membrane distillation (VMD) [64]. Compared with traditional Cr separation method, VMD to separate Cr containing solution will not generate secondary pollution. The whole operation proceeded continuously with high flux and rejection.

Du also explored the feasibility of concentrating wastewater containing Cr by VMD [65]. feed solution concentrations of Cr (VI) and Cr(III) were 100 ppm, rejection get close to 99%, while Cr concentration in distillate was 0.06 ppm. A very high Cr rejection has been achieved.

Chitosan hollow fiber modules have been prepared and tested by Vincent and Guibal for solvent extraction of Cr (VI) from dilute solutions using Aliquat 336 as the carrier [66]. Chitosan membranes served as barrier between aqueous and organic phases: both passive and active transport from water to extractant occurred, depending on the experimental conditions and the characteristics of the membrane. The extraction system was prepared by connecting several hollow fibers in parallel.

It was confirmed that hollow fiber solvent extraction prepared was very efficient at removing Cr (VI) compounds from dilute acidic solutions. When the pH is below 5, the extraction efficiency is only slightly influenced by the pH or the acid concentration. For the treatment of very corrosive solutions containing high level of both H_2SO_4 and Cr (VI) compounds, the chemical resistance of the fiber must be improved. Among the several methods to strengthen the fibers, chitosan reacetylation proves to be the most efficient one for maintaining extraction performances and enhancing mechanical and chemical resistances.

In Kumbasar's work [67], selective separation and concentration of Cr (VI) compounds from acidic solutions containing various metal ions was conducted using emulsion liquid membrane (ELM) . The liquid membrane phase consists of kerosene as diluent, ECA 4360J as surfactant, trioctylamine (TOA) as extractant; the stripping phase contains 0.5 M ammonium carbonate ($(NH_4)_2CO_3$) as receiving chemical. It was possible to selectively extract 99% of Cr (VI) from the feed solutions containing ions of Cu, Zn, cobalt (Co), Ni, and cadmium (Cd).

The extraction percentage of Cr (VI) was found to be higher than 99% even for feed solution containing 600 mg L^{-1} Cr (VI). The selectivity of liquid membrane for Cr (VI) with respect to other ions was high within the first 4 min.

3.3 Manganese

3.3.1 Introduction [68]

Pyrolusite is the most important Mn ore with manganese dioxide (MnO_2) being the existing form of Mn. Globally, about 80% of the known Mn resources are in South Africa. Metallurgical or high grade Mn ores (>40% Mn) are usually processed into suitable metallic alloys [ie, Fe-Mn, Si-Mn] by pyrometallurgical processes. Fe-Mn is a commonly used alloy. In its production, a mixture of Mn ore, reducing agent [eg, carbon] and flux [e.g. caustic lime] are smelt at above 1200°C to enable reduction and alloy formation. Standard grade Fe-Mn alloy comprises more than 76% Mn and about 7% carbon, and can be achieved either in blast furnace or electric furnace.

The world's rapidly growing demand for Mn has made it increasingly important to develop processes for economical recovery of Mn from low grade Mn ores and other secondary sources. A simplified flow chart for Mn metallurgical processes and products based on major and potential Mn sources is illustrated in Fig. 3.25.

A pyrometallurgical pretreatment (eg, reductive roasting) followed by hydro-metallurgical processing plays an important role in the treatment of low grade manganese ores (<40% Mn). The final products are chemical manganese dioxide (CMD), electrolytic manganese (EM), or electrolytic manganese dioxide (EMD).

Polymetallic Mn nodules currently cannot be economically processed by pyro-processes because the energy consumption is high. The nodules contain Ni, Co, Cu, etc. in their oxides, which are often present in the lattices of Fe and Mn minerals. Breaking up these lattices

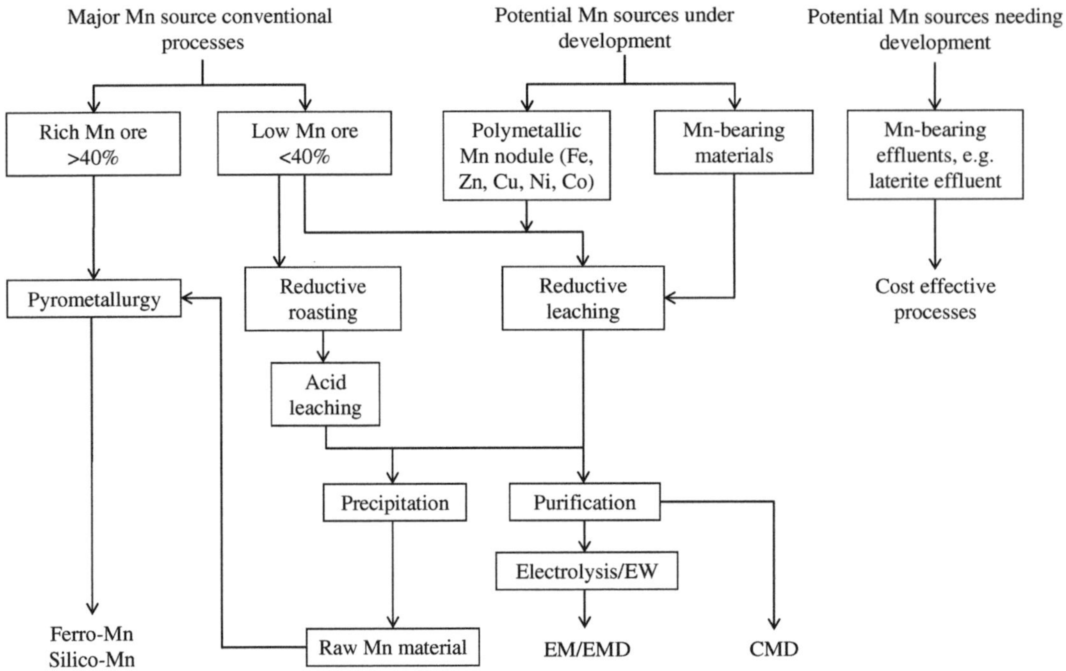

Figure 3.25: A simplified flow chart for manganese metallurgical processes and the manganese products based on major and potential manganese sources.

by hydrometallurgical reductive dissolution is an essential step for satisfactory recovery of valuable metals.

3.3.2 Purification of Manganese Sulfide ($MnSO_4$) Solution [69]

H_2SO_4 acid leaching of manganese dioxide (MnO_2) and ferrous disulfide (FeS_2) binary minerals in one step is widely adopted in electrolytic Mn production process. In order to recover Mn for providing feed to electrolysis step, manganese sulfate ($MnSO_4$) is to be separated from the slag also existing in the leaching slurry. With conventional filtration, the Mn recovery rate is very low, as the slag in slurry will form cake on the feed surface of filtration media to cause high resistance toward solution transfer.

YT metallic membranes were used in treating the leaching solution. The level of elemental sulfur (S) in the filtrate obtaining from H_2SO_4 leaching slurry by YT metal membrane filtration was reduced greatly.

The aforementioned $MnSO_4$ solution obtained by YT membrane filtration has favorable conditions for follow-up electrolysis. The current efficiency can be increased, thus realizing the target of energy saving. Additionally, the product (eg, EM, EMD) has higher quality or purity. This is because the elementary S and heavy metals' [eg, Cu, lead (Pb), Ni, Co] sulfide are efficiently retained in the feed. For example, the concentration of S in the filtrate was only 3×10^{-6} (mass fraction), as compared with 400×10^{-6} (mass fraction) in the original feed. According to the authors' comments, leaching of binary minerals will greatly improve the utilization rate of pyrolusite.

3.4 Conclusions

Ferrous metals compose the largest group in metal demand and consumption market. For ferrous metallurgy, the framework for membrane applications is very clear. They are concerned with pickling solution, oily wastewater, coking plant wastewater, desalination, and waste gas. The examples introduced include both lab-scale researches and industrial level applications. For the latter, membranes have helped obtain obvious benefits in environment and economy. Desalination is well established and widely employed application, as compared to other applications. Actually, usage of metal membrane in gas treatment has also demonstrated very promising results, due to the excellent stability of the membranes in the process. Investigations regarding Cr and Mn are relatively few and sporadic at present.

Glossary and Symbols

AEM	Anion-exchange membranes
Al	Aluminum
A^2/O	Anaerobic/anoxic/oxic
Bi	Bismuth
BPM	Bipolar membranes
Cd	Cadmium
CEM	Cation-exchange membranes
Cl_2	Chlorine
CMBR	Composite membrane bioreactor
CMD	Chemical manganese dioxide
Co	Cobalt
CO	Carbon monoxide
CO_2	Carbon dioxide
COD	Chemical oxygen demand
CrO_3	Chromium oxide
CrO_4^{2-}	Chromate
$Cr_2O_7^{2-}$	Dichromate
Cu	Copper
DC	Direct current
DD	Diffusion dialysis
ECO	Electrochemical catalytic oxidation
ED	Electrodialysis
ELM	Emulsion liquid membrane
EM	Electrolytic manganese
EMD	Electrolytic manganese dioxide
EU	European Union
EW	Electrowinning

Fe	Iron
FeAl	Ferro-aluminum
$FeCl_3$	Ferric chloride
Fe-Cr	Ferro-chrome
FeF_3	Ferric fluoride
Fe-Mn	Ferromanganese
Fe_2O_3	Ferric oxide
Fe-Si	Ferrosilicon
FeS_2	Ferrous disulfide
HCl	Hydrochloric
HF	Hydrofluoric
H_2	Hydrogen
KF	Potassium fluoride
KNO_3	Potassium nitrate
K_2SiF_6	Potassium fluorosilicate
GEF	Gas energy flocculation
$HCrO_4^-$	Bichromate
$HCr_2O_7^-$	Bi-dichromate
H_2CrO_4	Chromic acid
HNO_3	Nitric acid
H_2O_2	Hydrogen peroxide
H_2SO_4	Sulfuric acid
IEM	Ion-exchange membrane
MBR	Membrane bioreactor
MD	Membrane distillation
MF	Microfiltration
MnO_2	Manganese dioxide
$MnSO_4$	Manganese sulfate
MW	Molecular weight
O_2	Oxygen
NaOH	Sodium hydroxide
NF	Nanofiltration
NH_3	Ammonia
NH_3-N	Ammonia nitrogen
NH_4^+	Ammonium
$(NH_4)_2CO_3$	Ammonium carbonate
$(NH_4)_2S_2O_3$	Ammonium thiosulfate
NH_4SCN	Ammonium thiocyanate
Ni	Nickel
NiO	Nickel oxide
PAM	Polyacrylamide
Pb	Lead
PSF	Polysulfone
S	Sulfur
Si-Mn	Silicon-manganese
SIMS	Special Industry Membrane Separation
SS	Suspended solids
TOA	Tryoctylamine
TSNF	Two-stage nanofiltration
VMD	Vacuum membrane distillation
Zn	Zinc

References

[1] J. Zhou, China's steel industry demand trend analysis during twelfth five-year period, Macroecon. Manage. 5 (2011) 25–27.

[2] B. Wang, Study on the development strategy of Tangshan iron and steel industry, Master thesis, Tianjin University, 2002.

[3] T. Wu, C.L. Zhao, D.J. Liu, R. Gu, Exploration of wastewater zero-emission technologies in iron and steel industry, Environ. Eng. 33 (4) (2015) 146–149.

[4] M. Regel-Rosocka, A review on methods of regeneration of spent pickling solutions from steel processing, J. Hazard. Mater. 177 (2010) 57–69.

[5] J. Carrillo-Abad, M. García-Gabaldón, V. Pérez-Herranz, Study of the zinc recovery from spent pickling baths by means of an electrochemical membrane reactor using a cation-exchange membrane under galvanostatic control, Sep. Purif. Technol. 132 (2014) 479–486.

[6] F. Rögener, M. Sartor, A. Bán, D. Buchloh, T. Reichardt, Metal recovery from spent stainless steel pickling solutions, Resour. Conserv. Recy. 60 (2012) 72–77.

[7] Q.X. Zhang, Metallurgical Separation Science and Engineering, Science Press, Beijing, (2004).

[8] J. Dufour, J. L-Gálvez, C. Negro, F. López-Mateos, Optimization of iron and chromium recovery from mixed-acid pickling liquors, Global Symposium on Recycling, Waste Treatment and Clean Technology, 1 (2004) 361–370.

[9] D.C. Ge, X.J. Guo, Investigation of application of anion and cation exchange membranes and bipolar membrane in metallurgy. The Fourth National Symposium of Application of Membrane-based Separation in Metallurgical Industry, 2014, 100–102.

[10] K. Wang, W. Wang, W.H. Xing, Recovery of nitric acid from stainless steel pickling waste liquor by diffusion dialysis, Membr. Sci. Technol. 30 (2011) 62–65.

[11] H. Wang, A SIMS solution for comprehensive recovery of waste acid in a steel mill. The Third National Symposium of Application of Membrane-based Separation in Metallurgical Industry, 2009.

[12] K.M. Forsberg, A.C. Rasmuson, Recycling of waste pickle acid by precipitation of metal fluoride hydrates, Miner. Eng. 20 (2007) 950–955.

[13] M. Tomaszewska, M. Gryta, A.W. Morawski, The influence of salt in solutions on hydrochloric acid recovery by membrane distillation, Sep. Purif. Technol. 14 (1998) 183–188.

[14] M. Tomaszewska, M. Gryta, A.W. Morawski, Recovery of hydrochloric acid from metal pickling solutions by membrane distillation, Sep. Purif. Technol. 22 (2001) 591–600.

[15] M. Tomaszewska, A. Mientka, Separation of HCl from HCl/H2SO4 solutions by membrane distillation, Desalination 240 (2009) 244–250.

[16] K.H. Lum, S.J. Cook, G.W. Stevens, Zinc chloride and hydrochloric acid co-extraction from galvanizing pickling waste in the presence of iron (II). Results with hollow fiber membrane contactors, Ind. Eng. Chem. Res. 53 (2014) 4453–4461.

[17] F.J. Alguacil, S. Martinez, Solvent extraction of Zn(II) by Cyanex 923 and its application to a solid-supported liquid membrane system, J. Chem. Technol. Biotechnol. 76 (2001) 298–302.

[18] M.F. San Roman, I. Ortiz Gandara, R. Ibanez, I. Ortiz, Hybrid membrane process for the recovery of major components (zinc, iron and HCl) from spent pickling effluents, J. Membr. Sci. 415-416 (2012) 616–623.

[19] E. Paquay, A.M. Clarinval, A. Delvaux, M. Degrez, H.D. Hurwitz, Applications of electrodialysis for acid pickling wastewater treatment, Chem. Eng. J. 79 (2000) 197–201.

[20] C. Negro, M.A. Blanco, F. Lopez-Mateos, A.M.C.P. DeJong, G. LaCalle, J. van Erkel, D. Schmal, Free acids and chemicals recovery from stainless steel pickling baths, Sep. Sci. Technol. 36 (2001) 1543–1556.

[21] G. Csicsovszki, T. Kekesi, T.I. Torok, Selective recovery of Zn and Fe from spent pickling solutions by the combination of anion exchange and membrane electrowinning techniques, Hydrometallurgy 77 (2005) 19–28.

[22] J. Carrillo-Abad, M. Garcia-Gabaldon, E. Ortega, V. Perez-Herranz, Electrochemical recovery of zinc from the spent pickling baths coming from the hot dip galvanizing industry. Potentiostatic operation, Sep. Purif. Technol. 81 (2011) 200–207.

[23] B.K. Thomas, D.J. Fray, The effect of additives on the morphology of zinc electrodeposited from a zinc chloride electrolyte at high current densities, J. Appl. Electrochem. 11 (1981) 677–683.
[24] P. Diaz-Arista, O.R. Mattos, O.E. Barcia, F.J. Fabri Miranda, ZnFe anomalous electrode position: stationaries and local pH measurements, Electrochim. Acta 47 (2002) 4091–4100.
[25] N. Yi, W. Hu, Several treating methods for emulsion wastewater from cold rolling plants of iron and steel enterprises ( ), Metallurg. Power 5 (2004) 58–63.
[26] L. Zhang, G.F. Xue, J.F. Bao, Research advances in treatment of oily wastewater from cold rolling mill, Wisco Technol. 47 (2009) 7–10.
[27] J. Yue, S.C. Wan, H.Y. Wan, Progress in treatment technology of oily wastewater ( ), Pollut. Control Technol. 22 (2009) 52–55.
[28] Y. Guo, Z.D. Xiao, Explore and analysis on materialized reason of difficult treatment of emulsified oily wastewater, J. Educ. Inst. Jilin Province 31 (2015) 149–150.
[29] Y.L. Sun, Discussion on ultrafiltration membrane treatment technologies for emulsion wastewater in the cold rolling mill, Metallurg. Gansu 33 (2011) 77–80.
[30] Z.B. Fang, G.J. Xiao, G.Y. Yin, Application of inorganic membrane technology in oil emulsion treatment. The Third National (China) Symposium on Application of Membrane Separation Technologies in Metallurgical Industry, 2009.
[31] Z.G. Wang, Application of membrane separation technology in hydrometallurgy. The Fourth National (China) Symposium on Application of Membrane Separation Technologies in Metallurgical Industry, 2014.
[32] Y.H. Zhang, F.X. Gan, M. Li, Industrial test research on new combined process (principally based on MBR) for treatment of cold rolling emulsion wastewater. The Third National (China) Symposium on Application of Membrane Separation Technologies in Metallurgical Industry, 2009.
[33] C. Shu, Y.S. Li., A new idea of WISCO cold-rolled oily emulsion wastewater treatment. The Third National (China) Symposium on Application of Membrane Separation Technologies in Metallurgical Industry, 2009, 66–68.
[34] Y.B. Wen, H.M. Zhang, Y. Qian, Pilot study on the treatment of coking wastewater by AA/O method (AA/O), Water Supply Drain. China 8 (1992) 7–12.
[35] C.L. Yin, Z.J. Dan, Problems and solutions of wastewater treatment of coking plant, Water Supply Drain. 26 (2000) 35–37.
[36] Y.L. Yang, X.P. Bai, Technology for treatment of wastewater from coking and its development, Ind. Water Wastewater 32 (2001) 8–10.
[37] Y. Zhang, B.R. Jiang, Research on the technique of coking wastewater treatment in metallurgical industries, Indus. Saf. Environ. Prot. 28 (2002) 5–7.
[38] C.M. Zhong, M. Cai, P. Wang, C.Q. Liu, Investigation at the treatment of coking wastewater by submerged membrane bioreactor, J. Jiangxi Univ. Sci. Technol. 31 (2010) 5–8.
[39] J. Wang, B.C. Zhang, P. Zhang, Coking wastewater treatment and reuse by CMBR. The Third National (China) Symposium on Application of Membrane Separation Technologies in Metallurgical Industry, 2009.
[40] W. Zhang, L. Liu, R. Zheng, G. Wu, P. Yu, Y. Luo, Fenton oxidation and microfiltration for coking wastewater treatment and membrane fouling mechanism, Membr. Sci. Technol. 1 (2014) 104–110.
[41] Z.Z. Zhu, Z.L. Han, Z.S. Shen, J. Chen, Recovery of ammonia and phenol from wastewater of coking mill with membrane absorption, Indus. Water Treat. 5 (2006) 50–53.
[42] L. Zhu, Y.M. Zhao, F. Chen, Application of ion-exchange membrane electrolysis to pretreatment of coke plant wastewater, Environ. Sci. Technol. 12 (2009) 9–12.
[43] N. Yin, F. Liu, W.H. Xing, Study on the process of membrane treatment of coking desulfurization waste liquor, Membr. Sci. Technol. 30 (2011) 75–78.
[44] T.X. Li, Y.J. Zhou, Application and practice of membrane separation technology in treating coking wastewater. The Third National (China) Symposium on Application of Membrane Separation Technologies in Metallurgical Industry, 2009.

[45] X.W. Jin, E.C. Li, S.G. Lv, Z.F. Qiu, C. Zhou, Practical research on advanced treatment of coking wastewater by membrane process, Environ. Pollut. Control 11 (2013) 77–81.
[46] J.H. Feng, H.M. Li, Application of ceramic membranes to the oil removal from the coking residual ammonia water, Indus. Water Treat. 12 (2006) 31–33.
[47] J.H. Zhang, L.Y. Wu, Z.F. Liu, Operating standard making of desalination of wastewater in iron and steel company. The Third National (China) Symposium on Application of Membrane-based Separation in Metallurgical Industry, 2000, 172–176.
[48] D.W. Xia, L.M. Yin, J.Z. Tong, Overview on design of Rizhao Iron and Steel Group steelmaking wastewater reuse project. The Third Symposium on Application of Membrane-based Separation in Metallurgical Industry, 2009, 97–100.
[49] W.L. Zhong, H.S. Liang, W.C. Li, F. Liu, Y. Jin, Z.C. Wu, S.M. Liang, Application of RO technology to the project of steel plant wastewater reuse, Indus. Water Treat. 8 (2013) 83–85.
[50] W.L. Zhong, Application of domestic RO membrane in water treatment system of Shougang Group, China Steel 7 (2011) 29–31.
[51] Y.B. Jin, Discussion on application of ceramic membranes in deep desalination of industrial wastewater throughout the iron and steel enterprises. The Third National (China) Symposium on Application of Membrane based Separation in Metallurgical Industry, 2009.
[52] W.L. Zhong, L.X. Peng, Application of reverse osmosis technology in zero discharge of wastewater project in steelmaking company, Indus. Water Treat. 31 (2011) 81–83.
[53] K. Fan, D.S. Zheng, C.H. Ji, J. Guo, Application of membrane treatment technology in wastewater reuse in iron & steel company. The Third National (China) Symposium on Application of Membrane-based Separation in Metallurgical Industry, 2009.
[54] J.J. Dong, X.Q. Chen, Application of double-membrane water treatment process in the reuse system of metallurgical wastewater, Chin. Eng. Sci. 8 (2006) 92–94.
[55] Z.C. Zhang, Review on energy recovery and utilization of ferroalloy submerged arc furnace flue gas, Power Demand-side Manage. 13 (2011) 4–6.
[56] X.J. Zhang, Z.H. Wang, W.X. Zeng, Application of YT membrane separation technology in high temperature gas purification in closed ore furnace. The Fourth National (China) Symposium on Application of Membrane Separation Technologies in Metallurgical Industry, 2014.
[57] H.S. Zhu, New Technology and Theoretical Studies of Chrome Smelting, Master thesis, Kunming University of Science and Technology, 2004.
[58] N.F. Kizim, A.P. Lar'kov, E.Y. Sharova, Electrodialysis regeneration of chromium-containing solutions, J. Appl. Chem. USSR (Engl. Transl.) 60 (1987).
[59] A. Hafiane, D. Lemordant, M. Dhahbi, Removal of hexavalent chromiumby nanofiltration, Desalination 130 (2000) 305–312.
[60] M. Taleb-Ahmed, S. Taha, R. Maachi, G. Dorange, The influence of physico-chemistry on the retention of chromium ions during nanofiltration, Desalination 145 (2002) 103–108.
[61] N.K. Lazaridis, A.I. Zouboulis, M. Jekel, Removal of Cr(VI), Mo(VI), and V(V) ions from single metal aqueous solutions by sorption or nanofiltration, Sep. Sci. Technol. 38 (2003) 2201-.
[62] S.S. Chen, B.C. Hsu, C.H. Ko, Recovery of chromate from spent plating solutions by two-stage nanofiltration processes, Desalination 229 (2008) 147–155.
[63] X.Q. Zhang, L.S. Zhang, X.W. Lv, Research of treatment of Cr (VI) wastewater by nanofiltration, Chin. J. Environ. Eng. 63 (3) (2005) 25–27.
[64] Z.G. Ji, Progress in treatment of heavy metal wastewater by membrane distillation. The Fourth National (China) Symposium on the Application of Membrane Separation Technologies in Metallurgical Industry, 2014.
[65] J. Du, Separation of Solution Containing Cr by Vacuum Membrane Distillation, Master thesis, Chongqing Univ., 2002.
[66] T. Vincent, E. Guibal, Cr (VI) extraction using Aliquat 336 in a hollow fiber module made of chitosan, Indus. Eng. Chem. Res. 40 (2001) 1406–1411.

[67] R.A. Kumbasar, Studies on extraction of chromium (VI) from acidic solutions containing various metal ions by emulsion liquid membrane using Alamine 336 as extractant, J. Membr. Sci. 325 (2008) 460–466.
[68] W. Zhang, C.Y. Cheng, Manganese metallurgy review. Part I: Leaching of ores/secondary materials and recovery of electrolytic/chemical manganese dioxide, Hydrometallurgy 89 (2007) 137–159.
[69] D.S. Liu, J.X. Zhang, Q.L. Zheng, M. Jiang, L. Gao, Application of YT membranes in hydrometallurgy. The Fourth National (China) Symposium on Application of Membrane Separation Technologies in Metallurgical Industry, 2014.

CHAPTER 4

Heavy Nonferrous Metals

Z.-L. Xu

State Key Laboratory of Chemical Engineering, Membrane Science and Engineering R&D Lab, Chemical Engineering Research Center, East China University of Science and Technology, Shanghai, China

4.1 Nickel

4.1.1 Introduction

Nickel (Ni) is a kind of approximately silvery white metal which is ferromagnetic and ductile, and can conduct electricity and heat. Ni is mainly used to produce stainless steel (SS) and other corrosion-resistant alloys. It is also used as hydrogenation catalyst and in producing ceramic products, special chemical vessels, etc. There are three major types of the Ni deposits worldwide: copper–nickel sulfide ore, laterite, and weathering crust silicate nickel deposit. Among them, laterite and weathering crust silicate nickel deposit now account for three-quarters of the world's total reserves of Ni. However, at present, Ni extracted from sulfide ore accounts for about two-thirds of the total Ni and its compounds production [1].

Global Ni resources are mainly distributed in Australia, New Caledonia, Indonesia, the Philippines, Brazil, Russia, Cuba, South Africa, and Canada. China's Ni reserves are 3 million tons, ranking ninth in the major Ni resource countries in the world and only accounting for 3.7% of total global reserves. Therefore, the dependence on imports will not change in a fairly long period of time in China.

There exist several main approaches for Ni-containing product production. The first one is based on nickel sulfide ore, which employs pyrometallurgy-electrolytic refining to produce electrolytic nickel. It is the principal electrolytic Ni production method adopted in China. In the second method, oxidic Ni ore is raw material and hydrometallurgy is utilized to produce nickel. The third approach uses oxidic nickel ore and relies on blast furnace or electric furnace to produce ferronickel (FeNi) [2]. It is worth noting that due to the shortage of nickel sulfide ore, production of Fe-Ni using oxidic nickel ore has become a main direction of future Ni industry in China [2]. Oxidic Ni ore deposit which is customarily called laterite ore has gradually become the focus of the world's Ni industry development.

The Fe-Ni process comprises four steps: drying (ore preparation), calcination and prereduction, smelting, and refining. Oxidic nickel ore is preheated or partly prereduced in

Membrane-Based Separations in Metallurgy
http://dx.doi.org/10.1016/B978-0-12-803410-1.00004-9

rotary kiln, followed by addition of coke powder and reduction smelting; the final step is converting to produce Fe-Ni [3]. In practical operation, electric furnace smelting is adopted, which is simple but has high energy consumption and environmental pollution [4].

Vulcanizing agent is used in Ni matte smelting process, which is commonly conducted in blast furnace or electric furnace. The composition of Ni matte can be adjusted by changing the dosage of reducing agent (coke powder) and the vulcanizing agent (plaster). In this process, low concentration Ni matte (Ni and Co 20–30%) is obtained, and then sent to converter to be blown into high concentration Ni matte. Ni matte technology delivers high grade product and high recovery of Ni and cobalt (Co). Ni matte product is more flexible in adapting to subsequent refining process. This method is most suitable for processing garnierite ore, but has the disadvantages of high energy consumption and severe environmental pollution [4,5]. The two pyrometallurgical processes mentioned previously are shown in Fig. 4.1.

The limonite type of Ni laterite ore and garnierite which contains low content of magnesia, are usually treated by two hydrometallurgical processes. As shown in Fig. 4.2, one is reduction roasting-ammonia (NH_3) leaching process (RRAL) and the other is high-pressured acid leaching process (HPAL). The purpose of reduction roasting is to convert Ni in silicate and oxide into elementary Ni to the greatest degree and most of iron (Fe) into iron oxide black (Fe_3O_4). The next step is to transform metallic Ni and Co into complex compounds which can be dissolved in aqueous solution. Simultaneously, Fe was turned into coordinated compound of NH_3 that is soluble in water, followed by oxidization into ferric ion (Fe^{3+}) which will finally hydrolyze to produce ferric hydroxide ($Fe(OH)_3$) as precipitates. Meanwhile

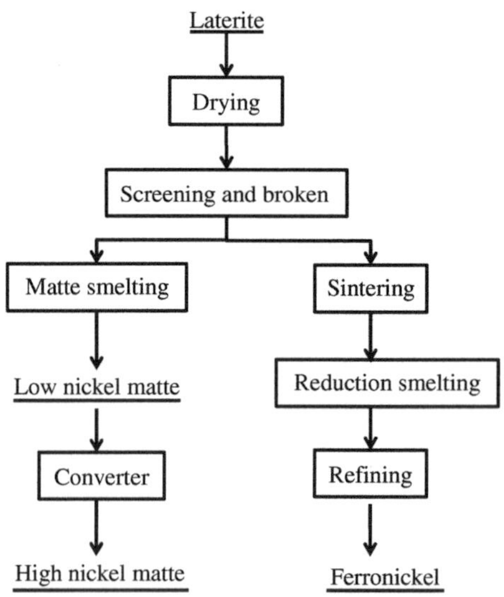

Figure 4.1: Two kinds of pyrometallurgical processes for producing different nickel compounds.

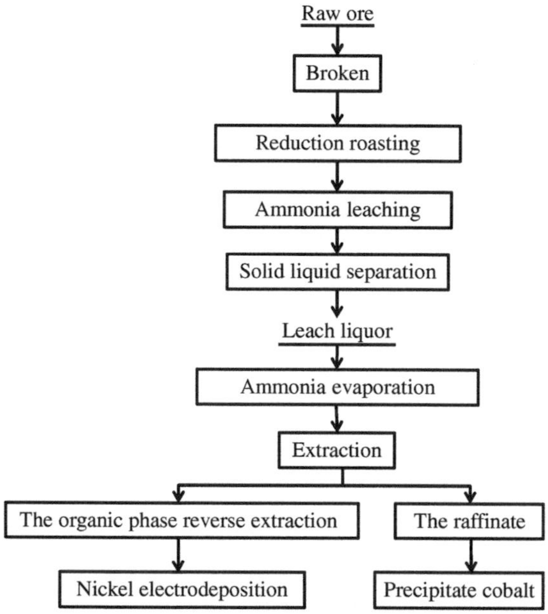

Figure 4.2: The flow chart of reduction roasting—ammonia leaching process.

most of Co would be removed in the form of precipitation. In this regard, the most serious weakness of the aforementioned process is the low Co recovery.

In the HPAL process, Ni, Co, Fe, aluminum (Al) in minerals are dissolved by dilute sulfuric acid (H_2SO_4). The next step of pH adjustment transfers impurities involving Fe, Al, and silicon (Si) into slag via hydrolysis, while Ni and Co remain in solution. The leaching liquor then reacts with hydrogen sulfide (H_2S), followed by reduction neutralization and precipitation to produce high quality nickel and cobalt sulfides (NiS and Co_xS_y) as final product. The flowchart of high-pressure acid leaching process is shown in Fig. 4.3. The most important advantage of this process is higher than 90% recovery for metals.

4.1.2 Improvement of Ni Electrowinning

Nel used membrane separation to facilitate the zero discharge of Ni electrowinning process [6]. The proposed process with nanofiltration (NF) is indicated in Fig. 4.4 In Ni electrowinning, a Ni-rich solution from the leaching section is sent to the Ni electrowinning unit with a concentration of approximately 90 g L^{-1} $Ni^{2+}(aq)$ at a pH of around 3.5 [7]. Energy expenses limit the recovery of Ni to 50%. Therefore, recycling the spent Ni electrolyte is very critical for profitable Ni production. The currently adopted process for treating spent Ni electrolyte before it is recycled back to leaching is shown in Fig. 4.4, comprising mainly four procedures. The process presented successful removal of water and excess sulfur (S) in the form of sodium sulfate (Na_2SO_4). The major disadvantage is

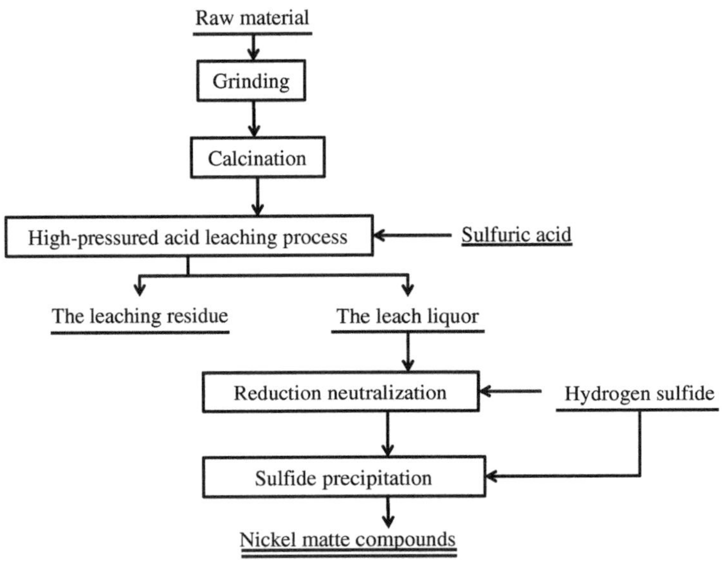

Figure 4.3: The flow chart of high-pressure acid leaching.

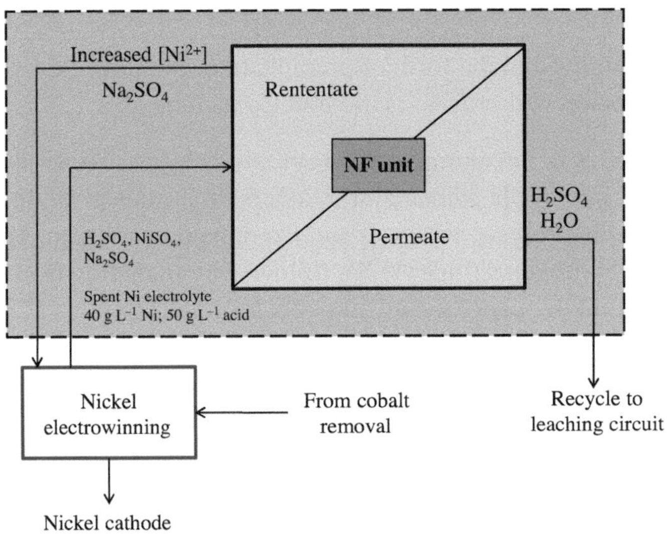

Figure 4.4: Proposed NF process for improving nickel electrowinning.

the consumption of large quantities of caustic soda to produce a relatively cheap Na_2SO_4 by-product. The process proposed by Nel et al. replaces the four units with a single NF unit. In the new system, the concentrated Ni in the retentate is recycled back to the Ni electrowinning (EW), together with the Na_2SO_4 to ensure conductivity within the EW cells. The permeate is fed back to the leaching circuit to reduce the overall H_2SO_4 consumption. The new process not only reduces capital costs, but also running costs.

Table 4.1: Metal ions content in Ni-P electroless plating wastewater

Metal Ion Type	Concentration, mol L^{-1}
Ni^{2+}	0.724
Na^{+}	0.171
Cu^{2+}	0.158
Zn^{2+}	0.077

In the experiment, the NF membrane was found to be very selective for Ni ions, with the rejection of Ni ions being between 54.4% and 98.2%. The rejection of hydrogen ions (H^+) ranged from −5.9% to 21.8%. The rejection of sodium ions (Na^+) varied between 16.6% and 72.4% at a crossflow velocity of 2.5 m s^{-1} and Na_2SO_4 concentration of 50 g L^{-1} in the feed solution.

4.1.3 Process Wastewater as Secondary Resource

Mohammad [8] investigated the possibility of applying NF for treatment of nickel–phosphorus (Ni-P) electroless plating waste using the real waste from industry. A crossflow cell (Osmonics, USA) was used to carry out the experiments. The NF membrane chosen is also supplied by Osmonics Corporation. The composition of the electroless plating waste investigated in this study is shown in Table 4.1. Although other metal ions also existed in the waste, the study focused solely on separation of Ni^{2+} and Na^+ ions. Results show that the rejections of Ni^{2+} are higher than 95%, compared to Na^+ ion rejections which are between 59% and 73%. The experimental data were also analyzed using two predictive models, namely Donnan steric pore model (DSPM) and ion transport model.

Changsha Lyrun Material Company is one of the production bases for electroplating spongy Ni [9]. In the electroplating process, saccharin is used as brightening agent and stress reducer. Upon directly discharged into environment, the wastewater with Ni^{2+} would present, under certain circumstances, toxicity to aquatic life, and in extreme cases, human life. The company has cooperated with the Hangzhou Water Treatment Center of National Bureau of Oceanography to develop the process for treating the rinsing water of Ni plating [9]. NF and Brinewater RO were employed in the first stage to dewater and concentrate the rinsing water, followed by concentration via Brinewater RO. The quality of the feed water is summarized in Table 4.2. The Ni concentration is in the range from 50 to 500 mg L^{-1}. The flowchart for treating the Ni electroplating wastewater is displayed in Fig. 4.5. The concentration of Ni in the treated water met the level required by application in plating. The recovery of water

Table 4.2: Content of components in electroplating nickel rinse water (pending)

	pH	SO_4^{2-}	Boric Acid	Ni^{2+}	Cl^-	Saccharin
Target concentration (mg L^{-1})	5.0 ± 0.2	411	50	50~300	58	0.5

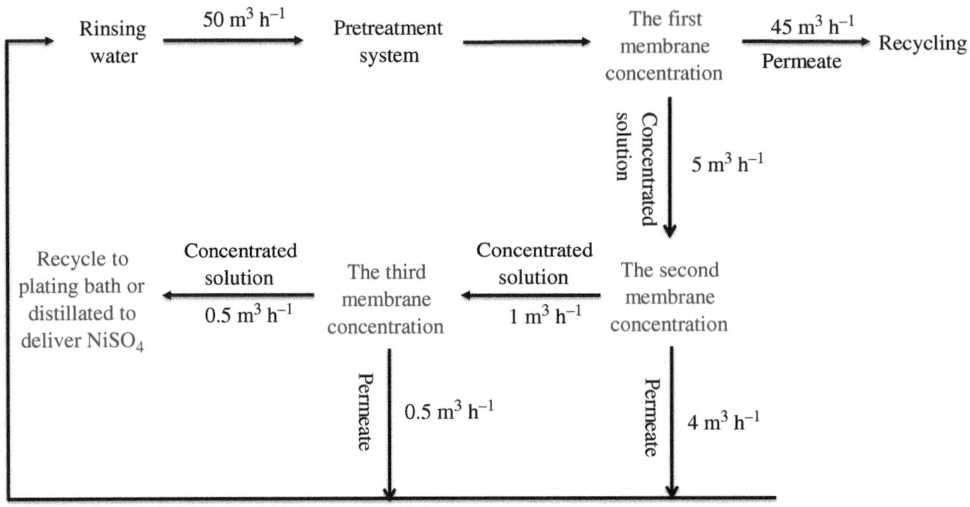

Figure 4.5: The process flow diagram of Changsha force yuan membrane separation system.

and Ni brings about a total net benefit of 2.86 million CNY year^{-1}. The application of RO in sponge-Ni electroplating has been industrialized.

In Shi et al.'s work [10], the method as shown in Fig. 4.6 involving sodium hydroxide (NaOH) and membrane filtration was used to treat wastewater. Inorganic membranes with average pore size of 0.5 μm were selected. The approach not only takes the advantages of chemical method, but also solves the challenging problem of colloidal sedimentation with the help of using membrane filtration. Generally, colloidal does not settle by gravity efficiently. The bulky sedimentation tanks are not necessitated any more with the new system. Nickel sulfate ($NiSO_4$) recovered is characterized by high purity which can be recycled to the production process. The concentration of Ni^{2+} in the effluent at permeate side is stably maintained at around 1 mg L^{-1}, which satisfied the national (China) standard for discharging. The edges of the method also include high flux, easy cleaning, and resistance to corrosion.

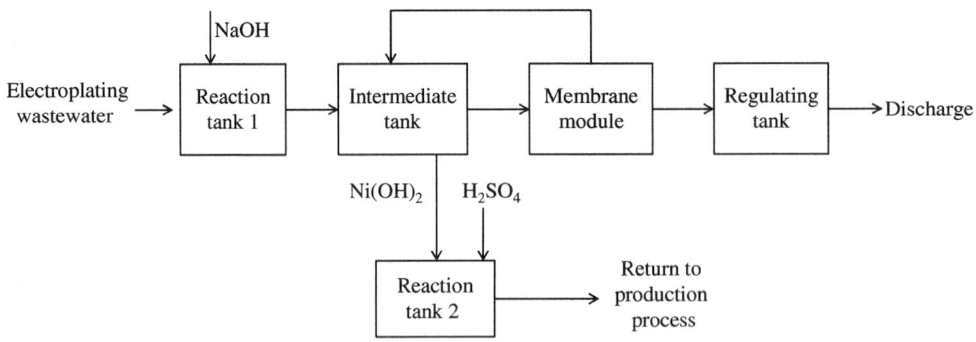

Figure 4.6: NaOH-membrane filtration treatment for process wastewater containing nickel.

Zhu et al. [11] treated the Ni electroplating wastewater using vacuum membrane distillation (VMD). The feed composition is as follows: COD_{Cr}, 40,000–50,000 mg L^{-1}; ammonia nitrogen (NH_3-N), 60,000–70,000 mg L^{-1}; total P, 1000–2000 mg L^{-1}, inorganic salt, 190–200 g L^{-1}; Ni^{2+}, 5000–5200 mg L^{-1}. The membranes utilized were hydrophobic polytetrafluoroethylene (PTFE) hollow fiber membranes. A 90 kPa vacuum is used at the distillate side. In the immersion VMD experiment, Ni electroplating water was first treated by ion exchange to remove partially the Ni^{2+}. The water stream was then heated to a certain temperature before being fed into VMD system. A test lasting 180 days was conducted. The pure water was produced at 2.4–2.6 kg ($h^{-1} \cdot m^{-2}$). Except for NH_3-N, the contents of the other components mentioned previously in the distillate reach the national (China) standard requirements (GB8978-1996).

Sofi Filtration Ltd, together with Department of Materials Science and Engineering at Aalto University in Finland, has developed a new type of high velocity, crossflow filter with a commercial trade name of Sofi Filter [12]. The system was applied in treating belt filter permeate at Ni processing plant. The operation temperature was set at 25°C; the overall process feed flow was 7 m^{-3} h^{-1}; the feed has a pH value of 10.5 and contains 10 μm metal hydroxide solids with a concentration of about 1000 mg L^{-1}. Membrane filtration in general is operated either in a dead-end or crossflow mode, of which the biggest problem is low filtration (flux) flow, solids' concentration, and fouling. Consequently, the membrane has to be either cleaned or replaced often. In Sofi Filter, the shearing effect takes place, that is, the feed flows tangentially along the membrane. This reduces the build-up of potential fouling layers on the membrane, and reduces concentration polarization, membrane surface adsorption, and cake layer formation.

The Sofi Filter is a compact unit consisting of an acid-proof steel vessel, a sintered metal filtrate element, ultrasonic cleaning unit, and (programmable logic controller) PLC unit. The modular structure design enables high capacity installations by economic combination of units for virtually all capacity needs. The filtration is done with overpressure, in other words the bubble point of the membrane being exceeded. The required feed pressure is low, varying between 0.05 and 0.2 MPa (0.5–2 bar).

4.2 Copper

4.2.1 Introduction

Copper (Cu) is a nonferrous metal which is closely related to human activities, being widely used in electrical, light industry, machinery manufacturing, construction industry, defense industry, and many other fields [13]. Cu reserves are concentrated in Chile, Peru, and Australia, followed by Mexico, the United States, China, Indonesia, Russia, Poland, Congo, and Zambia [14].

Nowadays, the countries from Latin America are the main suppliers of Cu resources and the manufacturing is controlled by mining companies from Europe, the United States, and Latin America. China has a great demand of Cu supply. Its production capacity is high, but still lagging behind the demands, leading to severe dependence on import [13]. Currently, China ranks first in the amount of Cu concentrate imported. The countries that rely heavily on Cu import are mainly from Asia, such as Japan, India, South Korea, etc. In 2013, China imported 10.059 million tons of Cu concentrate, up 28% year over year (Y-O-Y). Due to the demand from economic growth, while at the same time the limit by the mineral grade and the service life of mine, the demand for Cu import will keep increasing in the near future in China [15].

On a global scale, 85–90% of sulfide Cu ore is treated by pyrometallurgy process for extracting Cu. The major steps include matte smelting, copper matte converting, pyro-refining, and electrolytic refining [16]. A typical Cu pyrometallurgy process is illustrated in Fig. 4.7 [17]. Compared with hydrometallurgy, pyrometallurgy is more advantageous

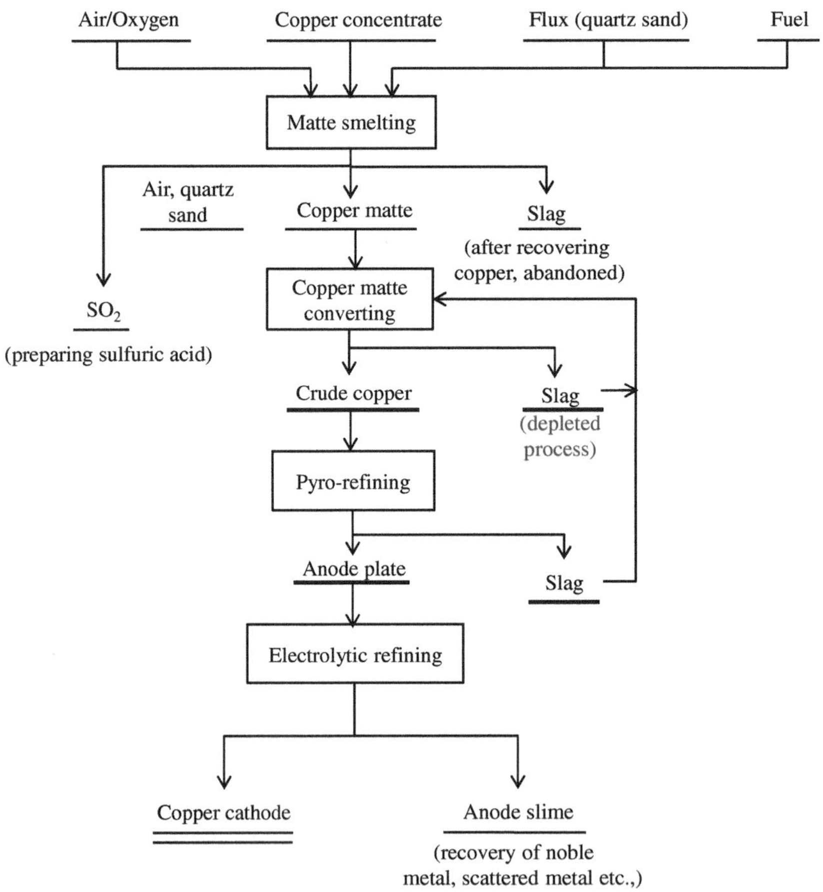

Figure 4.7: The typical copper matte pyrometallurgy process.

in terms of adaptability of raw materials to raw materials, scale of production, and ease in concentration/recovery of noble metals and rare metals. Conventional pyrometallurgical technologies such as flash smelting, bath smelting, have been progressively improved [18], leading to downsizing energy consumption and weakening pollution to environment.

In Cu matte converting, the priority is still given to Pierce-Smith (P-S) converter. After 1990s, continuous converting technique was successfully commercialized with a qualitative leap. In 1995, flash converting process was applied to Kenny Curt (K-C) smelters in the United States. The sulfur (S) recovery rate of the whole pyrometallurgical process (flash smelting and flash converting) improved from 95% to 99.9%. The basic procedures involved are as follows: Cu matte production in the flash furnace, water quench of Cu matte followed by milling and drying, conversion of Cu matte into crude Cu using oxygen-enriched air in another small-scale flash furnace, crude Cu delivered to anode furnace through sluice [18].

Generally, hydrometallurgy is used when processing of oxidic Cu and low grade Cu ore is concerned. A typical Cu hydrometallurgy flowchart is shown in Fig. 4.8 [16]. It comprises mainly leaching, extraction, reverse extraction. and electrodeposition. With the advancement of bio-heap leaching, pressurized leaching, etc., the application scope of copper hydrometallurgy has been expanded [17]. During the past decade, heap leaching has become one of the mainly adopted technologies for leaching of low grade Cu ore [17].

4.2.2 Assisting Selective Leaching Using Electromembrane

Sabba and Akretche applied the electromembrane process for in situ leaching of a typical Cu ore in NH_3 medium [19]. The ore is from Ougarta, a locality sited in the south west of Algeria. It contains Cu (average concentration is 6.5%) in multiple mineralogical forms, among which malachite ($Cu_2CO_3(OH)_2$) is the dominant one. The roasted Cu ore was mixed with ammonium (NH_4^-)/NH_3 solution and then fed into the central compartment of a five-compartment Plexigals cell between two (IEMs). The leaching mechanism is probably governed by the complexation reaction between both Cu (I) and Cu (II) and NH^3 solution. For all the experiment, Cu is recovered and enriched inside the catholyte compartment as both a free ions and amino complex (i.e., $Cu(NH_3)_2^{2+}$, $Cu(NH_3)_4^{2+}$). NH_4^+ crosses the membrane from an adjacent compartment to increase the dissolution giving rise to 98% of extraction yield after 400 h.

These recommendable extraction yields have been obtained without using pressure and oxidant in the course of an alkaline leaching. Although the kinetic of the process is slow, the number of operating units was reduced by cutting the filtration phase and purification–concentration steps. In addition, the results indicated that the recovery of metals could be highly selective by utilizing the discriminating function of the membrane.

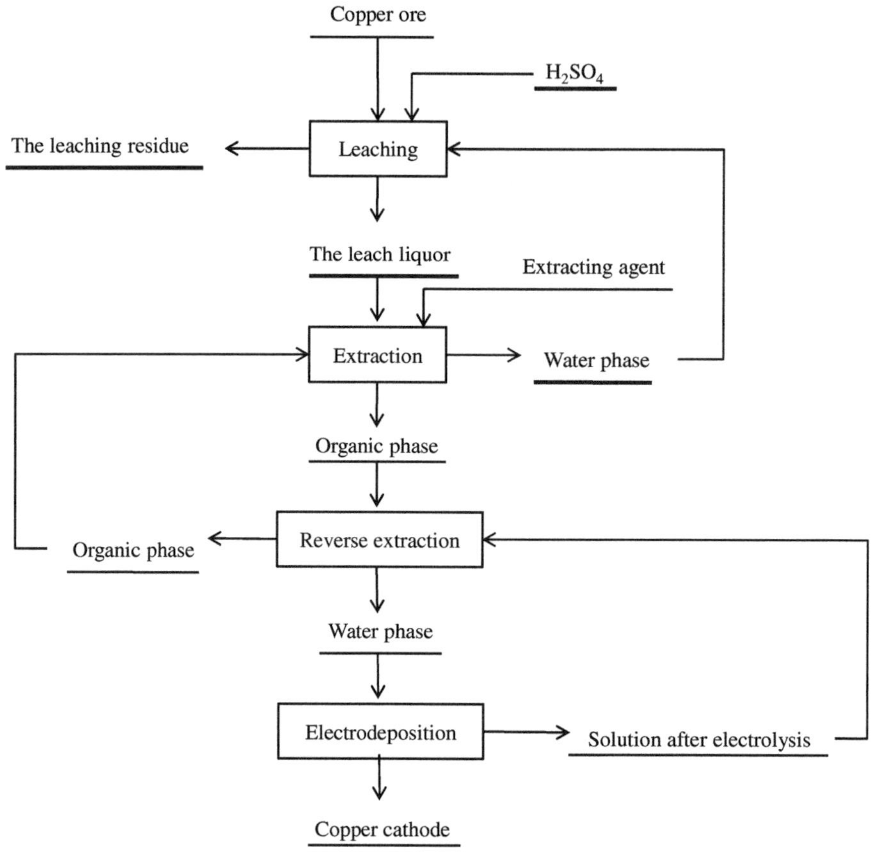

Figure 4.8: The copper hydrometallurgy process.

4.2.3 Improvement of Conventional Process Efficiency

Heap leaching by natural water precipitation (i.e., rain, snow) is widely adopted in many countries to capitalize on low-quality copper ore [20].

One typical example is the practice adopted at Mexicana de Cananea Mine. The leaching solution is characterized by very low concentration of copper (0.6 g L^{-1}), which results in inefficiency in follow-up conventional solvent extraction (SX) and EW. As a result, the Cu recovery rate is very low based on the traditional process, which also causes the high cost for wastewater treatment. The raffinate that can be returned to the leaching step for controlling the Cu concentration is limited. Almost half of the raffinate has to be discharged along with the copper (0.155 g L^{-1}). The wastewater is treated by lime to precipitate Cu and change the pH from 2 to 7. The slag was transferred to tailing tank.

Concentration of the leaching solution by NF before its delivery to the SX-EW procedure significantly changes the situation, as shown in Fig. 4.9. The Cu content in the wastewater of

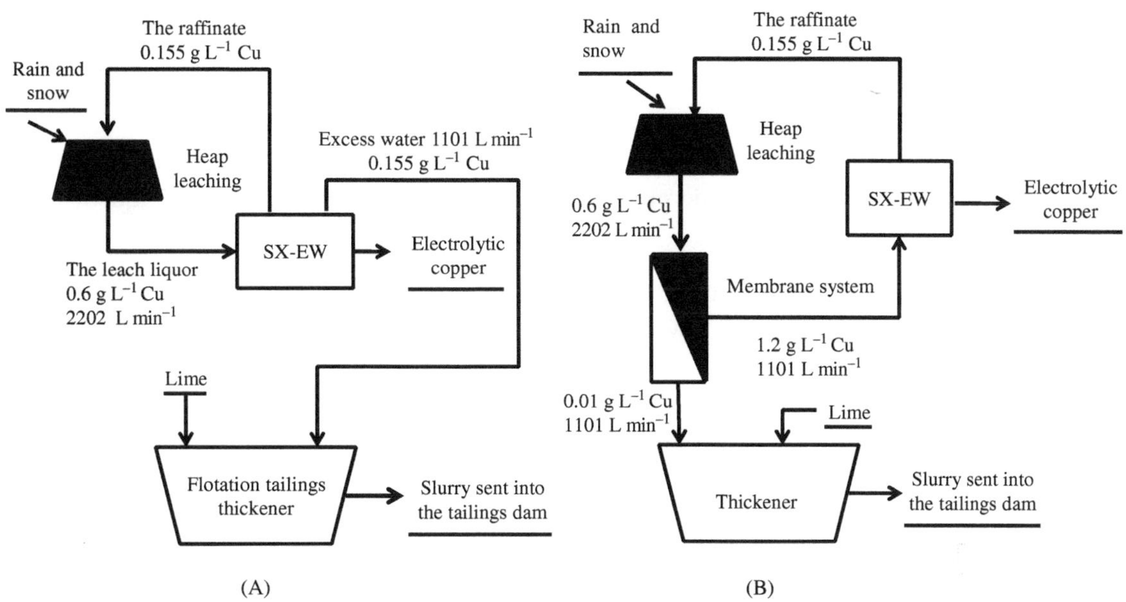

Figure 4.9: The comparison between (A) the original process and (B) the new process containing membrane separation.

the new process is only about 1/15 of that from the original process. So the recovery of Cu is enhanced and dosage of lime could be greatly reduced. Table 4.3 summarizes the economic analysis for the two processes. After a year and a half of operation, the system has handled nearly 7.5 million m^3 solution containing Cu of 0.6–0.8 g L^{-1}, and recovered Cu with values being more than US $10 million. The operating cost is US $0.11 m^{-3}, and it took less than 2 years to recoup their investment.

Instead, Huang [21] studied the concentration of copper in heap leaching solution using emulsion liquid membrane (ELM) technology. ELM is double water-in-oil-in-water

Table 4.3: The economic analysis of the original and new processes

Daily Operation Cost (Handling Capacity 1101 L min^{-1})	Lime Precipitation	Nanofiltration
Lime ($130 t$^{(1)}$) /U$	2536 (13 g lime L^{-1})	78 (0.4 g lime L^{-1})
Other consumption (power, flocculant, filter bag, etc.)/U$	320	190
Labor/U$	Part-time thickener operator	324
Membrane replacement costs/U$	––	202
Copper loss/U$	372 (0.155 g L^{-1} copper)	24 (0.01 g L^{-1} copper)
Net saving in operation cost/U$	––	2410
Investment/U$	––	1500 000
The payback period of investment/a	––	1.7

(W/O/W) emulsions stabilized by employing suitable surfactants. At the external W/O interface, Cu^{2+} and mobile carrier (indicated by H_2A_2) would react:

$$2H_2A_{2(O)} + Cu^{2+}_{(E)} = Cu(HA_2)_{2(O)} + 2H^+_E \tag{1}$$

where subscript "O" refers to organic phase and subscript "E" stands for external aqueous phase

The complex ($Cu(HA_2)_2$) generated would migrate into the organic phase. When arriving at internal O/W interface, the complex would react with H_2SO_4 from the internal phase, and release the Cu^{2+}:

$$Cu(HA_2)_{2(O)} + 2H^+_{(I)} = Cu^{2+}_{(I)} + 2H_2A_{2(O)} \tag{2}$$

where subscript "I" stands for the internal phase.

In Huang et al.'s work, Lix984N was used as immobilized carrier, kerosene solvent, H_2SO_4 as reverse extraction agent, and Spang80 as the surfactant. Based on the experiment which used simulated copper heap leaching solution with Cu^{2+} of 0.7 g L^{-1} and Fe^{3+} of 1.5 g L^{-1}, the separation factor for Cu/Fe by the ELM was 971. When treating the real Cu heap leaching liquid which contains Cu^{2+} of 0.406 g L^{-1} and Fe^{3+} of 0.558 g L^{-1}, the optimum operation conditions are shown in Table 4.4 and the separation results are summarized in Table 4.5. Visibly using ELM in handling this specific type of heap leaching liquid resulted in a Cu extraction rate of 98.46%, and a separation factor of copper/iron of 1393. The number of oil phase recycle has no obvious influence on extraction yield. According to the findings, the ELM technology is a good choice to deal with Cu heap leaching liquid containing high concentration Fe.

In the process of Cu EW, around 3.5 m^3 electrolysis barren solution is produced for production of each ton of Cu. The accumulation of impurity ions (eg, Fe^{3+}) in the

Table 4.4: The best conditions for treating the actual copper heap leaching liquid

Emulsifying Time	Lix984N in Kerosene	Span-80 in Kerosene	γ_{oi}	The Internal Phase	γ_{ew}	Leaching Time
20 min	0.10	0.09	1.0	4 mol L^{-1} H_2SO_4	1/3	15 min

Table 4.5: The results of the comprehensive experiment

The Feed		The Raffinate		The Stripping Liquor or Internal Phase		Extraction Yield (%)
Cu^{2+}	Fe^{3+}	Cu^{2+}	Fe^{3+}	Cu^{2+}	Fe^{3+}	98.46
0.406	0.558	0.0062	0.543	16.23	0.016	

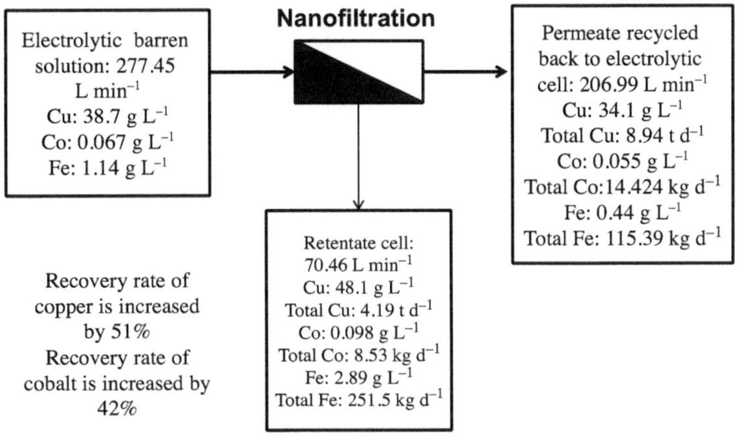

Figure 4.10: The effects of membrane separation to remove iron from electrolytic copper barren solution.

solution would affect the electrolysis efficiency, leading to higher current density, energy intensiveness, and degrading of product quality. The presence of Fe is mainly due to the leaching step, where the Fe is extracted along with Cu. On the other hand, direct discharge of the barren solution will lead to loss of Cu and Co, and serious pollution. Integration of membrane separation [22] has been found to efficiently solve the problem. As shown in Fig. 4.10, the separation process using special NF membrane rejects most of the Fe^{3+} in the barren solution, while allowing the passage of most of the Cu^{2+} and Co^{2+}. The permeate with Fe^{3+} at a concentration of only 0.44 g L^{-1} can be recycled back to the electrolytic cell. In addition, the modified process has greatly depressed the loss of Cu and Co.

4.2.4 Treatment of Various Process Waste Liquids

Waste liquids discharged from processes such as EW, metal fabricating, manufacturing of printed circuit boards for electroless Cu plating, and washing remediation of metal-contaminated soils associated with extractive metallurgy, generally contain metals and sometimes acid. This section will introduce the progress in applying membrane to process the various waste liquids for either reducing the loss of useful components or remove impurities to enhance the overall efficiency of extractive metallurgy.

In Cu hydrometallurgy, sometimes H_2SO_4 is used for extracting Cu. The barren solution of EW therefore is acidic. The pH of the solution is not high enough for leaching and SX-EW procedures which both involve application of acid. Zhang et al. investigated the application of diffusion dialysis (DD) in recovering acid of nonferrous metallurgical process [23]. Acid radical ion diffuses from high concentration side to low concentration side through the anion-exchange membrane (AEM), whereas the metal ions are trapped down in the retentate,

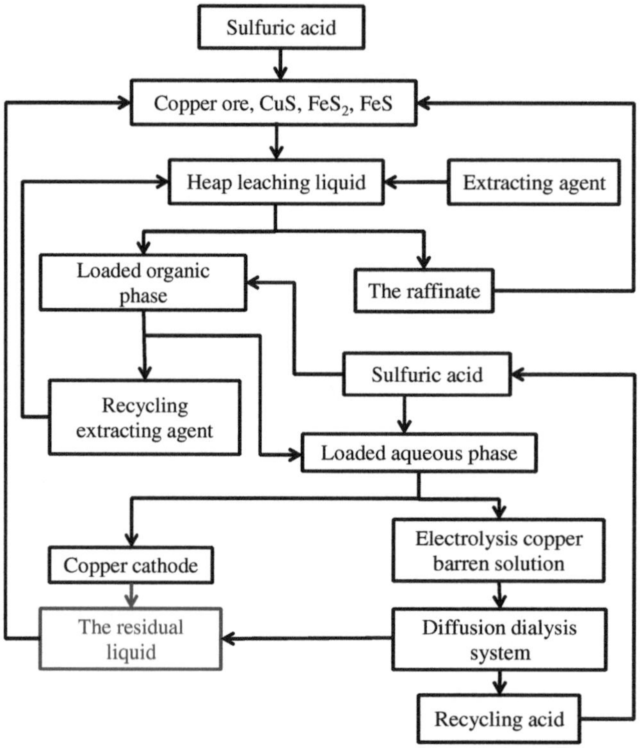

Figure 4.11: Copper hydrometallurgical process improved by involving diffusion dialysis for recovering acid.

realizing the separation of acid and metal ions. The overall hydrometallurgical process involving DD is shown in Fig. 4.11 [23]. Actually, in addition, the electrolytic effluent contains not only sulfuric acid of higher than 300 g L^{-1}, but also 20–30 g L^{-1} nickel sulfate. After removing most of the H_2SO_4 by DD, $NiSO_4$ could be retrieved by extraction process.

The traditional way to treat Cu mine acid wastewater is lime neutralization. In the high density slurry (HDS) method, the precipitate is stored as tailings, and the supernatant is discharged. Therefore, Cu has not been recovered in anyway and there might exist secondary pollution due to tailing leakage. If using S^{2-} to precipitate Cu in the acid, most Cu could be recovered. Nevertheless, new pollutants would be introduced, such as H_2S and S^{2-}.

Valenzuela et al. [24] investigated the recovery of Cu^{2+} ions from mining wastewater by ELM. The ELM in this research was prepared by dissolving LIX-860 (a salicylaldoxime) as an extractant and Span-80 as a surfactant inside kerosene. The double emulsion W/O/W was generated as shown in Fig. 4.12. In experiments conducted in a batch-type stirred tank at 30°C, the membrane allowed very efficient transfer of metal ions from the feed solution to the stripping liquor. According to the experimental results and the experimental design

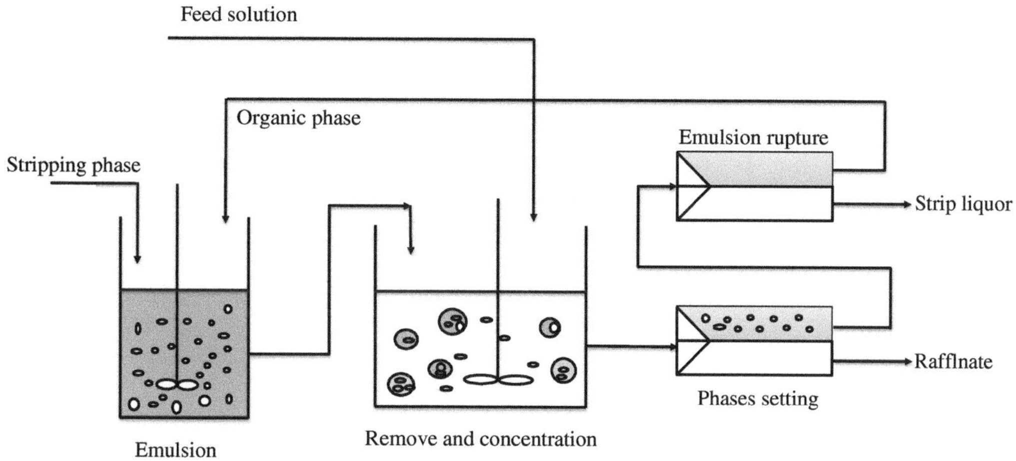

Figure 4.12: Schematic of double ELM process.

analysis, the statistically significant terms for Cu transport through the membrane are the extractant concentration, the stirring time, and the stirring speed of the double emulsion. Zhong et al. [25] applied NF to remove the heavy metal ions existing in acidic mining wastewater. Under optimum conditions the rejection of DK2540 NF membranes for heavy metals reaches 97%.

In Cai et al.'s work, the concentration of Cu^{2+} in the electroplating wastewater could be reduced from 340 ppm to less than 4 ppm after being treated by RO membrane [26]. The research also has shown that, metal concentrations in the treated water can be further reduced if transmembrane pressure drops. The results suggest that larger scale operations on treating electroplating waste water by RO membrane separation are possible and effective.

In Wu et al.'s research [27], the application of diaphragm electrolysis to recycle Cu from cyanide plating solution was investigated. The flow diagram of the process is shown in Fig. 4.13. The anode liquid inside the left chamber is Na_2SO_4 of which the concentration is 0.125 mol L^{-1}; the anode liquid inside the right chamber is sodium chloride (NaCl) of which the concentration is 0.25 mol L^{-1}. The current for electrolysis is 2.3 A dm^{-2}.

The reactions in the process include the following:

$$\text{Cathode reaction}: Cu(CN)_4^{3-} + e \rightarrow Cu + 4CN^-$$
$$\text{Anode reaction}: 4OH^- - 4e = 2H_2O + O_2$$

For the aforementioned process, the theoretical mass of Cu recovered = Cu electrochemical equivalent (1.186 g A^{-1} h^{-1}) ×electrolysis time (h) × the current (A).

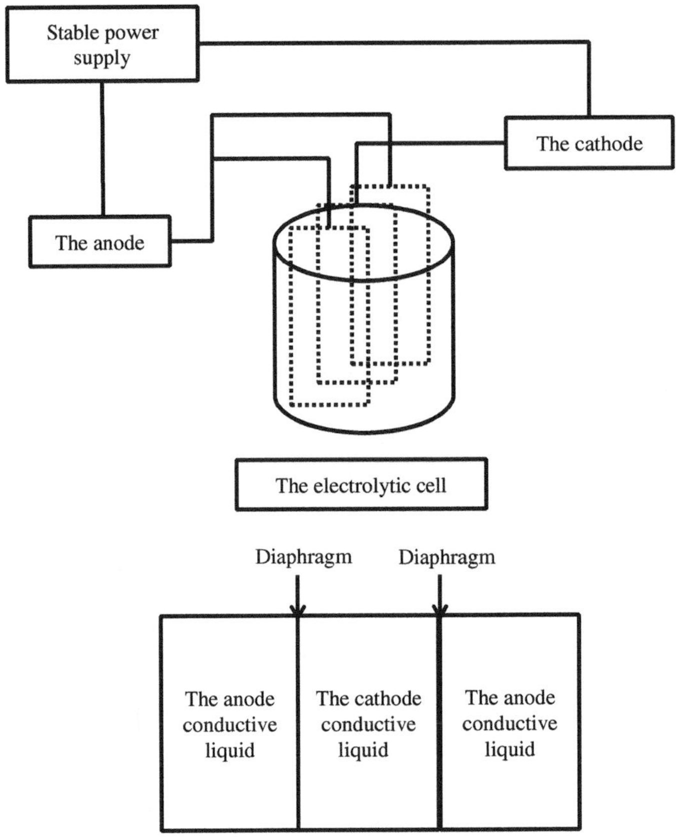

Figure 4.13: The schematic of cyanide copper diaphragm electrolysis.

Sun et al. [28] studies the recovery of HNO_3 from Cu deplating solution using DD. The Cu stripping solution was supplied by a plant at Wenzhou and the AEM was purchased from Shandong Tianwei Membrane Technology Co. Ltd. In the single-stage DD experiment, the optimum conditions include a ratio of 1:1 for feed/permeate flow rates, 90 mL h^{-1} volume flow rate and system temperature of 19°C. Under these conditions, the acid recovery rate reached 86.9% and the concentration of recovered acid is 6.04 mol L^{-1}. As for the three-stage DD, the concentration of the acid recovered was higher under the same conditions, and more feed could be disposed per unit time if the same concentration of acid as that in single-stage operation was to be achieved.

In Cu etching process using H_2SO_4, the effluent usually contains copper sulfate ($CuSO_4$), Na_2SO_4, and residual H_2SO_4 [29]. As shown in Table 4.6, the concentrations of these components are relatively high. Special Industrial Membranes (SIMS) supplied by HWA Technology Ltd that can sustain 20 wt.% H_2SO_4 was employed to separate $CuSO_4$ and Na_2SO_4. As shown in Table 4.7, the membrane has obviously higher retention for $CuSO_4$, as

Table 4.6: Typical composition of the circuit board etching liquid

Component	Cu^{2+} ($g{\cdot}L^{-1}$)	Na^{+} ($g{\cdot}L^{-1}$)	SO_4^{2-} ($g{\cdot}L^{-1}$)	H^{+} ($g{\cdot}L^{-1}$)
Concentration	13.27	23.45	96.96	0.66

Table 4.7: The efficiency in separation of $CuSO_4$, Na_2SO_4, H_2SO_4

Concentrate Factor of 2.2	Cu^{2+} ($g{\cdot}L^{-1}$)	Rejection Cu^{2+} (%)	Na^{+} ($g{\cdot}L^{-1}$)	Rejection Na^{+} (%)	SO_4^{2-} ($g{\cdot}L^{-1}$)	Rejection SO_4^{2-} %	H^{+} ($mol{\cdot}L^{-1}$)	Rejection H^{+} (%)
Feed 20 L	13.27		23.45		96.96		0.66	
Permeate 11 L	0.62	95.33	11.5	50.96	40.03	58.71	0.45	68.18
Retentate 9 L	22.64		31.09		141.29		1.01	

compared with those of Na_2SO_4 and H_2SO_4. The concentration factor obtained 2.2 times and the copper ion concentration in the feed finally reached 22.64 g L^{-1}.

Etching of the circuit board using acidic copper dichloride ($CuCl_2$) aqueous solution [30] was sensitive to the accumulation of cuprous ion (Cu^+). Usually the etchant waste is regenerated by oxidization using hydrogen peroxide (H_2O_2) and simultaneous addition of hydrochloric acid (HCl). Capenhurst method (CEER method) employs diaphragm electrolysis to regenerate etching liquid, recovering Cu at the same time [31]. A cation-exchange membrane (CEM) is used as a diaphragm. In the process of electrolysis, Cu^+ is oxidized into Cu^{2+} by releasing an electron at anode, whereas elementary Cu is generated at cathode.

The reactions in the electrolytic cell are summarized as follows:

Anode reaction

$$\begin{aligned} &Cu^+ \rightarrow Cu^{2+} + e \\ &CuCl_2 \rightarrow Cu^{2+} + 2Cl^- + e \\ &Cu^+ + CuCl^{2-} \rightarrow 2Cu^{2+} + 2Cl^- + 2e \end{aligned} \tag{3}$$

Cathode reaction

$$\begin{aligned} &Cu^+ \rightarrow Cu^0 - e \\ &CuCl_2 \rightarrow Cu^0 + 2Cl^- - e \\ &Cu^+ + CuCl^{2-} \rightarrow 2Cu^0 + 2Cl^- - 2e \end{aligned} \tag{4}$$

Overall reaction

$$2Cu^+ + CuCl^{2^-} \rightarrow 2Cu^0 + 2Cu^{2+} + 4Cl^- \tag{5}$$

The CEER method has achieved industrial-level application. For example, it has been installed in a printing circuit board factory in Scotland, the United Kingdom. The system operates at an electric current of 5000 A, regenerating the required volume of etchant and recovering 5 kg of Cu per hour.

Yang [32] also investigated the method of combining membrane separation and electrodeposition to regenerate $CuCl_2$ etching liquid and recover Cu as well. Thermodynamics analysis of copper chloride (Cu-Cl) complexes based on the E-pH diagram of Cu(II)-Cu(I)-NH_3-HCl-H_2O system showed that Cu(II) were dominated by the form of $CuCl^+$, while Cu(I) by $CuCl_3^{2-}$ in the experiment.

In some circumstances, discharges from the various processes have chelating agents such as ethylenediaminetetraacetic acid (EDTA), nitrilotriacetic acid (NTA), and citrate [33,34]. For electroless Cu plating rinse effluent, for example, it typically has 50–100 ppm Cu(II) and 300–500 ppm EDTA. The present chelating agents may cause the low efficiency of chemical precipitation such as hydroxide and sulfide. Moreover, the high buffer capacity provided by the chelating agents requires excessive amounts of chemicals to neutralize alkalinity.

In Juang and Huang's work [35], the extraction of Cu(II) from an equimolar EDTA solution was conducted across microporous hollow fibers to an organic phase containing Aliquat 336 (aquaternary amine) was investigated. It was shown that Cu(II) can be effectively co-extracted with EDTA in the form of chelated anions from water using Aliquat 336, as long as the pH is not far beyond 7 [36]. A mass-transfer model was proposed considering aqueous-layer diffusion, membrane diffusion, and organic-layer diffusion on the basis of extraction chemistry and the transport properties of the relevant geometry. The present hollow-fiber extraction process is found to be governed by aqueous layer and membrane diffusion. Although the role of interfacial chemical reaction is experimentally considered to be insignificant here, there have been no certain criteria to judge whether the resistance of interfacial reaction is important or not in dispersion-free SX processes. The calculated time profiles of Cu(II) concentrations were in reasonable agreement with the measured ones (average standard deviation, 12%).

4.2.5 Integrated or Hybrid Systems and Industrial Level Applications

SIMS and DD dominated processes have been implemented by Zijin Copper Mining Company (China) for treating acidic stripping solution and acid mining drainage (AMD), respectively, targeting recovery of copper and water, and actualizing zero liquid discharge (ZLD). [23,37]. For the former, the integrated process has already been introduced as shown in Fig. 4.11. The flow chart for the latter is illustrated in Fig. 4.14 [38].

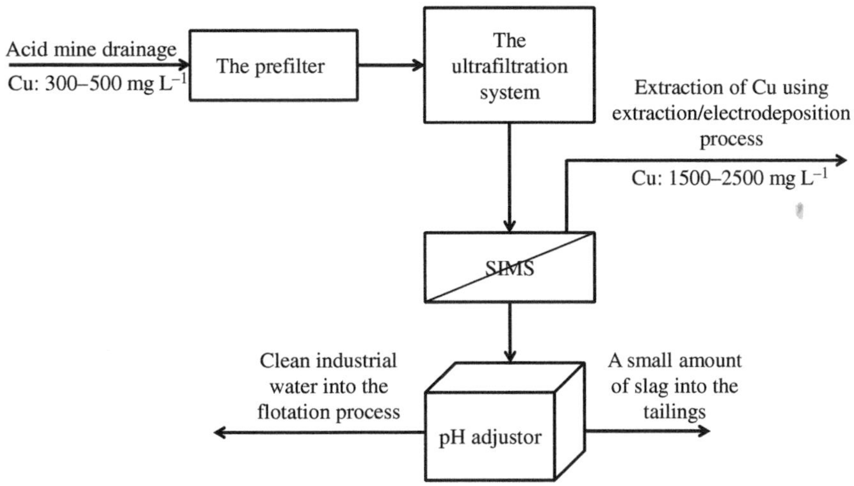

Figure 4.14: Schematic diagram of ZLD and copper recovery using special industrial membrane.

The Cu AMD is a very common and thorny problem in may mining sites. For the system adopted by Zijin Mine Company [38], a prefilter is used to intercept larger suspended solids, while ultrafiltration (UF) system retains the smaller suspended solids (eg, ferric oxide (Fe_2O_3)). Inside the SIMS system, metal ions are separated from the acid and obtain higher concentration. After pH adjustment, the permeate which contains a small amount of acid is recycled or discharged. The concentrated solution is transferred to SX-EW process for producing Cu. The cost of Cu AMD processing by SIMS is around 4 CNY ton^{-1} (inclusive equipment depreciation, cost of membrane replacement, electricity, etc.). According to the current Cu price, generally the investment can be recouped in a year.

The benefit analysis based on a process treating 4000 m^{-3} d^{-1} of the wastewater is given as follows:

1. Recovery of 600–700 t Cu and 1.1 million m^3 of water per year;
2. Completely solving the problem of environmental pollution;
3. Saving a lot of lime every year;
4. Reducing the dependence on the tailings, saving the cost of land resources and the construction of the tailing dams;
5. Lessen the secondary pollution.

In effect in the 1990s, the comprehensive treatment of Cu AMD technique using physical membrane separation [ie, reverse osmosis (RO)] has already been applied successfully at Mexico Cananea Copper Mine in North America [38]. This is one of the largest scale Cu mines.

At the Cu mine of Seville Cobre Los Crucas in Spain, the mining well water has high levels of arsenic (As), fluoride (F), and boron (B). Application of SIMS system as demonstrated

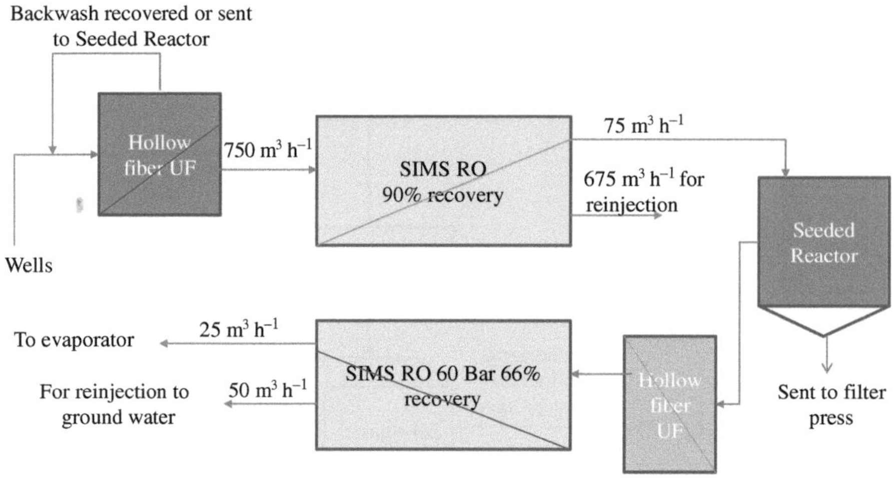

Figure 4.15: Process flow diagram for recovery 95–97.5% water from copper mine.

schematically in Fig. 4.15 [37] delivers ultrahigh concentration of Cu and 95–97.5% recovery of water. The concentration in the first stage of membrane treatment is processed by seeded reactor with the sludge being sent to filter press and the supernatant further being processed by second stage of membrane process comprising UF and RO. The final concentrate is evaporated and the water purified is reinjected into ground water.

Application of SIMS process has also been employed by Phelps-Dodge Company for separation and recovery of Cu and acid from mining drainage wastewater at open-air copper mining sites [37]. The system is composed mainly of integrated membrane separations as shown in Fig. 4.16. RO retains both the acid and inorganic salts, whereas NF only rejects divalent ions. By integrating the two technologies in the system, the acidic wastewater can be almost completely recycled. The effluent with 29 Kppm L^{-1} Cu is directly fed to the smelting process. The other effluents are sent to rinsing step, dissolution process or mixed with the original feed, depending on the contents of Cu and acid.

Cu foil wastewater treatment in Liaoning Benxi Copper Company (China) is conducted using hybrid system involving membrane separation [39]. The feed water quality as shown in Table 4.8 is highly acidic and contains various metal ions. In addition, a large amount of gelatin is present. By controlling pH value of the wastewater, the gelatin becomes suspended, whereas the metal ions remain dissolved in the solution. The double-layer filter and activated carbon filter in the following step are able to remove the suspended solids and colloid, making the turbidity and silt density index (SDI) of the water satisfy the requirements of RO. The permeate of the RO (I) procedure which is characterized by Cu ≤ 0.5 mg L^{-1} and electrical conductivity ≤ 30 μS cm^{-1} can be used as production process water. The concentrated water of RO(I) is pumped to the RO (II) for enrichment of copper. Then fresh

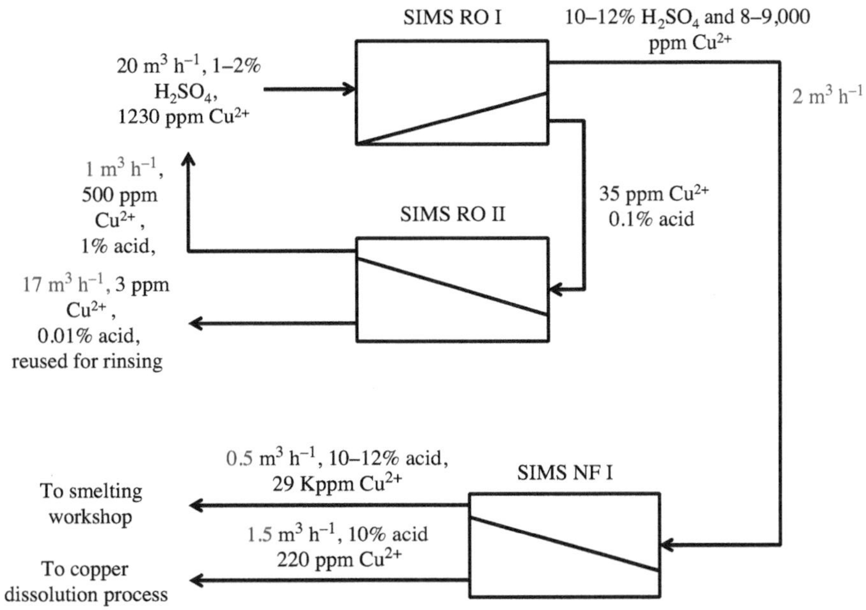

Figure 4.16: Flowchart of SIMS process at mining site of Phelps-Dodge Company.

Table 4.8: The quality of copper foil wastewater (mg L^{-1})

Acid	Cu^{2+}	Zn^{2+}	Cr^{3+}	Se^{2+}
50–80	50–80	5–10	5–10	1–2

water from RO(II) is cycled to the feed of RO(I), whereas the retentate with concentrated Cu is suitable for recovery by SX-EW. The concentrate from RO(II) is produced at 2–5 $m^3\ h^{-1}$ with Cu of 2800–1120 mg L^{-1}. The aforementioned processes are schematically displayed in Figs. 4.17 and Fig 4.18.

The economic analysis about the aforementioned overall process from feed to EW is given hereinafter:

Electricity consumption: 350,000 kWh^{-1} $year^{-1}$, the operating cost is 175,000 CNY $year^{-1}$ (0.5 CNY ($kW^{-1}\ h^{-1}$)).
Alkali consumption: 500 kg (in terms of NaOH), 350 CNY $year^{-1}$.
RO consumable: 30,000 CNY $year^{-1}$.
Water recycle: 400,000 t $year^{-1}$.
The total operation cost: 205,350 CNY $year^{-1}$ = 0.513 CNY t^{-1} water.
The Cu recovered a year: 26,600–42,560 kg $year^{-1}$.

The concentration of Cu in pickling wastewater of a Cu finishing process is generally low [20]. A factory employed a hybrid process which consists of RO, NF, and ion exchange to

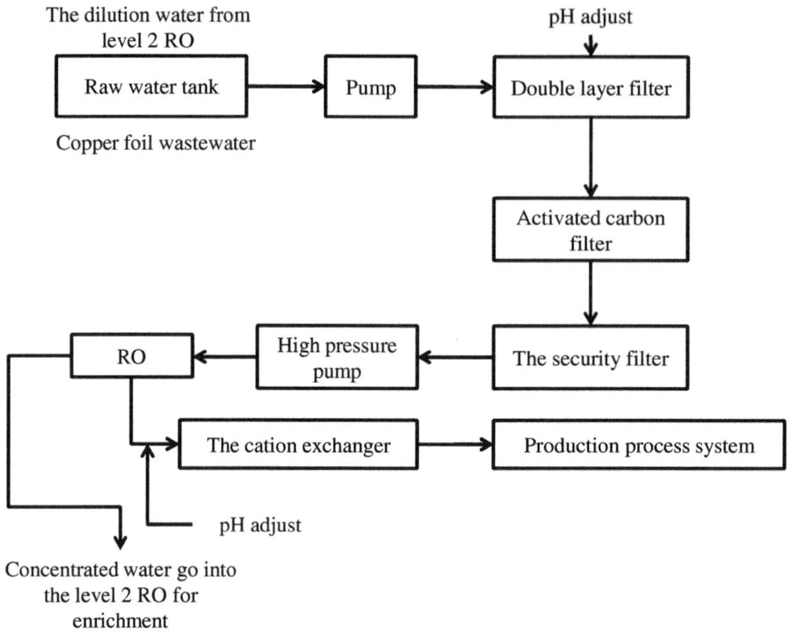

Figure 4.17: Copper foil wastewater level 1 reverse osmosis treatment process flow diagram.

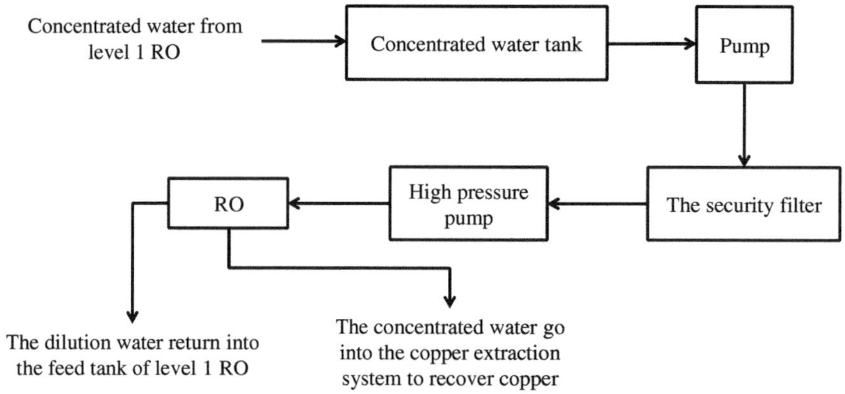

Figure 4.18: Copper foil wastewater level 2 reverse osmosis treatment process flow diagram.

recover acid and Cu, as shown in Fig. 4.19. The system helps save US $560,000 per year for the company and all investment was recouped within 1 year.

4.3 Zinc and Lead

4.3.1 Introduction

Lead (Pb) and zinc (Zn) are generally symbiotic in the ores. Pb and Zn resources are widely distributed in the world in more than 50 countries. According to statistics of

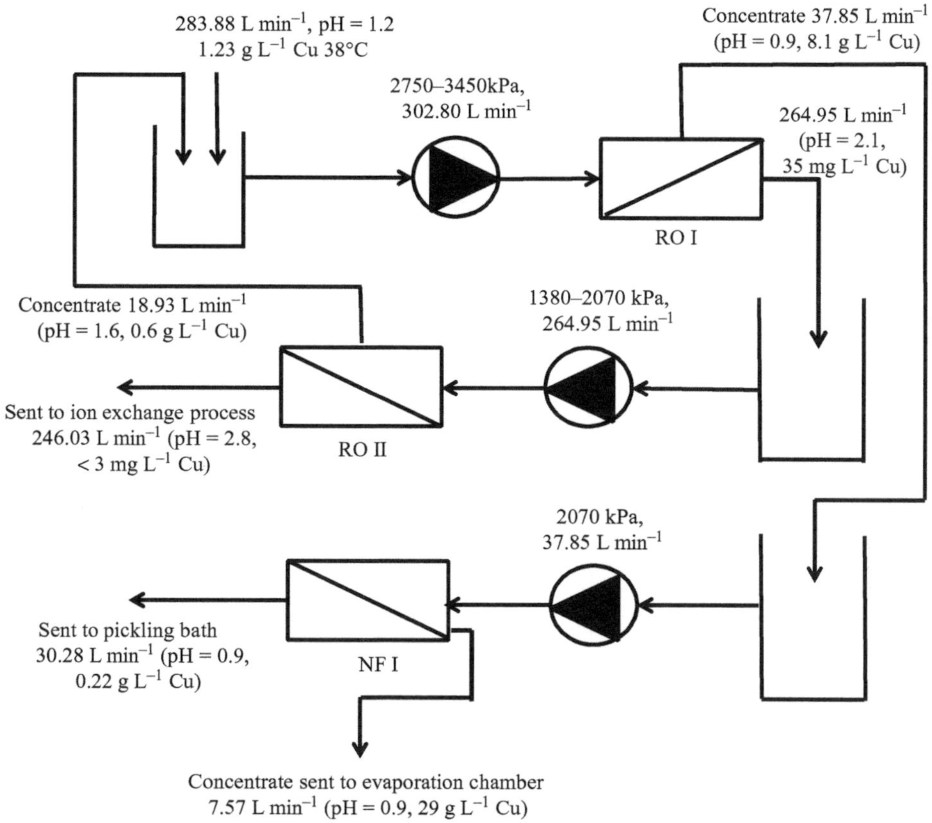

Figure 4.19: The process of recycling acid and copper from pickling wastewater.

America Geological Survey in 2003, the reserve of Pb resources in the world that has been proven is 1500 million tons, and the reserve base 140 million tons; the identified reserve of Zn resources is 1900 million tons, and the reserve base is 460 million tons [40]. Pb resources distributed in Australia, China, the United States, and Kazakhstan account for 60.3% of the world's total reserve, and 71.2% world's reserve base. The Zn resources are mainly distributed in Australia, China, Peru, the United States, and Kazakhstan. The amount of Zn reserves located in these countries occupies 67.2% of the world's total and 70.9% of the reserve base.

The world's and China's production and consumption of Pb and Zn, both refined and regenerated, are summarized in Table 4.9–4.10. As can be seen, the average annual growth rates for all these items in China are obviously higher than the counterparts' worldwide from 1998 to 2008.

The major approach for Pb smelting is pyrometallurgy, whereas hydrometallurgy is still at experimental stage. Among the various pyrometallurgical methods, the priority is given to two-step sintering roasting-blast furnace smelting [41]. Initially, powdered lead sulfide (PbS)

Table 4.9: Comparison of production of lead and zinc in 1998 and 2008, and the average growth rate

	Refined Lead Production (Million Tons)		Production of Regenerated Lead (Million Tons)		Refined Lead Consumption (Million Tons)		Refined Zinc Production (Million Tons)		Refined Zinc Consumption (Million Tons)	
Year	The World	China	The World	China	The World	China	The World	China	The World	China
1998	6.00	0.7569	3.25	0.2416	6.0742	0.5302	7.986	1.486	7.9855	1.1278
2008	8.67	3.2064	4.67	0.957	8.6683	3.1349	11.531	3.913	11.346	4.0186
Average annual growth rate	3.8%		3.7%		3.6%	19.4%	3.7%	10.2%	3.6%	13.5%

Table 4.10: Proportion of Chinese consumption of refined zinc and lead in world total in 1998 and 2008

Year	Refined Lead (%)	Refined Zinc (%)
1998	8.7	14.1
2008	36.3	35.04

concentrate is desulfurized and sintered into block by high temperature oxidation roasting. Secondly, crude Pb is generated by using reduction smelting in blast furnace. Due to its simplicity, stable production, high recovery, the sintering roasting-blast furnace smelting process is widely applied for many years. The crude Pb produced by this method accounts for 70–80% of the world's output, and the situation is quite similar in China. Nevertheless, the method has been challenged by several problems: serious pollution due to smoke and dust with heavy metal, the massive return circulation of material, the difficulty of utilization of low concentration by-product sulfur dioxide (SO_2) in making acid [42].

To solve the problem, several new lead smelting technologies have been developed, including Qaureaa–Schuhman–Lurgi (QSL) method by Lurgi Company in German, Kivcet method, ISA method by MountIsa Company in cooperation with Commonwealth Scientific and Industrial Research Organization, Shui Kou Shan (SKS) method, and Kaldo method. Nowadays, only ISA method has achieved extensive application, with around 11 plants adopting this method inclusive those situated in China [43]. Oxygen bottom blown smelting—blast furnace reduction lead smelting process (also called SKS method) was developed by China [41]. The production cost of SKS method is around 85% of that with traditional processes, and lower than those associated with the aforementioned new technologies. Currently in China, SKS method has become the mainstream technology of Pb metallurgy and put to use by more than ten lead smelting factories [42].

Zn smelting can be carried out using either pyrometallurgy or hydrometallurgy [44]. The latter is currently the principal method with production accounting for about 85% of the

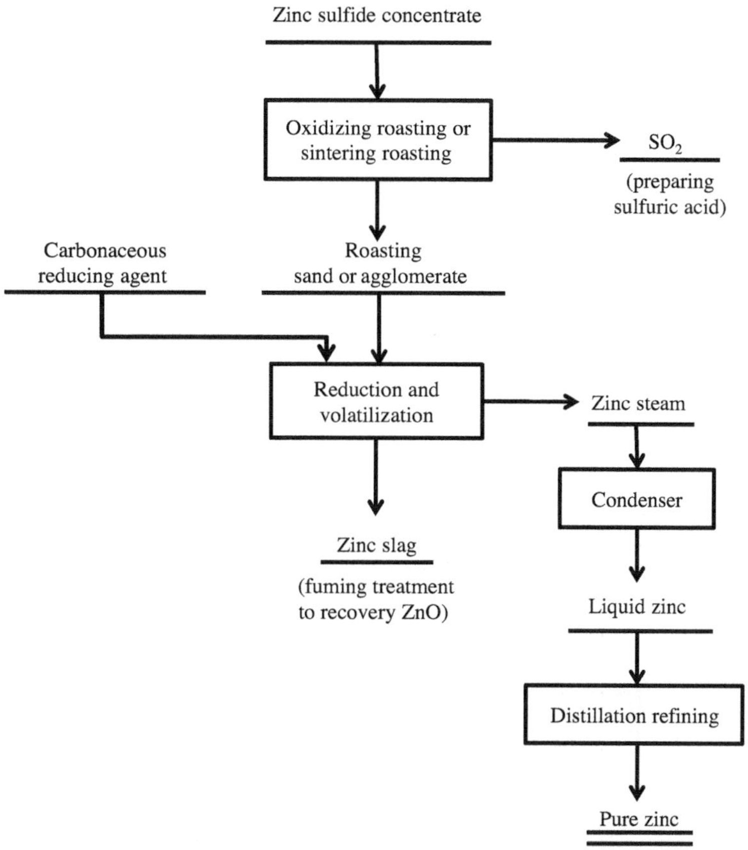

Figure 4.20: Zinc pyrometallurgy process.

world total output. Airtight blast furnace smelting process is the main Zn pyrometallurgy method in the world [45]. At first, Zn concentrate is treated using oxidizing roasting or sintering roasting to convert zinc sulfide (ZnS) into zinc oxide (ZnO) that is easier to be reduced by carbonaceous reducing agent. Due to the low boiling point, Zn reduced at temperature higher that its boiling point would evaporate from the furnace charge in vapor state. As a result, Zn is separated from other components in the furnace charge. Along with furnace gas, Zn steam flows into the condenser where liquid zinc forms. Then refining crude Zn is carried out. Based on the difference in boiling point between Zn and metal impurities, distillation is adopted to purify Zn. Finally the refined Zn with purity higher than 99.99% is cast into Zn ingots. The Zn pyrometallurgy process is shown in Fig. 4.20.

The Zn hydrometallurgy involves roasting, leaching, purification, and electrodeposition procedures [44]. Generally Zn hydrometallurgy can be classified into traditional method, hot acid leaching, and oxygen (O_2)-assisted direct leaching. The flowsheets of traditional and hot acid leaching methods are shown in Fig. 4.21 [45]. With the assistance of (O_2), direct O_2 leaching process can be operated without roasting. It is widely regarded as a

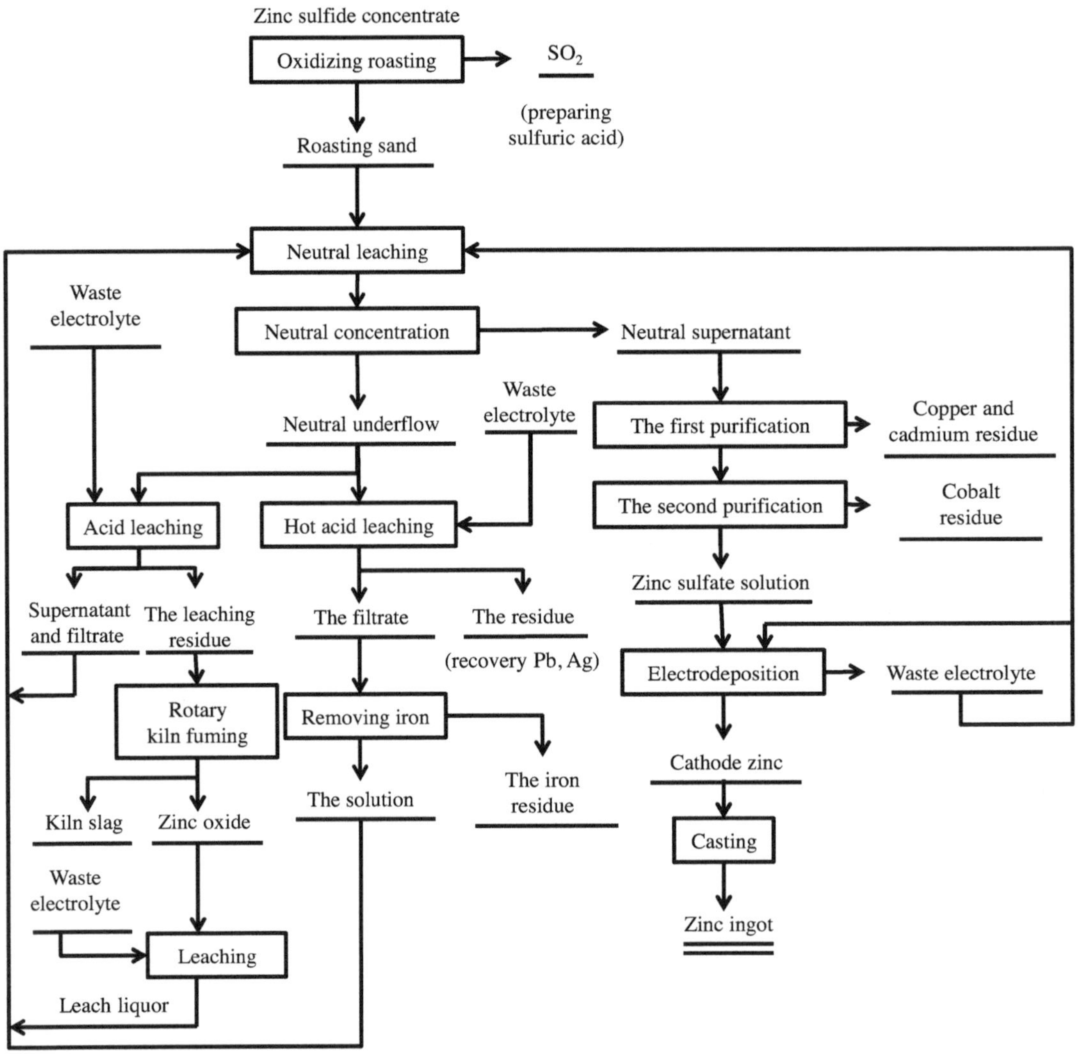

Figure 4.21: The traditional method and hot acid leaching method.

major breakthrough for Zn smelting, hence entitled the Third Generation of Zinc Smelting Technology. The direct O_2 leaching process can be further divided into pressurized (1100–1300 kPa) leaching and atmospheric pressure leaching. The process diagram of the latter which is safer is shown in Fig. 4.22 [41].

4.3.2 New Process for Zinc Sulfate Purification [46]

As aforementioned, Zn hydrometallurgy is the principal choice for world Zn production. Zinc sulfate ($ZnSO_4$) is an important intermediate product. The separation of its aqueous

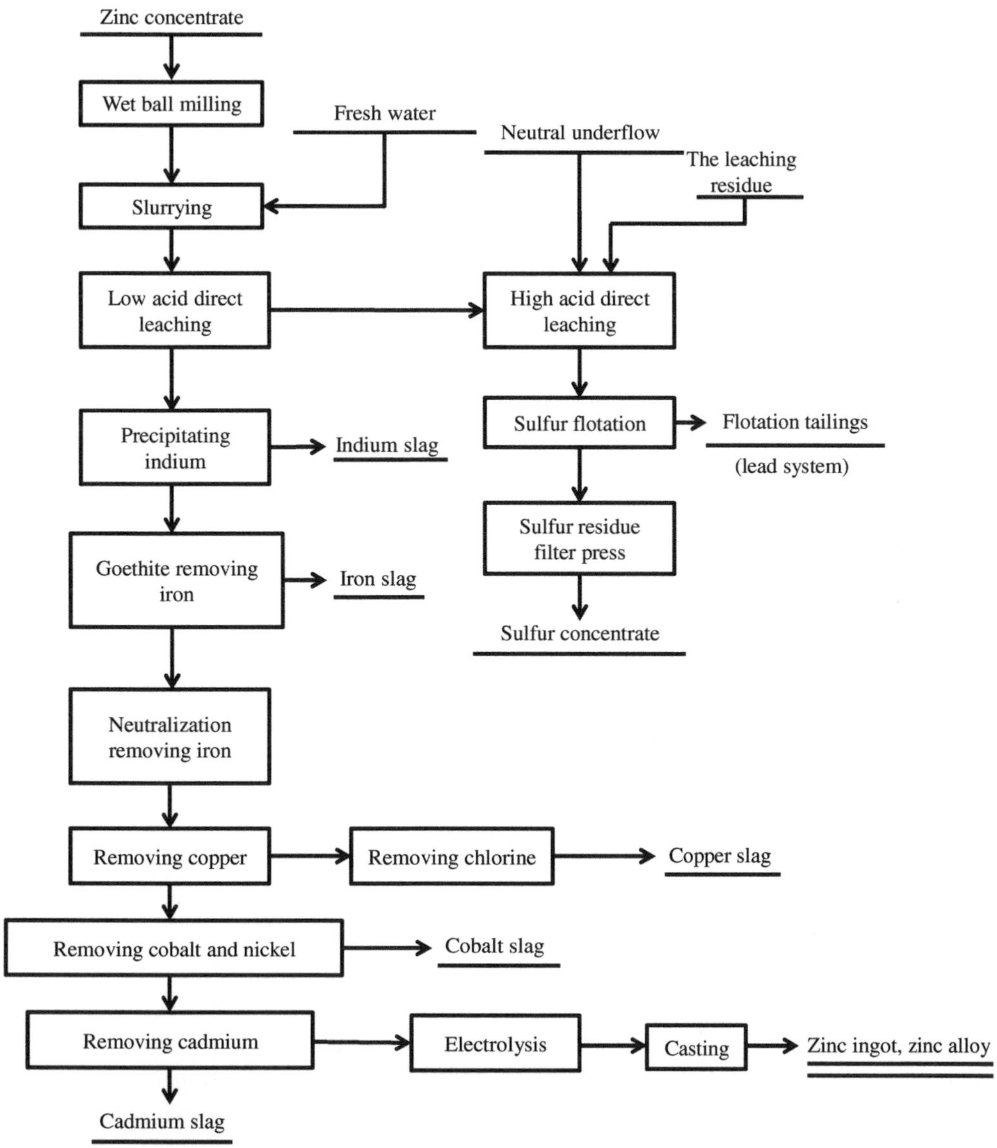

Figure 4.22: Direct pressure leaching with atmosphere leaching process.

solution from the solid and follow-up purification play key roles, directly affecting the Zn product quality, environmental management, and investment costs. In traditional Zn hydrometallurgical process, the solid/liquid separation is realized using natural sedimentation and plate pressure filtration. The disadvantages of these approaches are system blockage, solution penetration, and incomplete separation, which not only degrade product quality, but also cause waste of resources and environmental contamination.

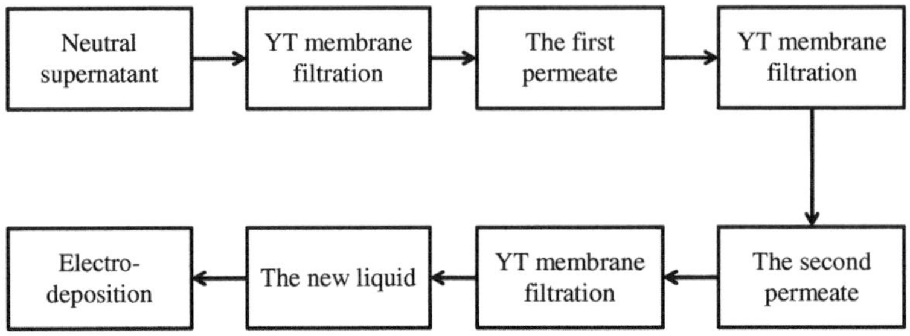

Figure 4.23: The new technology of YT membrane zinc sulfate purification process flow diagram was shown.

Liu et al. proposed a new process that uses membrane separation as the key component for $ZnSO_4$ separation and purification [46]. The process diagram is displayed in Fig. 4.23. Hydrometallurgical process is characterized by harsh environment due to the presence of particle abrasion, acidic or basic corrosion, etc. The YT special membranes made of metal alloys were chosen for the system, since they have excellent temperature resistance, acid and alkali and organic solvents, abrasion and corrosion resistance, and sufficient mechanical strength. In the first stage of the design, a YT membrane-equipped crossflow filtration device was used to remove suspended solid impurities from the supernatant. The next step is reverse antimony (Sb) low temperature purification to remove impurities of Cu, cadmium (Cd), etc., followed by fine filtration. Thereafter, second stage high temperature purification to take away Co and Ni, and rapid and precise filtration were employed.

The application of YT membrane filtration has achieved following benefits: (1) the concentration of solid impurities is lower than 5 ppm in the permeate; (2) the first stage intensive purification efficiently prevents redissolution and leakage of Cd, and the permeate contains lower than 0.8 mg L^{-1} of Cd^{2+}; (3) the second stage intensive purification generates filtrate meeting the industrial standard for new liquid, with Cu < 0.10 mg L^{-1}, Co ≤ 0.6 mg L^{-1}; (4) $ZnSO_4$ solution is purified by a two-step operation, greatly shortening overall process compared with the original one; (5) the total Zn powder consumed is cut down by 40%. In addition, reduction of electrical power consumption in EW also resulted due to the higher quality of the new liquid.

4.3.3 Arsenic Removal From Nonferrous Pyrometallurgy Process [47]

In nonferrous metal smelting, toxic As-containing materials are almost present throughout the whole process [48]. The approaches for dearsenization include mainly pyrogenic process, wet process, and combined process. Volatilization roasting, burning, and vacuum are technologies employed in pyrogenic process. They are relatively mature and have strong adaptability, simple process, and easy operation. The disadvantages of these methods are low recovery rate of As, the bulky dust collecting system, high investment, and serious pollution. The wet process via acid leaching, alkaline leaching, and chlorination leaching are characterized by tremendous volume of leaching liquid, many procedures, extended process, and high cost. The combined process involves mainly soda sintering and water leaching. It will consume a large amount of expensive additives and is seldom used industrially.

In nonferrous pyrometallurgical process, trivalent arsenic (As(III)) is the major existing form in the flue gas released from smelting furnace. The vapor pressure of arsenic oxide (As_2O_3) is higher relative to those of other metal oxides. Considering this difference, implantation of membrane separation in the pyrometallurgical process to filter the high temperature flue gas is expected to actualize separation of arsenic oxide from other metal oxides and enrichment of the former.

The process schematic diagram explaining the aforementioned design is shown in Fig. 4.24. The first step of both routs treating the gas of high temperature inside YT membrane filter operates at 300–800°C to remove the powdery dust. The first route is a better choice for dealing with flue gas with low content of S. Within the process, the As containing permeate from the first filter is pumped to the second filter set at around 100°C. At this temperature, vapor state As_2O_3 condenses into solid that is then trapped by YT membrane. The purified gas has very little solid (<10 mg m^{-3}) and can be directly released into atmosphere. As for flue gas with high concentration of S, second route is selected. After being treated by the

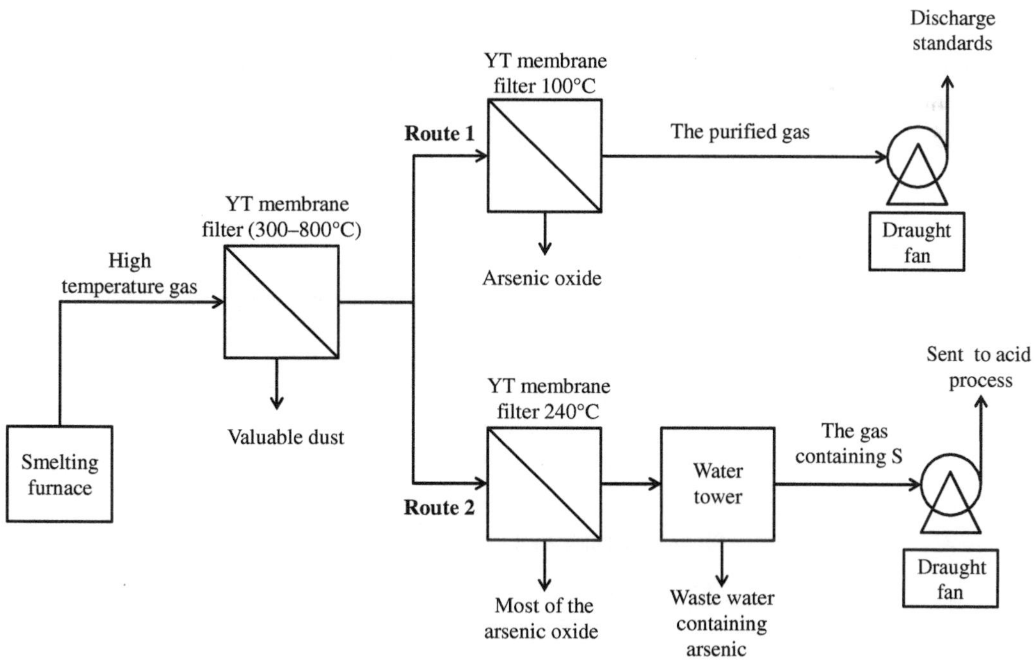

Figure 4.24: The YT-membrane involved separation process for removing arsenic.

second filter at 240°C, most of the As_2O_3 is intercepted as condensate. The permeating gas goes through a water scrubber to get rid of residual arsenic. The sulfur containing effluent (powdery dust < 10 mg m^{-3}) from the scrubber can be used to produce H_2SO_4.

The aforementioned process has been applied to treat high temperature flue gas of Pb anode slime smelting furnace, achieving 62.46% of dearsenization, realizing the separation of As_2O_3 from other valuable metal oxide and enrichment of Sb_2O_3 and As_2O_3. The recoveries of Sb_2O_3 and As_2O_3 are higher than 98 and 75%, respectively. Optimized three-stage process with more stable and efficient temperature has also been developed. The rapid cooling is the key to induce more effective separation of As_2O_3 from other valuable metal. The dearsenization rate is raised up to 80%, completely satisfying the requirements of industrial level production.

4.3.4 Facilitating Chlorine Metallurgy

Chlorine metallurgy has been widely applied, not only in nonferrous primary ore and concentrate, but also secondary resources. Particularly, it has great potential in utilizing complicated sulfide ore or oxide ore [49]. Tecnicas Reunidas Company in Spain has conducted an effective research on base metal chlorine metallurgy and developed a series of new technologies addressing the extraction of complex sulfide ores of base metal. Most of them count on chloride extraction and electrolysis to obtain final pure metal product. The Metclor process flow diagram is shown in Fig. 4.25 [20]. On the basis of Metclor electrolyzer, the Cuprex, Ledclor, Zinclor processes have been developed. The metal was deposited at anode, whereas chlorine (Cl_2) gas was generated at cathode. Part of the barren solution from the anodic cell that is removed of the targeted metal was fed to the cathodic cell to top-up the NaCl depleted during electrolysis. The balance of the solution is enriched in the metal and mixed with the feed for extraction.

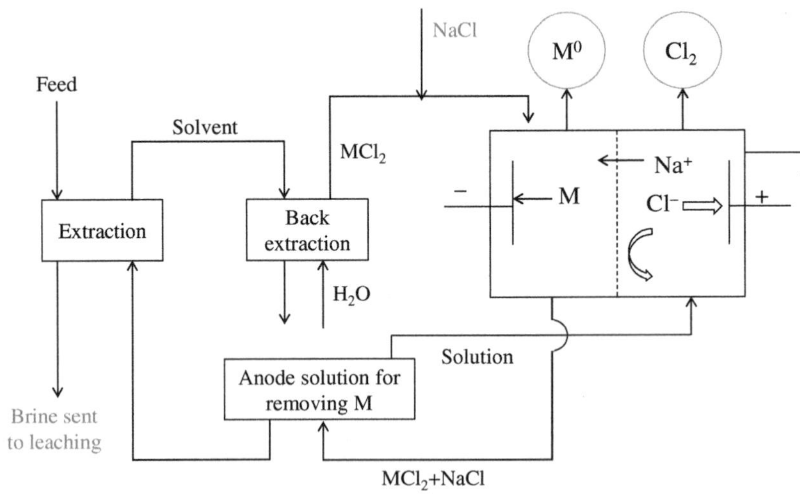

Figure 4.25: The flow diagram of Metclor process for base metal production.

In addition, these processes have the following common features: (1) handling complex concentrate; (2) application of ferric chloride ($FeCl_3$) to leach the ore; (3) high yield of metal, Cu ≥ 98%, Pb > 98%, Zn > 95%; (4) beneficial recovery of by-products of S, gold (Au), and silver (Ag); (5) the product has high purity, up to 99.99%; (6) the main chemical agents can be recycled, and Cl_2 gas is used for regenerating oxidizer; (7) under normal operation conditions, the current efficiencies of both anode and cathode approach 95%.

4.3.5 Industrial Level Applications in Process Wastewater Treatment

Shaoguan Smelter Plant in China adopted an advanced system in treating wastewater from production process [50]. Conventionally, the wastewater is treated by two-step chemical precipitation method using lime and ferrous sulfate ($FeSO_4$). Water quality and scaling analyses show that the water discharged reaches the emission standard, but is still liable to scaling which is detrimental to equipment. The quality of water is tabulated in Table 4.11. Therefore, an advanced water treatment system has been installed for further removing inorganic salts, so as to eliminate scaling phenomenon.

Roughly, three major stages are incorporated in the system: dual-media filtration, UF, and NF. The total volume of the feed water disposed by the system is 800 and 540 $m^3\ h^{-1}$ water is purified or recovered, with an approximately 70% recovery. NF is the key technology for desalination. Due to the charge loaded on the membrane, NF can achieve high desalinization at low feed pressure with membrane characterized by high molecular weight cut-off (MWCO). Spiral-wound membrane modules are used in the system. The water recovery and desalinization in NF procedure are 75 and 80%, respectively, and the pressure head at outlet is 0.1 MPa. The pretreatment process of NF is UF to get rid of suspended particle, bacteria, organic matter of high molecular weight (MW) and so on. Kristal UF membrane produced by

Table 4.11: Water quality of the original wastewater and that treated by two-step precipitation

Brine Composition	Wastewater From Production Line	Water Treated by Two-Step Precipitation
Total Cd	0.5	0.01
Total As	1.0	0.4
Total Zn	20	1.8
Total Hg	0.05	0.03
Total Pb	35	0.05
SS	50	20
Ca^{2+}	50	300–500
SO_4^{2-}	600	1000
CO_3^{2-}	30	10
Cl^-	900	400
pH	7–8	7–8
Water temperature	40–50°C	20–40°C

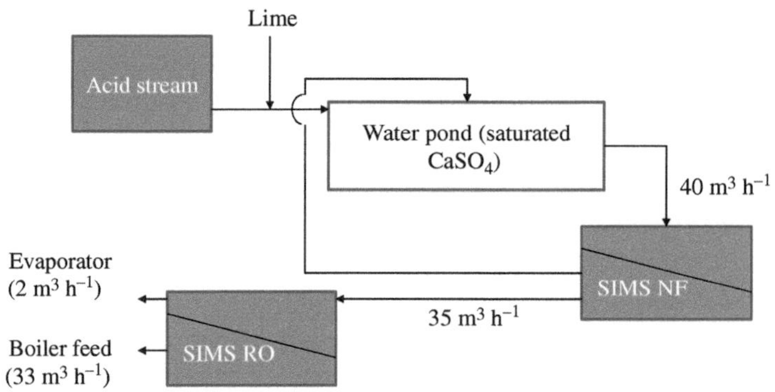

Figure 4.26: Flowchart of the advanced water treatment based on integrated membrane system.

Hyflux Company is selected and water recovery is higher than 90%. Before UF, a multimedia filtration is employed. The media in the top layer of the filter is anthracite coal, whereas that in the bottom layer is quartz sand. The system went online in Sep. 2007 and exhibits a stable operation till the date of report in Oct. 2009.

A Zn refinery getting an investment of 0.5 billion Australia dollars has been built in Townsville, Australia [37]. A zero discharge water treatment system mainly relying on membrane-based separation was installed. The plant was located near the Great Barrier Reef. Therefore, "zero discharge" of any pollutant is undoubtedly desired. In addition, it has also been found after construction that there exists up to 30 ppm of B in the plant effluent. It is well known that Boren is a kind of contaminant that cannot be easily removed by conventional precipitation technology. The flowchart for the water treatment process with zero discharge is displayed in Fig. 4.26. The water generated by RO is qualified as boiler feed. The system was equipped with the SIMS membrane supplied by HWA Technology Ltd.

4.4 Other Researches About Heavy Metal Wastewater Treatment

In some cases, the composition of the wastewater is complicated due to the coexistence of several types of heavy metal ions and many other components. The reason may be that the effluent from different processes will more or less mix together. Yet another reason is that some components are purposely added into the process to facilitate the efficiency improvement. The researches reviewed in this part do not specifically address a certain type of heavy metal and its extraction process according to the information of the publications. They are briefly classed into investigations on (1) basic unit operations and (2) integrated or hybrid processes or systems. The focus of the review will be on heavy metal removal and water recovery. Actually, there also exist organic components in many wastewater resources. However, researches about membrane for tackling this type of pollution are few and will be introduced in Chapter 10.

4.4.1 Basic Unit Operations

In Zhang and Li's review, researches about using RO to treat wastewater were introduced [51]. Its advantage is higher efficiency in removing heavy metal. Depending on the quality of the feed, recycling ratio extends from 75% to 95%. Zhong et al. [52] conducted the research about the processing and recycling of acidic mining drainage wastewater containing various heavy metals. The feed has gone through secondary treatment. The membrane employed is Korean Saehan RE4040-BL. Under optimum conditions, these ultralow pressure RO (ULP-RO) membranes exhibited 97% for heavy metal ions and 98% for total conductivity.

Wastewater containing copper and cadmium can be produced by several industries. The application of both RO and NF technologies for the treatment of wastewater containing Cu and Cd ions to reduce fresh water consumption and environmental degradation was investigated. Synthetic wastewater samples containing Cu^{2+} and Cd^{2+} ions at various concentrations were prepared and subjected to treatment by RO and NF in the laboratory. The results showed that high removal efficiency of the heavy metals could be achieved by RO process (98% and 99% for Cu and Cd, respectively). NF, however, was capable of removing more than 90% of the Cu^{2+} existing in the feed water. The effectiveness of RO and NF membranes in treating wastewater containing more than one heavy metal was also investigated. The results showed that the RO membrane was capable of treating wastewater with an initial concentration of 500 ppm and reducing the ion concentration to about 3 ppm (99.4% removal), whereas the average removal efficiency of NF was 97%. The low level of the heavy metals concentration in the permeate implies that water with good quality could be reclaimed for further reuse.

A new configuration of electrodeionization (EDI) with a bipolar membrane (BPM) with a segment electrode was proposed by Shang et al. for the treatment of simulated dilute Ni^{2+}-containing wastewater [53]. Under the experimental conditions, Ni^{2+} -free diluate and a solution concentrated by 27 times (820 ± 50 mg L^{-1}) was obtained. After 98.5 h of experimental running at the current densities of 16.5 mA cm^{-2}, the system allows the almost complete removal of the Ni^{2+} and sulfate anions (SO_4^{2-}) from the dilute solutions.

The metal in the concentrated retentate in NF or RO process can be recovered via pH adjustment and precipitation, fulfilling the dual purpose of retrieving both water and metals.

4.4.2 Integrated or Hybrid Processes

Low-pressure reverse osmosis (LPRO) was utilized by Ujang to concentrate the heavy metals [54]. Addition of EDTA into the wastewater brought about the rejection coefficients of more than 99% for both Zn^{2+} and Cu^{2+}. In 1980, Michaels pioneered in proposing for the first time the use of water soluble polymers for heavy metal capture. With this approach [55], the coordination between polymer and heavy metal ions leads to formation of larger complex molecules that can be efficiently rejected by UF. As a result, the heavy metal removal efficiency is improved.

Since the complexations between various polymer and heavy metal ions are largely reversible, the metal ions can be released from the rejected complexes under certain conditions, whereby the heavy metals are recovered [55].

Generally, polymers have selectivity in forming complexation with heavy metal ions. For example, poly(sodium 4-styrenesulfonate) (PSS) preferentially forms complexes with Cd^{2+} [34], poly(allylamine) (PALA) with Ni^{2+} [56], poly acrylamide-co-1-(2-hydroxyethyl) aziridinyl with Cu^{2+}, to name just a few. Therefore, better separation effect can be achieved in treating heavy metal in wastewater on condition that the polymers for complexation are chosen specifically.

The advantages associated with combination of complexation using water soluble polymer and UF are the simultaneous recovery of heavy metal and reclamation of water. Nevertheless, it also confronted with the challenge of high cost and low efficiency. This technology is in the stage of R&D without any industrialized application.

In another promising development, Bhattacharya et al. [57] attached multifunctional polymeric complexing agents, such as polyacrylic acid or polyamino acids to membrane surfaces for high capacity metal capture [58]. The advantage expected is that there will be no secondary pollution due to organic agent permeating through the membrane.

During the same period, micellar-enhanced ultrafiltration (MEUF) was proposed by some researchers, which couples the surfactant and UF. The surfactants used in MEUF are mainly synthesized, such as the cetylpyridiniumchloridemonohydrate (CPCI). Some, for instance the phosphatidylcholine, are naturally obtained. Xu et al. [59] took advantage of MEUF to remove Cd^{2+} and Pb^{2+} in aqueous solution. The results indicated that the rejection coefficients were higher than 99%. Natural surfactant of phosphatidylcholine used in Ahmadi et al.'s work [60] has following order of affinity toward five heavy metal ions in a mixing state: Cu > Cd~Zn > Ni. As for metal ions tested separately, the affinity took the following order: Ni > Cu~Zn > Cd. MEUF is simple and has high efficiency in tackling heavy metal pollution, whereby wastewater containing low concentration of heavy metal can be processed with recovering both metal and water. Notwithstanding most of the synthesized surfactants are toxic, hence causing secondary pollution in application. Search for nontoxic substitute will be the future direction.

Coupling of ULP-RO/NF was applied by Zhong [61] to dispose AMD wastewater. The feed was collected from a mine located in Anhui Province, which was already processed by

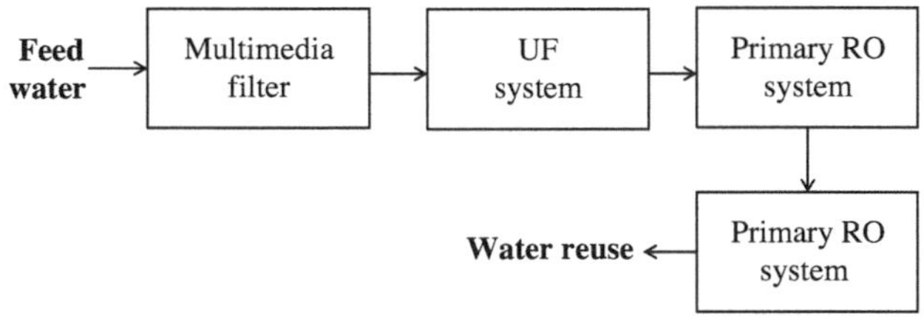

Figure 4.27: Main process for dual-membrane separation.

secondary treatment. The metal ions include Cu^{2+} (78.83 mg L^{-1}), Ni^{2+} (16.56 mg L^{-1}), Zn^{2+} (18.28 mg L^{-1}) Pb^{2+} (9.96 mg L^{-1}), and Na^{+}, whereas the anions are SO_4^{2-}, Cl^{-}. Compared with ULPRO, the ULPRO + NF obviously enhanced, whereas NF + ULPRO depressed the water recovery rate. The coupled process also improved the rejection to heavy metal ions and other salts, leading to higher quality of output water.

In Wu et al.'s work, a dual-membrane system as shown in Fig. 4.27 was employed to treat the nickel electroplating wastewater [62]. The electroconductivity of the feed water was around 6000 μs cm^{-1}, while the value acceptable for reuse was below 100 μs cm^{-1}. With a conductivity removal efficiency as high as 98.3%, water produced by the system can be used directly for plating workshop cleaning. Most of the wastewater can be recycled, thus saving the cost on fresh water purchase and sewage discharge. The water reuse rate associated with the aforementioned process was more than 60%, reducing the sewage discharge by 5160 t day^{-1}. The direct economic benefit was payment cut-down for purchasing fresh water by 3.27 million CNY $year^{-1}$. Based on the aforementioned advantages, the payback period for the total investment was relatively short.

The schematic of complexation UF integrated with electrolysis is shown in Fig. 4.28 [63]. For many years, electrochemistry method has been employed in refining metals or recovery of heavy metal existing in the electroplating wastewater. In this work, flat-sheet graphene with high performance and low price was selected as electrodes. Under optimum conditions, the recovery of Cu and Zn by EW both reached more than 60%. The energy consumptions were 11.5 kW h^{-1} kg^{-1} for Cu and 14.0 kW h^{-1} kg^{-1} for Zn.

It is more suitable for solutions with high concentration of metal ions that are characterized by higher current efficiency. As far as wastewater is concerned, the concentration of metal ions present is generally low, and therefore the current efficiency is low and power consumption is high when traditional electrochemistry approach is adopted. 3D electrode had very high specific surface area, allowing relatively easier mass transfer between solution and electrode [63]. This is one of the reasons for achieving high current efficiency for dilute solution. Martin et al. [63] applied 3D graphene electrode to remove Cu^{2+} and Zn^{2+}. The concentration of Cu^{2+}

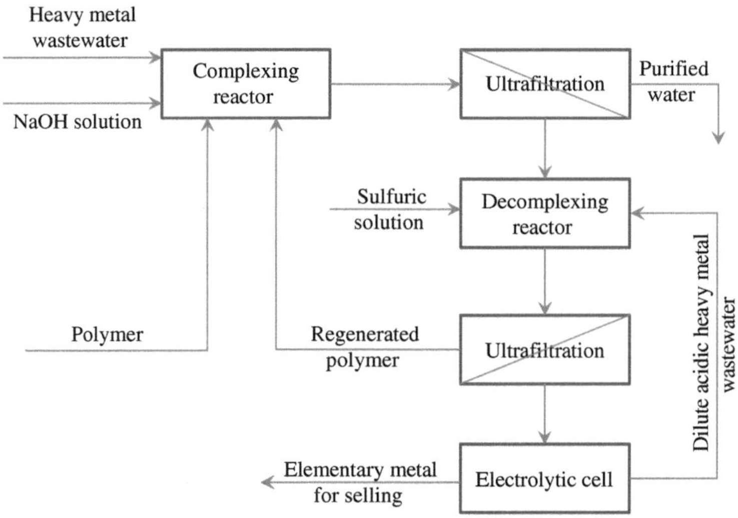

Figure 4.28: The schematic of complexation–ultrafiltration integrated with electrolysis.

declined from 150 to 0.05 mg L^{-1}, while that of Zn^{2+} from 200 to 1.1 mg L^{-1}. The current efficiencies for Cu^{2+} and Zn^{2+} removal were 68 and 65%, respectively.

Application of chemical precipitation/microfiltration (MF) to treat mixed electroplating wastewater was investigated by Broom et al. [64]. Its removal efficiencies for Cd, chromium (Cr), Pb, and Hg were 98.4, 98.9, 91.3, and 99.0%, respectively.

A design that integrates DD, UF (or NF, RO) is introduced by Zhang and Li [23]. The DD system is composed of a series of cells that contain dialysis (feed) and diffusion (permeate) chambers separated by an AEM. Driven by concentration difference across the membrane, both the acid and the salt have the tendency to transport through the membrane. As the AEM exhibits selectivity toward anions, the acid radicals pass the membrane favorably. According to the electroneutrality rule, part of the cations will move along with anions to the permeate side. Due to the smaller diameter of hydrated H^+, its transfer through the membrane will be preferentially taking place. Based on the aforementioned mechanism, acid will be enriched in the permeate side. As shown in Table 4.12, acid recovery is generally higher than 80% and rejection toward metal higher than 90%. The acid can be recycled after being concentrated. The retentate in DD process being removed of acid is further treated by UF, NF, or RO to get rid of heavy metals. Regulation of pH value by adding NaOH causes the development of metal hydroxide in the form of colloid. Thereafter, application of flocculent precipitates the colloid. Follow-up treatment using UF will reduce the heavy metal concentration to be significantly lower than the discharge standard.

Table 4.12: Efficiency of diffusion dialysis in treating different waste acid

Waste Acid	Feed Concentration	Recovery of Acid (%)	Rejection of Metal (%)	Application Field
HCl-$FeCl_2$	HCl: 15–20%; Fe^{2+}: 50–90 g L^{-1}	82.5	93	Steel pickling
H_2SO_4-$CuSO_4$	H_2SO_4: 20%; Cu^{2+}: 50 g L^{-1}	82	91	Hydrometallurgy
H_2SO_4-$FeSO_4$	H_2SO_4: 18%; Fe^{2+}: 0.0008 g L^{-1}	83	92	Battery
HCl-$AlCl_3$	HCl: 5 mol L^{-1}; Al^{3+}: 20 g L^{-1}	82.5	92.6	Foil
HNO_3-$Al(NO_3)_3$	HNO_3: 0.8 mol L^{-1}; Al^{3+}: 6 g L^{-1}	83.3	93.1	Polycrystalline silicon
HCl-HF-Fe^{2+}	HCl-HF: 3.8 mol L^{-1}; Fe^{2+}: 0.0014 mol L^{-1}	84	91	
H2SO_4-$ZnSO_4$	H_2SO_4: 15%; $ZnSO_4$: 150 g L^{-1}	83.5	92.8	Galvanization

4.5 Remediation of Heavy Metal Polluted Soil

4.5.1 Introduction

Soil contamination, which has been known to threaten human health and the environment, has been caused by a number of anthropogenic inefficiencies. Metallic contaminants in soil result from discharges of mining and metal finishing processes. Metallurgy is not the only source for soil contamination. Frequently, the contaminants are organic compounds, which result from leaky underground fuel tanks, or discharges of chlorinated cleaning solvents, such as trichoroethylene (TCE), trichloroethane (TCA), and perchloroethane (PCA). Other contaminants that are found in the contaminated soil are recalcitrant organic compounds, such as glycol de-icing agents, or heavier organics, such as polyaromatic hydrocarbons (PAHs), pentacholorphenols (PCPs), and polychorinated biphenyls (PCBs).

An important purpose of remediating contaminated soil is to restore the soil to beneficial future use. To meet a cleanup goal, identifying the future use of the site will determine what cleanup levels will be set. The technologies for soil treatment include the following: capping, solidification/stabilization, vitrification, chemical deoxidation, incineration, bioremediation, leaching/extraction, soil washing/soil flushing. Depending on the site-specific conditions, the choices for dealing with soil contamination vary.

In soil remediation, membrane separation is generally used in conjunction with other technologies, specifically, the soil leaching/extraction and soil washing/flushing [65]. Membranes, per se, do not change the hazardous nature of the contaminants but separates the contaminants from the bulk of the water. It can greatly reduce the volume of contaminated water, and render the bulk of the water substantially free of contaminants. The latter stream so that the water can either be reused or legally discharged. Membrane techniques in this context that have been considered for soil remediation are RO, NF, UF, MF, and pervaporation (PV).

4.5.2 Conventional Operations

Several complications may arise when the soil wash water contains highly toxic organic and metal compounds, which either cannot be discharged, or corrupt the activated sludge process, or cause uncontrolled foaming in the publicly owned treatment works (POTWs) because of the presence of surfactants. Membrane technologies are likely to find applications in treating this wash water, regardless of the nature of the contaminants.

Rochem Environmental's (Torrance, CA) high-pressure (1000 psig) Disk Tube technology involves RO to remove organics and metal contaminants from landfill leachates with an efficiency of more than 98% [66]. Geckeler and Volchek [67] showed that for low contaminant concentrations (below 50 ppm), if the pH of the solution is controlled within a narrow range, excellent binding and subsequent removal by polysulfone UF membranes (MWCO 10 or 20 kDa), could be achieved in the presence of sodium salts.

A MF process, known as the DuPont/Oberlin process, was developed to remove suspended metallic impurities from landfill leachates, electroplating rinse water, and heavy metal-laden groundwater using DuPont's polyolefinic Tyvek filter material. This technology was demonstrated at the Palmerton Zinc Superfund site in Palmerton, Pennsylvania [65]. Removal of suspended Zn solids, which was the primary objective, was achieved with better than 99% removal efficiency. It is claimed by the developer that the technology was best suited for suspended solid with concentration below 5000 ppm.

An integrated soil washing process, known as the GHEA process, was introduced by Gotlieb et al. [68] to treat soils highly contaminated with coal tar, PCBs, chlorinated organics, nitroaromatics, and heavy metals. It was demonstrated that water containing surfactants and other additives could carry away the contaminants from the soil matrix efficiently, and the residual hazards were at a level of 1 ppm, in both in situ and ex situ methods.

In waters contaminated with dilute concentrations of organics or metals, a more efficient way to treat these would be to remove the contaminants through the membrane. An organic extractant can be used to extract the organic pollutants without actually mixing it with the aqueous phase. Hutter and Vandegrift [69] developed such a membrane-assisted extraction-stripping process for removing benzene and a host of chlorinated organic compounds from groundwater contaminated at some Department of Energy (DOE) sites. A process similar in principle but used for the removal of toxic metals from water using selective complexing agents, is marketed by Commodore Separation Technologies (Kennesaw, GA). PV is another emerging technology that also removes contaminants, more specifically volatile organic compounds (VOCs), through a nonporous or dense hydrophobic membrane, cleaning up the aqueous stream in the process [70]. A crossflow PV system developed by Zenon Environmental was evaluated under Environment Protection Agency (EPA)'s SITE program for cleaning a TCE-contaminated site in California. The process removed 99% of TCE, nevertheless the treated water failed to meet the stringent criteria of 5 μg L^{-1} established by the Safe Drinking Water Act [65]. Cox and Baker [71] described a similar pervaporation PV process, which was successfully tested in an industrial site in California for dealing a methylene chloride-bearing stream. The permit to this facility for discharging the treated water into the local POTW required achieving an effluent concentration of 34 μg L^{-1}. Jiang et al. [72] applied PV in the laboratory work for cleaning a solution simulating simulated a Superfund effluent laden with several VOCs such as TCA, toluene, and surfactants.

A great deal of research has also been conducted on isolating metal ions from aqueous solutions by using liquid membranes (LM) that employ metal-complexing ligands.

4.5.3 Unconventional Application of Ion-Exchange Membranes for In situ Soil Remediation

An unconventional in situ treatment method applies membranes in the soil itself with electrokinetic (EK) transport for removal of contaminants, especially toxic metals [65]. The significance of this method is its low operation cost and potential applicability to a wide range of contaminants. The potential of this technique for waste remediation has been demonstrated in several studies [65].

During EK processing, hydroxides are inevitably precipitated by the hydroxyl ions generated through the electrolysis of water in a cathode compartment, and these precipitates prevent the removal of contaminants. In order to avoid precipitation in the cathode compartment, various techniques have been proposed and implemented, which include [65] (1) using enhancing agents such as acetic acid or a hydroxyl ion (OH^-) selective membrane in the cathode reservoir to prevent precipitation or solubilize precipitates of cationic metal contaminants near the cathode; (2) conditioning the anode and/or the cathode reservoirs to control pH and zeta potential, enhance desorption, and increase electroosmotic flowrate to increase mobility of contaminants; (3) mixing complexing agents such as ammonia (NH_3), citrate, and EDTA into soil, which will compete with soil for metal contaminants and form soluble complexes. Among these technologies proposed thus far, the one involving the prevention of metal precipitation has received the greatest attention and experimental evaluation. If an acidic agent is used in cathode electrolyte to prevent (OH^-) from flowing into a soil bed, the current efficiency decreases due to the introduction of new ions into the treated systems. In order to overcome this decrease in current efficiency, EK processing enhanced by electrodialysis (ED) with IEM was developed.

Hansen et al. [73] applied AEM and CEM in an EK process to remove Cu, Cr, mercury (Hg), Pb, and Zn (concentration several hundred ppm each) from soils contaminated by chlor-alkali plants, and Cu rolling mills. The results showed that cleanup objectives were achieved using less than 20 mA current on sand and loamy sand over a period of several weeks. The experimental apparatus consists of four principal parts: soil cell, electrode compartments, electrolyte solution reservoirs, and power supply. Each electrode compartment contained a sufficient volume of electrolyte solution to avoid sudden changes of pH in the electrolyte solution.

EK soil remediation technology enhanced by an IEM, was experimentally evaluated by Kim et al. [74]. The efficacies of two different types of electrode configurations, rectangular and cylindrical, were investigated. The schematics showing the standard ED demineralization process and the electrodialytic (IEM-enhanced EK) soil process is given in Fig. 4.29. In

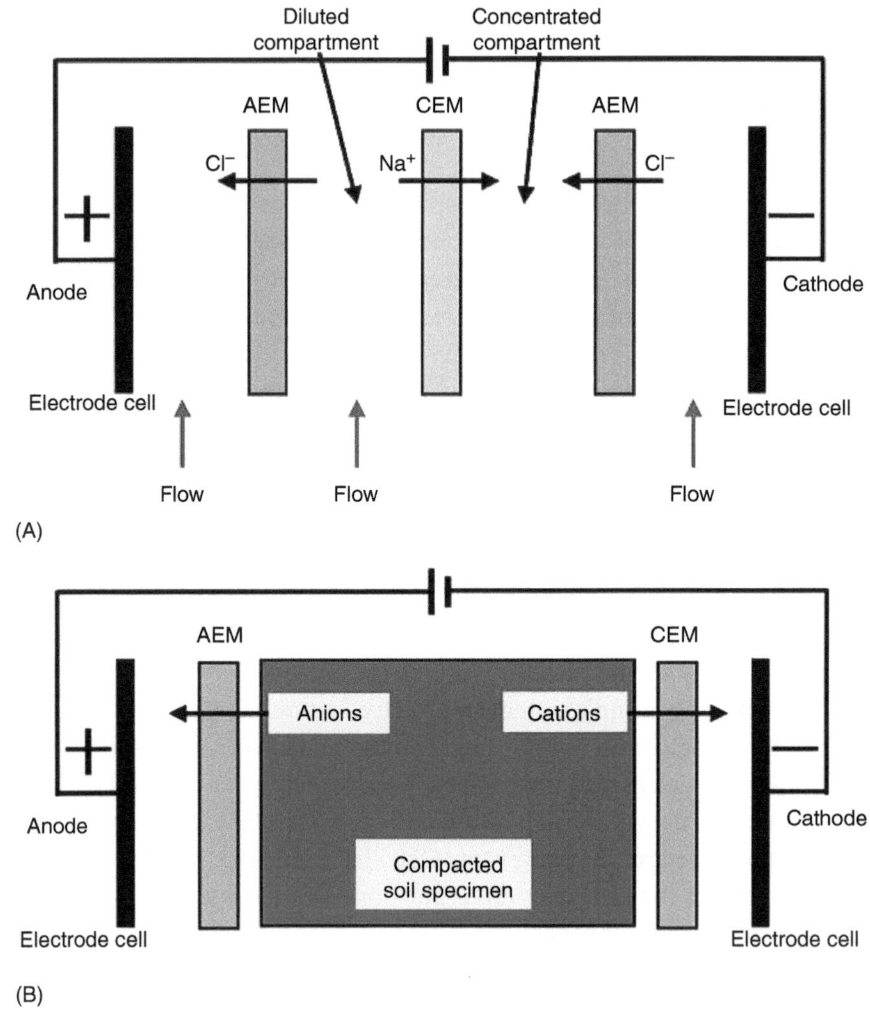

Figure 4.29: Schematic diagrams.
(A) Standard electrodialysis demineralization process and (B) electrodialytic (IEM-enhanced EK) soil process (AEM, anion-exchange membrane; CEM, cation-exchange membrane).
(copied from [74]).

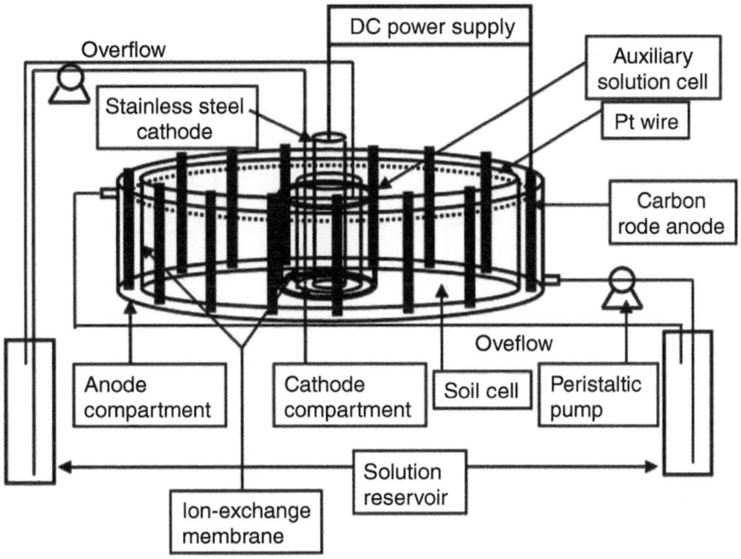

Figure 4.30: The schematic showing the location of the ACS inside EK system. *(copied from [74]).*

particular, this study focused on a fouling problem and its settlement by employing an auxiliary solution cell (ASC). The location of the ACS inside EK system is presented in Fig. 4.30. It was found that the effectiveness of the EK technology was increased by an enhancement scheme using an IEM. This is probably the prevention of metal precipitation in the region near the cathode originating from OH^- generated by the electrolysis of water in the cathode. Part of the OH^- move from the cathodic electrolyte into the ASC through the holes developed purposely on the CEM. The metal contaminants therefore precipitate in the ASC before moving on to the cathode electrolytes. The experimental results also indicate that the cylindrical electrode configuration can be implemented in practical situations to improve the treatability of cathode effluent containing a high level of contaminants.

4.6 Groundwater Remediation

Asarco Refinery is a legacy refinery with very serious ground water pollution problem after 100 years of operation [37]. Lime neutralization and precipitation system as demonstrated in Fig. 4.31 for groundwater remediation was installed in 1985 at an investment of more than US $1 million and required huge operating cost. In addition, the supernatant effluent did not meet the standard for discharge into aquatic system, due to the presence of heavy metals. For solving the challenge, the Wastewater Reclamation Project was initiated in 1993 and the corresponding flowchart is revealed in Fig. 4.32. It provided membrane separation before

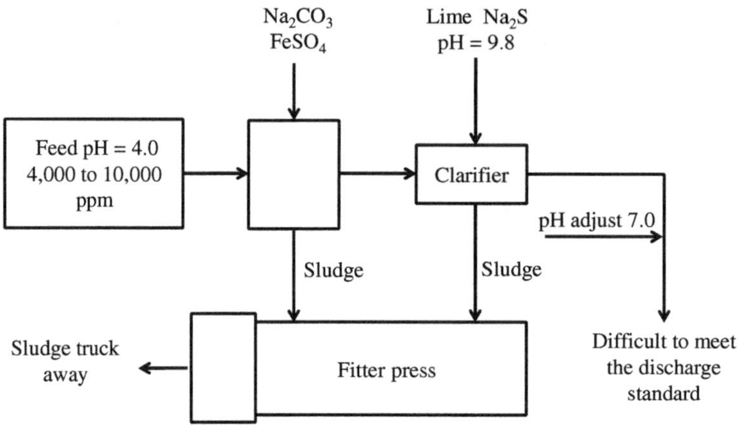

Figure 4.31: The process treating the ground water using lime precipitation method at Asarco legacy mine.

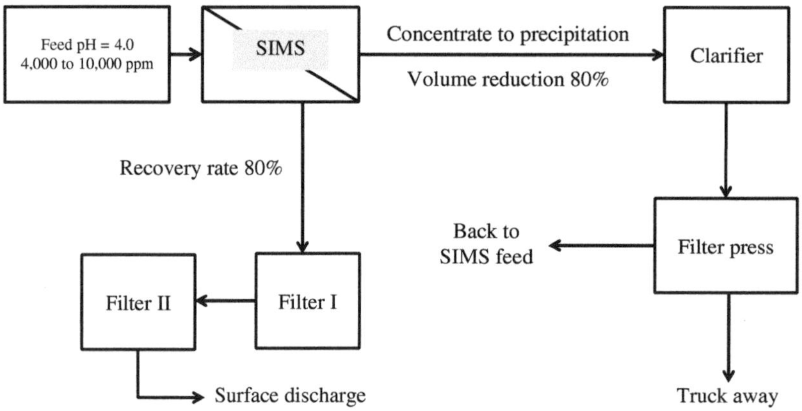

Figure 4.32: Process flow diagram of membrane system followed by precipitation. (Zinc metallurgy).

lime treatment. Table 4.13 summarizes the qualities of the feed and the permeate in terms of metal ions inside the membrane system. Undoubtedly, the total rejection of the metal ions is higher than 99%, which is the major reason for the environmental benefit achieved. With this innovation, the volume of liquid to be precipitated was reduced from 6 to 1 $m^3\ h^{-1}$. In addition as shown in Table 4.14, addition of membrane process has led to recommendable saving in capital and operation costs.

A Ni refinery company has caused serious ground pollution at mining sites due to the leakage [20]. The ammonium sulfate ($(NH_4)_2SO_4$) concentration in the ground water is ~2%. A two-stage RO process was then applied to repair the aquatic system, as shown in Fig. 4.33. In the

Table 4.13: Feed and permeate water qualities [33]

Project	As	Cd	Zn	Pb	Cu	Fe	Mn	Total Metal	Metal Rejection
Feed (mg L^{-1})	10.1	14.5	33.5	3.07	0.073	0.986	3.33	67.9	
Perm (mg L^{-1})	0.081	0.05	0.01	0.05	0.01	1.10	0.05	0.583	99.14%

Table 4.14: Capital and operating costs comparison between old and new processes [33]

Process	Capital	Chemicals (\$ m^{-3})	Sludge Disposal (\$ m^{-3})	Precipitation	Total Sludge Generated (kg m^{-3})	Total Operation Cost (\$ m^{-3})
Precipitation	1 million	2.61	26.53	Yes	19.24	29.15
Membrane	0.3 million	0.15	1.35	No	2.88	1.82

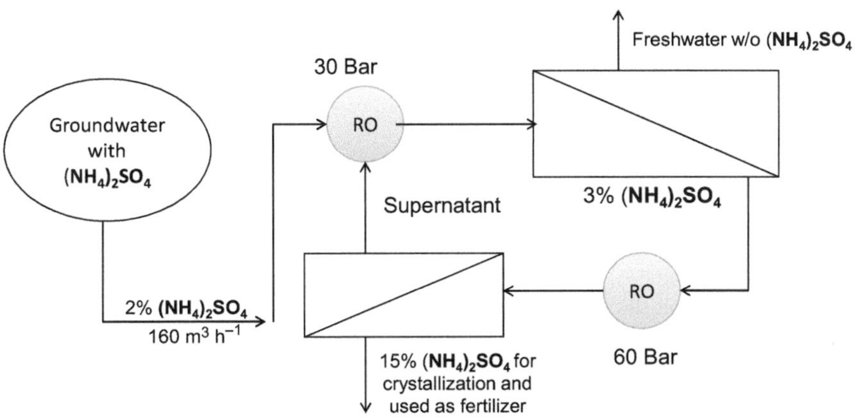

Figure 4.33: Treatment of $(NH_4)_2SO_4$ contaminated groundwater by a Nickel Refiner at Australia.

first stage of RO operated at 30 bar, $(NH_4)_2SO_4$ was concentrated to 3%; in the second stage of RO at 60 bar, the concentration reached 15%. Crystallized $(NH_4)_2SO_4$ can be used as a fertilizer. The system has been running for more than 9 years since the installation.

4.7 Conclusions

This chapter is mainly about the investigation on applications of membrane-based separations in nonferrous extractive metallurgy. The four heavy metals, Ni, Cu, Pb, and Zn, being in demand in huge amount are given more attention. As for pyrometallurgy, the segment that currently involves membrane technology is waste gas treatment. Due to the harsh conditions caused by high temperature and fine particles, inorganic membranes become the sole choice.

In hydrometallurgy, almost all the segments, such as leaching, concentration, electrolysis, wastewater treatment, have witnessed the benefits rendered by membrane application. The bases for all these merits in terms of environment and economy mainly include recovery of metals and water, and improvement of process efficiency. As for the last point, a very typical example is removal of impurities from electrowinning. Industrial-level application has already been realized and extended to countries in Asia, America, and Australia. One regret is that some researches do not specify their research background, which causes the confusion in research classification. A more direct understanding of the research object will lead to intimate combination of laboratory work and demand from industry, facilitating the problem solving.

Glossary and Symbols

AEM	Anion-exchange membranes
AMD	Acid mining drainage
Ag	Silver
As	Arsenic
As_2O_3	Arsenic oxide
ASC	Auxiliary solution cell
Au	Gold
B	Boron
BPM	Bipolar membrane
Cd	Cadmium
CEM	Cation-exchange membrane
Cl_2	Chlorine
Co	Cobalt
Co_xS_y	Cobalt sulfides
CPCI	Cetylpyridiniumchloridemonohydrate
Cu	Copper
Cu^+	Cuprous ion
Cr	Chromium
$Cu_2CO_3(OH)_2$	Malachite
$Cu(NH_3)_2^{2+}, Cu(NH_3)_4^{2+}$	Copper amino complex
$CuCl_2$	Copper dichloride
Cu-Ni	Copper-nickel
$CuSO_4$	Copper sulfate
DD	Diffusion dialysis
DOE	Department of Energy
DSPM	Donnan steric pore model
ED	Electrodialysis
EDI	Electrodeionization
EDTA	Ethylenediaminetetraacetic acid
EK	Electrokinetic
ELM	Emulsion liquid membrane
EPA	Environment Protection Agency
EW	Electrowinning
Fe^{3+}	Ferric ion
Fe-Ni	Ferronickel
Fe_2O_3	Ferric oxide

Fe_3O_4	Iron oxide black
$FeSO_4$	Ferrous sulfate
$Fe(OH)_3$	Ferric hydroxide
H^+	Hydrogen ions
HCl	Hydrochloric acid
HDS	High-density slurry method
Hg	Mercury
H_2O_2	Hydrogen peroxide
HPAL	High-pressured acid leaching process
H_2S	Hydrogen sulfide
H_2SO_4	Sulfuric acid
IEM	Ion exchange membrane
K-C Smelter	Kenny Curt smelters
LM	Liquid membranes
MEUF	Micellar-enhanced ultrafiltration
MF	Microfiltration
MW	Molecular weight
MWCO	Molecular weight cut-off
Na^+	Sodium ions
NaCl	Sodium chloride
NaOH	Sodium hydroxide
Na_2SO_4	Sodium sulfate
LP-RO	Low-pressure reverse osmosis
NF	Nanofiltration
$(NH_4)_2SO_4$	Ammonium sulfate
Ni	Nickel
Ni-P	Nickel-phosphorus
NiS	Nickel sulfide
$NiSO_4$	Nickel sulfate
NTA	Nitrilotriacetic acid
O_2	Oxygen
PAHs	Polyaromatic hydrocarbons
PALA	Poly(allylamine)
Pb	Lead
PbS	Lead sulfide
PCA	Perchloroethane
PCBs	Polychorinated biphenyls
PCPs	Pentacholorphenols
PEI	Polyazirine
PLC	Programmable logic controller
POTWs	Publicly owned treatment works
P-S converter	Pierce-Smith converter
PSS	Poly(sodium 4-styrenesulfonate)
PTFE	Polytetrafluoroethylene
PV	Pervaporation
QSL method	Qaureaa–Schuhman–Lurgi method
RO	Reverse osmosis
NH_3	Reduction roasting-ammonia
RRAL	Leaching process
S	sulfur
Sb_2O_3	Antimony oxide

Si	Silicon
SIMS	Special Industrial Membranes
SKS method	Shui Kou Shan method
SO_2	Sulfur dioxide
SO_4^{2-}	Sulfate anions
SS	Stainless steel
SX	Solvent extraction
TCA	Trichloroethane
TCE	Trichoroethylene
UF	Ultrafiltration
ULP-RO	Ultra-low pressure reverse osmosis
VMD	Vacuum membrane distillation
VOCs	Volatile organic compounds
W/O/W	Water-in-oil-in-water
Y-o-Y	Year over year
ZLD	Zero liquid discharge
Zn	Zinc
$ZnSO_4$	Zinc sulfate

References

[1] M.B. Liu, L.W. Duan, Y. Gao, W.Z. Yin, Present situation and utilizing technique of nickel ore in China, China Min. Mag. 20 (2011) 98–102.
[2] R.J. Luo, Current situation and prospects of nickels resources in China, Urban Fin. China 11 (2013) 53–55.
[3] Q.G. Chen, Study on the development and utilization of nickel oxide ore resources, Hydrometallurgical 27 (2008) 7–9.
[4] C.J. Li, Y.M. Chen, X.R. Huang, H. Yu, Process status and progress of nickel ore, Utilization Miner. Resour. 6 (2012) 8–11.
[5] Q.H. Li, J. Wang, Z.H. Liu, Progress in the development of the world's nickel laterite resources and hydrometallurgical technologies, Conserv. Utilization Miner. Resour. 6 (2012) 8–11.
[6] D.W. Nel, P. van der Gryp, H.W.J.P. Neomagus, D. Bessarabov, Application of membrane technology in a base metal refinery, J. S. Afr. Inst. Min. Metall. 113 (2013) 363–374.
[7] F.K. Crundwell, M.S. Moats, V. Ramachandran, T.G. Robinson, W.G. Davenport, Extractive Metallurgy of Nickel Cobalt & Platinum-Group Metals, Elsevier, Holland, (2011) 537-549.
[8] A.W. Mohammad, R. Othaman, N. Hilal, Potential use of nanofiltration membranes in treatment of industrial wastewater from Ni-P electroless plating, Desalination 168 (2004) 241–252.
[9] Y.T. Lou, Y.T. Chen, S.G. Wang, J.C. Dong, Application of membrane process to electroplating nickel effluent treatment, Membr. Sci. Technol. 22 (2002) 43–47.
[10] H.W. Shi, A.G. Chen, C.B. Xia, J.Y. Feng, Nickel-containing electroplation wastewater treatment with sodium hydroxide-membrane filtration, Environ. Pollut. Control 24 (2002) 93–95.
[11] H.L. Zhu, S.B. Ding, J.M. Li, Y.M. Guo, Submerged vacuum membrane distillation for treatment of electroplating wastewater, The Fourth National Symposium on Application of Membrane Separation in Metallurgical Industry, 2014, 172–175.
[12] V. Hakala, Practical filtration-metallurgy: new membrane filter tackles metallurgical water demands, Filtr. Separat. 50 (2013) 41–43.
[13] W. Wang, Y.K. Li, N. Feng, Global copper resource pattern analysis, Resour. Indus. 15 (2013) 27–32.
[14] Land and Resources Information Center Annual Review of World Mineral Resources, Geological Publishing House, Beijing, R. 2012.
[15] Y.N. Liu, The status and analysis of copper imports in China, China Natl. Cond. Strength 3 (2014) 58–60.

[16] Beijing Nonferrous Metallurgy Design and Research Institute Heavy Non-ferrous Metal Smelting and Design Manual, Copper–Nickel Volumes, Metallurgical Industry Press, Beijing, R. 1996.

[17] X.P. Zhang, Introduction of Metallurgical Engineering, Central South University Press, China, 2007.

[18] S.P. Chen, Z.L. Wu, B.B. Lan, Summarize on the technology of copper pyrometallurgy, Copper Proj. 4 (2010) 44–49.

[19] N. Sabba, D.E. Akretche, Selective leaching of a copper ore by an electromembrane process using ammonia solutions, Miner. Eng. 19 (2006) 123–129.

[20] Q.X. Zhang, Metallurgy Separation Science and Engineering, Science Press, China, R. 2004.

[21] W.F. Huang. Experimental research Emulsion liquid membrane technology to recover the copper from copper heap leaching liquid, The Third National Symposium on Application of Membrane Separation in Metallurgical Industry, 2013, 212–215.

[22] Z.G. Wang, The application membrane separation technology in hydrometallurgy, The Fourth National Symposium on Application of Membrane Separation in Metallurgical Industry, 2014, 54–61.

[23] X.M. Zhang, Z.G. Li. The application of diffusion dialysis in nonferrous metal industry used in waste acid recovery, The Fourth National Symposium on Application of Membrane-based Separation in Metallurgical Industry, 2014 91–93.

[24] F. Valenzuela, C. Fonseca, C. Basualto, O. Correa, C. Tapia, J. Sapag, Removal of copper ions from a waste mine water by a liquid emulsion membrane method, Miner. Eng. 18 (2005) 33–40.

[25] C.M. Zhong, X.H. Fang, Z.L. Xu, Study on removing heavy metal ions of acid mine drainage by using nanofiltration membranes, Environ. Sci. Technol. 30 (2007) 10–12.

[26] X. Chai, Pilot scale membrane separation of electroplating wastewater by reverse osmosis, J. Membr. Sci. 123 (1997) 235–242.

[27] J. Wu, C. Liu, Application and research of membrane separation technology used in electroplating wastewater treatment and metal recycling, The Fourth National Symposium on Application of Membrane Separation in Metallurgical Industry, 2014, 163–164.

[28] Y.H. Sun, L.S. Xiao, G.Q. Zhang, Diffusion dialysis recycling HNO_3 from copper stripping solution, The Fourth National Symposium on Application of Membrane-based Separation in Metallurgical Industry, 2014, 94–98.

[29] H. Wang, Separation and recovery of copper from copper etching liquid—the separation of copper sulfate and sodium sulfate, sulfuric acid, The Fourth National Symposium on Application of Membrane Separation in Metallurgical Industry, 2014, 220–221.

[30] D.C. Ge, X.J. Gou, Discussion on application of anion and cation membrane and bipolar membrane in metallurgical field, The Fourth National Symposium on Application of Membrane Separation in Metallurgical Industry, 2014, 100–102.

[31] D.S. Flett, Ion Exchange Membrane, E. Horwood Ltd, Chichester, U.K, 1983.

[32] X. Yang, Theory and technology research of diaphragm electrowinning sheet-form copper from acidic cupric chloride etchant waste, Master Thesis, Central South University, 2011.

[33] R.M. Spearot, J.V. Peck, Recovery process for complexed copper-bearing rinse waters, Environ. Prog. 3 (1984) 124–128.

[34] B.E. Reed, S. Arunachalam, B. Thomas, Removal of lead and cadmium from aqueous waste streams using granular activated carbon columns, Environ. Prog. 13 (1994) 60–64.

[35] S.J. Ruey, H. Pyng, Hollow-fiber membrane extraction of copper(II) from aqueous ethylenediamine tetraacetic acid solutions with Aliquat 336, Ind. Eng. Chem. Res. 39 (2000) 1409–1415.

[36] R.S. Juang, Y.J. Chen, I.P. Huang, Amine-based extraction recovery of copper(II) from aqueous solutions in the presence of EDTA, Equilibrium Stud. Sep. Sci. Technol. 34 (1999) 3099–3112.

[37] H. Wang, SIMS Special industrial membrane technology, The Fourth National Symposium on Application of Membrane Separation in Metallurgical Industry, 2014, 206–213.

[38] H. Wang, Copper acid mine drainage AMD composite membrane treatment technology—Copper SIMS zero discharge project, The Third National Symposium on Application of Membrane Separation in Metallurgical Industry, 2013, 84–85.

[39] B.W. Qiao, Copper foil wastewater zero discharge treatment, The Third National Symposium on Application of Membrane Separation in Metallurgical Industry, 2013, 115–117.

[40] Z.X. Dai, Distribution, type and exploration criteria for lead and zinc resources in the world, World Nonferrous Metals (2005) 15–23.

[41] R.G. Li, Current status and development of China's lead and zinc smelting process, Chin. Nonferrous Metall. 39 (2010) 13–20.

[42] D.W. He, H.N. Zhou, L. Liu L., Technological advances and pollution prevention of lead smelting, China Metal Bulletin (2011) 18–19.

[43] J.M. Jiang, Domestic status and development trend of lead smelting technology, Saving Energy Nonferrous Metall. 29 (2013) 4–8.

[44] D.J. Yang, Y.S. Liao, Y.F. Xu, Overview of zinc metallurgical process, Yunnan Metall. 31 (2002) 22–26.

[45] X.P. Zhang, Introduction of Metallurgical Engineering, Central South University Press, China, R. 2007.

[46] D.S. Liu, X.J. Zhang, Q.L. Zheng, YT membrane application in hydrometallurgical industry, The Fourth National Symposium on Application of Membrane Separation in Metallurgical Industry, 2014, 38–41.

[47] Z.H. Wang, X.J. Zhang, W.X. Zeng, M. Jiang, L. Gao, YT membrane separation technology in the non-ferrous metallurgical industry arsenic removal application, The Fourth National Symposium on Application of Membrane Separation in Metallurgical Industry, 2014, 34–37.

[48] W.Z. Tian, B.Z. Chen, The state of processing and utilization about the arsenic material in non-ferrous metallurgical industry, Hunan Nonferrous Metals 20 (2004) 11–19.

[49] S.M. Li, Research progress on chlorine metallurgy, Non-ferrous Min. Metall. 26 (2010) 34–36.

[50] D.W. Xia, L.M. Yin, J.Z. Tong, The design and operation of ShaoGuan smelter wastewater depth treatment, The Third National Symposium on Application of Membrane Separation in Metallurgical Industry, 2009, 101–104.

[51] X.M. Zhang, Z.G. Li, The application of diffusion dialysis in nonferrous metal industry used in waste acid recovery, The Fourth National Symposium of Application of membrane separation technique in metallurgical industry, 2014, 91–93.

[52] C.M. Zhong, Z.L. Xu, X.H. Fang, X.J. Wang, Treatment and reuse of acid mine drainage by ultra-low pressure reverse osmosis membrane, Technol. Water Treat. 33 (2007) 77–80.

[53] G.H. Shang, G.Q. Zhang, C.J. Gao, Treatment of simulated dilute Ni^{2+}-containing wastewater by electrodeionisation with a bipolar membrane: feasibility and current density distribution, Desalination 353 (2014) 1–7.

[54] Z. Ujang, G.K. Anderson, Application of low-pressure reverse osmosis membrane for Zn and Copper removal from wastewater, Water Sci. Technol. 34 (1996) 247–253.

[55] H. Strathmann, Selective removal of heavy metal ions from aqueous solutions by ultrafiltration of macro-molecular complexes, Sep. Sci. Technol. 15 (1980) 1135–1152.

[56] I. Korus, M. Bodzek, K. Loska, Removal of zinc and nickel ions from aqueous solutions by means of the hybrid complexation ultrafiltration process, Sep. Sci. Technol. 17 (1999) 111–116.

[57] D.B. Bhattacharya, L.G. Bachas, J. Hestekin, S. Richie, S.K. Sikdar, High capacity functionalized membrane sorbents for heavy metal ion sequestration, J. Membr. Sci. 141 (1998) 121–135.

[58] S.K. Sikdar, D. Grosse, I. Rogut, Membrane technologies for remediating contaminated soils: a critical review, J. Membr. Sci. 151 (1998) 75–85.

[59] Z.L. Xu, H.M. Xu, X.D. Zhai, Treatment of waste streams containing Pb^{2+} and Cd^{2+} by micellar-enhanced ultrafiltration, Membr. Sci. Technol. 22 (2002) 15–20.

[60] S. Ahmadi, I.K. Tseng, B. Bactchelor, Micellar enchanced ultrafiltration of heavy metal using lecithin, Sep. Sci. Technol. 29 (1994) 2435–2450.

[61] C. M. Zhong, Study on acid mine drainage treatment by reverse osmosis, nanofiltration and coupling of RO-NF membrane processes, East China University of Science and Technology, Doctoral thesis, 2007.

[62] Wu J, Liu C, Research of application of membrane separation technology in electroplating wastewater treatment and metal recovery, The Fourth National Symposium of Application of Membrane Separation Technique in Metallurgical Industry, 2014, 163–167.

[63] Z.L. Xu, Y.F. Zhang, Research progress in treatment of heavy metal wastewater using the hybrid process of complexation-ultrafiltration-electro deposition, Membr. Sci. Technol. 23 (2003) 141–150.
[64] G.P. Broom, R.C. Squires, M.P.J. Simpson, The treatment of heavy metal effluents by crossflow microfiltration, Membr. Sci. Technol. 87 (1994) 219–230.
[65] S.K. Sikdar, D. Grosse, I. Rogut, Membrane technologies for remediating contaminated soils: a critical review, J. Membr. Sci. 151 (1998) 75–85.
[66] R.A. Griffiths, Soil-washing technology and practice, J. Hazard. Mater. 40 (1995) 175–189.
[67] K.E. Geckeler, K. Volchek, Removal of hazardous substances from water using ultrafiltration in conjunction with soluble polymers, Environ. Sci. Technol. 30 (1996) 725–734.
[68] I. Gotlieb, J.W. Bozzelli, E. Gotlieb, Soil and water decontamination by extraction with surfactants, Sep. Sci. Technol. 28 (1993) 793–804.
[69] J.C. Hutter, G.F. Vandegrift, Decontamination of groundwater by using membrane-assisted solvent extraction, in: G.F. Vandergrift, D.T. Reed, I.R. Tasker (Eds.), Environmental Remediation: Removing Organic and Metal Ion Pollutants, 509, ACS, America, 1992, pp. 47–60.
[70] W. Ji, S.K. Sikdar, S.T. Hwang, Modeling of multicomponent pervaporation for removal of volatile organic compounds from water, J. Membr. Sci. 93 (1994) 1–19.
[71] G. Cox, R. Baker, Pervaporation: a new membrane technology effectively treats small volumes of volatile organic compound-contaminated wastewater, Ind. Wastewater January/February (1998) 35–37.
[72] J.S. Jiang, L.M. Vane, S.K. Sikdar, Recovery of VOCs from surfactant solutions by pervaporation, J. Membr. Sci. 136 (1997) 233–247.
[73] H.K. Hansen, L.M. Ottosen, S. Laursen, A. Villumsen, Electrochemical analysis of ion-exchange membranes with respect to a possible use in electrodialytic decontamination of soil polluted with heavy metals, Sep. Sci. Technol. 32 (1997) 2425–2434.
[74] W.S. Kim, S.O. Kim, K.W. Kim, Enhanced electrokinetic extraction of heavy metals from soils assisted by ion exchange membranes, J. Hazard. Mater. 118 (2005) 93–102.

CHAPTER 5

Light Nonferrous Metals

L.-S. Xiao

School of Metallurgy and Environment, Central South University, Changsha, Hunan Province, China

5.1 Lithium

5.1.1 Introduction

Lithium (Li) is a kind of light metal, which is widely applied in energy storage, aerospace, nuclear fusion power generation, etc. [1]. Global Li resources are rich, 13 million tons calculated using elementary Li, according to the America Geological Survey Data issued in 2014 [2]. Li resources are mainly distributed in Chile, China, Brazil, Bolivia, the United States, Canada, Australia, and Zimbabwe. The Li reserves in Chile account for almost 58% of the world's total, followed by those in China, Australia, and Argentina in sequence; 90% of the global Li output is delivered by these four countries.

Naturally occurring Li mainly exists in the brine of salt lakes (surface and underground) and granitic pegmatite deposits. Hereinto, the amount with salt lake occupies 66% of the world's Li reserve, and over 80% of the world's Li reserve base. Usually, Li in salt lake has paragenetic relationship with the salts of potassium (K) and sodium (Na), and the minerals of halogens [eg, bromine (Br), iodine (I), etc.]. Therefore, extraction of Li can generate other valuable side products such as K, Na, magnesium (Mg), Br, I, etc. [2–6]. Salt lake containing Li mainly distributes in Qinghai and Tibet regions of China, Nevada of the United States, and Chile, Bolivia, and Argentina of South America. Granite pegmatite type Li deposits are widely distributed in Canada, Australia, Zimbabwe, Russia, the United States, and China [1,7]. Extraction of Li principally utilizes spodumene, lepidolite, petalite, etc.

Driven by the demand due to new energy market, Li exploitation in recent years experiences prosperity. The growth in global Li production is primarily attributed to the extraction from brine in Argentina and Chile. An average growth rate of around 11% per annum is registered with the output of Li from brine [8]. At present, the top three Li producers in the world, SQM, FMC, and Chemeall, supply 94% of salt lake Li in global market.

The process for Li extraction from minerals relies heavily on sulfate digestion process. Briefly, potassium salt consisting of potassium sulfate (K_2SO_4) (70%) and sodium sulfate (Na_2SO_4 30%) are mixed with lepidolite. The mixture is sintered at high temperature, followed by water quench and multistage countercurrent leaching. Sodium hydroxide (NaOH) is used to adjust

Membrane-Based Separations in Metallurgy
http://dx.doi.org/10.1016/B978-0-12-803410-1.00005-0

the pH value of leachate and subsequently sodium carbonate (Na_2CO_3) for precipitating lithium carbonate (Li_2CO_3). The solid precipitant is rinsed and dried to obtain industrial level Li_2CO_3. This technology is relatively mature, but with serious pollution and high cost.

The technologies widely used for extracting Li from salt lake brine include carbonate precipitation, solvent extraction, adsorption, calcinations and leaching, and solar pond and freshwater extraction. Carbonate precipitation is the most mature and reliable one [9]. The concentration of the brine relies on solar energy for evaporation. Nevertheless, it is not economical for brine comprising high ratio of Mg/Li. This is because an excessive amount of Na_2CO_3 is needed to remove Mg with the presence of supersaturated magnesium chloride ($MgCl_2$). Solvent extraction is classified into unitary and synergistic extractions. It is efficient in recovering Li from its low grade salt lake brine. The key of both is suitable extractant. Tributyl phosphate (TBP), an extractant suitable for brine with high Mg, is widely applied on a large scale. The lithium iron chloride complex ($LiFeCl_4$) in brine is extracted to form $LiFeCl_4 \cdot$ TBP complex is then washed by acid and back extracted by hydrochloric acid (HCl). Its disadvantages are the serious system corrosion and solution loss of extractant. In adsorption separation, Li ions will be preferentially adsorbed by ion-exchange resins or inorganic adsorbents and subsequently eluted [10]. Inorganic adsorbents have high selectivity, large adsorption capacity, and high recovery rate; therefore they are the main research direction in exploiting salt lake brine Li resource [11,12].

Relative to mineral extraction, brine extraction has advantages of low energy consumption, high recovery, and low pollution. In addition, the product from brine has a higher content of Li_2CO_3 with sound stability and low solubility. According to a survey by Yaksic in 2008 [13], salt lake brine ranks first in cost competitiveness for obtaining Li, compared with other resources including mineral, sedimentary clay, oil and gas, and seawater. Therefore, utilization of salt lake brine is currently the mainstream of Li exploitation.

As one of the solutions to the drain of the natural Li ion reserves, the recovery of Li metals from disposed Li batteries or from solutions (such as sea water and waste water) containing Li metals is attracting increasing attention [14].

The subsequent step after obtaining Li_2CO_3 and lithium chloride (LiCl) is metal extraction. Elementary Li can be produced by molten salt electrolysis and thermal reduction. The process for LiCl–potassium chloride (KCl) molten salt electrolysis is schematized in Fig. 5.1. This is the only method till present that has been applied in industrial production [15].

5.1.2 Lithium Recovery From Various Solutions

The components in salt lake brine are very complex, with the coexistence of Mg^{2+}, Li^+, boron (B), Ca^{2+}, SO_4^{2-}, etc. As aforementioned, the high Mg^{2+}/Li^+ ratio and low concentration of targeted component will elevate the cost of production. In most resources, the concentration

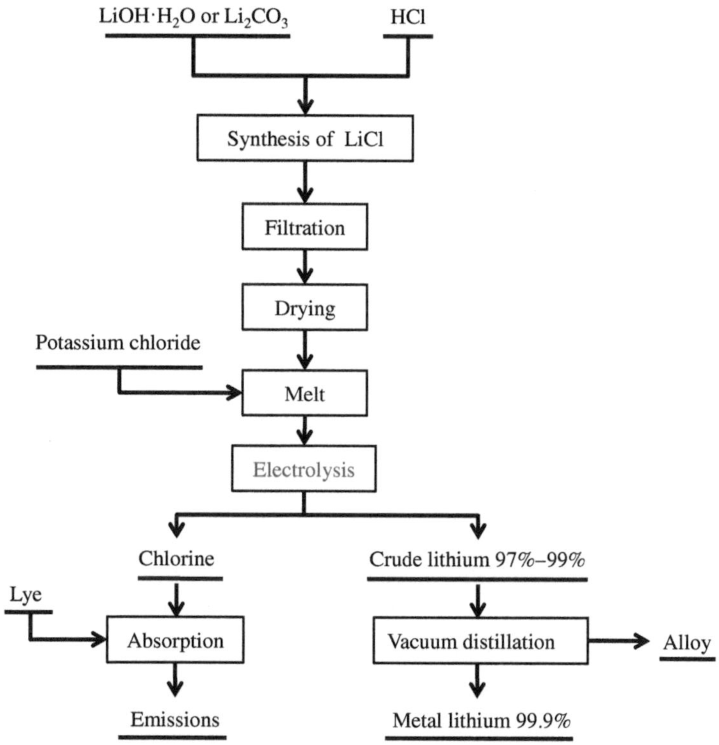

Figure 5.1: Flow chart for the production of metallic lithium by electrolysis.

of Li^+ is low, which further presents difficulty for Li extraction. Kang et al. investigated separation of Mg^{2+}/Li^+ by nanofiltration (NF) method [16]. The low concentration salt lake brine was used as feed water and spiral-wound DK NF membranes (DK4040F, GE Osmonics, USA) were employed. These membranes were negatively charged.

Under the experimental operating conditions, membrane separation permeate factors (SF) of Mg^{2+}/Li^+ of three types of feed waters with different compositions were all less than 0.1 and the Mg^{2+}/Li^+ mass ratios were reduced from 48.50, 42.31, and 28.30 in the feed to 4.04, 3.21, and 1.86 in the permeate, respectively, as shown in Table 5.1. It is also found that the contents of Ca^{2+} and SO_4^{2-} were almost zero in the permeate. At least 73.8% B were retained in the feed. The presence of Ca^{2+} may lead to scaling of membrane and the suggestion by the

Table 5.1: The composition of permeate water

Raw Water	Ions Concentration in the Permeate (g L^{-1})							
	Mg^{2+}	Li^+	Ca^{2+}	SO_4^{2-}	Cl^-	B	Mg^{2+}/Li^+	α
A	0.93	0.23	0.00	0.05	4.11	0.05	4.04	0.062
B	0.61	0.19	0.04	0.00	3.16	0.06	3.21	0.027
C	0.67	0.36	0.00	0.03	4.28	0.46	1.86	0.026

Note: $\alpha = [\rho(Mg^{2+})/\rho(Li^+)]_{permeate}/[\rho(Mg^{2+})/\rho(Li^+)]_{retentate}$.

authors is to remove most of the Ca^{2+} in the step of salt pond treatment. The permeate with enriched Li^+ is used for the production of Li_2CO_3, whereas the retentate is used to extract relevant inorganic salts.

A positively charged polyamide (PA) composite NF hollow fiber membrane was used for Li^+ and Mg^{2+} separation [17]. It was fabricated by the interfacial polymerization of 1,4-bis(3-aminopropyl)piperazine (DAPP) and trimesoyl chloride (TMC) on the polyacrylonitrile (PAN) ultrafiltration (UF) fiber support. The salt rejection order of the membrane was $MgCl_2 > MgSO_4 > NaCl \geq LiCl$. The mass ratio of Mg^{2+}/Li^+ decreased from initial 20:1 in the feed of 2.0 g L^{-1} $MgCl_2$ and LiCl mixture to 7.7:1 in the permeate. The operating pressure was 0.3 MPa.

In Liu et al.'s work, a novel membrane electrolysis method, as schematically shown in Fig. 5.2, was developed to extract Li from high Mg^{2+}/Li^+ ratio salt lake brine [18–20]. The mechanism for lithium iron phosphate ($LiFePO_4$) and iron phosphate ($FePO_4$) are used to extract Li from brine. $FePO_4$ is the anode placed in the brine and $LiFePO_4$ cathode in the Mg^{2+}-free electrolyte. By applying external potential, intercalation reaction takes place at the anode with Li^+ inserting and depositing inside the $FePO_4$ crystal structure to form $LiFePO_4$; whereas, at the cathode Li^+ is released from $LiFePO_4$ into the Mg^{2+}-free electrolyte. The basic chemistry for the process is expressed by following equations:

$$\text{Cathode}: \text{LiFePO}_4 \rightarrow \text{Li}^+ + \text{FePO}_4 + \text{e}$$

$$\text{Anode}: \text{FePO}_4 + \text{Li}^+ + \text{e} \rightarrow \text{LiFePO}_4$$

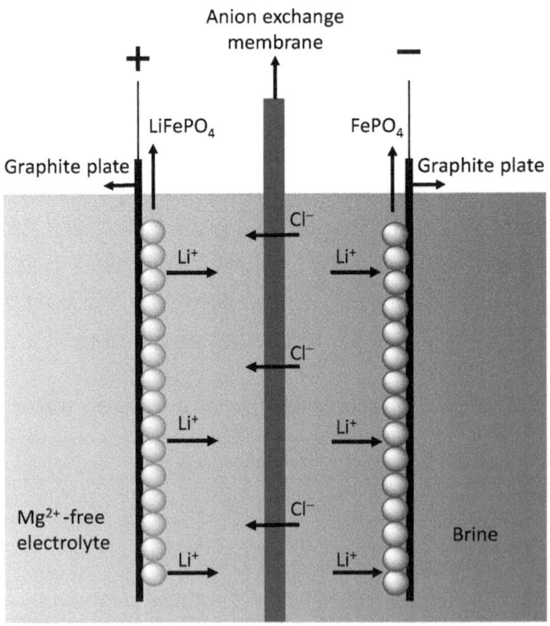

Figure 5.2: The schematic diagram of the electrolytic cell for extraction of Li^+ using $FePO_4$.

By this way, Li^+ can be preferentially removed from the salt lake brine. The mass ratio of Mg/Li inside anode was 0.3, in contrast with 493 of the brine solution. The side effects of Na^+, K^+, and Mg^{2+} at anode may be controlled by adjusting electrolytic voltage [19,20]. The disadvantages of the method are the low Li^+ exchange capacity and stability of electrode.

Porous mixed matrix nanofiber (MMN) membrane (roughly microfiltration (MF) membrane) using Li^+ ion sieves (LIS) particle (Li^+ adsorbent $H_{1.6}$ manganese $(Mn)_{1.6}O_4$, abbreviated as MO) embedded polysulfone (PSF) matrix as a base material was fabricated for Li^+ membrane adsorber (MA) [21]. The MMN was prepared via electrospinning and the lithium manganate ($LiMnO_2$) was the precursor for adsorbent.

The particles were well distributed inside the porous network as revealed by Electron Probe Micro-analyzer combined with Energy Dispersive X-ray Spectrometer (EPMA-EDS) mapping for elements Mn and O (Fig. 5.3). The adsorption sites of the particles mostly (96.6%) remained accessible to Li^+ even after their incorporation into the MMN. The convective flow through MMN facilitated the fast Li^+ transport to the MO surface. At fixed volumetric flow rate, the accumulated volume of treated seawater steadily increased while the transmembrane pressure (TMP) (<2 kPa difference) were stable for the entire operation (Fig. 5.4A). The *quick* breakthrough of other cations *under continuous operation* (Fig. 5.4B) *suggests* the selectivity of the MMN towards Li^+; the $C_p/C_f = 0.95$ occurs at $t_{ex} = 11.8$ h for Li^+. Similar *pattern, as shown in* Fig. 5.4C were obtained for breakthrough curves during the operation *which monitors* the Li^+ uptake (qU) in each cycle. Meanwhile, the recoveries (qR) measured from the acid solution (i.e., stripped Li^+ from the MMN) per cycle were occasionally higher than qU (Fig. 5.4D). This maybe due to the coelution of residual Li^+existing in the MO. Enrichment or elution of Li^+ was successfully realized by repeated Li^+ desorption using a small volume of acid solution.

The breakthrough study showed that a balance between kinetics and dynamic Li^+ adsorption capacity was actualized at optimal feed/MMN contact time by adjusting the feed flow rate

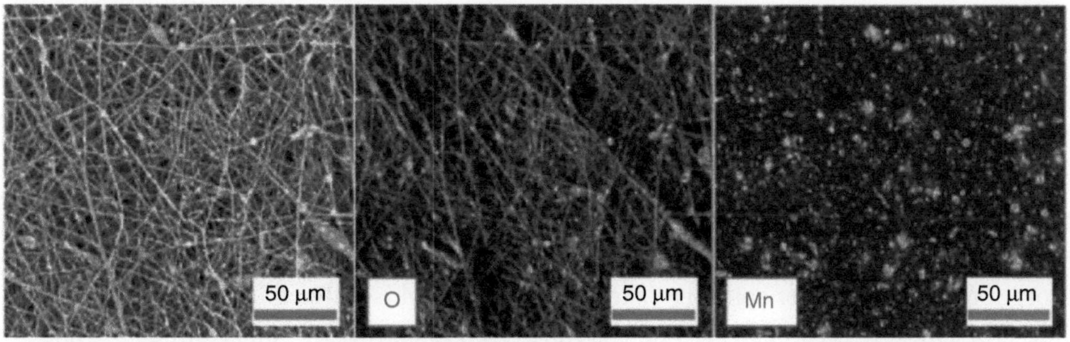

Figure 5.3: Field emission-electron probe micro-analyzer (FE-EPMA) with EDS mapping of Mn and O elements present on the surface of the MMN.

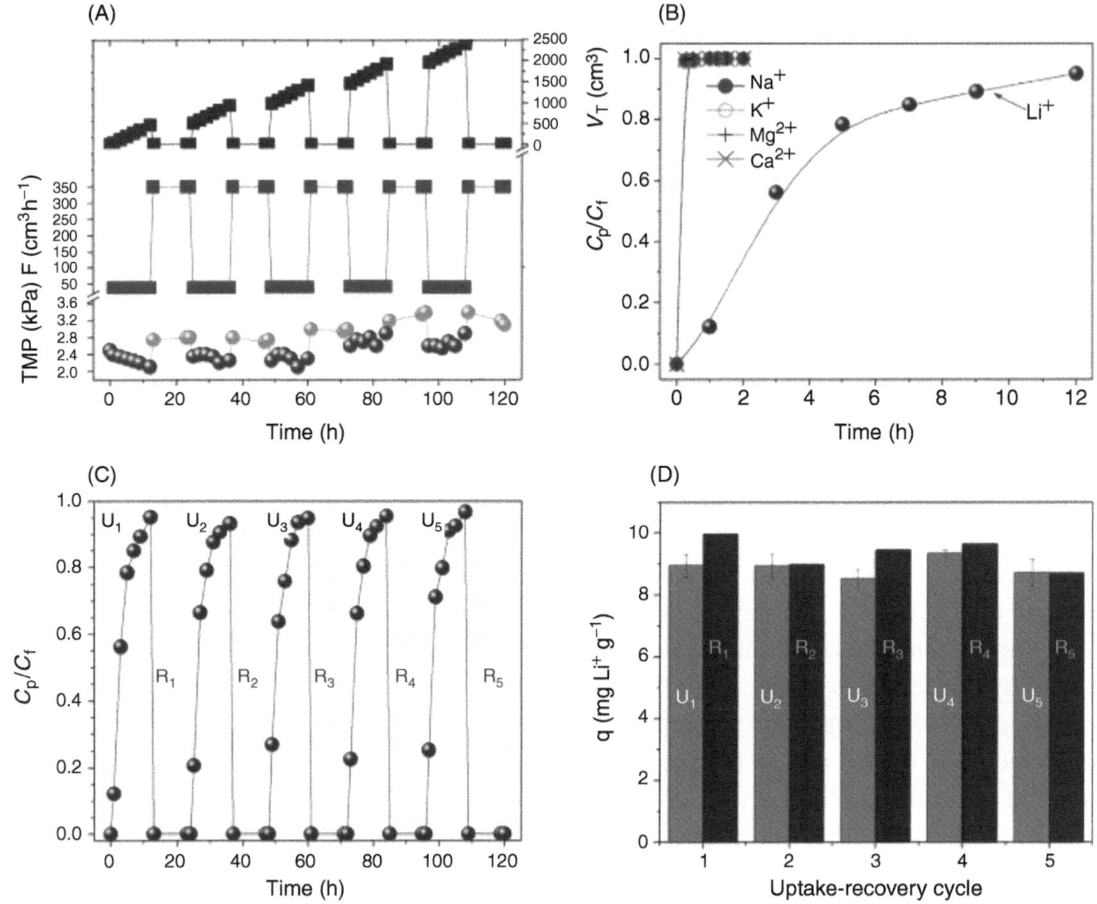

Figure 5.4: Continuous operation (Run 1) of flow-through MMN membrane adsorber for Li^+ recovery from seawater.

(A) Monitored TMP, VT profiles, controlled flow rate of feed (F), or acid leaching solution (B) breakthrough curves of Li^+ and other cations (Na^+, K^+, Mg^{2+}, Ca^{2+}) present at high levels in seawater (U1); (C) breakthrough profiles for five U-R cycles; and (D) Li^+ uptake and recovery using fresh 0.5 M HCl solution with VA¼100 mL in each cycle.

or MMN membrane thickness. The closeness of q*U*, q*R*, and Thomsan rate constant (k_{th}) values from each cycle indicate recommendable consistency and regenerability of the MMN performance.

Electrodialysis (ED) involving bipolar membrane (BPM) was proposed by Hwang et al. in their work using $Li_{1.6}Mn_{1.6}O_4$(LMO) absorbent for enriching Li^+ via adsorption and desorption [14]. The water splitting characteristic of BPM is utilized instead of directly using HCl solution. Hydrogen (H^+) ions are provided to extract Li^+ ions from LMO, and

Li^+ ions are transferred to another section using positive ion exchange membranes (IEMs). In the space between the two IEMs, LMO was filled and reacted with the H^+ ions. The result is the desorption of Li^+ ions from the LMO. Thereafter, the targeted ions transfer to the concentrate compartment through the cation exchange membrane (CEM) driven by the electric potential.

Vacuum membrane distillation (VMD) crystallization was investigated by Quist-Jensen et al. as a potential technology for extracting Li from nonconventional resources (eg, seawater, wastewater) [22]. To precipitate LiCl from single salt solutions, a concentration above 14 M is required. The osmotic effect associated with feed of high salinity has been avoided by using VMD. Additionally, only the application of VMD, among all the MD configurations proposed, achieved supersaturation required for LiCl crystallization. This is mainly because VMD has eliminated temperature polarization (TP) and distillate liquid, hence leading to higher driving force even at high feed concentrations. The resultant LiCl crystal morphology was investigated and it was found that cubic or orthorhombic polymorphic structures depended on the operative conditions. For example, crystal mean diameter and growth rate decrease with increase in flow rate.

5.1.3 Lithium Hydroxide Production

The traditional methods for lithium hydroxide (LiOH) production are precipitation and ED. In Jiang et al.'s work [23], BPM-based ED was investigated for generating high purity LiOH. First, an ordinary ED was employed as pretreatment for concentrating the brine. The brine discharged was precipitated using Na_2CO_3. The Li_2CO_3 solution was evaporated to obtain Li_2CO_3 powder which is then redissolved to get feed solution for the BPM ED.

As shown in Fig. 5.5, the BPI ED membrane stack was composed of two CEMs and two BPMs. The effective area for each membrane is 7.07 cm^2. Both electrodes were made of ruthenium (Ru) coated titanium (Ti). The whole system has two acid and two base compartments. Constant current operation with a density of 20–60 mA cm^{-2} was employed. $LiCO_3$ was circulated through acid compartment. Li^+ ions transport through the CEMs to the base compartment, where it combines with OH^- dissociated inside BPM and evolved at cathode to form LiOH.

The concentration factor (CF) of the ED pretreatment was around 8. The purity of Li_2CO_3 powder with secondary crystallization reached 95.3%. The current efficiency ranges from 91% to 95%, while the energy consumption is between 18 and 42 kW·h kg^{-1}. The BPMs were supplied by Tukuyama Co. Ltd. of Japan, while the CEMs and the anion exchange membranes (AEMs) were provided by Beijing Tingrun Membrane Technology Development Co. Ltd.

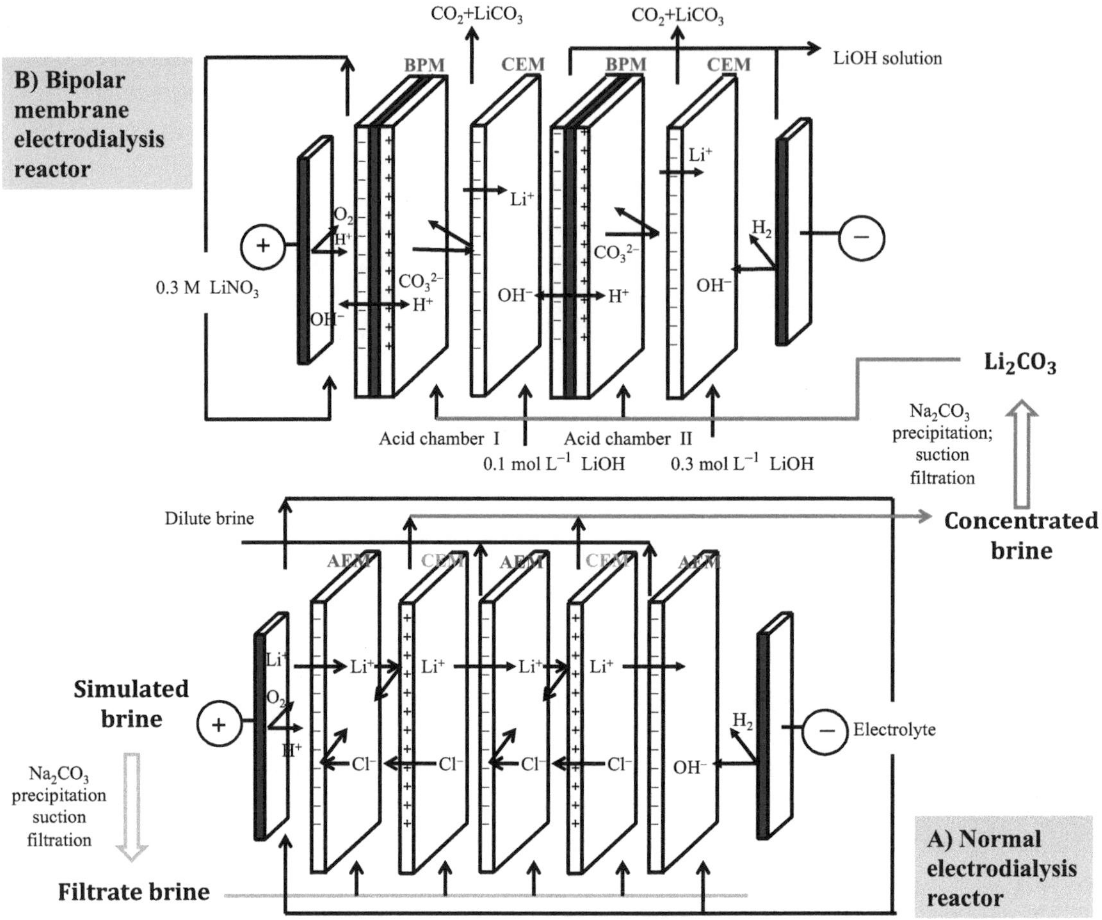

Figure 5.5: Integrated figure of normal electrodialysis membrane reactor and electrolytic-bipolar electrodialysis membrane reactor.

5.1.4 Processing of Powder Material for Lithium Ion Battery

In the metal working process, particularly for ultrafine powder materials, there often exist many impurities. The ultrafine powder synthesized using hydrothermal method for Li ion battery contains 5% ammonium (NH_4^+) salts of sulfuric acid (H_2SO_4) and phosphoric acid (H_3PO_4). The traditional methods, such as plate-and-frame filtration, high-speed centrifugation, are not efficient due to the fine particles. The problems are low interception and high material loss, and difficulty in removing impurities.

A new process demonstrated in Fig. 5.6 using ceramic membranes was introduced in Wang's report [24]. This is a consolidated three-stage filtration and cascade cyclic water reuse design. With the high filtration accuracy of the ceramic membranes, the following advantages were achieved: high rejection of powder materials (>99.99%); high purity (>99.99%) powder;

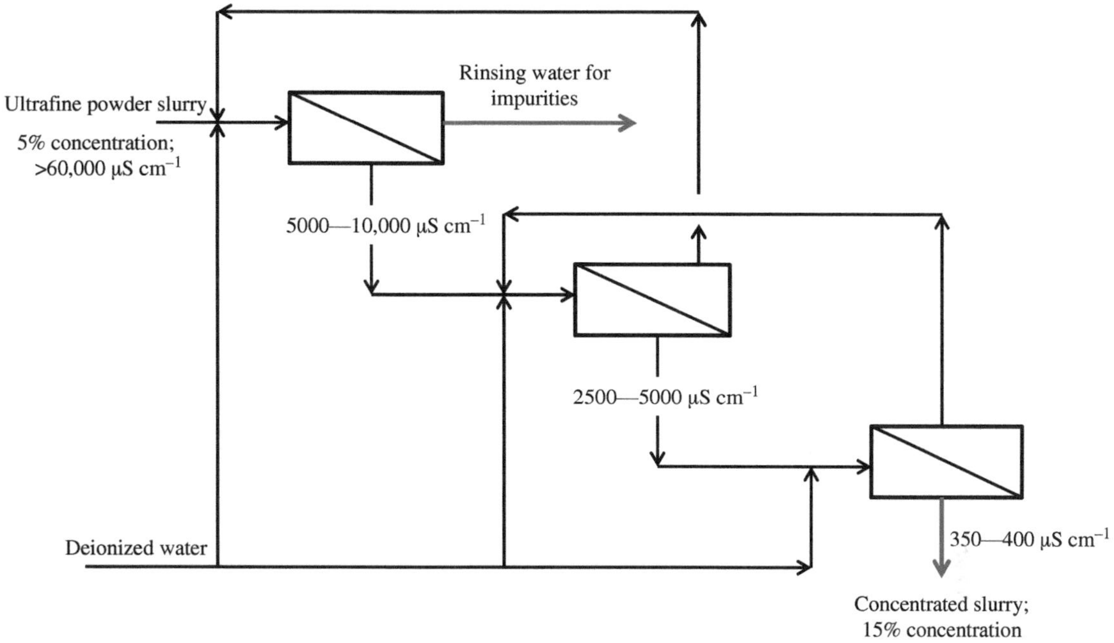

Figure 5.6: The three-stage process flow for washing Li battery ultrafine powder using ceramic membranes and with a design of cascade reuse of rinsing water.

Table 5.2: The effect of washing ultrafine Li-ion battery powder using ceramic membranes

Slurry	Solid Mass Fraction (%)	Conductivity (μS m^{-1})	Impurities (wt.%)	pH
The initial	5	60,000	5	8
After cleaning	15	350-400	0.2	7

save in water. Water consumption was 40–45 m^3 ton^{-1} of product. In addition, the operation of new design is simple and continuous; washing and filtration can be conducted simultaneously. As indicated in the figure, the ion conductivity was reduced from 5000–10,000 to 350–400 μS cm^{-1}; the effect of washing using ceramic membrane is shown in Table 5.2 and the impurity in the product was only 0.2%.

5.2 Aluminum

5.2.1 Introduction

Aluminum (Al) is one of the leading metals, second only to steel in demand by economic development. It has the features of low density, "resistance to corrosion," and good ductility. Al and its alloys have very versatile applications. For example, they are construction materials for airplane, automobile, rocket, etc.; pure Al is used for manufacturing high voltage cable [25].

Bauxite is the main source of Al in the world; the other resources with much less reserve include alunite and urtite [26]. The Al resources that have proven reserve in China are all bauxite, and rank the ninth in the world [27,28]. There are three types of bauxite deposits in China: sedimentary, accumulation, and red clay. The first one is the dominant type.

To obtain elementary Al [29], the classic way requires production of alumina (Al_2O_3) first. There are three approaches for obtaining Al_2O_3: bayer process, sintering process, and the combination of the two. The flowchart for the Bayer process is given in Fig. 5.7. It involves principally ore crushing, slurry grinding, high temperature leaching or digestion, dilute separation and rinsing, seed decomposition, calcination of aluminum hydroxide ($Al(OH)_3$),

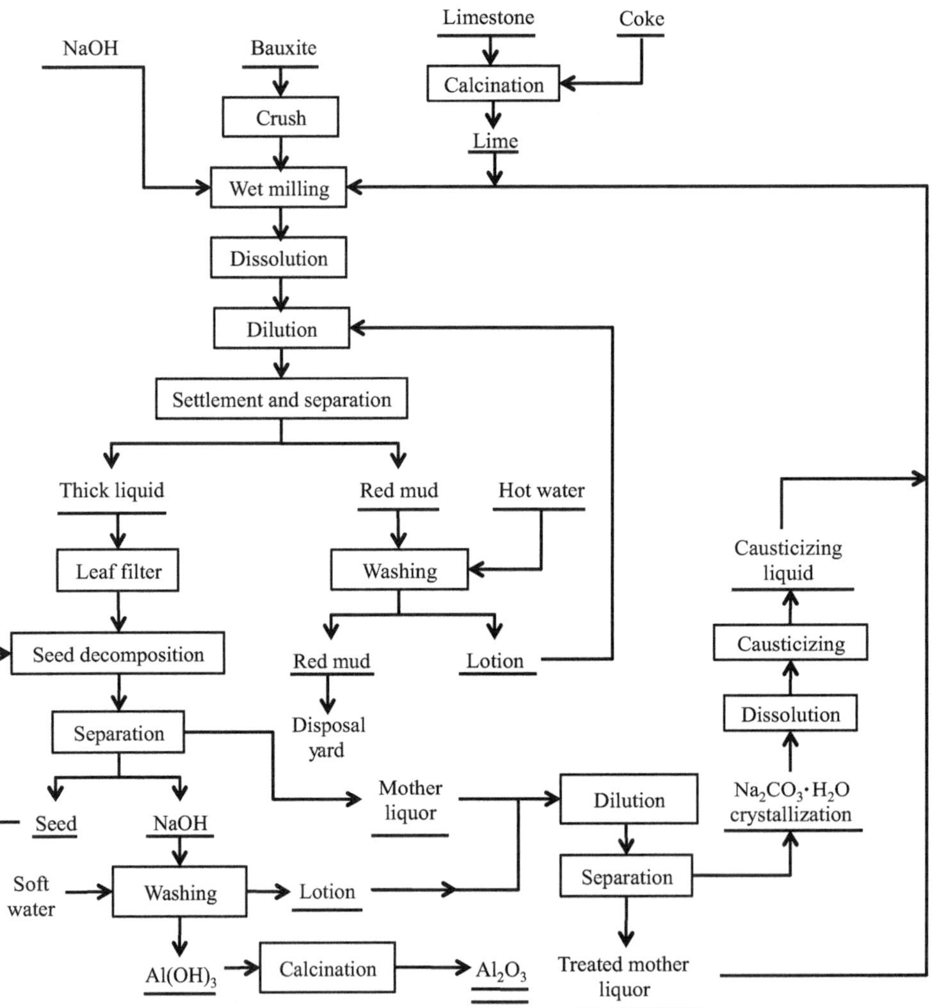

Figure 5.7: Bayer process for alumina production.

mother liquor concentration, and monohydrate soda causticization. The major chemistry in the Bayer process is summarized as follows [30,31]:

$$\text{Hot ore digestion}: Al_2O_3 \cdot 3H_2O + 2NaOH \rightarrow 2NaAl(OH)_4$$

$$Al(OH)_3 \text{ precipitation}: NaAl(OH)_4 + \beta Al(OH)_3(\text{seed}) \rightarrow (\beta+1)Al(OH)_3 \downarrow + NaOH$$

At low temperature, $Al(OH)_3$ is added as a seed into sodium aluminate ($Na_2O{\cdot}Al_2O_3$) solution to precipitate $Al(OH)_3{\cdot}3H_2O$. The mother liquor with a certain mole ratio of (sodium oxide) Na_2O/Al_2O_3 is reused after being treated as shown in Fig. 5.7 for dissolving bauxite at high temperature.

The essence of soda–lime sintering process [29] of which the flow chart is given in Fig. 5.8 is conversion by sintering of mixture with bauxite and a certain amount of soda and lime (or limestone) into grog composed mainly of water soluble $Na_2O{\cdot}Al_2O_3$, sodium citrate ($Na_2O{\cdot}Fe_2O_3$) which is liable to hydrolysis, and di-calcium silicate ($2CaO{\cdot}SiO_2$) which is insoluble in water and alkali. Digestion by water or alkali and follow-up separation delivers *sodium aluminate* ($NaAl(OH)_4$) and NaOH. Then after carbon dioxide (CO_2) carbonation decomposition, $Al(OH)_3$ as precipitant and mother liquor mainly containing Na_2CO_3 are obtained. Finally, calcination of $Al(OH)_3$ produces Al_2O_3, and the mother liquor after being concentrated is recycled to wet grinding.

The cryolite-alumina molten salt electrolysis is the key process for producing Al, which is shown in Fig. 5.9 [29]. The first step includes the melting of electrolyte and maintaining its temperature at 950–970°C using Joule heat generated by direct current in electrolytic tank. Cryolite is instrumental in lowering the melting point. The second is electrolysis of Al_2O_3 and reduction of Al at cathode; simultaneously, oxygen (O_2) evolves at anode, which oxidizes anodic carbon into carbon monoxide (CO) and CO_2. Yet another key step is vacuum extraction of liquid state Al which is then clarified, purified, and poured into ingots.

5.2.2 Modification of Carbonation Process [31]

The seed precipitation of the Bayer process generally takes 48–70 h and requires a large quantity of $Al(OH)_3$ seeds (seed ratios of 1–2). Even so, only up to 50% of $Na_2O{\cdot}Al_2O_3$ in the green liquor (leaching solution) is transformed into $Al(OH)_3$. Therefore, a substantial amount of $Na_2O{\cdot}Al_2O_3$ remain in the recycle of the process, which limits the digestion efficiency.

Carbonation using sodium bicarbonate ($NaHCO_3$) combined with membrane electrolysis for the Bayer process was investigated by Yu et al. [31], as demonstrated in Fig. 5.10. The validity of carbonation by $NaHCO_3$ has already been tested for precipitating $Al(OH)_3$ from synthetic green liquor of the sintering process [32] and the reaction is given as follows:

$$NaAl(OH)_4 + NaHCO_3 \rightarrow Al(OH)_3 \downarrow + Na_2CO_3 + H_2O$$

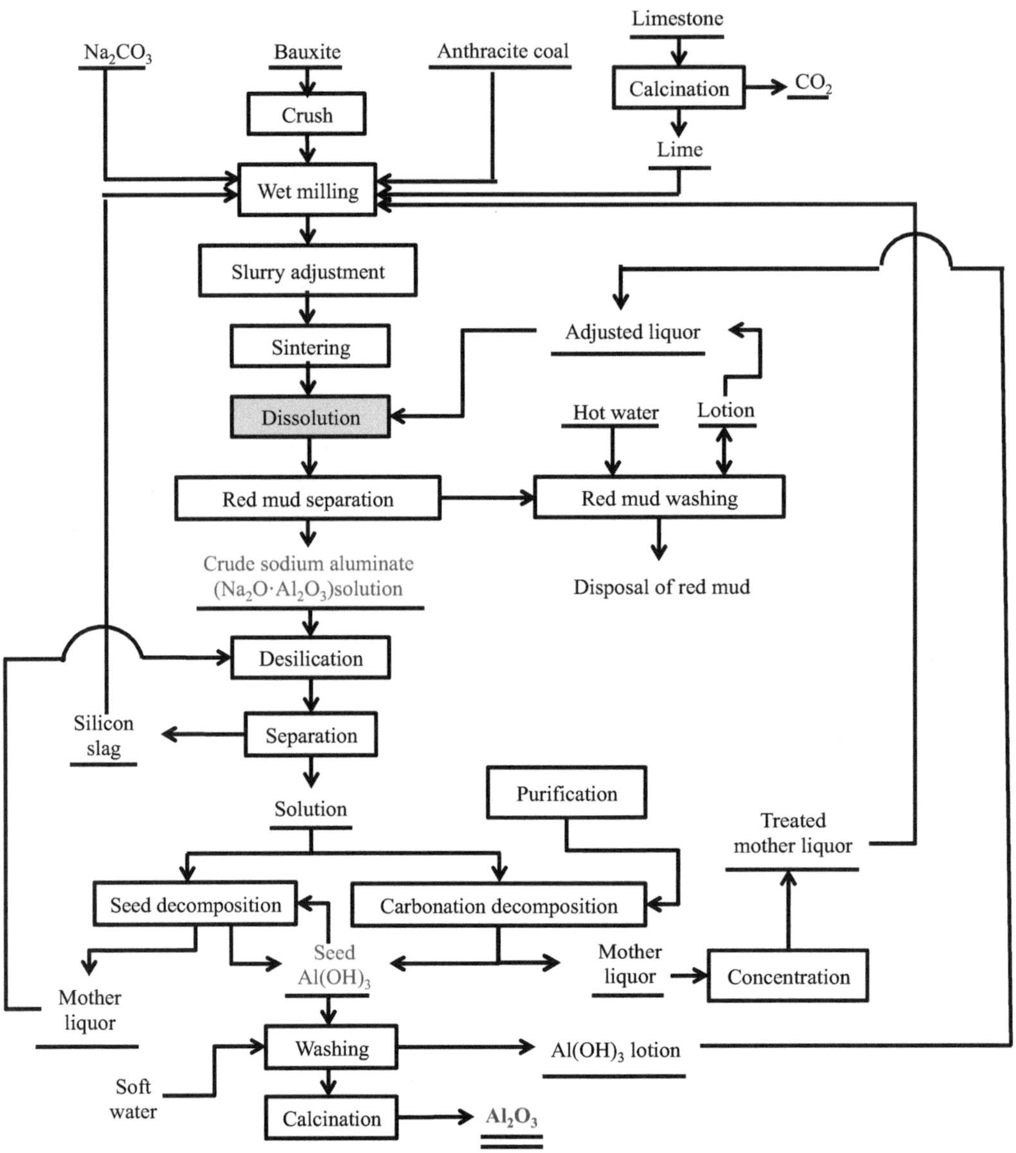

Figure 5.8: Flow diagram of sintering process for alumina production.

The kinetics of $NaHCO_3$ carbonation and precipitation is fast and, this facilitates more thorough precipitation of $Al(OH)_3$. Application of $NaHCO_3$ carbonation in the Bayer process undoubtedly requires the causticization of the Na_2CO_3 liquor resulted for obtaining sufficient digestion capability. The products of the membrane electrolysis are a NaOH solution, $NaHCO_3$ solution, hydrogen (H_2), and oxygen (O_2). The reaction occurring during the electrolysis is as follows:

$$Na_2CO_3 + 2H_2O \rightarrow NaHCO_{3(anode)} + NaOH_{(cathode)} + H_2 \uparrow_{(cathode)} + 1/2\ O_2 \uparrow_{(anode)}$$

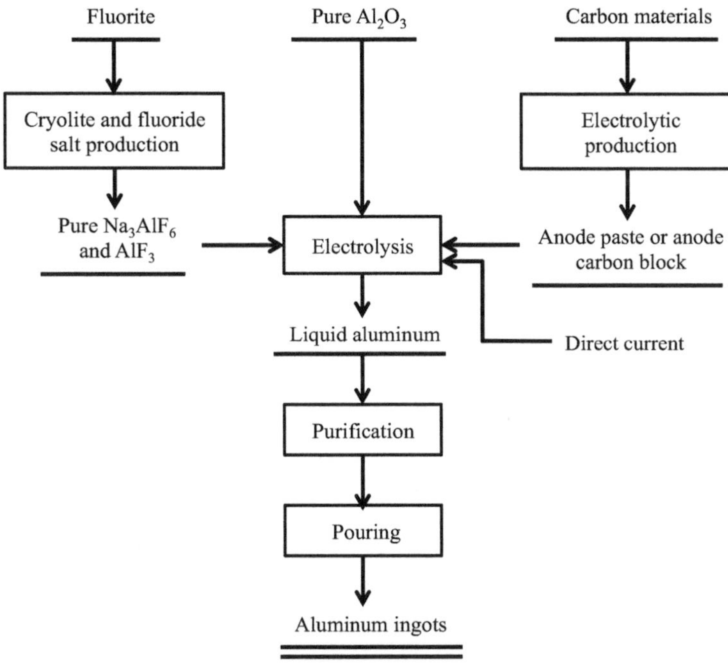

Figure 5.9: The flow chart of cryolite-alumina molten salt electrolysis process.

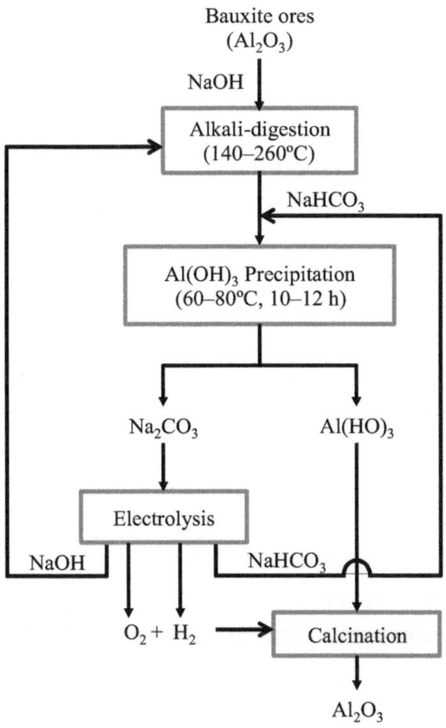

Figure 5.10: Simplified flowchart of modified Bayer process comprising membrane electrolysis for causticization of Na_2CO_3.

The NaOH solution can be used for leaching bauxite ores. H_2 obtained can be used for the calcination of $Al(OH)_3$; O_2 can be used for oxygen-enriched air combustion.

The aforementioned proposed process was demonstrated in a pilot unit with a capacity of 10 kg alumina (Al_2O_3) per day using diaspore ore as a raw material. Despite a substantial saving of energy due to utilization of gaseous products, an independent evaluation still indicated that the costs for the proposed process were about 15% higher than those of the Bayer process. This is attributed to the high electricity charges. The proposed process may be competitive when sufficient and low-cost renewable electricity is available and particularly when additional factors, such as environmental costs and product quality, are considered [31,32].

In the sintering process for Al_2O_3 production, as shown in Fig. 5.6, the carbonation decomposition mother liquor generated in $Al(OH)_3$ precipitation contains Na_2CO_3, some NaOH and small amounts of Al_2O_3 and SiO_2. It is recycled to preparation of raw slurry after concentration which is actualized by evaporation, an energy intensive process. If technology solely relying on latent heat of the mother liquor for evaporation is available, the energy consumption will be significantly reduced. (MD) can efficiently utilize low quality thermal resources and has been investigated [33]. The schematic for the MD system is given in Fig. 5.11. The pH value of the distillate liquid was maintained at 13, indicating that the alkali rejection by the membrane was very high.

5.2.3 Treatment of Red Mud Wastewater

Al_2O_3 production will deliver a large amount of red mud, of which each ton carries 3–4 m^3 wastewater. The composition of the water is very complicated. It mainly comprises Na_2CO_3, NaOH, and $NaAl(OH)_4$. The anions and cations include $Al(OH)_4^-$, OH^-, CO_3^{2-}, SO_4^{2-}, Si_3^{2-}, Cl^-, Na^+, Ca^{2+}, Mg^{2+}, Fe^{2+}, Fe^{3+}, etc.

Liu explored the recovery of alkali and water from the red mud wastewater using IEM technology [34]. The process flow is shown in Fig. 5.12. The red mud liquid is strong alkaline electrolyte, suitable for IEM process. However, some components will form colloidal substances during the IEM separation process. For example, with decreasing Na^+ concentration and alkalinity, $Al(OH)_4^-$ will hydrolyze to produce $Al(OH)_3$ precipitate, which will block membrane surface pores. Another case is regarding Fe^{3+}, which may precipitate near anode. In addition, the suspended matter can easily block passage with diameters from a few angstroms to 20 Å. Therefore, pretreatment by de-alkalization is considered to remove the negative impact from the above components [34], etc. 65–70% Na_2O was transferred from red mud to the solution; 98% Al_2O_3 is removed; the precipitation rate of SiO_4 is 87%. In summary, the liquid becomes diluted but with Na_2O_T (e.g. Na_2O, sodium peroxide (Na_2O_2)) being concentrated by 2.5–3.5 times.

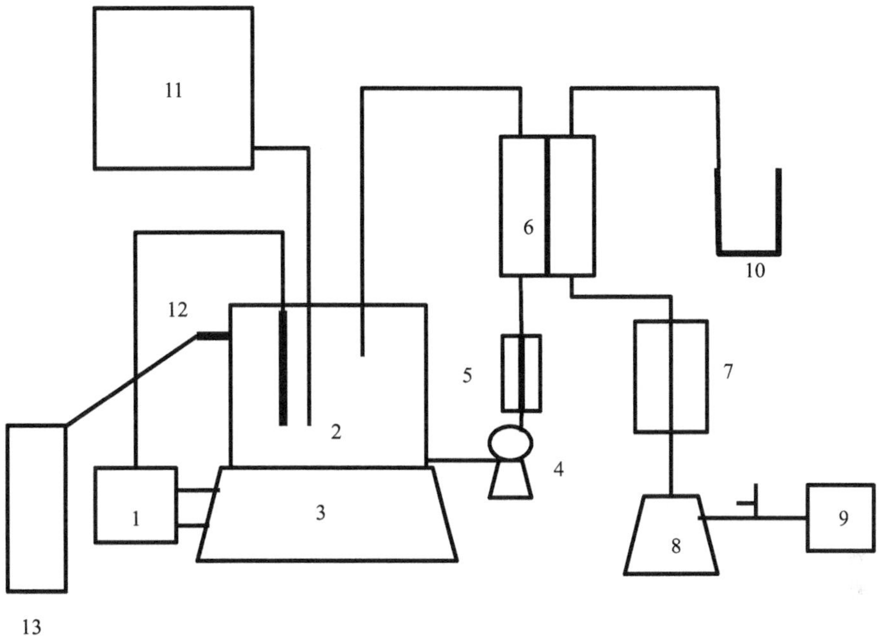

Figure 5.11: Schematic diagram of continuous vacuum membrane distillation device. (1) Temperature controller; (2) circular groove; (3) heater; (4) circulating pump; (5) meter; (6) vacuum distillation membranes; (7) condenser; (8) receive bottles; (9) vacuum pump; (10) pressure gauge; (11) high trough; (12) the overflow mouth; (13) metering collection tank.

ED experiment showed that in the extremely dilute $NaAlO_2$ solution, Al_2O_3 exists in the form of $Al(OH)_4^-$ which can pass through the AEM along with OH^- to the concentrate compartment. Meanwhile, Na^+ passes through CEM and enters the concentrate compartment. With increasing NaOH concentration and caustic ratio, the Al_2O_3 becomes more stable. In contrast, the $Al(OH)_4^-$ in the fresh water chamber may hydrolyze, as Na^+ and OH^- are gradually depleted. Therefore, the pH value of the diluent must be controlled carefully.

Table 5.3 shows the analysis results for liquids from concentrate and diluent chambers. It is found that concentrated alkali solution with 73.9 g L^{-1} of Na_2O and diluted water with 0.45 g L^{-1} of Na_2O could be obtained. The latter can be reused as process water. The alkali recovery is greater than 83.7% and the water recovery is higher than 85.8%.

A pilot line has been installed and semiindustrial experiment was completed. A problem observed in the long-term (3–3.5 months) test is that the AEM or CEM showed severe deterioration in ion exchange capacity after being used in the treatment of practical red mud liquid. Besides, calcite type calcium carbonate ($CaCO_3$) deposited on membrane surface.

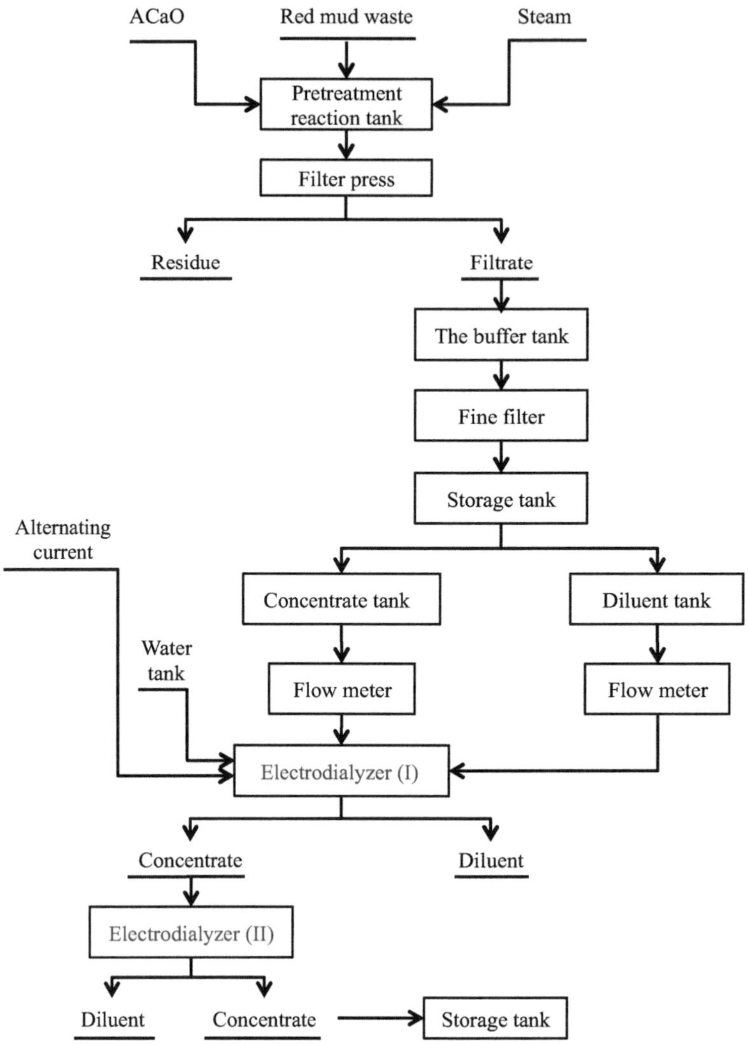

Figure 5.12: Recovery alkali from red mud waste using membrane technology.

Table 5.3: The experimental material liquid analysis results

	Composition (mg L^{-1})					
	Na_2O_T	SiO_2	Fe_2O_3	Al_2O_3	Na_2OC	SO_4^{2-}
Raw liquid	7020	65	32	3930	820	1250
Refined liquid	6070	80	<0.05	360	1280	510
Concentrate	73,850	160	40	600	4900	7940
Diluent	450	6	2	200	60	41

Table 5.4: The composition of electrodialysis solution (g L^{-1})

ID	d_{NT}	d_{NH}	Al_2O_3	SiO_2 ($\times10^{-4}$)	Fe ($\times10^{-4}$)	SO_4^{2-}	Cl^-	Ca^{2+} ($\times10^{-3}$)	Mg^{2+} ($\times10^{-4}$)
1	9.30	1.80	0.11	9.9	2.10	0.69	0.048	0.9	1.2
2	10.42	1.40	0.32	5.9	3.90	0.63	0.053	1.68	1.2
3	9.49	1.68		7.4	1.40	0.49	0.051		
4	9.89	1.38			0.75	0.60	0.064		

d_{NT} is the total alkalinity of the solution calculated based on Na_2O.
d_{NH} is the concentration of $NaHCO_3$ of the solution calculated based on Na_2O.

Table 5.5: The desalination index comparison of sodium carbonate and sodium bicarbonate material liquid

The Membrane Stack Voltage (V)	Liquid Composition	Liquid Concentration (ρ/g L^{-1})	The End Concentration of Diluent Chamber (ρ/g L^{-1})	W $(kW.h)^{-1}$	η (%)	f (%)
24.5	Na_2CO_3	11.02	0.43	0.461	67.8	96
24.5	$NaHCO_3$	11.25	1.12	0.607	52.7	90
18.5	Na_2CO_3	10.07	0.60	0.360	70.1	94
18.5	$NaHCO_3$	12.37	1.37	0.655	46.8	88.9

η is the current efficiency of diluent chamber and f is the total desalination rate of process.

Zhang et al. used a combined carbonation and ED to remove and recover Al, alkali, and water resources [35]. The samples were provided by Guizhou Aluminum Plant and the composition is shown in Table 5.4. The major content is alkaline (12 g L^{-1} of Na_2O equivalent) and Al (8 g L^{-1} of Al_2O_3 equivalent). Undoubtedly, this situation causes resource waste and environmental pollution upon leakage without treatment.

Table 5.5 compares the desalination index of Na_2CO_3 and $NaHCO_3$ in treating the wastewater. As seen, the removing of Na_2CO_3 is more efficient than that of $NaHCO_3$. Therefore, in the experiments, the reduction of both Al and $NaHCO_3$ contents should be taken into account. According to the test results, the terminus of carbon decomposing step should be set at 1–1.5 g L^{-1} of $NaHCO_3$ (in the amount of Na_2O).

The influence of feed water concentration and flow rate on the energy consumption is studied. As we can see, under the same level of feed concentration and at the limiting current, the desalination rate changed little at the limiting current flow. This is because the concentration affects more strongly the limiting current than the flow rate. For the more concentrated solution with increasing the flow rate, the limiting current increased significantly; as the desalination rate was a linear function of the current, the negative effect of flow rate enhancement on the desalination rate was negated.

Table 5.6 is the comparison of a single-stage/single-pass and a dual-stage/dual-pass processes. Although the unit energy consumption of the single-stage/single-pass design is slightly lower,

Table 5.6: The comparison of two-stage-two-process and one-stage-one-process

ID	ρ_{di}/g L^{-1}	Q (m^3 h^{-1})	I_1	I_2	f (%)	η (%)	W(kW.h)$^{-1}$
1	6.5	3.2	159	118	65.4	84.9	1.55
			138		32.6	85.9	1.31
2	6.5	3.2	129	95	62.3	87.5	1.13
			112		29.7	83.4	1.04
3	2.4	3.2	52	42	60.1	82.4	1.19
			51		31.1	78.7	1.31
4	2.4	3.2	39	40	47.8	78.1	0.87
			39		22.2	73.1	0.73

ρ_{di} is the feed concentration of diluent chamber, Q is the flow of diluent chamber, I_1 is the current of the first stage, and I_2 is the current of the second stage.

the desalination rate of the latter is higher by about 50%. Therefore, the desired number of ED units can be reduced, which in turn reduces the investment cost. In this regard, single-stage/single-pass process is a better choice when the water to be treated is in huge amount, whereas when treating a small quantity of water, design dual-stage/dual-pass process should be chosen.

5.2.4 α-Alumina Powder Washing

Washing of α-Al_2O_3 powder by separation employing ceramic membranes was reported by Wang [24]. The raw material in slurry state contains up to 3 wt.% of Na_2O. The particle size estimated based on particle number distribution is 0.7 μm, which makes it hard to hold up particles and efficiently remove of Na_2O using traditional methods, such as centrifuge. After washing the slurry 11–15 times using the system relying on ceramic membranes, the ion conductivity of the feed solution decreased from 5700–53,000 to 250–280 μS cm^{-1}. The purified slurry had only 0.02–0.28 wt.% of Na_2O. The contents of other impurities, ferric oxide (Fe_2O_3) and silica (SiO_2), were both <0.01 wt.%.

5.2.5 Treatment of Wastewater From Aluminum Electrolysis Plant

The wastewater in electrolytic Al plant is a mixture from several resources: (1) industrial wastewater from reserve tank and carbon manufacturing process, (2) rainwater, and (3) domestic wastewater. Table 5.7 displays the quality of the integrated wastewater. If water recycle and reuse in electrolytic Al production is required, total dissolved solids (TDS) ≤20 mg L^{-1} and pH in the range of 6–9 are to be achieved. This necessitates the desalination, which has not been incorporated into large-scale wastewater treatment in Al manufacturing plant at present.

Table 5.7: Raw water quality

Items	Values
COD (mg L^{-1})	25–50
Fluoride (mg L^{-1})	3–15
Oil (mg L^{-1})	5–15
Conductivity (μS cm^{-1})	500–3000
Suspended solids (SS) (mg L^{-1})	50–200
Volatile phenol (mg L^{-1})	0.2–0.5
pH	6–9
TDS (mg L^{-1})	250–1500

Addressing the feed water quality and requirements for reuse, Zhong et al. employed reverse osmosis (RO) for desalination [36]. The membrane selected is antifouling PA membrane provided by VONTRON Company. The pretreatment of RO is composed of sewage regulation pool, integrated biochemical pool, flotation, filtration tower (sand tank), absorption tower (activated carbon filter), and fiber ball filter. Its design in terms of technical specifications and device quantities was to ensure that the silting density index (SDI) of the feed to the RO was <5.

The first stage project has a total treatment capacity of 300 m^3 h^{-1}, which is shared equally by three systems. The practical operation shows that the final water product of the overall system has a TDS ≤20 and a pH value of 6–9. The water recovery rate is around 93%. The retentate or concentrate is transferred to processing system outside the plant. Installation of the system is expected to save 6700 m^3 new water intake per day and 2.44 million m^3 per annum. The cost for producing each ton of water is roughly (CNY) 5.0, based on the major cost elements such as fixed assets discount, energy consumption, chemical agents, and manpower.

5.3 Conclusions

In this chapter, the applications of membrane separation technologies in Li and Al metallurgy are introduced. For Li metallurgy, the sector that attracts most attention regarding using membrane separation is the extraction of Li ions from various liquid resources, such as salt lake brine, seawater, waste water, etc. The most challenging issue with these resources is the high content of impurities. The membrane processes that have been considered and also present a certain level of Li^+ enrichment include NF, ED, MD, and (MA). For Al, membrane separation has been united with other methods (eg, chemical treatment) to improve traditional hydrometallurgical process efficiency. The research about using ED in red mud liquid treatment has reported the problem of membrane degradation and fouling.

Glossary and Symbols

AEM	Anion exchange membrane
Al	Aluminum
Al_2O_3	Alumina
$Al(OH)_3$	Aluminum hydroxide
BPM	Bipolar membrane
Br	Bromine
C_f	Feed concentration (mg/L)
C_p	Permeate concentration (mg/L)
$CaCO_3$	Calcium carbonate
$2CaO{\cdot}SiO_2$	Dicalcium silicate
CEM	Cation exchange membrane
CF	Concentration factor
CO	Carbon monoxide
CO_2	Carbon dioxide
DAPP	1,4-Bis(3-aminopropyl)piperazine
ED	Eectrodialysis
EPMA-EDS	Electron probe micro-analyzer-energy dispersive X-ray spectrometer
Fe^{2+}	Ferrous ion
Fe^{3+}	Ferric ion
$FePO_4$	Iron phosphate
H^+	hns
H_2	Hydrogen
HCl	Hydrochloric acid
H_3PO_4	Phosphoric acid
H_2SO_4	Sulfuric acid
I	Iodine
IEMs	Ion exchange membranes
K	Potassium
k_{th}	Thomsan rate constant
KCl	Potassium chloride
K_2SO_4	Potassium sulfate
Li	Lithium
Li_2CO_3	Lithium carbonate
LiCl	Lithium chloride
$LiFeCl_4$	Lithium iron chloride complex
$LiFePO_4$	Lithium iron phosphate
$LiMnO_2$	Lithium manganate
LiOH	Lithium hydroxide
LIS	Li^+ Ion sieves
LMO	$Li_{1.6}Mn_{1.6}O_4$
MA	Membrane adsorber
MF	Microfiltration
Mg	Magnesium
$MgCl_2$	Magnesium chloride
MMN	Mixed matrix nanofiber
Mn	Manganese
MO	$H_{1.6}Mn_{1.6}O_4$
Na	Sodium
Na_2CO_3	Sodium carbonate

$NaAlO_2$; $NaAl(OH)_4$; $Na_2O{\cdot}Al_2O_3$	Sodium aluminate
$NaHCO_3$	Sodium bicarbonate
Na_2O	Sodium oxide
Na_2O_2	Sodium peroxide
$Na_2O{\cdot}Fe_2O_3$	Sodium citrate
NaOH	Sodium hydroxide
Na_2SO_4	Sodium sulfate
NF	Nanofiltration
NH_4^+	Ammonium
O_2	Oxygen
PA	Polyamide
PAN	Polyacrylonitrile
PSF	Polysulfone
qR	Li recovery capacity(mg/g)
qU	Li uptake capacity(mg/g)
RO	Reverse osmosis
Ru	Ruthenium
SDI	Silting density index
SF	Separation factors
SiO_2	Silica
TBP	Tributyl phosphate
TDS	Total dissolved solids
Ti	Titanium
TMP	Transmembrane pressure
TP	Temperature polarization
UF	Ultrafiltration
VMD	Vacuum membrane distillation

References

[1] D.Z. Lin, Use of lithium and its resource exploitation, Chin. Saf. Sci. 14 (9) (2004) 72–77.

[2] U.S. Geological Survey. Mineral Commodity Summaries, 2014.

[3] U.S. Geological Survey. Mineral Commodity Summaries, 2008.

[4] U.S. Geological Survey. Mineral Commodity Summaries, 2009.

[5] U.S. Geological Survey. Mineral Commodity Summaries, 2010.

[6] U.S. Geological Survey. Mineral Yearbook, 2006.

[7] X.P. Wang, X.X. Cai, W.J. Cui, Exploitation and utilization of global lithium resources: trends and our responses, Chin. Mining Mag. 23 (6) (2014) 10–13.

[8] F. Gao, J.P. Zheng, Z. Nie, J.H. liu, P.S. Song, Brine lithium resources in salt lake and advances in its exploitation, Acta Geoscientica Sin. 32 (4) (2011) 484–492.

[9] H. Zhong, Y.F. Zhou, H.A. Yin, Progress in technology for extracting lithium from Li-bearing brine resources, Multipurpose Util. Miner. Resour. (1) (2003) 23–28.

[10] W.D. Wang, X. Cao, Production process and current situation of lithium carbonate extraction from salt lake brine in China, J. Salt Lake Res. 18 (4) (2010) 52–60.

[11] N. Chen, H. Zhong, H. Yan, Present situation of the process and technique of lithium recovery from brine around the world, J. Salt Chem. Ind. 43 (3) (2014) 1–4.

[12] Z.J. Yang, L. Xiang, Progress on the extraction of lithium from the salt lake brine, J. Salt Chem. Ind. 34 (6) (2005) 27–29.

[13] X.P. Wang, X.X. Cai, W.J. Cui, Exploitation and utilization of global lithium resources: trends and our responses, Chin. Mining Mag. 23 (6) (2014) 10–13.

[14] C.W. Hwang, M.H. Jeong, Y.J. Kim, W.K. Son, K.S. Kang, C.S. Lee, T.S. Hwang, Process design for lithium recovery using bipolar membrane electrodialysis system, Sep. Purif. Technol. 166 (2016) 34–40.
[15] X.L. Di, Q.S. Pang, Q. Li, Comparative analysis of productive technology for metallic lithium, J. Salt Lake Res. 13 (2) (2005) 45–52.
[16] W.Q. Kang, L.J. Shi, Y.J. Zhao, D.Y. Zhang, H.T. Zhang, M. Wang, Preliminary test of separation of Mg^{2+}/Li^+ in salt lake brine by nanofiltration, Inorganic Chem. Ind. 46 (12) (2014) 22–24.
[17] X.H. Li, C.J. Zhang, S.N. Zhang, J.X. Li, B.Q. He, Z.Y. Cui, Preparation and characterization of positively charged polyamide composite nanofiltration hollow fiber membrane for lithium and magnesium separation, Desalination 369 (2015) 26–36.
[18] X.H. Liu, X.Y. Chen, L.H. He, Z.W. Zhao, Study on extraction of lithium from salt lake brine by membrane electrolysis, Desalination 376 (2015) 35–40.
[19] X.H. Liu, X.Y. Chen, Z.W. Zhao, X.X. Liang, Effect of Na^+ on Li extraction from brine using $LiFePO_4/FePO_4$ electrodes, Hydrometallurgy 146 (3) (2014) 24–28.
[20] Z.W. Zhao, X.F. Si, X.H. Liu, L.H. He, X.X. Liang, Li extraction from high Mg/Li ratio brine with $LiFePO_4/FePO_4$ as electrode materials, Hydrometallurgy 133 (2) (2013) 75–83.
[21] M.J. Park, G.M. Nisola, E.L. Vivas, L.A. Limjuco, C.P. Lawagon, J.G. Seo, H. Kim, H.K. Shon, W.J. Chung, Mixed matrix nanofiber as a flow-through membrane adsorber for continuous Li^+ recovery from seawater, J. Membr. Sci. 510 (2016) 141–154.
[22] C.A. Quist-Jensen, A. Ali, S. Mondal, F. Macedonio, E. Drioli, A study of membrane distillation and crystallization for lithium recovery from high-concentrated aqueous solutions, J. Membr. Sci. 505 (2016) 167–173.
[23] C.X. Jiang, Y.M. Wang, T.W. Xu, The study of producing of lithium hydroxide by electrodialysis method, The Fourth Symposium on the Application of Membrane-Based Separation in Metallurgical Industry, 2014, pp. 72–76.
[24] Z.G. Wang, Application membrane separation technology in hydrometallurgy, The Fourth Symposium on the Application of Membrane-Based Separation in Metallurgical Industry, 2014, pp. 54–61.
[25] A.G. Yuan, S.L. Sun, J. Li, Geological characteristics and exploitation of bauxite Situation in Henan, Resour. Ind. 11 (4) (2009) 92–95.
[26] S.F. Liao, R.T. Liang, Chinese Bauxite Geology, Guizhou Science and Technology Press, Guiyang.
[27] Z.G. Wang, Status situation and sustainable development of aluminum mineral resources in China, China Mining Mag. 11 (5) (2002) 28–30.
[28] USGS Mineral Commodity Summaries. Available from: http://minerals.Usgs.Gov/minerals/pubs/commodity/aluminum/index.html#mcs 2014-1-16.
[29] X.P. Zhang, Introduction of Metallurgical Engineering, Central South University Press, Changsha, (1998).
[30] Y.L. Zhang, Y. Tang, Q. Qiu, Y.M. Chen, Y.Z. Sun, P.Y. Wan, X.J. Yang, Electrochemically enhanced adsorption of aluminum from sodium carbonate solution by activated carbon fibers, Ind. Eng. Chem. Res. 52 (40) (2013) 14449–14455.
[31] Z. Yu, Y. Chen, Y. Niu, Y. Tang, P. Wan, Z. Lv, X.Y. Yang, Efficient and sustainable production of alumina by electrolysis of sodium carbonate, Angew. Chem. Int. Ed. 50 (49) (2011) 11719–11723.
[32] Y. Li, Y. Zhang, C. Yang, Y. Zhang, Precipitating sandy aluminium hydroxide from sodium aluminate solution by the neutralization of sodium bicarbonate, Hydrometallurgy 98 (1–2) (2009) 52–57.
[33] Q.X. Zhang, Metallurgy Separation Science and Engineering, Science Press, Beijing, (2004) pp. 318-319.
[34] Z.L. Liu, Y.C. Luo, P. Tan, Waste liquor of red mud by ion—exchange membrane separation and recovery of alkali, Light Metals (4) (1997) 10–14.
[35] Q.X. Zhang, A.P. Luo, C.G. Shi, Q.C. Zeng, Industrial test research on red mud backwash of alumina factory by membrane technology (I), J. Central South Univ. Technol. 28 (2) (1997) 134–137.
[36] W.L. Zhong, Y. Jin, Y. Xiang, Y.S. Huang, F. Liu, Z.C. Wu, S.M. Liang, The application of high recovery RO system in the large-scale project of aluminum wastewater reuse, The Fourth Symposium on Application of Membrane-Based Separation in Metallurgical Industry, 2014, pp. 153–156.

CHAPTER 6

Refractory Metals

L.-S. Xiao

School of Metallurgy and Environment, Central South University, Changsha, Hunan Province, China

6.1 Molybdenum

6.1.1 Introduction

Naturally occurring molybdenum (Mo) is principally found as various metal sulfide or oxide compounds, existing in either single or paragenic ores. Its reserve in the earth is very low. The major source for Mo production in China and many other countries is molybdenite (MoS_2). The global distribution of Mo resources is in an extremely imbalanced state. They are concentrated in the United States, China, Chile, Canada, and Russia, of which the total Mo reserves account for about 88% of the world's total and ultimate production 90% of the global output [1–3]. Specifically, the total reserve and the product in China are the largest among these five countries.

Mo products include primary products, chemical products, metal products, and deep processing products. They are generated at different steps of the Mo metallurgy, of which a schematic for the simplified overall process is shown in Fig. 6.1 [4]. These various forms of products in pure metals, alloys, compounds, etc. have been widely used in many industries. The one-off consumption of Mo is mainly associated with steel industries, where the Mo compounds (eg, ferro-molybdenum, abbreviated as Fe–Mo) are used as additive or alloy agents to enhance the properties such as strength, hardness, and ductility of metal materials. About 80% of world Mo production is consumed by steel alloys. In recent years, the secondary consumption of Mo received attention and took up increasing proportion in its overall demand. These applications include fabrication of solar panels, aircraft engine, liquid crystal display (LCD), armor or shield of nuclear fusion reactor, etc. [1–5].

At present, the Mo consumption structure and the share for each category are somewhat similar between developed and developing countries. In western developed countries the demand from steel industry is around 75% of the total. The applications as catalyst, pure Mo, super alloy, and other Mo chemicals put forward Mo demand about 8, 6, 4, and 7% of the total, respectively [5]. In China the consumption of Mo generated by steel industry, Mo metal, super alloy such as nickel (Ni)-based alloy with 18% of Mo, tungsten (W)–Mo alloy, Mo–rhenium (Re) alloy, Mo chemicals, such as molybdenum trioxide (MoO_3) and molybdate

Membrane-Based Separations in Metallurgy
http://dx.doi.org/10.1016/B978-0-12-803410-1.00006-2

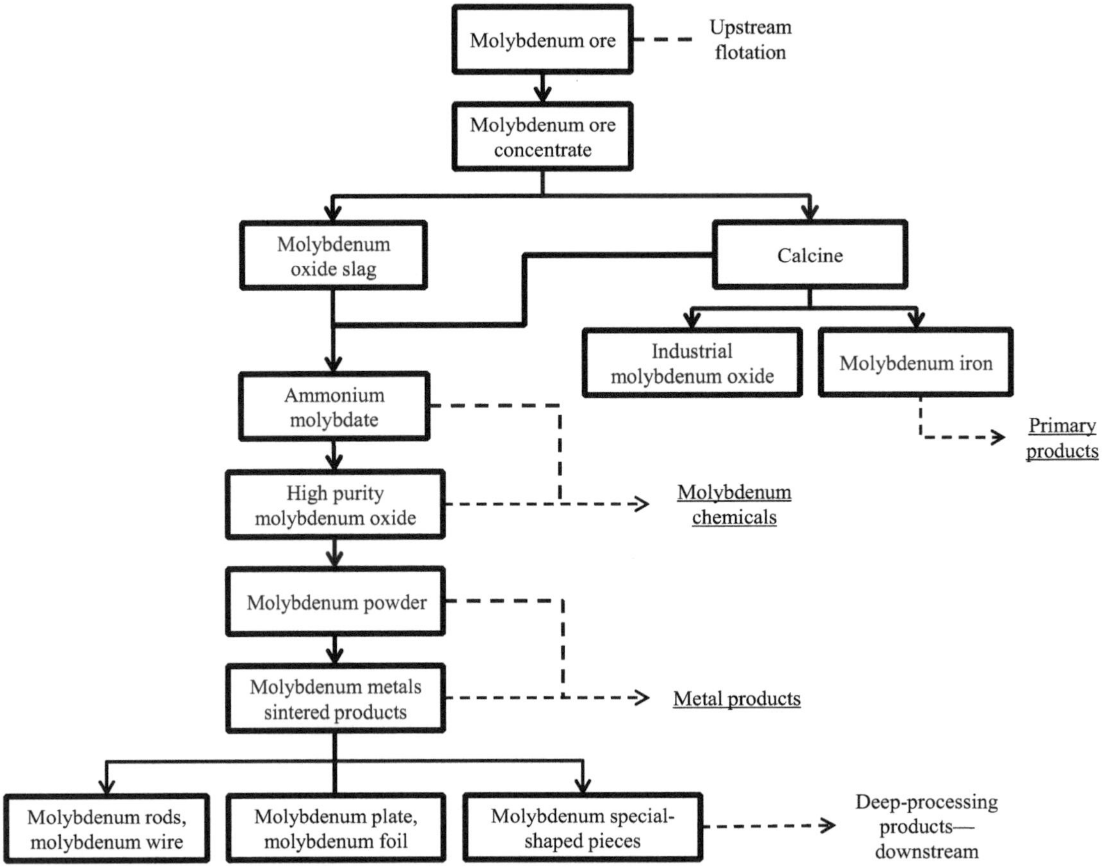

Figure 6.1: Flow diagram of a general molybdenum metallurgical process.

(chemical formula: M indicate metal; MO indicates acid radical) used as catalysts in chemical and petroleum industries, MoS_2 as lubricated in aerospace and mechanical industries, and the other applications are 75, 5, 3, 14, and 3%, respectively.

Depending on the final product, either pyro- or hydro-metallurgy, or their combination is employed in Mo smelting. Fe–Mo alloy widely used in steelmaking as mentioned previously is produced [4,6] by either "block method" using carbon as a reducer in electric furnace smelting, or metal hot reduction furnace method, which uses the silicon (Si) and aluminum (Al) as reducing agents. The second approach characterized by energy efficiency is more widely applied. A typical Fe–Mo production process is shown in Fig. 6.2 [4]. The raw materials for this Fe–Mo smelting comprise industrial MoO_3, reducing agent of Si and Al, flux of fluorite, heat repair agent of saltpeter and steel scrap for adjusting iron alloy grade [4].

Ammonium molybdate ($(NH_4)_2MoO_4$) is a very important intermediate product in Mo smelting process. It is mainly used as a basic raw material for producing high purity Mo

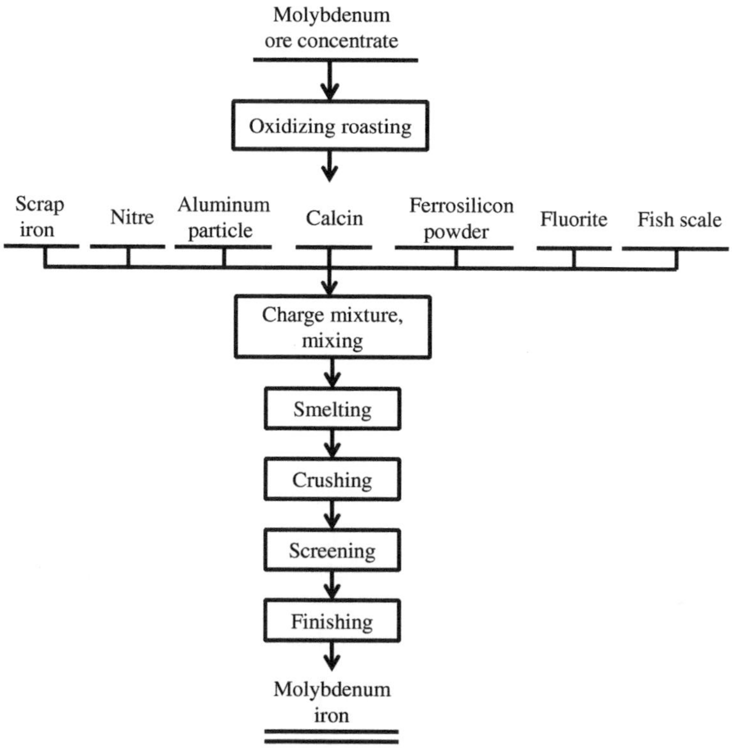

Figure 6.2: Production process of ferro-molybdenum (Fe–Mo) alloy.

products (eg, MoO_3) and other chemicals, for example, catalyst, pigments. The production process of $(NH_4)_2MoO_4$ is shown in Fig. 6.3 [6]. The oxidizing roasting approach generates sulfur dioxide (SO_2) which can pollute environment. With addition of lime in the oxidizing roasting step, the sulfur (S) in the raw material is converted into gypsum instead of SO_2. The second method adapts conveniently to processing low grade Mo concentrate, but has higher production cost [4].

MoS_2 component of raw material reacts rapidly inside heated nitric acid (HNO_3) with concentration in the range of 25–50% to be transformed into molybdic acid (H_2MoO_4) in the state of slag [4]. Abandoning the roasting step gets rid of the air pollution due to SO_2. Similarly, acid/alkali leaching under hyperbaric oxygen atmosphere makes the roasting of concentrate unnecessary. The advantages associated with this method are that it has high metal recovery and achieves comprehensive utilization of the multinary components (eg, Re) in raw material. However, it demands equipment with high pressure resistance and high corrosion resistance. For example, the autoclave uses titanium lining, which renders high cost of investment. In general, it is not easy for small and medium-sized manufacturers to adopt this technology [4].

The properties $(NH_4)_2Mo_7O_4$ determine the purity and quality of subsequent Mo products [7,8]. Therefore, calcine obtained by roasting MoS_2 concentrate and filtration cake of

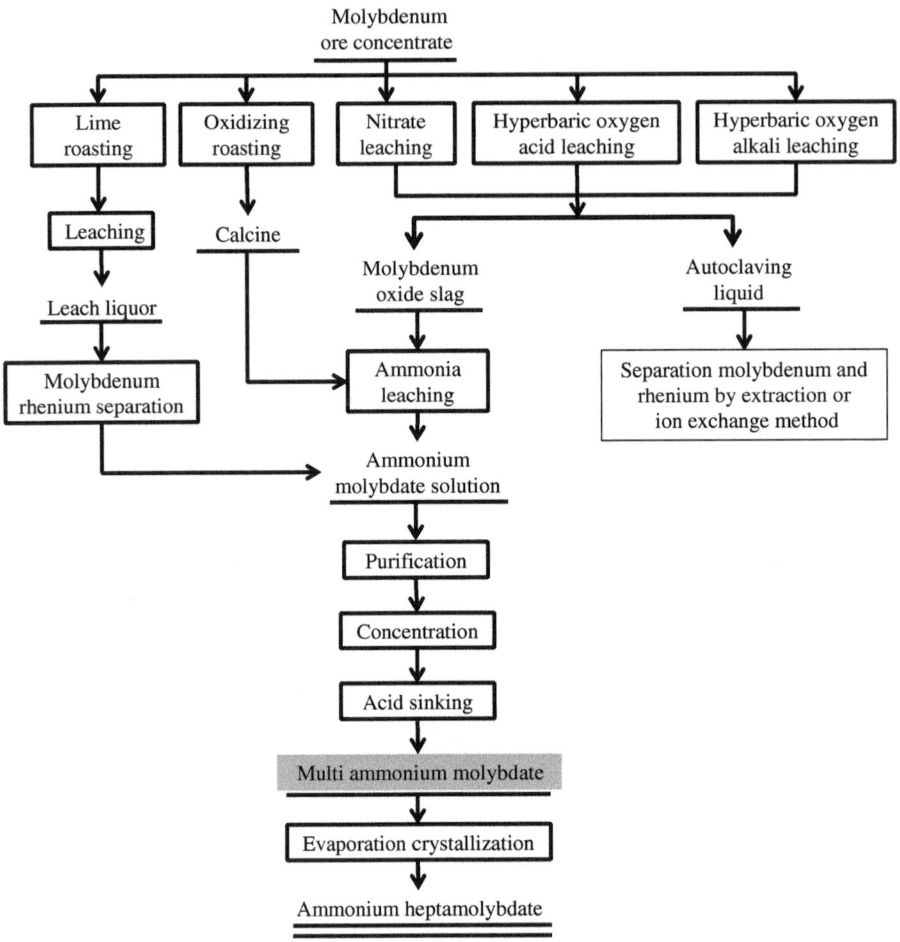

Figure 6.3: Flow chart of production process for ammonium molybdate ($(NH_4)_2[Mo_7O_4]$) and ammonium para-molybdate ($(NH_4)_6[Mo_7O_{24}]$).

H_2MoO_4 generated in oxidizing acid/alkali leaching process need purification. The classical method is the combination of ammonia leaching and acid precipitation, which is adopted by most manufacturers in China currently. It ensures stable product quality, nevertheless, renders the problem of delivering a lot of "three wastes."

The industrial production of Mo powder employs generally the calcinations of ammonium para-molybdate ($(NH_4)_6Mo_7O_{24}{\cdot}4H_2O$) into molybdenum trioxide (MoO_3). Thereafter, hydrogen (H_2) is used to reduce MoO_3 into elementary Mo. Direct reduction from ($(NH_4)_6Mo_7O_{24}{\cdot}4H_2O$) by H_2 can also produce Mo powder. However, the complicated atmosphere containing NH_3, water vapor, etc. in reduction process will give rise to serious corrosion of equipment and hence contamination of products. As a result, it is not favored by plants having stringent requirements on product quality [4].

6.1.2 Concentration of Ammonium Molybdate Crystallization Mother Liquor

Wang et al. investigated the preconcentration of mother liquor delivered in $(NH_4)_2MoO_4$ acid precipitation crystallization using electrodialysis (ED) [9]. The flow chart for the process is given in Fig. 6.4.

The feed solution containing 160 g L^{-1} ammonium nitrate (NH_4NO_3) is the mother liquor generated by precipitating $(NH_4)_2MoO_4$ using HNO_3. It has been treated by ion exchange to reclaim Mo before being sent to ED step. The fresh water collected in the ED system was recycled to MoS_2 metallurgical process shown in Fig. 6.3 for $(NH_4)_2MoO_4$, etc. production. The concentrated solution was further treated by evaporation. Thereafter, gypsum ($CaSO_4{\cdot}2H_2O$) addition and spray drying were carried out to produce calcium (Ca) containing NH_4NO_3 compound fertilizer.

At the start of the experiment, the dilute solution chamber, concentrate solution chamber, and polar solution chamber in the ED system were all filled with mother liquor removed of Mo. During the experiment, the variations of specific gravities of dilute and concentrate streams were used, supplemented by change in ammonium ion (NH_4^+) concentration measured by Nessler's reagent colorimetric method, as indicators for determining the impact of concentration on the ED efficiency. It was shown that with single stage/single pass ED concentration and current density of 126.5 A m^{-2}, the current efficiency was maintained above 80% before the concentration ratio of concentrate/dilute solutions reached 2. Meanwhile, the final content of NH_4NO_3 increased from 160 to 179 g L^{-1} in the concentrate chamber, whereas in dilute

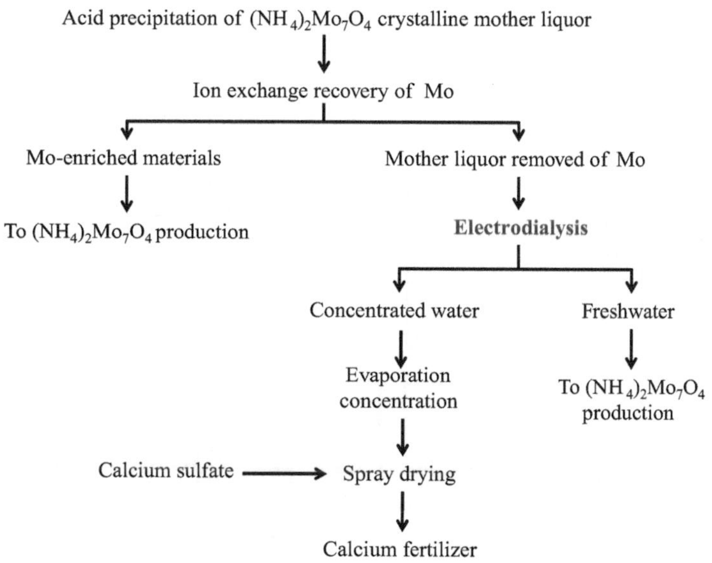

Figure 6.4: The process for treating mother liquid generated in acid precipitation crystallization of ammonium molybdate.

chamber decreased to 95 g L^{-1}. The edge of ED is the higher energy efficiency compared with evaporation concentration.

6.1.3 Comprehensive Utilization of Wastewater from Ammonium Molybdate

A huge amount of acidic wastewater is released from the pretreatment of Mo roasting sand (or calcine) by hydrochloric acid (HCl) (not shown in Fig. 6.3). The waste solution is light yellow-green transparent liquid with a pH value of 0.5–1. Apart from Mo, it also comprises heavy metals, alkaline earth metals, as well as a large amount of NH_4NO_3.

Fig. 6.5 is the schematic of setup for the aforementioned waste water treatment [10]. Nanofiltration (NF) separation is the key step of the whole process. It retained Mo in the feed to generate a solution that has higher Mo concentration and can be recycled to the Mo metallurgical process. NF has relatively stringent requirements on feed and therefore the wastewater must be pretreated before entering membrane system. The permeate from NF step was neutralized by alkali, desalinated using reverse osmosis (RO), and finally reused in production. [10].

In effect for a long time it is difficult to manage this type of acidic wastewater. Direct emissions cause not only the loss of a great deal of Mo resources, but also serious environmental pollution. In this regard, technologies that can efficiently treat $(NH_4)_2Mo_7O_4$ wastewater, thus achieving recovery of valuable metals is particularly important, at least for the part of economic development and environmental protection. The application of NF

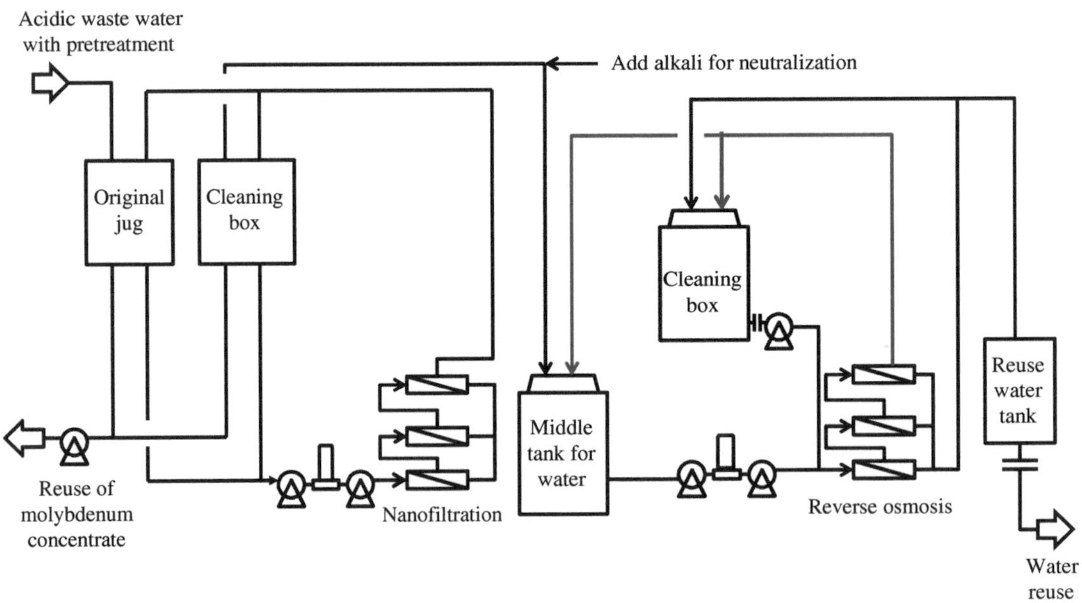

Figure 6.5: Process flow diagram for ammonium molybdate wastewater treatment.

membrane separation as shown earlier has partially met these requirements, which will bring about substantial benefit.

The tail liquid of sodium molybdate (Na_2WO_4) production via alkali wet process has around 6–10 gL^{-1} Mo. Nevertheless, the monovalent salt content is relatively high, which was around 20–25 gL^{-1} in the amount of chloridion (Cl^-). In order to remove Cl^-, chemical reaction, flocculation, precipitation, etc. are adopted in the conventional treatment process. Yu et al. relied on NF to take away Cl^- [11]. In tackling the feed with 8 gL^{-1} of Mo and 23 gL^{-1} of Cl^-, two stages of NF achieved higher than 99% efficiency in Mo recovery and Cl removal.

6.2 Titanium

6.2.1 Introduction

Titanium (Ti) can easily combine with oxygen (O), hence usually exists as oxide or acid salt compounds in nature. It is also proved that the minerals of Ti generally symbioses with those of iron. Ilmenite and rutile are the major Ti minerals that have utilization value with the up-to-date technologies [12,13].

China ranks first globally in Ti resources amount, possessing about 64% of the world's mining reserves [14]. However, the ore is characterized of low grade and complex composition. Based on the industrial application requirements, primary and secondary ores are classified, both of which occur naturally in China. Hereinto, ilmenite, a kind of primary Ti ore, accounts for 97% of the total Ti resource; rutile type resources are mostly, ca. 86%, secondary ore of low grade [15].

The major products of Ti are elementary Ti and titanium white (or titania, titanium dioxide (TiO_2)). Fig. 6.6 presents the process schematic for Ti metallurgy [13]. Mineral dressing and enrichment of natural Ti ores generate rutile and ilmenite concentrates. The content of Ti (in the amount of TiO_2) in natural rutile concentrate is commonly higher than 93–95% (mass fraction), while that (based on TiO_2) in ilmenite concentrate is in the range of 42–64% (mass fraction). Therefore, rutile concentrate is a better choice for producing titanium chloride ($TiCl_4$) and titanium sponge. However, due to the limited amount of its reserve, rutile type ore is not the major raw material.

Ilmenite concentrate can be utilized in two distinctive ways. The first is conversion into acid soluble Ti-rich material (ie, high Ti slag) by electric furnace smelting or HCl leaching, followed with producing titanium white by H_2SO_4 method [16]. The production of titanium white will discharge by-product of ferrous sulfate ($FeSO_4$) and dilute sulfuric acid (H_2SO_4) that can cause serious pollution [17]. The Ti compound obtained by HCl leaching has TiO_2 content above 80 wt.%, and can reduce raw material or mine consumption, the acid

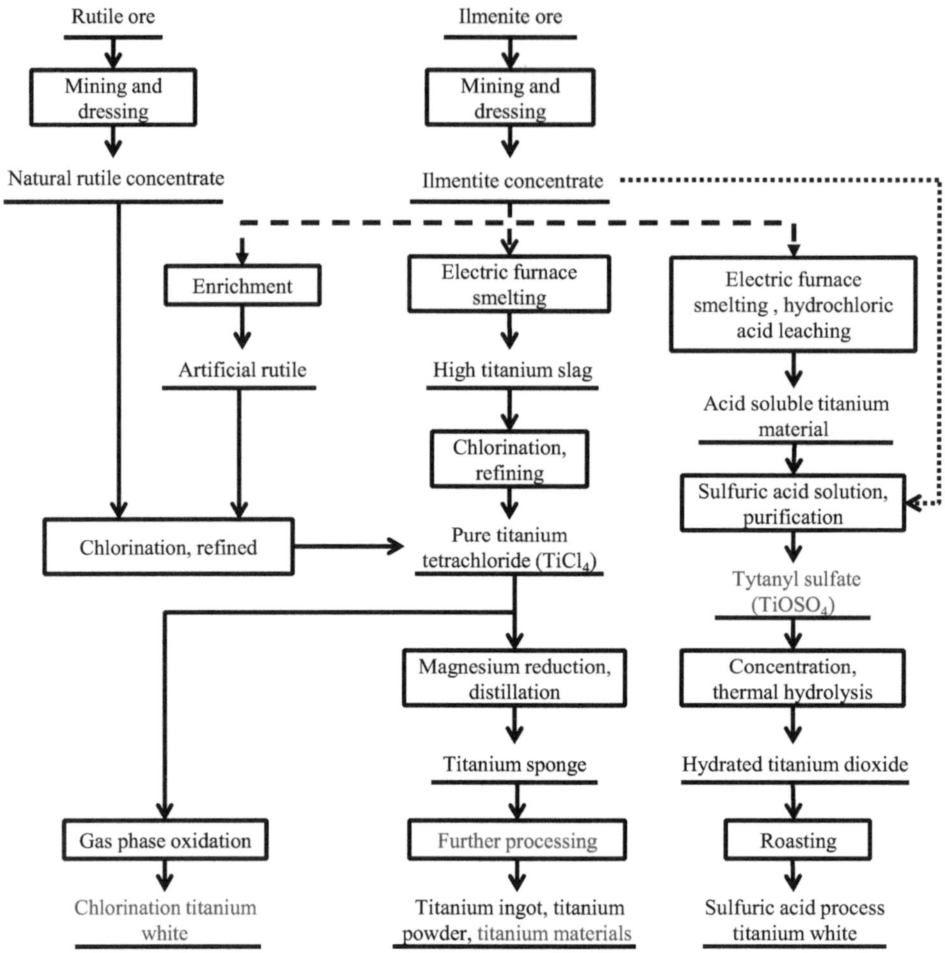

Figure 6.6: Flow chart for the process of extraction of titanium resources and fabrication of titanium products.

consumption and energy consumption when used as a raw material for titanium white production.

Another way for ilmenite exploitation is change into high titanium slag or artificial rutile. The subsequent procedure involves chlorination, impurity removal, and refining to obtain highly purified $TiCl_4$. In the chlorination step, reaction of Ti-rich material, petroleum coke and chlorine (Cl_2) in the chlorination furnace takes place and the resultant chemical is coarse $TiCl_4$ gas. Gas phase oxidation of refined $TiCl_4$ leads to the generation of chlorinated titanium white (ie, TiO_2), whereas magnesium (Mg) reduction–vacuum distillation process produces Ti sponge, an intermediate for manufacturing elementary Ti ingot, Ti powder, etc. Another product of reduction–distillation step is magnesium chloride ($MgCl_2$), which is then split into

Mg and Cl_2 through electrolysis. Cl_2 gas is returned to the chlorination process to be used as raw materials for $TiCl_4$ production and metal Mg is cycled to reduction distillation step [18].

6.2.2 Recovery of Waste Acid Released in Hydrolysis Step

One of the commercial processes to produce TiO_2 is the "sulfate" process utilizing ilmenite mineral or titanium slags as feedstock [17]. For every ton of TiO_2, around 6–8 tons of H_2SO_4 (20 wt.%) dissolving inorganic salts (eg, ferric sulfate ($FeSO_4$), titanyl sulfate ($TiOSO_4$), aluminum sulfate ($Al_2(SO_4)_3$)) are generated [19]. Therefore, a major concern of TiO_2 industry is controlling acid discharge to prevent environment havoc.

Xu et al. carried out research on recycling titanium white waste acid (original concentration of 175 g L^{-1}) using membrane dialysis [20]. The membrane is laboratory-made anion exchange membrane (AEM). The driving force of dialysis is the concentration difference across the membrane. Fresh water and acidic wastewater are fed to the chambers on respective side of the membrane. The concentration of H_2SO_4 and inorganic salts in waste acid is much higher than that in the pure water, generating the tendency for the ions infiltration into the pure water across the membrane.

According to the experimental results shown in Table 6.1, it is confirmed that the AEM can actualize the recovery of H_2SO_4 and its separation from other inorganic constituents in the wastewater. With static operation of batch system, the concentration of H_2SO_4 in dialysate was reduced to 25.1 L^{-1} and in diffusate reached 149.9 g L^{-1}. The separation factors (the ratio of dialysis coefficient) for separating H_2SO_4 from $FeSO_4$, and $TiOSO_4$ were not lower than 26.0 and 51.0, respectively.

It was also found that the contents of the components in the feed have observable impact on the dialysis efficiency. It is feasible to enhance the permeability of one component by increasing the content of other component(s). Particularly, the presence of inorganic salt at certain range of concentration enhanced the dialysis of the acid. Throughout the experiment, the AEMs employed demonstrated good resistance to acid.

Table 6.1: Efficiency of membrane dialysis in recovery of titanium white sulfuric acid[a]

	H_2SO_4	$FeSO_4$	TiO_2
Original wastewater (g L^{-1})	175	102	75
A Chamber (g L^{-1})	149.9	101.2	74.8
B Chamber (g L^{-1})	25.1	0.8	0.2
Dialysis coefficient $\times 10^{-3}$ (m h^{-1})	17.95	0.68	0.35

[a]Membrane ion exchange capacity is 2.78 mmol g^{-1} (dry); a chamber wastewater 100 mL, B chamber deionized water 100 mL; effective membrane area 9.02 cm^2

Li et al. [21] designed a simple, compact, and light weight dialysis membrane module in spiral and wind configuration and used it in recovering H_2SO_4 in simulated titanium white waste acid. It was reported that module achieved an acid recovery of 80% when the flow rate ratio of water/acid was 0.9–1.2, and the handling capacity 0.12–0.24 L m^{-2} h^{-1}. This module has a higher unit treatment capacity than small scale plate and frame module.

Some other researchers have investigated the concentration and recovery of titanium white sulfuric acid by membrane distillation (MD). In Li et al.'s work [22], the synthetic and actual waste sulfuric acid generated in TiO_2 hydrolysis was condensed by vacuum membrane distillation (VMD) using microporous commercial polypropylene (PP) flat membrane [21]. The pure H_2SO_4 could be concentrated from around 18.3% to 65.5 wt.%, but the acid containing $FeSO_4$ only to 31–32 wt.%. Further experimental design was conducted to reveal that Ti salts in the waste solution have no effect on membrane performance in terms of wetting and rejection deterioration. Therefore, it was thought that membrane wetting due to the $FeSO_4$ crystallization caused the difference.

Addressing the problem, several pretreatments have been proposed. For example, 8-stage solvent extraction was used to remove $FeSO_4$ from the solution, with an acid recovery rate of 91.4%. The concentration of the treated acidic wastewater could then be raised to around 65.1 wt.% by VMD. Nevertheless the tedious extraction will greatly enhance the operation cost in practice [18].

Considering the characteristics of $FeSO_4$ saturation concentration, Feng et al. made some preliminary investigation on simultaneous acid concentration and $FeSO_4$ crystallization by MD [19]. The synthetic titanium white waste acid was concentrated by direct contact membrane distillation (DCMD) using laboratory prepared polyvinyldenefluoride (PVDF) hollow fibers. Tentative efforts were made to decouple the effects of different factors on wetting and DCMD performance. $FeSO_4$ crystals were found homogeneously distributed on the outer surface of the hollow fibers while treating $H_2SO4/FeSO_4$ solution. Inhomogeneously distributed crystals inside the membrane and on the lumen surface were identified. Elemental analysis indicated that liquid intrusion was a factor nonnegligible in causing membrane wetting. Rejection coefficients nevertheless were dependent on the solute type. For H_2SO_4 and $H_2SO_4/FeSO_4$, the rejections were maintained at high level throughout the DCMD process before cleaning. In contrast, concentration of sodium chloride (NaCl) solution witnessed obvious deterioration in rejection when approaching saturation. This difference refers to the influence of the amount and property of the solute crystallizable in feed solution. At 60°C, the H_2SO_4 can only be enriched to around 30 wt.%.

In Zhao et al.'s work [23], the research was to study the clarification of titanium white waste acid by ceramic microfiltration (MF) membrane. The main focus was the effect of high concentration $FeSO_4$ and flocculant of polyacrylamide (PAA) on the filtration. The influence of operating parameters such as trans-membrane pressure, cross-flow velocity, concentration, and temperature on filtration process was also investigated. The results showed that hydrated TiO_2 in waste acid can be successfully recovered by ceramic membrane (MF) and an excellent clarification effect was achieved. The high concentration of $FeSO_4$ in waste acid complicates the effect of temperature on the filtration process. When the temperature decreased to less than 293 K, crystallization of $FeSO_4$ may occur in the membrane pores, hence it is difficult to regenerate the membrane. Flocculation by modified PAA, of which the concentration was less than 5 ppm, slightly increased the membrane flux. Further addition of flocculants might cause adsorption fouling of the membrane and resulted in decrease of flux. The membrane flux was also affected by other parameters. Analyses of filtration resistances indicated that the cake resistance was the main component of the fouling resistance and was 68–80% of the total resistance. The analyses gave more quantitative understanding into the effects of the operating parameters on the filtration.

6.2.3 Recovery of Waste Acid in Material Processing or Fabrication

In fabrication of Ti material, surface cleaning by acid to remove impurities is necessary for follow-up treatment (eg, electroplating). The acid is typically a mixture of HNO_3 and hydrofluoric acid (HF). Generally, the acid pickling solution is considered spent when the acid concentration decreases by 75–85%, which is accompanied by metal content increase even up to 150–250 g L^{-1} [24]. Spent baths must be dumped, conventionally after 1–2 cycles of usage, because the efficiency of cleaning decreases with increasing content of dissolved metal in the bath.

Specifically in Ti surface cleaning, the acid solution is discarded when the concentration of Ti reached 20 g L^{-1}. The main components in this wastewater are HNO_3 (4.5–5 mol L^{-1}), HF (3 g L^{-1}), $Ti^{(IV)}$ (mainly in titanium tetrafluoride (TiF_4), $[TiF]_6^{2-}$) (18–24 g L^{-1}). Due to the existence of $[TiF_6]^2$, the use of membrane dialysis is difficult to achieve a high separation efficiency. Shandong Tianwei Membrane Technology Co., Ltd. developed a specialized dialysis membrane [25], and tested its performance in recovering waste acid at pilot scale. The schematic for recovery process is shown in Fig. 6.7. The acid recovery rate is more than 75%, and Ti content in the diffusate is less than 5 g L^{-1}. The residue reacted with soda to form sodium fluoride titanium that can be extracted and recovered. The consumption of soda is decreased as a result of acid recovery. No scientific information about the membrane was revealed in the work.

6.2.4 Purification of Titanium Tetrafluoride [26]

Titanium tetrachloride ($TiCl_4$) is liquid at ambient temperature and its boiling point is 136°C. It is an important intermediate in spongy Ti and chlorination oxidation Ti white production process. The raw material for $TiCl_4$ is high Ti slag; thus the crude $TiCl_4$ usually contains

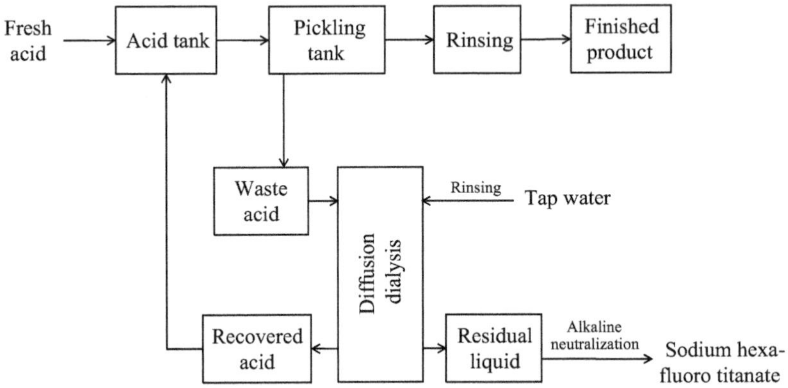

Figure 6.7: HNO_3 recovery process diagram.

impurities such as $FeCl_2$, calcium chloride ($CaCl_2$), silicon tetrachloride ($SiCl_4$), vanadium oxychloride ($VOCl_3$). The former two chemicals are solid at room temperature, which renders the filtration the first step in crude $TiCl_4$ purification. Titanium–aluminum (Ti-Al) alloy MF was employed and found to be better than many other filtration media. Experimental results showed that filtration rate reached 2.0 $m^3(m^2{\cdot}h)^{-1}$ under 0.14 MPa and the filtrate satisfied the industrial production requirements. Backwash pressured used in 0.18 MPa.

6.3 Tungsten

6.3.1 Introduction

More than 30 countries and regions of the world own tungsten (W) resources, but with uneven distribution. Most W deposits are distributed in regions around the Pacific. Another important region is the Mediterranean and the part of Europe on the verge of Atlantic Ocean. The countries with the largest amount of W resources are China, Canada, Russia, the United States, and Bolivia. The major W ore in the United States are Pine Creek deposit, Springer deposit, Straub Lee deposit, Emerson and Clay Mexes deposit, and the first four are all scheelite deposit. The northwest region and the junction of Yukon region are most abundant in W resources in Canada [27].

China's W reserve accounts for more than 60% of the world total. The characteristics of W resources in China include diversity in deposit types, existence of many symbiotic and associated components, high comprehensive utilization value, dominance of lean ores, and the difficulty in dressing. The exploitation degree of W resources in China is higher and has long been relying on wolframite [28–30]. With the development of technologies for scheelite and mixed ore beneficiation, development and utilization of scheelite and mixed W resources is initiated [31,32].

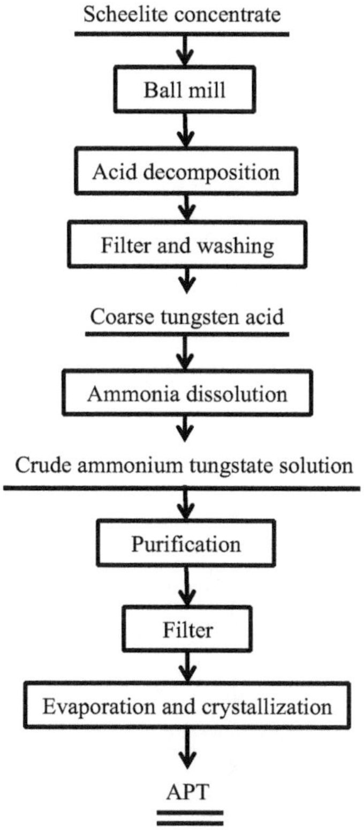

Figure 6.8: Flow chart for acid method in tungsten metallurgy.

W is widely used in alloy, electronic, and chemical industries. Currently, ammonium para-tungstate ($(NH_4)_{10}(H_2W_{12}O_{42})\cdot 4H_2O$, briefly APT) is a major product of W, which is an intermediate product for fabricating metal W powder, tungsten oxide (WO_3), ammonium meta-tungstate ($(NH_4)_6H_2W_{12}O_{40}$, briefly AMT), etc. APT production mainly adopts following processes: the classic white W concentrate process (also called acid method), as shown in Fig. 6.8 [33]), mainly used for scheelite smelting, W concentrate alkali pressure boiling—solvent extraction—evaporation crystallization process (also called the alkali pressure cooking—extraction process), of which the flow chart is given in Fig. 6.9 [34], and W concentrate alkali cooking—ion exchange—evaporation crystallization process (also called alkali pressure cooking—ion exchange process), as displayed in Fig. 6.10 [34].

Due to the short technological process, high metal yield, low consumption of auxiliary materials, high product quality, low impact on environmental conditions, and simple operation, etc., alkali cooking—ion exchange process gets rapid development since its invention. At present, it is employed by 90% of W smelting industry, becoming the mainstream technology of APT production in China [35].

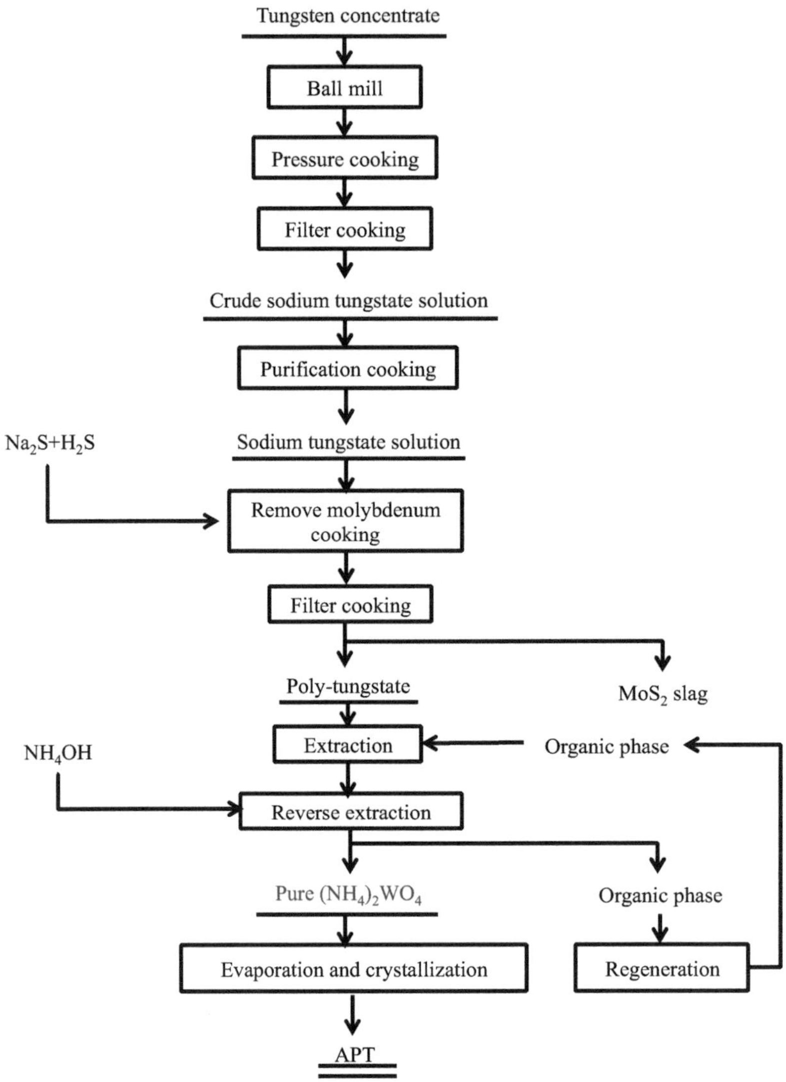

Figure 6.9: Flow chart for the alkali pressure cooking—extraction process in tungsten metallurgy.

6.3.2 Free Alkali Recovery

Ion-exchange membrane (IEM) electrolysis technology has been widely used in many industries [36,37]. Its usage in the recovery alkali from sodium tungstate (Na_2WO_4) solution has been investigated [26], and achieved industrial applications. Xiao et al. [38] made the first attempt in applying it in China and conducted a series in-depth research on the influence of anode, electrolytic tank structure, tank assembly, etc. [39–41]. On the basis of these previous works, industrial scale-up experiment was performed using practical Na_2WO_4 solution as

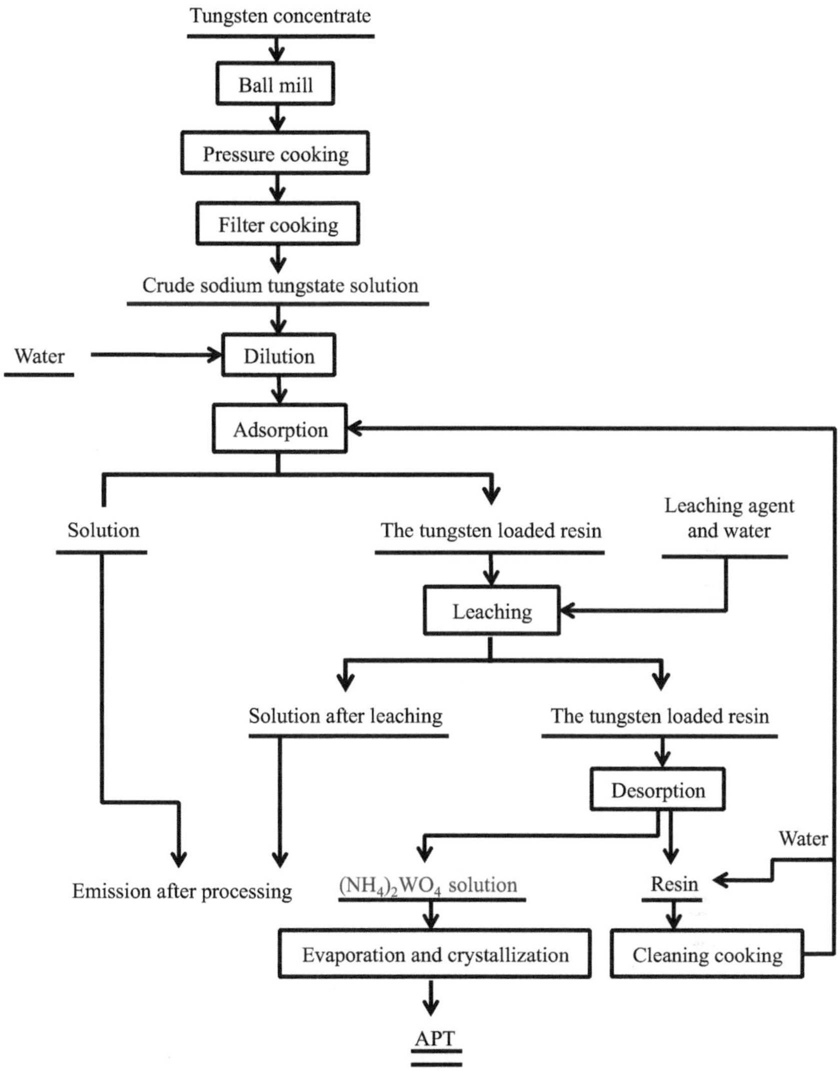

Figure 6.10: Flow chart for the alkali pressure cooking—ion exchange process in tungsten metallurgy.

feed. The experimental setup is shown in Fig. 6.11 and the principle for alkali recovery is shown in Fig. 6.12. The overflow liquid from anodic circulation chamber is Na_2WO_4 solution containing 10–15 mg L^{-1} sodium hydroxide (NaOH), whereas that for cathodic circulation chamber has about 4.5 mol L^{-1} NaOH.

The results show that higher temperature improves electrolyte conductivity and brings down the overpotential of electrode reaction, thereby reducing the cell voltage, and lowering the energy consumption. However, increase in temperature intensifies the reverse transfer of ions. In addition, rise in temperature itself will consume more energy. The other factors interfering

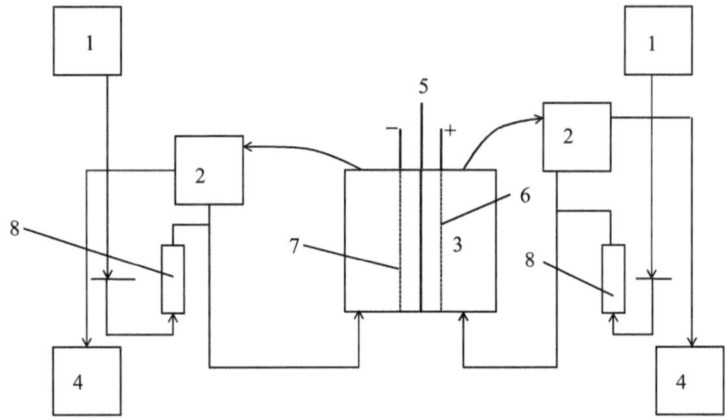

Figure 6.11: Experimental setup for Membrane electrolysis of sodium tungstate ($Na_2WO_4 \cdot 2H_2O$).

1, high slot; *2*, cycle overflow tank; *3*, electrolyzer; *4*, fluid product lower tank; *5*, cation-exchange membrane; *6*, anode; *7*, active nickel cathode; *8*, flow meter.

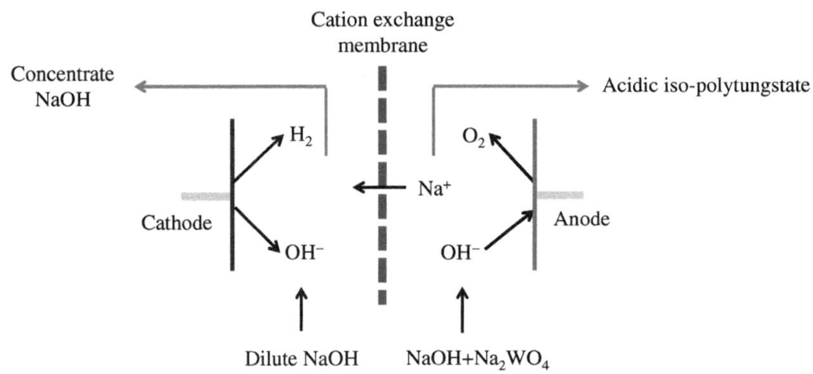

Figure 6.12: Principle of membrane electrolysis for recovery free alkali.

with selection of temperature mainly refer to the endurance capacity of the membrane, electrolytic tank frame, and the channels.

Continuous operation mode is in favor of the stable operation of the alkali recovery process. At 60–65°C and with a current density of 1000 A m^{-2}, the current efficiency is higher than 88% and energy consumption is lower than 2250 kWh ton^{-1} of NaOH, which makes the process practically valuable for industrial application.

Luo et al. found that AEM-based electrodialysis that allows preferential transport of OH^- and reject bivalent WO_4^{2-} also demonstrated feasibility of recovering free alkali [42].

Zhang and Xiao reported the recovery of alkali from Na_2WO_4 solution by diffusion dialysis (DD) in a review [26]. The feed solution was sheelite alkali decomposition solution, inside which W content in WO_3 equivalent is 91.25 g L^{-1} and NaOH 1.36 mol L^{-1}. The recovery rate of alkali was up to 50%.

Recovery of alkali can reduce the acid consumption for neutralization in the follow-up stages, and hence the amount of resultant inorganic salts and their impact on environment. With electrolysis, the H^2 evolved from cathode can be utilized as a reducing agent in W powder production.

6.3.3 Utilization of Ammonium Para-Tungstate Crystallization Mother Liquor

The APT mother liquor released in process shown in Fig. 6.10 generally has Cl^- concentration in the range of 15–70 gL^{-1}. This condition makes it impossible to recycle the mother liquor to the ion exchange step of the main stream production. pH adjustment by alkali combined with artificial scheelite precipitation is traditionally applied to assist W recovery.

A series of membrane-based methods for recovering W and ammonium chloride (NH_4Cl) from mother liquor of APT crystallization were investigated [43–45]. The composition of the mother liquor of APT crystallization is tabulated in Table 6.2. In principle, NF or ultrafiltration (UF) membranes were used to separate W compounds and chloride (Cl^-) ions present in the mother liquor; the retentate solution with W is transferred directly to the ion exchange step for the production of $(NH_4)_{10}[H_2W_{12}O_{42}]$. Alternatively, the concentrate is mixed with leaching solution in mainstream process. The residual NaOH in the leaching solution can depolymerize the poly-tungstate to generate tungstate (WO_4^{2-}) ions. The NH_4Cl solution at the permeate side may be used for other purposes after appropriate treatment. In addition, some other impurities (e.g. phosphorus (P), arsenic (As), Si) were also transporting along with NH_4Cl. The concentration factor of W in the retentate was around 4.

In Zhang et al.'s work [44], NF treatment of APT mother liquor was targeted. The experimental process setup is designed as demonstrated schematically in Fig. 6.13. As revealed in Table 6.3, NF is effective in separating W compounds and chloride in APT crystal liquor and concentrating W. The highest de-chlorination efficiency reached more than 90%; the concentration of WO_3 in retentate was improved to 81–113 g L^{-1} and only a trivial loss (1%) of W to the permeate side. Obviously, separation by NF membrane is efficient within a wide range of pH.

Table 6.2: Composition of mother liquor of APT crystallization

No.	WO_3 (g L^{-1})	Cl^- (g L^{-1})	SO_4^{2-} (g L^{-1})
1	22.92	28.36	8.7
2	13.55	18.79	19.8

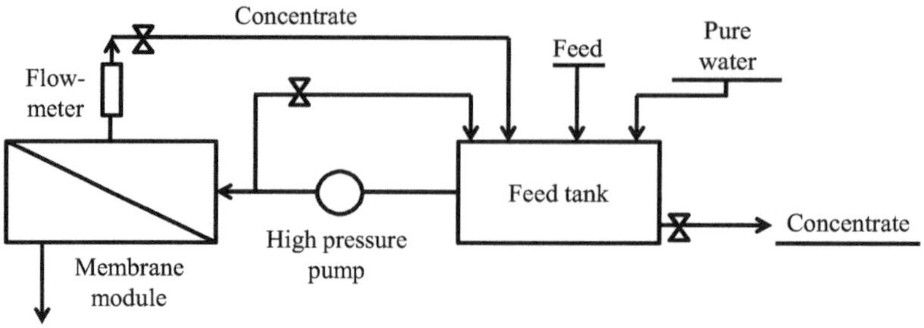

Figure 6.13: Separation system using NF as the key step for treating mother liquor of APT crystallization.

Table 6.3: Experimental results for dealing APT mother liquor using NF

No.	Feed pH	V_f (L)	Concentration in feed (g L^{-1})		Concentration in retentate (g L^{-1})		P (kg cm^{-2})[a]	J (L m^{-2} h^{-1})[a]	ξWO_3 (%)[a]	η_{Cl^-} (%)[a]
			WO_3	Cl^-	WO_3	Cl^-				
1	5.0	80	21.39	33.32	113.17	13.83	20	47.1	0.15	92.2
2	10.0	69	24.44	33.85	81.80	4.43	20	29.3	1.7	96.0

[a]ξWO_3 is the tungsten loss rate, η_{Cl^-} is the chloride ion removal rate, J is the membrane flux, and P is the operating pressure.

In addition, the research also mentions that NF membrane has high rejection toward sulfate (SO_4^{2-}) group, and hence cannot literally discriminate sulfate (SO_4^{2-}) group and W compounds. With higher concentration of SO_4^{2-}, the flux is lower, leading to increased W loss.

UF separation treatment of APT mother liquor was examined by Zhang et al. [45]. The pattern of performance as a function of pH value is quite similar to that observed in using NF introduced previously. The molecular weight cut off (MWCO) of the UF membrane was 1000~2000 Dalton. The solution was pretreated by pH adjust to convert all the W components to metatungstate with higher molecular weight (MW) and charge number, so that W and NH_4Cl may be separated more efficiently. The de-chlorination rate is generally higher than 85%, and the concentration of WO_3 in the feed reached about 50 g L^{-1}. It was confirmed that UF separation delivered a lower recovery for W, as compared with NF, but achieved division of W and SO_4^{2-}. When Cl^- removal rate reached 90%, the SO_4^{2-} removal rate was 50~60%. The rise of operating pressure and temperature was beneficial to the improvement of the permeation flux, but has little effect on rejecting W compounds. On the whole, the loss of W is dependent on the content of SO_4^{2-}.

Compared with conventional technologies (eg, neutralization, crystallization, freeze drying, solvent extraction and spray drying, ion-exchange), the advantages of hydraulic pressure driven membranes for treating APT crystallization mother liquor are low chemical reagent consumption, high W recovery rate, simple operation, etc.

6.3.4 Other Applications

6.3.4.1 Preparation of sodium meta-tungstate

Zhang et al. investigated sodium meta-tungstate ($Na_6[H_2W_{12}O_{40}]$) production by membrane electrolysis using CEM [26]. The principle for production is shown in Fig. 6.12. Metatungstic acid $H_2W_{12}O_{40}$ in anodic chamber is resulted from condensation of WO_4^{2-} and reaction with H^+. Simultaneously, the process actualized the recovery of alkali (ie, NaOH), evolution of (H^2) at cathode, and (O^2) at anode. H_2 may potentially be used as a reducing agent in W powder fabrication. The energy consumption is 2.9–3.9 kWh kg^{-1} NaOH (the concentration of W is 100 g L^{-1}) when the recovered NaOH was 200 g L^{-1} in the solution.

This design is featured with nonpollution. It was installed and operated in the pilot plant of a Hungary tungsten company (currently owned by GE, USA) [38]. As compared with the processes relying on conventional technologies, the membrane-based process obviates the sodium sulfate (Na_2SO_4) discharge and saves the usage in new chemicals (eg, acid, alkali). Verbann and Brown carried out similar investigation, with difference being pH < 1 for the anodic chamber [26]. Consequently, super-saturation of WO_4^{2-} in a metastable state was resulted. It continuously evolved out from the solution.

6.3.4.2 Preparation of AMT

Zeng et al. put forward the application of bipolar membrane (BPM) based ED technology for the preparation of AMT [46]. Guan et al. also studied this method [47]. The principle of the method in generation of AMT inside ED system is schematized in Fig. 6.14. A series of

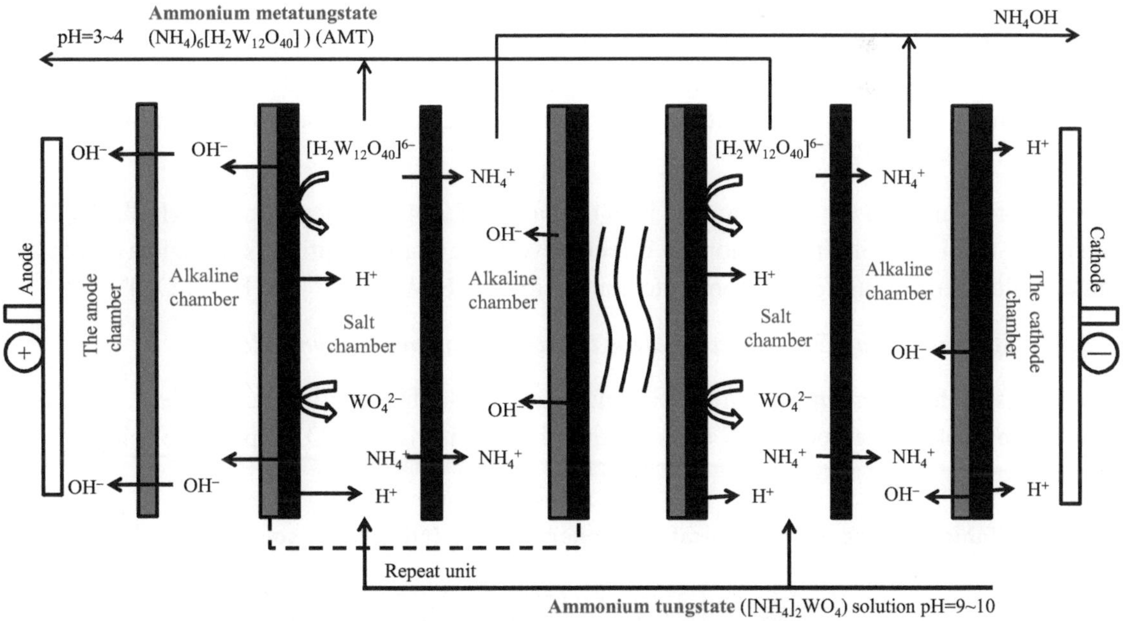

Figure 6.14: Principle of the preparation of ammonium metatungstate *(AMT)* using bipolar membrane electrodialysis (BPM-ED) method.

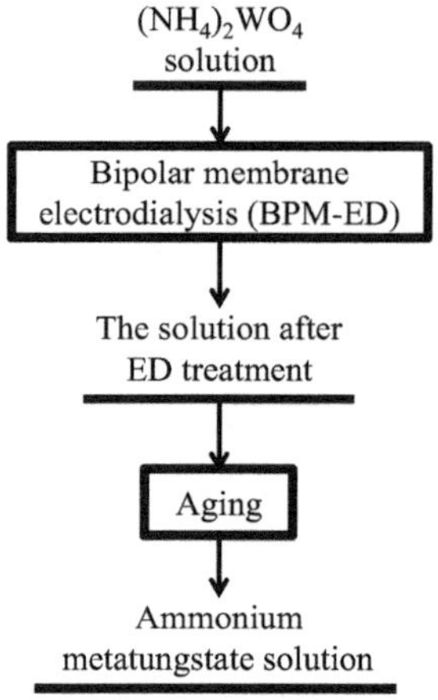

Figure 6.15: Flow chart of the preparation of ammonium metatungstate solution using bipolar membrane electrodialysis *(BPM-ED)* method.

the tungstate (WO_4^{2-}) ions with the proton (H^+) generated inside BPM to form meta-tungstate ($[H_2W_{12}O_{40}]^{6-}$) ions. The water molecule is split by BPM into H^+ and hydroxyl ions (OH^-) which further migrate from the junction part through the cation and anion exchange layers into the solutions.

The flow chart of the process involving BPM-ED in preparation of AMT is shown in Fig. 6.15 and the main chemical components inside the original $(NH_4)_2WO_4$ solution are summarized in Table 6.4. The optimum conditions revealed in the research involve current density of 600–750 A m^{-2}, the initial concentration of NH_4HCO_3 in alkaline chamber at 1.50 mol L^{-1}, the concentration of W (in the amount of WO_3) at 150–200 g L^{-1}, and the pH of salt chamber at 2.8–3.0. Under these conditions, the direct recovery of WO_3 reached 99.97%, which is much higher than that obtained with conventional thermal decomposition method [48] and the current efficiency is 72.50%.

The pH of solution in the salt chamber is in the range of 2–4 and theoretically $[H_2W_{12}O_{40}]^{6-}$ is the major form of W in the solution. In effect, nevertheless, the solution is in a meta-stable state, which need aging for the complete conversion of $[H_2W_{12}O_{42}]^{10-}$ to $[H_2W_{12}O_{40}]^{6-}$.

6.3.4.3 Splitting of Na_2SO_4 by Electrodialysis to produce acid and alkaline [26]

There exists a huge amount of Na_2SO_4 in raffinate of W hydrometallurgical process. ED was reported by Tokuyama Soda Co. Ltd as a method for splitting Na_2SO_4 to produce acid and

Table 6.4: The main chemical components of the $(NH_4)_2WO_4$ solution (g L^{-1})

WO_3	NH_4OH	P	Na	SiO_2	As	Sn	Mo
389.20	70.00	<0.007	0.050	0.095	0.0165	0.0001	0.012

alkaline. The membrane employed was Neosepta IEM. The overall replacement reaction is given below:

$$Na_2SO_4 + NH_4OH = 2NaOH + (NH_4)_2SO_4$$

6.4 Vanadium

6.4.1 Introduction

Vanadium (V) is an important and indispensable material to the development of modern industry, modern national defense, and modern science and technology. V is widely distributed in the earth's crust and there are 65 kinds of V minerals, mainly patronite, vanadinite, sulvanite, V-Ti-Fe mineral, V-Ti magnetite, etc. [49]. Ninety-eight percent of the world's proven reserves of V resources were symbiotic V-Ti magnetite, mainly distributed in Russia, the United States, South Africa, China, Norway, etc.

According to the statistics by American Mining Bureau, the V resources reserve base in the world is 160 million tons, which can be mined for 150 years, following the current production scale. South Africa accounts for 46% of the V reserve base, Russia 23.6%, the United States 13.1%, China 11%, and the sum of the other countries is less than 6% [50]. China is rich in V resources, mainly vanadium titanium magnetite, and has a lot of V containing stone coal resources [51].

The methods for extracting V from V-Ti magnetite are V slag blowing method, V containing steel slag method, and soda residue method [52–54]. In V slag blowing process, atomization method or converter is used for blowing molten pig iron, producing semi-steel and V slag comprising 12–16% V_2O_5. The V slag is the raw material for V extraction. Presently, 70% of raw materials for V industry in China are V slag, a by-product of iron and steel production, followed by stone coal [51,55]. For the second method, V containing molten iron is blown into steel in converter following basic single slag method, with products being V steel slag possessing 5% V_2O_5 for V extraction and molten steel. Soda residue approach adds sodium carbonate (Na_2CO_3) directly to molten iron V at temperature between 1400 and 1600°C. In this step, V in the molten iron turns into sodium vanadate (Na_3VO_4), sodium slag and semisteel are obtained, and sulfur and phosphorus in molten iron are removed simultaneously.

Usage of stone coal for extraction of V leads to different situations [56]. The traditional process is sodium roasting method comprising sodium roasting, V precipitation with acid, water leaching, alkali dissolution, metavanadic acid precipitation, etc. [57]. The method has good technological adaptability and the cost of vanadium oxide (V_2O_5) production is low. Nevertheless, it releases calcined smoke with a substantial amount of gases such as HCl, Cl_2, and wastewater with high salinity. In addition, the V recovery is low, only about 50% usually. Therefore, many areas do not support its application [58,59]. To reduce the

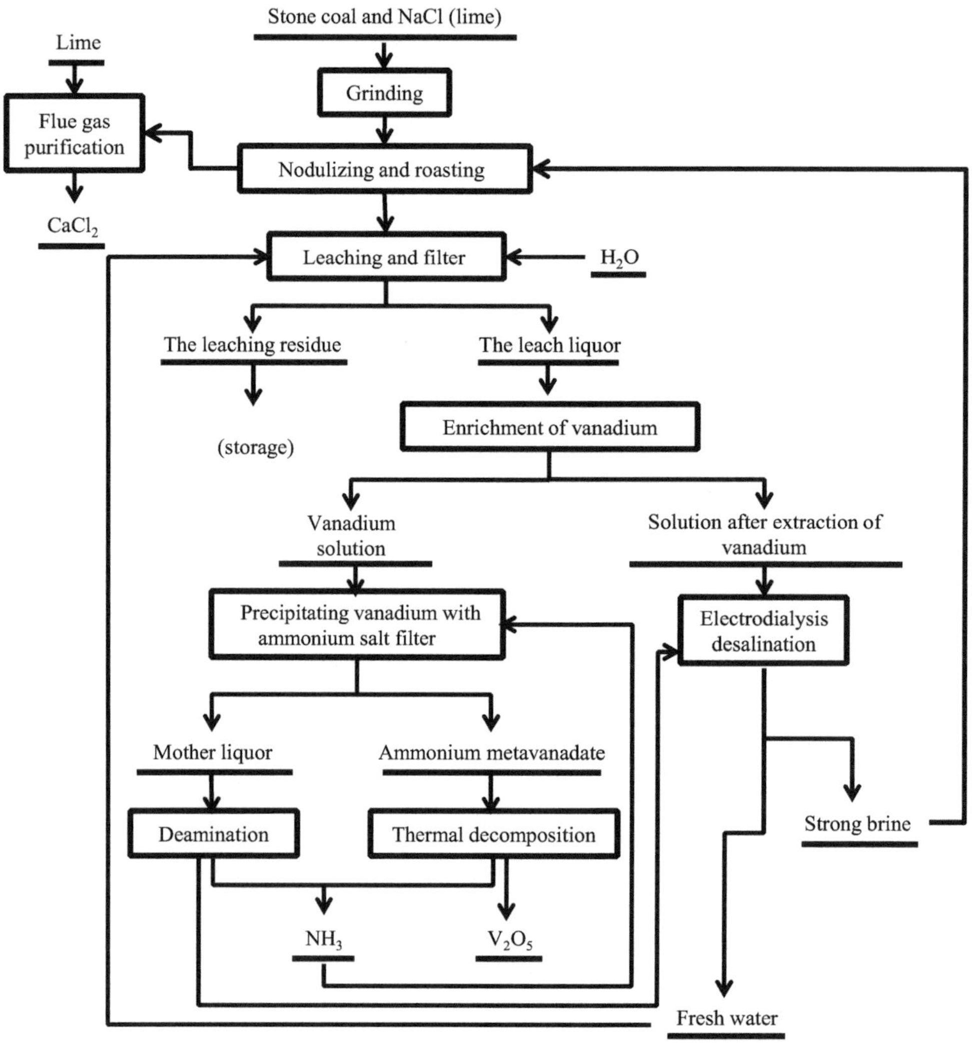

Figure 6.16: The improved process flow of vanadium extraction from stone coal using NaCl roasting.

environmental impact, an improved process based on the traditional one is proposed, as shown in Fig. 6.16 [60]. The key innovation lies with the immobilization of Cl_2 inside calcine and its follow-up recovery. Application membrane technology will be introduced in the forthcoming section.

There also exist some other methods to extract V from stone coal, which are the combination of stone coal oxidizing roasting—alkaline leaching—ion exchange and the stone coal acid leaching—solvent extraction [59]. These methods are progressively replacing the traditional one, attributing to their high extraction yield and low pollution.

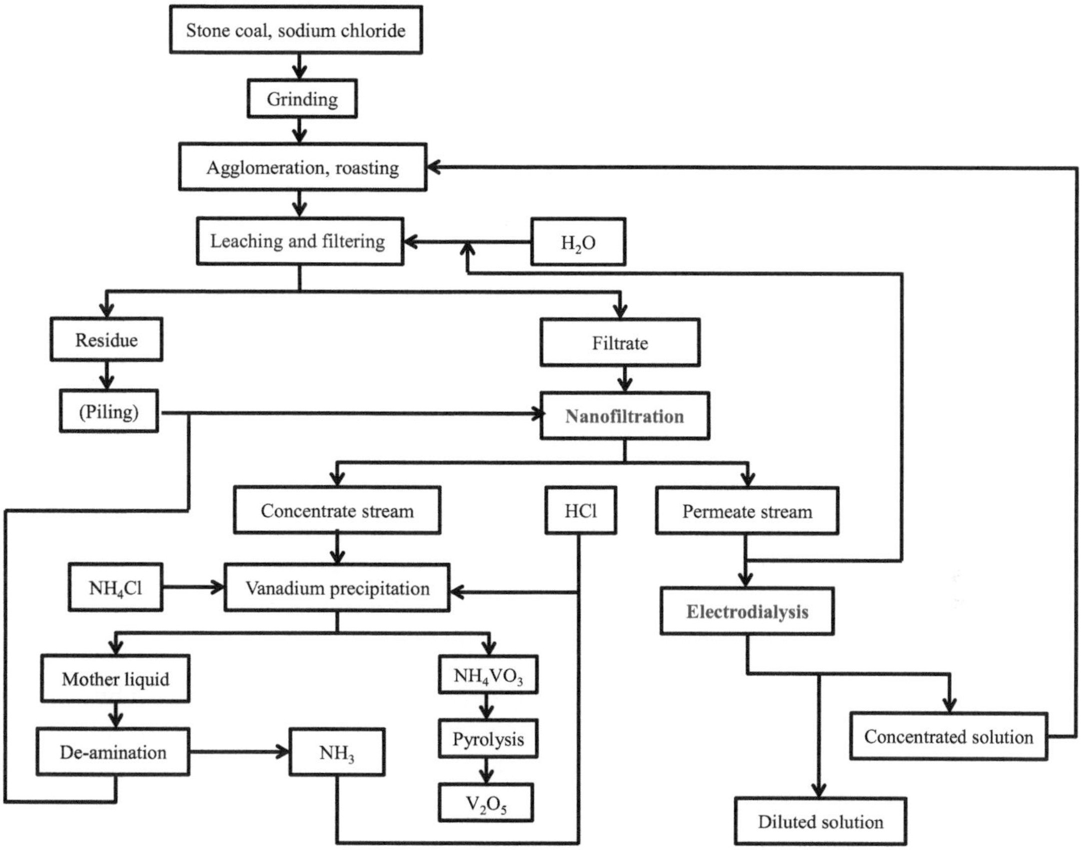

Figure 6.17: The process flow chart of vanadium recovery from the leaching solution of stone coal roasted with NaCl.

6.4.2 New Process for Vanadium Extraction From Stone Coal

Hazardous gases (ie, HCl, Cl_2) are emitted in the traditional process of extracting V from stone coal via NaCl roasting-water leaching [61]. Fig. 6.17 demonstrates an innovated process facilitated by membrane separation [61]. The leaching solution comprising $NaVO_3$ and unreacted NaCl flows to NF system, where VO^{3-} was rejected, while Cl^- was driven to the permeate. Membrane ED is then applied to treat the retentate solution; the concentrated solution released from ED was reused in roasting, and freshwater in calcine leaching. With the innovation, Cl component is completely recycled and zero discharge of wastewater is actualized.

With hydraulic pressure of 10.0–12.5 kg cm^{-2} and pH value 5.16–6.15, the content of V (calculated base on V_2O_5) was raised to above 35 g L^{-1} by NF concentration. At pH value of 6.15, highest V rejection was acquired, which is 92%. 2.2 times of the amount NH_4Cl equivalent for precipitating V was fed to the solution. The resultant ammonium vanadate (NH_4VO_3) precipitates were put to calcinations of 500°C for 1 h, producing V_2O_5 of 98.6% purity.

6.5 Zirconium

6.5.1 Introduction

The content of zirconium (Zr) in the earth's crust is the 20th among all the naturally occurring elements. Countries which have most abundant Zr resources are Australia and South Africa. They own together more than 75% of Zr reserves of which the data are available. Zr occurs naturally in more than 140 minerals, and the principal commercial source of Zr is zircon, a silicate mineral. Zircon is mainly used for producing zirconium oxychloride ($ZrOCl_2$), zirconium dioxide (ZrO_2), Zr metal, and other products [62]. Zr and its compounds are very important to many aspects of the economic development. For example, they are consumed in massive amount by ceramic industry, steel and nonferrous metallurgy, and glass industry. Other applications include nuclear power, space and aeronautic industry, etc.

In recent years, China's imports and consumption of zircon sand increased gradually. The estimated importing amounts of zircon sand in 2011 are 800,000 tons [63]. It has become the world's largest importer and consumer, and meanwhile one of the largest producers of ZrO_2 (primary products). Ninety percent of the world's ZrO_2 is made in China [64].

Zircon sand is the raw material for manufacturing Zr compounds and some useful by-products. The major component of Zr is (Zr, hafnium (Hf))SiO_4. Removing silica (SiO_2) and preparing $ZrOCl_2$ is the first step of Zr metallurgy based on zircon for obtaining other Zr compounds and Zr metal. $ZrOCl_2$ is an important basic material for the metallurgical, nuclear, and chemical industries [65,66]. An important by-product during $ZrOCl_2$ metallurgy is sponge (Hf). With further processing, they are both the core materials of nuclear power and nuclear military industry applications, and chemical equipment materials, etc.

The methods for decomposing zircon sand and preparing $ZrOCl_2$ are sintering-based and chlorination-based processes [67–69]. As shown in Fig. 6.18 [70], the essential steps of the former are alkali decomposition or sintering, acid leaching, and crystallization. The sintering-based process using NaOH is the dominant selection of large-scale enterprises in China.

Chlorination process is classified into direct chlorination and carbide chlorination processes. Due to the usage of electric arc furnace in removing Si, the second approach entails high power consumption, and special requirements on production condition and equipment. Therefore, its application is quite limited. In direct chlorination process, also known as boiling chlorination process, or fluidized chlorination process, fluid is used to suspend solid particles, strengthening the contact between gas–solid, liquid–solid, and gas and –liquid–solid. The process flow for chlorination method is displayed in Fig. 6.19 [70]. The hydrolysis of zirconium tetrachloride ($ZrCl_4$) generates $ZrOCl_2$. Chlorination method has the advantages of continuous and automatic operation, short process, relatively low energy consumption,

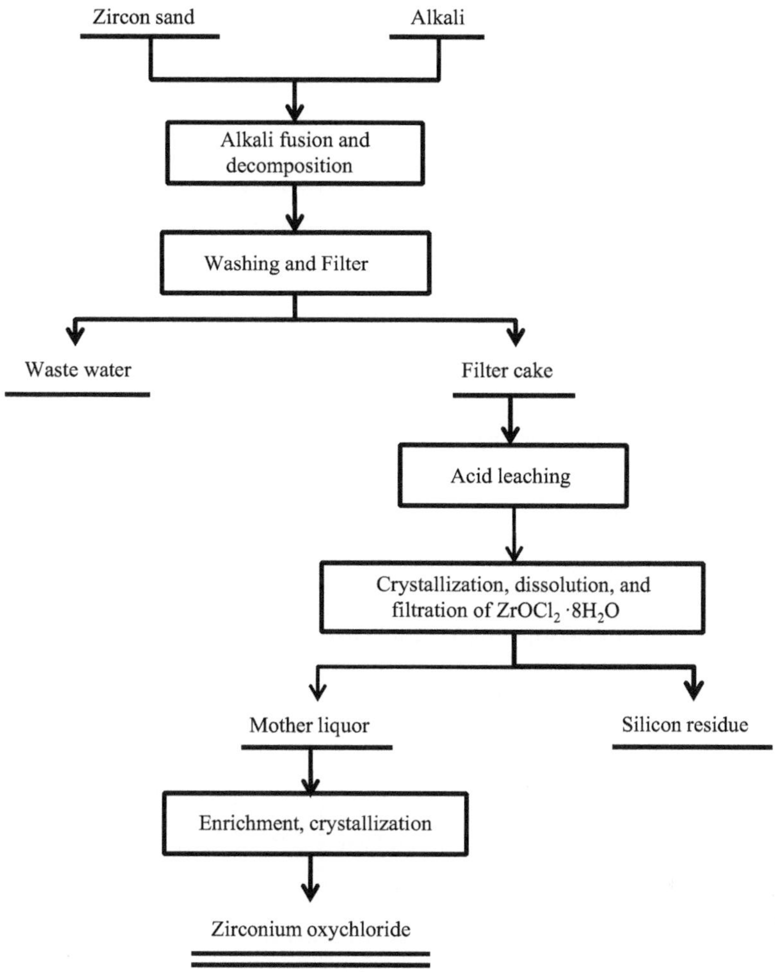

Figure 6.18: Flow chart of sintering process for producing zirconium oxychloride.

and stable and sound product quality. Therefore, it is adopted by most sponge Zr industry for zircon metallurgy.

6.5.2 Treatment of Wastewater in Zirconium Oxychloride Production

$ZrOCl_2$ is produced with discharging wastewater containing salt content up to 40,000 mg L^{-1}. Undoubtedly its direct discharge will seriously contaminate groundwater. An ED/RO integrated membrane system was developed by Shandong Tianwei Membrane Technology Co. Ltd. for tackling this waste [71]. The flow sheet for the process entailed by the system is illustrated in Fig. 6.20. The salt components in the wastewater were enriched by ED, followed with reuse of the concentrated water. RO treated freshwater from ED diluent chamber was qualified as process water.

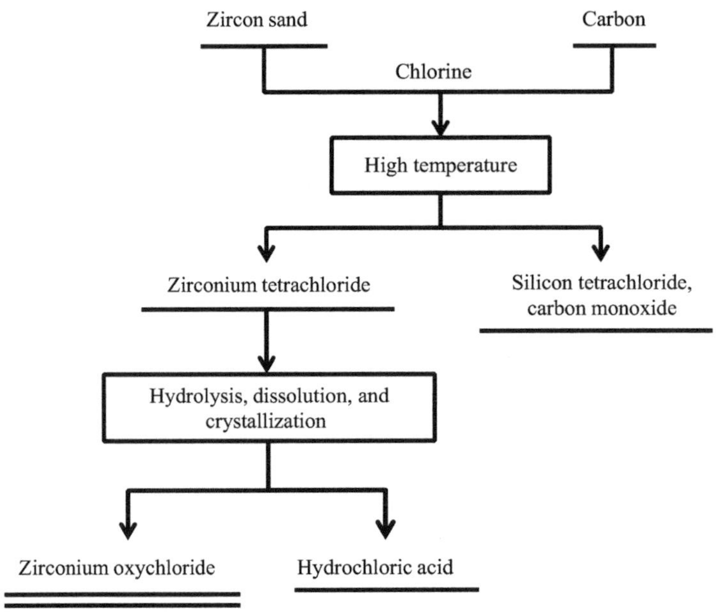

Figure 6.19: Flow chart of boiling chlorination process producing zirconium oxychloride.

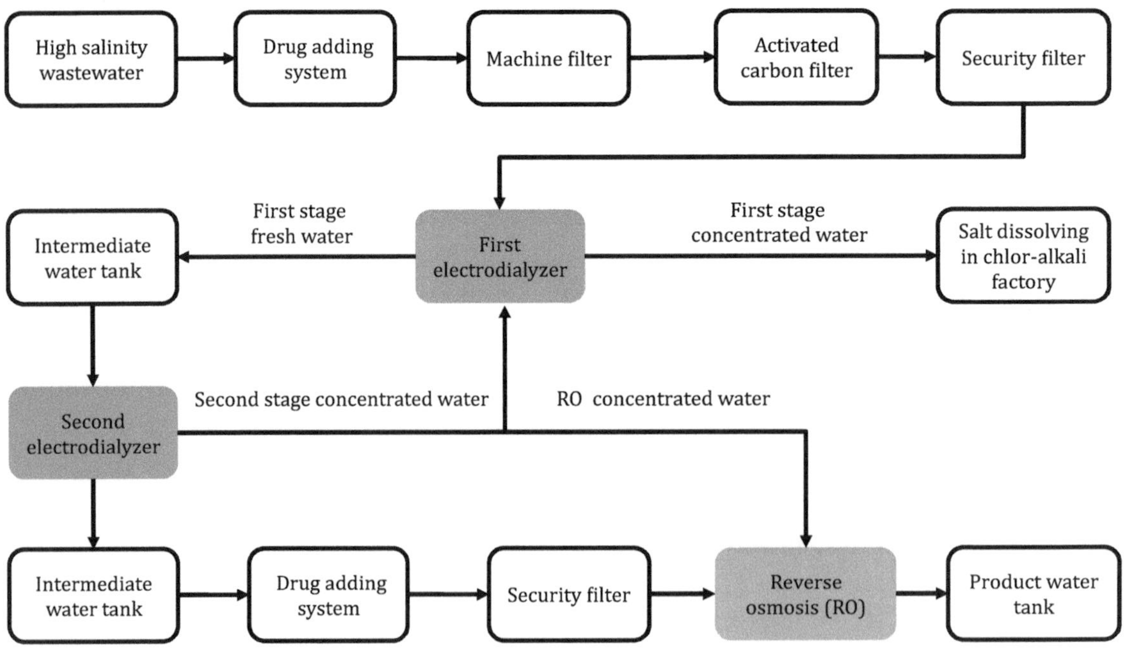

Figure 6.20: Process for treating saline wastewater in some chemical plant in China.

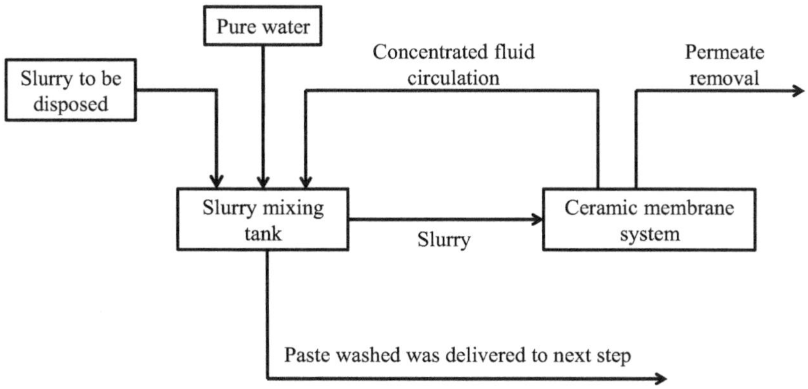

Figure 6.21: Technological flow chart for washing nanopowders using ceramic membrane module.

The benefits brought about by the ED/RO integration are zero discharge of sewage, and the creative application of concentrated water in chloro-alkali plant for salt dissolving [71].

6.5.3 Rinsing of Zirconium Oxide Nanoparticles

According to Xiao et al.'s review, ceramic membranes have been used in washing ZrO_2 [72]. In the production of ultrafine powder materials by wet chemistry, the impurities of ions should be removed from the product slurry by thorough washing to obtain highly purified nanoparticles. The schematic for the general process flow for separating water soluble impurities and nanosized particle product is displayed in Fig. 6.21. The membrane used in the system is microporous ceramic membrane with a pore size of 50 nm. The pH value of slurry changed from 10 to 6.5. The operation cost for the aforementioned process is summarized in Table 6.5.

Table 6.5: The running cost calculation using inorganic membrane washing nanozirconia

Items	Quota	Unit price (¥)	Total amount (¥)	Notes
Electricity (kWh t^{-1})	10	0.5	5.0	Air-compression station
Air (m^3 t^{-1})	0.2	0.5	0.1	
Pure water (m^3 t^{-1})	3	2	6.0	
Washing fee (kg t^{-1})	0.1	0.5	0.05	The acid and alkali consumption
Maintenance costs (¥ t^{-1})	0.66	—	0.66	
Operation cost				11.81 ¥ t^{-1}

6.6 Conclusions

Refractory metals include Ti, Zr, Hf, V, niobium (Nb), tantalum (Ta), Mo, and W. Their common properties are the extraordinary resistance to heat and wear. In this chapter, some basic information about the properties, applications, and metallurgical processes for Mo, Ti, W, and V and how membrane-based separation technologies are involved in their extractive metallurgy and manufacturing processes are introduced. Depending on the chemical nature of the raw materials, and hence the production processes selected, membrane technologies that can identify suitable utilization for metallurgy of respective metal are different.

Symbols and Nomenclature

AEM	Anion exchange membrane
Al	Aluminum
$Al_2(SO_4)_3$	Aluminum sulfate
As	Arsenic
BPM	Bipolar membrane
Ca	Calcium
$CaCl_2$	Calcium chloride
$CaSO_4 \cdot 2H_2O$	Gypsum
Cl^-	Chloridion, chloride ion
Cl_2	Chlorine
DCMD	Direct contact membrane distillation
DD	Diffusion dialysis
ED	Electrodialysis
Fe-Mo	Ferro-molybdenum
$FeSO_4$	Ferric sulfate
H^+	Protons
H_2	Hydrogen
HCl	Hydrochloric acid
Hf	Hafnium
HF	Hydrofluoric acid
H_2MoO_4	Molybdic acid
HNO_3	Nitric acid
H_2SO_4	Sulfuric acid
$H_2W_{12}O_{40}$	Metatungstic acid
$[H_2W_{12}O_{40}]^{6-}$	Metatungstate ions
IEM	Ion-exchange membrane
LCD	Liquid crystal display
MD	Membrane distillation
MF	Microfiltration
Mg	Magnesium
$MgCl_2$	Magnesium chloride
Mo	Molybdenum
MoO_3	Molybdenum trioxide
MoS_2	Molybdenum sulfide
MMO	Molybdate (chemical formula: M indicates metal; MO indicates acid radical)
MW	Molecular weight

MWCO	Molecular weight cut off
Na_2CO_3	Sodium carbonate
NaCl	Sodium chloride
$Na_6[H_2W_{12}O_{40}]$	Sodium metatungstate
NaOH	Sodium hydroxide
Na_2SO_4	Sodium sulfate
NF	Nanofiltration
NH_4^+	Ammonium ion
NH_4Cl	Ammonium chloride
$(NH_4)_{10}(H_2W_{12}O_{42})\cdot 4H_2O$ or APT	Ammonium paratungstate
$(NH_4)_6H_2W_{12}O_{40}$ or AMT	Ammonium metatungstate
$(NH_4)_6Mo_7O_{24}\cdot 4H_2O$	Ammonium paramolybdate
NH_4NO_3	Ammonium nitrate
$(NH_4)_2MoO_4$	Ammonium molybdate
NH_4VO_3	Ammonium vanadate
$(NH_4)_2WO_4$	Ammonium tungstate
Na_3VO_4	Sodium vanadate
Na_2WO_4	Sodium tungstate
OH^-	Hydroxyl ions
P	Phosphorus
PAA	Polyacrylamide
PP	Polypropylene
PVDF	Polyvinyldenefluoride
W	Tungsten
Re	Rhenium
RO	Reverse osmosis
S	Sulfur
Si	Silicon
$SiCl_4$	Silicon tetrachloride
SiO_2	Silica
SO_2	Sulfur dioxide
SO_4^{2-}	Sulfate
Na_2MoO_4	Sodium molybdate
Ti	Titanium
Ti-Al	Titanium-aluminium
$TiCl_4$	Titanium tetrachloride
TiF_4	Titanium tetrafluoride
TiO_2	Titanium dioxide, titania, titanium white
$TiCl_4$	Titanium chloride
$TiOSO_4$	Titanyl sulfate
UF	Ultrafiltration
V	Vanadium
VMD	Vacuum membrane distillation
V_2O_5	Vanadium oxide
$VOCl_3$	Vanadium oxychloride
W	Tungsten
WO_3	Tungsten oxide
WO_4^{2-}	Tungstate
Zr	Zirconium
$ZrCl_4$	Zirconium tetrachloride
ZrO_2	Zirconium dioxide
$ZrOCl_2$	Zirconium oxychloride

References

[1] X. Wu, L. Cao, Z.W. Zhang, Annual review of molybdenum in 2010, China Molybdenum Ind. 34 (2010) 1–8.

[2] L.J. Jiang, Y. Liu, W.Z. Zhang, Annual review of molybdenum in 2011, China Molybdenum Ind. 36 (2012) 1–9.

[3] X. Wu, L. Cao, Z.W. Zhang, Annual review of molybdenum in 2012, China Molybdenum Ind. 37 (2013) 1–7.

[4] H. Huang, F.L. Chen, Y. Jiang, The current situation of domestic molybdenum resources and molybdenum smelting analysis, Yunnan Metall. 43 (2014) 66–70.

[5] W.Z. Zhang, Q.S. Xu, Present situation of molybdenum resource development of our country and development trend, Express Inf. Mining Ind. 25 (2006) 1–4.

[6] T.G. Xiang, Molybdenum Metallurgy [M], Revision, Central South University Press, Changsha, (2009).

[7] Z.Z. Luo, X.Q. Yang, L.B. Liao, Progress in molybdenum smelting and processing technology of China, China Molybdenum Ind. 32 (2008) 14–18.

[8] J. Chen, D.J. Li, D.C. Liu, Progress in research of molybdenite smelting technology, Yunnan Metall. 38 (2009) 33–36.

[9] X.W. Wang, M.Y. Wang, D.W. Feng, X.Y. Sun, J. Hu, Y.F. Peng, Investigation on electrodialysis pre-concentration of the mother liquor from ammonium molybdate crystallization by acid precipitation, The Third Symposium (China) on the Application of Membrane Separation Technologies in Metallurgical Industry, 2009, 136–139.

[10] Z.P. Cheng, D.H. Lu, Q.M. Di, New technology for comprehensive utilization of wastewater from ammonium molybdate-Membrane technology, The Third Symposium on the Application of Membrane Separation Technology in Metallurgical Industry, 2009, 141–143.

[11] J. Yu, Z.P. Cheng, Y.K. Tong, Investigation on removal of impurity chloridion from molybdate salt by nanofiltration, The second National (China) Symposium on Application of Membrane Separation Technologies in Metallurgical Industry, 2006, pp. 296–298.

[12] W.Z. Zou, X.L. Guo, X.Y. Xie, Titanium Manual[M], Chemical Industry Press, Beijing, (2012).

[13] Y.Q. Liu, H. Ding, D. Xie, Titanium resource and the environmental effects of sponge titanium production in China, Frontiers of Earth Science 21 (2014) 281–293.

[14] Cao Jian-fei. Titanium sources and their tapping and utilization, Chem. Min. Geol. 18 (2) (1996) 127–134.

[15] X. Wu, J. Zhang, The distribution and characteristics of titanium resources in China, Titan. Ind. Prog. 23 (2006) 8–12.

[16] J.B. Li, S.Z. Chen, Application of new acid dissolved rich titanium product in titanium dioxide production by sulfuric acid method, China Coat. 27 (2012) 55–60 67.

[17] J. Yang, K. Li, A.T. Tang, Comprehensive utilization of Ilmenite resources: present status and future prospects, Mater. Rev. 17 (2003) 44–46.

[18] Y.J. Li, X.Y. Zhang, Process status and development trend of titanium sponge, Dev. Appl. Mater. 28 (2013) 102–107.

[19] X. Feng, L.Y. Jiang, Y. Song, Titanium white sulfuric acid concentration by direct contact membrane distillation, Chem. Eng. J. 285 (2016) 101–111.

[20] T.W. Xu, Z.M. Liu, Experiment study on the recovery of sulfuric acid from titanium white waste by membrane separation, Water Treat. Technol. 25 (1999) 204–208.

[21] C.R. Li, T.W. Xu, X. Zhang, Volume type of diffusion dialyzing membrane component design and manufacture and recycling of waste acid, Paper Collection of the Third National Meeting of Membrane Separation Technique in Metallurgical Industry, 2014, 51–55.

[22] Q. Li, Q.X. Zhang, G.Q. Zhang, K.G. Zhou, Study on direct concentration of waste sulfuric acid from titanium oxide hydrolysis by vacuum membrane distillation, Rare Metal Cemented Carbide 146 (2001) 1–5.

[23] M. Regel-Rosocka, A review on methods of regeneration of spent pickling solutions from steel processing, J. Hazard. Mater. 177 (2010) 57–69.

[24] Y.J. Zhao, Y. Zhang, W. Xing, H.N.P. Xu, Treatment of titanium white waste acid using ceramic microfiltration membrane, Chem. Eng. J. 111 (2005) 31–38.

[25] X.M. Zhang, Z.G. Li, Application of diffusion dialysis in non-ferrous metallurgical industry of waste acid recycling [J], The Fourth Symposium (China) on Application of Membrane Separation Technologies in Metallurgical Industry, 2014, pp. 91–93.
[26] Q.X. Zhang, L.S. Xiao, Application of membrane technology in metallurgy of rare metals, Chin. J. Rare Metals 27 (2003) 1–6.
[27] S.C. Liu, W.D. Yan, S. Xi (Information Center of Ministry of Land and Resources, P. R. China) (Eds.), The World Mineral Resources Annual Review (2011–2012), Geological Press, Beijing, 2012, pp.113–118.
[28] H.G. Li, Metallurgy of Rare Metals, Metallurgical Industry Press, Beijing, (2009).
[29] Z.W. Zhao, Y. Liang, X.H. Liu, H.G. Li, A.L. Chen, X.X. Liang, Sodium hydroxide digestion of wolframite by reactive extrusion, Chin. J. Nonferrous Metals 21 (2011) 2946–2951.
[30] H.Y. Shi, Z.W. Zhao, Thermodynamics analysis on caustic sodium decomposition of wolframite, China Tungsten Ind. 21 (2006) 24–27.
[31] M.Y. Wang, M.X. Jia, Y.W. Xiao, The present situation and countermeasures of sustainable development of tungsten resources China, Non-ferrous Metal Eng. 2 (2014) 18.
[32] Z.W. Zhao, J.T. Li, X.Y. Chen, X.H. Liu, Technology status and development of scheelite metallurgy in China, Ferrous Metals Sci. Eng 4 (2013) 11–14.
[33] J.N. Cui, Development and comparison of tungsten smelting technology, Rare Metals Cemented Carbides 32 (2004) 51–55.
[34] H.Q. Li, Z.Y. Tang, L.H. He, The application and progression exchange technology in of tungsten metallurgy, China Tungsten Ind. 29 (2014) 34–39.
[35] Z.R. Hu, The birth and development of tungsten smelting process of ion exchange method, China Tungsten 12 (1997) 41–43.
[36] C.J. Gao, Advances in liquid membrane separation, Proceeding of the Second National Symposium Membrane and Membrane Process, 1996, 2–6.
[37] D. Fang, W.Y. Yang, Preparation, Properties and Application of Perfluorinated Ion-Exchange Membrane [M], Chemical Industry Press, Beijing, (1993) 172–273.
[38] L.S. Xiao, Q.X. Zhang, Industry experimental study on recovery of free alkali in sodium tungstate solution by membrane electrolysis, Membr. Sci. Technol. 21 (2001) 14–16.
[39] Q.X. Zhang, A.P. Luo, Research of single anode in the removal of alkali in sodium tungstate solution, J. Central South Inst. Min. Metall. 24 (Tungsten Collection) (1994) 126–135.
[40] L.S. Xiao, W. Wang, Q.X. Zhang, Tungstate solution research on membrane electrolysis anode system of choice, The First National Proceedings of the Seminar on Application of Membrane Separation Technology in Metallurgical [C], 1999, pp. 92–95.
[41] L.S. Xiao, Q.X. Zhang, G.Q. Zhang, Study on ion-membrane electrolytic cell structure of sodium tungstate solution [A], The First National Proceedings of the Seminar on Application of Membrane Separation Technology in Metallurgical [C], 1999, pp. 96–98.
[42] A.P. Luo, Q.X. Zhang, Y.L. Liu, Removal of free alkali from crude sodium tungstate by electrodialysis, The First National Symposium (China) on Application of Membrane Separation Technologies in Metallurgical Industry, 1999, pp. 130–133.
[43] G.Q. Zhang, L.S. Xiao, Q.X. Zhang, A method for recovery of tungsten and ammonium chloride from the mother liquor of ammonium paratungstate, Chinese patent, Application number: 200510032251.9.
[44] G.Q. Zhang, G.H. Shang, L.S. Xiao, Treating mother liquor of amnion paratungstate crystallization by nanofiltration, J. Rare Metals 31 (2007) 346–350.
[45] G.Q. Zhang, G.H. Shang, L.S. Xiao, Pressure driven membrane process of secondary tungsten acid ammonium crystallization mother liquor (II)—ultrafiltration separation, The Second National Workshop on Membrane Technology in Metallurgy Application, 2006, 309–313
[46] C.W. Zeng, G.Q. Zhang, W.J. Guan, Preparation of ammonium metatungstate solution using bipolar membrane electrodialysis, China Tungsten Ind. 27 (2012) 32–34.
[47] W.J. Guan, G.Q. Zhang, C.W. Zeng, Study on preparation of ammonium metatungstate solution using bipolar membrane electrodialysis[C], Papers of the Fourth Session of National Membrane Separation Technique in Metallurgical Industry Symposium, 2014, 111–116.

[48] W.J. Guan, G.Q. Zhang, C.W. Zeng, Study on preparation of ammonium metatungstate solution using bipolar membrane electrodialysis, The Fourth National (China) Symposium on Application of Membrane Separation Technologies in Metallurgical Industry, 2014, 111–116.
[49] Y.K. zhang, Vanadium (V)—metals applications in modern military, Metal World 5 (1994) 22.
[50] J. Li, Y.L. Zhang, Summary of vanadium mineral resources and extraction process of vanadium, Hubei Agric. Mechanization (1) (2009) 60.
[51] Q.X. Wang, H.L. Ma, Status and prospects of V_2O_5 product research & production and Chinas's vanadium resources, Conserv. Util. Miner. Resour. 5 (2009) 47–50.
[52] Y.J. Pan, Non-Ferrous Metals Extraction Technique Manual, Metallurgical Industry Press, Beijing, (1999).
[53] D.X. Huang, Extracting Vanadium and Smelting Steel [M], Metallurgical Industry Press, Beijing, (2002).
[54] L.Q. Duan, Q.H. Tian, X.Y. Guo, Review on production and utilization of vanadium resources in China, Hunan Nonferrous Metals 22 (2006) 17–20.
[55] H.Y. Zhao, Analysis of vanadium resources utilization and demand for vanadium in China, Conserv. Util. Miner. Resour. 2 (2014) 54–58.
[56] X.D. Hui, Y.X. Wang, Z.X. Wu, The current research and application of process of vanadium extraction from stone coal, China Nonferrous Metall. 2 (2010) 10–16.
[57] D.Q. Zang, Y.G. Li, Y.Y. Zhang, The current research situation of vanadium and titanium resources and its utilization at home and abroad, Sichuan Nonferrous Metals (2011) 1–6.
[58] W.L. Liu, X.W. Wang, M.Y. Wang, Mineral decomposition process of vanadium recovery from stone coal by low temperature sulphating roasting, Chin. J. Nonferrous Metals 19 (2009) 943–948.
[59] X.W. Wang, M.Y. Wang, State-of-art development of vanadium extraction from stone coal, Iron Steel Vanad. Titan. 33 (2012) 8–14.
[60] X.W. Wang, G.Q. Zhang, M.Y. Wang, A stone coal sodium roasting method of extracting vanadium, Chinese patent, ZL200810030460.3 [P].2008.
[61] X.W. Wang, W.L. Liu, G.Q. Zhang, Recovery of vanadium by nanofiltration from the leaching solution of stone coal roasted with sodium chloride, J. Process Eng. 9 (2009) 289–292.
[62] R.Q. Wu, Rationally developing zirconium resources to meet domestic demand, China Metal Bull. 41 (2012) 16–19.
[63] Y. Wen, Zirconium and hafnium resource in China, Rare Metals Lett. 27 (2008) 46–47.
[64] L.P. Wu, W.Z. Fei, Properties and application prospect of overview of zirconium oxide, Sichuan Chem. 16 (2013) 25–27.
[65] D.M. Jiang, L.J. Wang, X.K. Che, The Preparation Technology and Application of Zirconium Oxychloride [M], Metallurgical Industry Press, Beijing, (2012).
[66] B.K. Xiong, The application status of zirconium in the nuclear power industry, Rare Metal 24 (2005) 45–47.
[67] B.K. Xiong, Overview of zirconium and hafnium overview production technology, Rare Metal 24 (2005) 42–43.
[68] W.D. Chen, G.Q. Yan, J.X. Zhao, Research on the decomposed process of zircon sand by NaOH and $CaCl_2$, Titan. Ind. Prog. 30 (2013) 38–40.
[69] Z.J. Li, Y.X. Huang, H.C. Yao, Research on the preparation of zirconium oxychloride using zirconium carbide alkali fusion, Rare Metal 30 (2006) 415–418.
[70] S.M. Huang, W.D. Chen, Y.F. Wang, Current situation and developing strategy of the zirconium oxychloride industry in China, Titan. Ind. Prog. 31 (2014) 1–7.
[71] Y.F. Lou, W.H. Zhang, Application of homogeneous charged membrane technology in hydrometallurgy, The Fourth Symposium on the Application of Membrane Separation Technology in Metallurgical Industry, 2014.
[72] Z.W. Chen, G.Y. Yi, W.R. Liu, Application of inorganic ceramic membrane in the production of nano powder material, The Third Symposium on the Application of Membrane Separation Technology in Metallurgical Industry, 2009.

CHAPTER 7

Scattered and Rare Earth Metals

L.-S. Xiao

School of Metallurgy and Environment, Central South University, Changsha, Hunan Province, China

7.1 Introduction

Scattered metals include gallium (Ga), indium (In), thanllium (Tl), germanium (Ge), selenium (Se), tellurium (Te), and rhenium (Re) [1]. They are widely recognized as the supporting materials for developing high tech industries, for example, electronic computer communication, aerospace, energy, and medicine & health sectors.

This group of metals has very low abundance in the earth's crust. There are totally 195 minerals that comprise scattered metals [1,2]. They generally occur in the relevant minerals with the isomorphous state and seldom form individual ores with mining value. To date only very few individual Ge ores, Se ores, and Te ores are identified, nevertheless with small deposit scale. Some basic information about the scattered metals is summarized in Table 7.1.

Most of the elementary or compound products of scattered metals are generated as by-product in metallurgical processes of main metals or recovered from chemical or metallurgical process intermediates or wastes, and coal dust [1]. Table 7.2 displays the main metal ores and other resources for scattered metals.

There are totally 17 kinds of rare earth elements [1–3], which are lanthanum (La), cerium (Ce), praseodymium (Pr), neodymium (Nd), promitnium (Pm), samarium (Sm), europium (Eu), gadolinium (Gd), terbium (Tb), dysprosium (Dy), holmium (Ho), erbium (Er), denebium (Tm), ytterbium (Yb), lutetium (Lu), scandium (Sc), and yttrium (Y). They are named "the treasure-house of new materials in the 21st century" by scientists nowadays [2]. The basic information about rare earth metals is summarized in Table 7.3.

There are 18 kinds of important industrial rare earth minerals; hereinto the most important ones are monazite (also called cerium lanthanum phosphate ore), bastnaesite, xenotime, etc. Rare earth metal resources mainly distribute in China, Russia, the United States, Australia, and India [4]. China's rare earth reserves are 36 million tons (based on 2009 data), accounting for about 36% of the global reserve, and its output is about 120,000 tons, taking up about 97% of the world's total production. The rare reserves in the United States are 13 million tons, about 13% of the global reserves; nevertheless, the output in the United States is zero.

Membrane-Based Separations in Metallurgy
http://dx.doi.org/10.1016/B978-0-12-803410-1.00007-4

Table 7.1: Summary of the properties of scattered metals

Metal	Symbol	Atomic Number	Isotope	Physical and Chemical Properties
Gallium	Ga	31	Ga-69, Ga-71	Gray blue or silver white, melting point of 29.8°C, boiling point of 2204°C. Easily hydrolyzed, very stable in the air.
Indium	In	49	In-113, In-115	Silver white with little light blue, strong plasticity and ductility. From room temperature to the melting point, indium has a slow with oxygen in air to form a thin oxide film (In_2O_3).
Thallium	Tl	81	Tl-81, Tl-205,Tl-203	Silvery white, soft, inelastic, easy to melt. Having black oxide film coated on the surface in air, soluble in nitric acid and sulfuric acid, difficult to react with hydrochloric acid, insoluble in water.
Germanium	Ge	32	Ge-70, Ge-72, Ge-73,Ge-74, Ge-76	Gray white, luster, quality hard. Insoluble in water, dilute hydrochloric acid, caustic solution, dissolved in aqua regia, concentrated nitric acid or sulfuric acid, with both sexes, therefore dissolved in molten alkali peroxide, alkali, alkali metal nitrates or carbonate, stable in air.
Selenium	Se	34	Se-74, Se-76, Se-77, Se-78, Se-80, Se-82	Gray with metallic luster. Insoluble in water and alcohol, soluble in sulfuric acid, nitric acid, alkali.
Tellurium	Te	52	Te-120, Te-122	Good heat transfer and conductive properties, medium density, low melting and boiling point. Black or silver white. Soluble in sulfuric acid, nitric acid, aqua regia, potassium cyanide, potassium hydroxide; insoluble in cold water and hot water, carbon disulfide.
Rhenium	Re	75	Re-185, Re-187	High melting point of 3180°C. Silver white metal or gray to black powder. Insoluble in hydrochloric acid, soluble in nitric acid.

Table 7.2: Main metal deposits and other resources for scattered metals

Symbol	Main Metal Ores and Other Resources
Ga	Zinc and aluminum ores, dcoal dust
In	Hematite, lead zinc ore and tungsten ore
Tl	Nonferrous metal sulfide
Ge	Hematite, lead zinc, coal
Se	Sulfide, anode mud of electrolytic copper, berzelianite, clausthalite, manganese nodule
Te	Anode of electrolytic refining of copper and lead
Re	Molybdenum and copper metallurgy

Table 7.3: Summary of the properties of the rare earth element

Metal	Symbol	Atomic Number	Isotope	Physical and Chemical Properties
Scandium	Sc	21	Sc-45	Silvery white, soft, soluble in acid
Yttrium	Y	39	Y-89	Ash black powder, metal luster
Lanthanum	La	57	La-138, La-139	Silvery white, soft, malleability, chemical properties and lively, soluble in acid
Cerium	Ce	58	Ce-136, Ce-138, Ce-140, Ce-142	Gray metal, luster crystal, chemical properties and lively, malleability
Praseodymium	Pr	59	Pr-141	Silvery white, soft, malleability
Neodymium	Nd	60	Nd-142, Nd-143, Nd-144, Nd-145, Nd-146, Nd-148	Silvery white, chemical properties and lively, malleability
Promethium	Pm	61	Pm-145	Artificial radioactive element
Samarium	Sm	62	Sm-144, Sm-147, Sm-148, Sm-149, Sm-150, Sm-152, Sm-154	Gray metal, radioactive
Europium	Eu	63	Eu-151, Eu-152	Silver-white metal
Gadolinium	Gd	64	Gd-152, Gd-154, Gd-155, Gd-156, Gd-157, Gd-158, Gd-160	Silver-white metal
Terbium	Tb	65	Tb-159	Colorless crystalline powder, toxic
Dysprosium	Dy	66	Dy-156, Dy-158, Dy-160, Dy-161, Dy-162, Dy-163, Dy-164	Its oxide is white, and salt is slightly yellow
Holmium	Ho	67	Ho-165	Silvery white, soft, malleability
Erbium	Er	68	Er-162, Er-164, Er-166, E-167r, Er-168, Er-170	Silver-white metal. Salts and oxides show pink to red
Thulium	Tm	69	Tm-169	Silver-white metal, malleability
Ytterbium	Yb	70	Yb-168, Yb-170, Yb-171, Yb-172, Yb-173, Yb-174, Yb-175	Silver-white metal, glossy
Lutetium	Lu	71	Lu-175, Lu-176	Silvery white, soft, malleability, chemical properties and lively, soluble in acid

The production in Russia is also zero, although it holds 19 million tons of the reserves. India's rare earth reserves are 3.1 million tons and the output is 2700 tons. Japan has no output of any products, but imports rare earth metal raw material and primary products.

The process of manufacturing rare earth metal products generally covers steps of separation, leaching, smelting, and refining [2,5–7]. The most critical ones are smelting and refining. Solvent extraction, ion exchange, and extraction chromatography are the main technologies used in smelting and refining.

Solvent extraction generally has the advantages of high selectivity and high separation efficiency. Many rare earth deposits, for example the baiyuneboite, contain several or even more than ten types of rare earth minerals, and in most cases many concomitant impurities [eg, uranium (U), thorium (Th), niobium (Nb), fluoride (F)] [2]. Therefore, separation of single metal from the mixture is a complicated and tedious process. Conventionally, more than hundred or even hundreds stages of cascade extractions are necessitated, largely due to the small separation factor neighboring metals. The consequence is the high consumption of organic extractant and serious environmental pollution.

Ion exchange process can generate single and highly purified rare earth metal products in a relatively easy manner. Nevertheless, it is not suitable for continuous operation and is limited by long production cycle and low capacity. Besides, the eluants or stripping agents, such as ethylene diamine tetraacetic acid (EDTA) and 2-hydroxyethylethylenediaminetriacetic acid (HEDTA), are expensive, causing lack of competitiveness for the products.

Extraction chromatography synergizes the strength of solvent extraction and ion exchange methods. Notwithstanding the gradient elution adopted brings about inconvenience in operation and the consumption of acid is large.

7.2 Extraction Using Liquid Membrane for Mixed Rare Earth Enrichment

Pool leaching—oxalic acid ($H_2C_2O_4$) or ammonium bicarbonate (NH_4HCO_3) precipitation is widely used for extracting mixed rare earth minerals from ion adsorption rare earth ores, which is unique to southern China. The method is characterized of a long process flow and high concentration of impurities [eg, calcium (Ca), magnesium (Mg), manganese (Mn)] in the leaching solution; even the pool leaching alone will take around 3–5 days.

Separation by emulsion liquid membrane (ELM) has been investigated for quite a long time. The disadvantages of using the traditional carriers, P507 and P204, are the strong acidity required in the internal aqueous phase, and the high content of impurity metal ions (eg, Ca(II)) transported to the stripping solutions. The concentration of Ca(II) can be up to 1–10 g L^{-}, which is unfavorable for further solvent extraction to separate rare earth metals.

In Qin et al.'s work for enriching the rare earth metals from the leaching solution [8], the focus was related to a novel extraction agent, Cyanex272. The actual leaching solution of the medium Gd & rich Eu type rare earth ore in Southern China was used as feed. Its content of rare earth metals (Re_2O_3 equivalent) was 1.7 g L^{-1}.

The experimental results showed that ELM with Cyanex 272 as the carrier was feasible in rare earth ore mentioned earlier. The optimum liquid membrane formulation and operation conditions are as follows:

Liquid membrane phase: 3% (v/v) Cyanex272 as an extractant, 2% (w/v) LMS-2 for surfactant, sulfonated kerosene as a diluent.
Internal phase: 2–4 mol L^{-1} HCl, $R_{oi} = 2:1$, $R_{ew}=1:30$, $t = 15$ min. (R_{oi} is oil and internal phase ration in volume; R_{ew} is emulsion phase and external aqueous phase ratio in volume)

Temperature: Room temperature.
Feed solution: $[R_E^{3+}] = 1 - 2\,g\,L^{-1}$, [NaAc]=6–10 g L^{-1}, pH = 5.0. (Note: R_E represents rare earth metal; Ac indicates acid radical)
Stirring speed: 250–300 r min^{-1}.

A single-stage batch-enrichment delivered a recovery or extraction efficiency of more than 98%. The rare earth metal concentration in the internal aqueous phase is up to 90 g L^{-1}. The primary nonrare earth impurities like Ca, Mg, and aluminum (Al) barely entered the internal phase.

7.3 Researches and/or Applications Addressing Specific Metal

7.3.1 Indium

In is one of the most important materials for modern electronic industry. Its production mainly relies on lead (Pb) and zinc (Zn) metallurgy, where it is the by-product of the mainstream process.

7.3.1.1 Recovery of indium from metallurgical process

Solvent extraction has been generally applied to separate In(III) ions from other ions in the solutions. Jia et al. investigated a coupled supported liquid membrane-ultrafiltration (SLM-UF) process to extract In [9]. The schematic of the SLM-UF system is shown in Fig. 7.1. The feed is acid leaching solution containing In(III) and trivalent iron ion (Fe(III) ions of 1 and

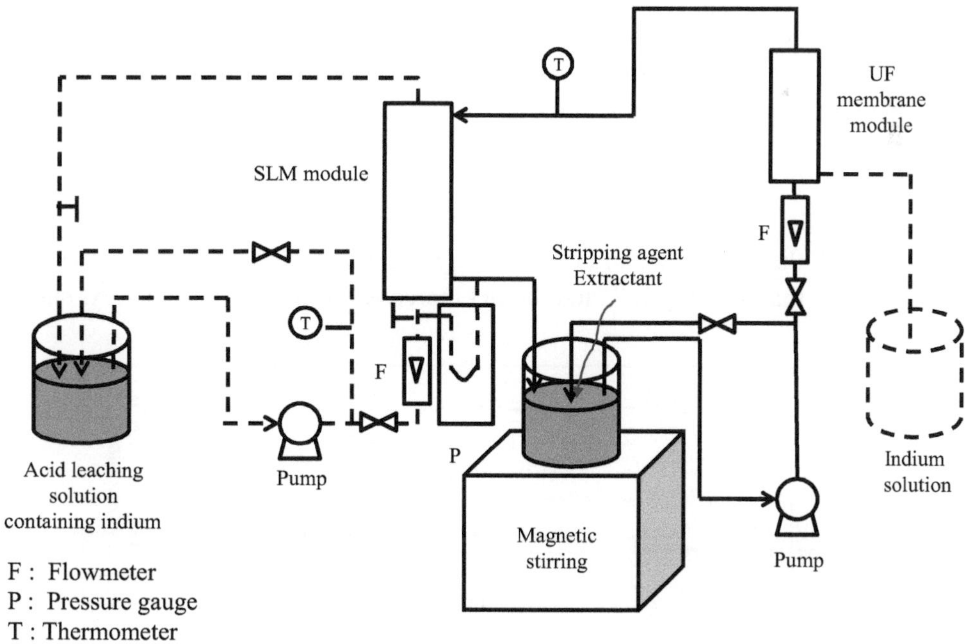

Figure 7.1: The schematic diagram of supported liquid membrane-ultrafiltration *(SLM-UF)* system for indium recovery.

9 g L^{-1}, respectively. The solid membrane used as a support was microporous polyvinyl difluoride (PVDF) hollow fiber membrane, while the liquid organic phase was composed of kerosene as a diluent and di-(2-ethylhexyl) phosphoric acid (D2EHPA) as an extractant. The two components have a volume ratio of 3/7.

Feed was circulated in the shell channel. At the interface of feed and organic phase, In(III) was extracted and form complex with extractant. The complex transported through the liquid phase and released In(III) into stripping phase at the lumen channel. The two streams were in counter-current flow. In addition, the stripping solution contains dispersed organic phase for strengthening the mass transfer. The function of UF is for oil and water separation and recovery. The aqueous phase was enriched of In(III) and the organic phase was reused in compensating and regenerating the liquid trapped in SLM.

It was found that the packing density SLM module had an optimal value of 35%, which resulted in an extraction rate of 89% and membrane flux of 2.7 g m^{-2} h^{-1} and a separation factor of In/Fe 426. About 98% of In(III) in the feed was extracted and 89% recovered in the stripping phase. The negative impact of SLM aging was obviously observed; the recovery rate at third operation was only 56%, in contrast with 71% when the organic phase in both SLM and stripping solution was refreshed.

7.3.1.2 Treatment of indium nanoparticles

Nanosized indium oxide (In_2O_3) is generally produced by liquid phase method, which can ensure homogeneous composition and narrow distribution in particle size. Washing after precipitation or crystallization of the particle is necessary. Xiao et al. introduced the application of ceramic membrane separation in washing In_2O_3 nanopowder [10]. The system setup is the same as that in Fig. 6.21. The effective membrane area of the inorganic ceramic membrane within the system is 0.42 m^2 and the pore diameter is 50 nm. The operation performance of one of the systems installed in a plant located at Hunan Province is shown in Fig. 7.2. The Cl(I) ions passed through the membrane and its concentration in the In_2O_3 slurry decreased from 0.13 mol L^{-1} to less than 100 mg L^{-1}, and the pH increased from 1 to around 7. As shown in

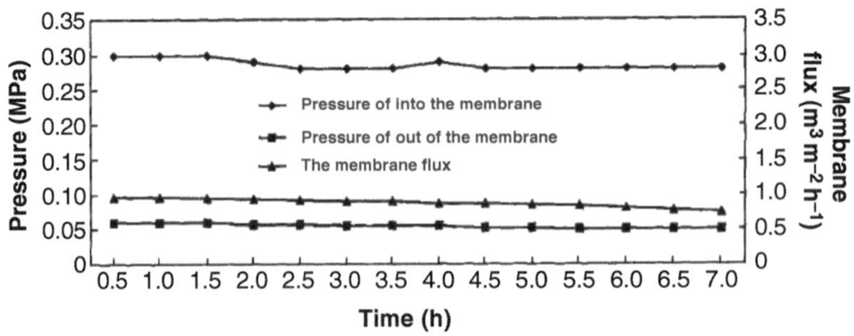

Figure 7.2: The pressure and flux change during the washing of nanosized indium oxide using inorganic ceramic membrane.

Fig. 7.2, the inlet and outlet pressures of the membrane module are stable, and the fluctuation of the trans-membrane flux was trivial. The average liquid flux is 0.96 $m^3\ m^{-2}\ h^{-1}$.

7.3.2 Rhenium

Re and its alloys are widely applied in aerospace, electronic industry, and petrochemical industry. According to the statistics by US Geological Survey Bureau in 2013, high-temperature alloy consumes the most of Re product, which takes up around 80% of total Re consumption. Catalyst is the second largest consumption areas.

The content of Re is up to 0.16% in the molybdnum (Mo) concentrate obtained from porphyry copper. Additionally, the smelting fume of some copper (Cu) ore, platinum (Pt)-group metal ores, Nb ore, sphalerite, the wastewater generated in treating low-grade Mo minerals, and Re containing wastes (eg, waste alloys) can all be used as feedstock. Selection of Re extraction methods depends on the resources and the product quality requirements. The major steps in Re metallurgy consist of following steps: Re containing solution production from feedstock, solution separation and purification to enrich Re, reduction by hydrogen (H_2) to yield pure Re power, and fabrication into production via methods such as powder metallurgy.

7.3.2.1 Recovery of rhenium [11]

The circulation water and overflow of thickener from Cu metallurgy contain 0.05–0.1 mg L^{-1} Re(IV), (VII). A hybrid separation combining complex formation (CF) and UF was employed to recover Re. High molecular weight polyelectrolyte was added to the solution to form complex with Re, which was then retained and concentrated by UF. Fig. 7.3 is the process for recovering Re. The complex was treated by sodium sulfite (Na_2SO_3) to regenerate polyelectrolyte and release Re(IV), (VII).

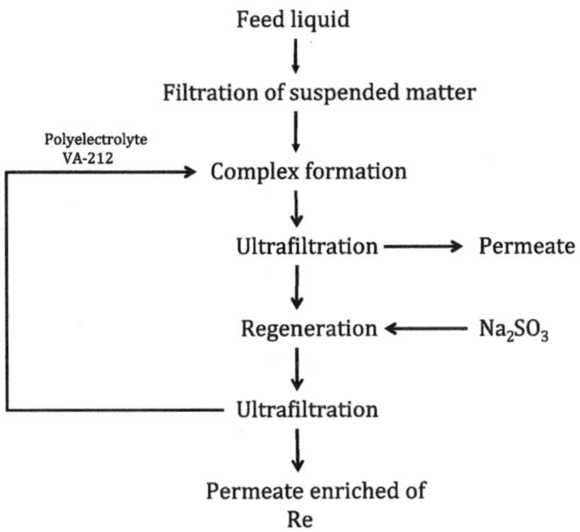

Figure 7.3: Principal flowsheet for recovery of Re by CF-UF method.

7.3.3 Lanthanum

In the rare earth elements family, the abundance of La in the crust is 32 ppm, accounting for 14.1% of the total rare earth. It ranks after Yr and Ce as the third most abundant rare earth metal. La is the most active one in rare earth metal family; thus can be used as the reducing agent in metal thermal reduction process, to prepare other rare earth metals [12]. The applications of La have also covered the modifier in steel production, H_2 storage material, magnetic refrigeration materials, and usage in nuclear industry [13].

Molten salt electrolysis and metal thermal reduction are the two methods of preparing elementary La [13,14]. Regarding lanthanum chloride—potassium chloride ($LaCl_3$—KCl) binary molten salt system, it has been almost abandoned due to evolution of chlorine (Cl_2) and easy moisturization of $LaCl_3$ that will cause low efficiency in electrolysis. The major disadvantage of lanthanum fluoride—lithium fluoride (LaF_3—LiF) or LaF_3—LiF—barium chloride (BaF_2) molten salt system is the system corrosion.

Metal thermal reduction uses principally lithium (Li) and Ca as reducing agents [13]. The reaction for heat reduction of $LaCl_3$ by Li is $LaCl_3 + 3Li = La + 3LiCl$. Inert protective atmosphere is necessary for the reaction. Vapor pressure of La metal at 1000°C is far lower than those of Li and lithium chloride LiCl, whereby 99.974% La is obtained via vacuum distillation [13,15]. Ca thermal reduction uses LaF_3 as a raw material and reaction is also taking place inside protective environment. After casting, cooling, and smashing of reaction products and removing the slag, vacuum remelting is used for extracting high purity La [13].

7.3.3.1 Recovery of lanthanum

Xu et al. investigated the effect of the ions concentration, electrolyte, and pH on the rejection efficiency of (La(III)) iond by ceramic membrane with 5 nm apertures [16]. The raw feed solution is La(III) wastewater discharged from processes in metallurgical, electronic, and chemical industries. The temperature for the experiment is 25 ± 2°C, the trans-membrane pressure is 0.2 MPa, and the surface velocity of feed solution is 3 m s^{-1} and pH 6.2.

The experimental results showed that the interception rate of La(III) decreases with the increase of ion concentration and the membrane flux declines slightly. The phenomenon of rejection is due to the presence of electric double layer (EDL). In addition, the dissociation and chemical adsorption of water on the surface of the ceramic membrane induces hydroxylation, which can significantly change the membrane surface chemistry. When the mass concentration of La(III) in the feed rises from 9.1 to 245.0 mg L^{-1}, the interception rate of La(III) decreases from 82.4% to 66.2%, and the membrane flux declines by 5%.

The effect of sodium chloride (NaCl) concentration on the flux and rejection is shown in Fig. 7.4. Apparently with increasing NaCl concentration to 100 mg L^{-1}, the interception rate

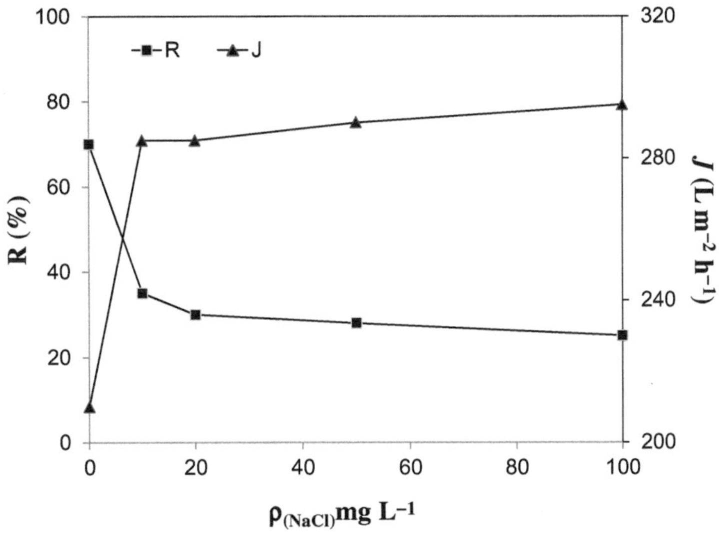

Figure 7.4: Effects of NaCl mass concentration on permeate flux and rejection.

of La(III) decreases from about 69% to 26%, and the membrane flux increases by 30%. This is because NaCl neutralizes and shields the negative charge on the surface. Therefore surface charge density was lowered and the repelling toward La(III) weakened apparently.

Variation of the pH of the solution not only changes the membrane surface charge, but also the existence form of La in water. Therefore, it has obvious effect on the membrane separation performance, as revealed in Fig. 7.5. When pH = 1, the La mainly exists in the form of La(III) ion and the interception rate of La^{3+} is 25%. When pH = 10, lanthanum hydroxide ($La(OH)_3$) is the principal existing state, and the rejection of overall La is more than 99.7%; under this condition, the membrane flux was up to 127 L m^{-2} h^{-1} at 0.1 MPa. When pH = 6–8, there exists a substantial amount of $La(OH)^{2+}$. It has the severe tendency to cause pore blockage of the membrane, resulting in the depression of membrane flux. Therefore, adjusting the pH value will be helpful for ceramic UF application in recovering La from wastewater. Using 1% HNO_3 solution to clean the membrane for 1 h, the membrane flux reached more than 95% of that of unused membrane.

One attempt investigated recovery of La compounds via MD from the sulfuric acid (H_2SO_4) after extraction of apatite phosphor-gypsum [17]. A modeled 12% H_2SO_4 solution saturated with calcium sulfate ($CaSO_4$) was used. The La compounds precipitated after cooling the treated acid solution, with a recovery ratio of 25%. The inorganic salts rejection coefficients were close to 100% in all the MD experiments. The process was continued till the acid concentration reached 40%. During concentration of the model solution, crystallization on the feed-side membrane surface was observed. This was due to the supersaturation of the small part of the feed solution remaining in the compartment in the periodical operation.

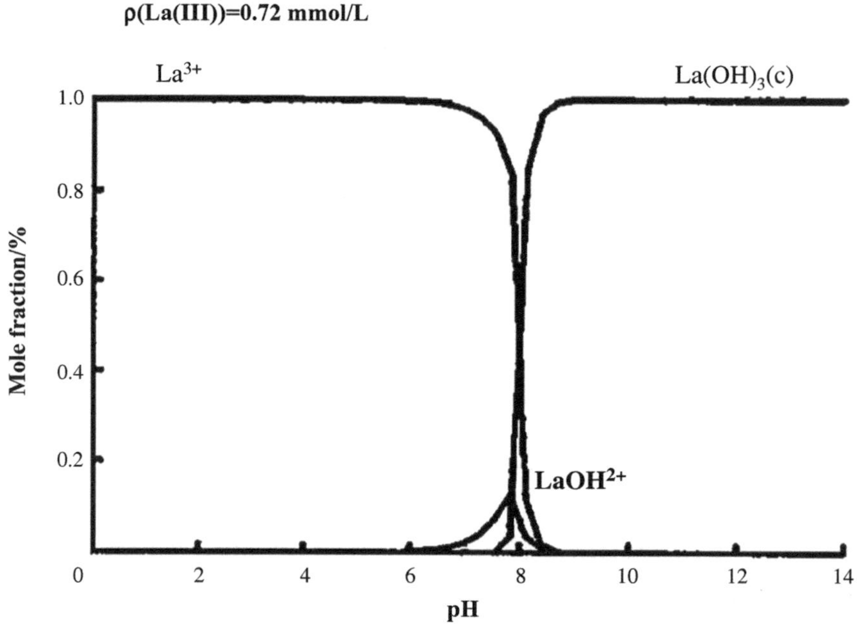

Figure 7.5: Speciation diagram of La in solution with different pH.

7.3.4 Europium

Eu is a silvery white metal that has important application in laser materials and nuclear power industry. It has abundance of 0.000106% in the earth's crust and is the rarest metal within the family of rare earth metals. It mainly occurs in eremite and hamartite. The two isotopes of Eu are Eu-151 and Eu-153. Eu doping is very critical for improving the life span and conversion efficiency of light emitting diode (LED). The technologies employed in producing highly purified europium oxide (Eu_2O_3) comprise extraction chromatography, reduction extraction, etc. [18]. The reduction making it easier for separation of Eu from other rare earth metals relies on mainly Zn and electrolytic reduction. To avoid Zn contamination of the product, the latter approach receives more attention [18]. The frequently-used method for producing elementary Eu is vacuum distillation of the mixture of Eu_2O_3 and metal La, during which Eu (VI) is reduced by La.

7.3.4.1 Production of europium compounds by membrane electrolysis reduction

In Long et al.'s work [19], membrane electrolytic reduction was investigated for reducing (Eu(II)) ions in the rare earth metal chloride solution. The membrane could be either anion exchange or cation exchange membranes (AEM or CEM). The anode was coated or plated with titinium (Ti) mesh, and Ti mesh or perforated Ti was used as a cathode.

The feed rare earth chloride solution with Eu(III) was put into the cathode chamber liquid; the anolyte is hydrochloric acid (HCl) or solution with HCl and NaCl mixtures. The anode chamber and the cathode chamber were separated by ion exchange membrane (IEM). Eu(III) was reduced to Eu(II) under electrolysis condition. Thereafter, the Eu(III) is severed from other rare earth metals via solvent extraction. Compared with other methods, a higher reduction efficiency or rate can be achieved with membrane electrolysis, which as claimed by the inventors in addition avoids the occurrence of oxidation of Eu(II). The half cell reactions are described as follows:

1. Using AEM: Cl^- in the anolyte transfers to the catholyte
   Anode reaction: $Eu^{3+} + e \rightarrow Eu^{2+}$, $H^+ + e \rightarrow 1/2\ H_2\uparrow$
   Cathode reaction: $Cl^- - e \rightarrow 1/2Cl_2\uparrow$
2. Using CEM
   Anode reaction: $Eu^{3+} + e \rightarrow Eu^{2+}$, $H^+ + e \rightarrow 1/2\ H_2\uparrow$
   Cathode reaction: $Cl^- - e \rightarrow 1/2Cl_2\uparrow$

7.3.5 Cerium

Ce is an active metal that is prone to self-ignition in air and can easily dissolve in acid. Its abundance in the earth's crust is around 0.0046%. The resources for Ce are mainly present in monazite and bastnaesite, as well as products of U, Th, and plutonium (Pu) splitting. Metal Ce is obtained by reduction of cerium oxide (CeO_2) using Ca or Mg, or melting electrolysis of cerium chloride ($CeCl_3$). The major applications of Ce comprise reduction agent, catalyst, alloying agents, additive in special glass.

7.3.5.1 Cerium oxidation by membrane electrolysis [11]

The major approach for separating Ce from other rare earth metals relies on oxidization of Ce(III) to Ce(IV). Electrolytic oxidation is an economical method for achieving this conversion in the rare earth nitrate or sulfate solution. The process will not introduce new metal impurities into solution and does not use chemical reagents. Currently, the technology is principally used for oxidization inside nitrate solution.

For smelting of some mixed-type rare earth concentrate, for example, in Baotou, China, extraction of single rare earth is dominated by using H_2SO_4 system. Membrane electrolytic oxidation of Ce in H_2SO_4 was first examined by Zhang and Xiao [11]. The challenge for the process lies with the low solubility and hence the small content of Ce in H_2SO_4. This situation causes high electric resistance and energy intensiveness.

Addressing this problem, the approach for oxidation of Ce(III) by electrolysis using AEM was studied by Wei et al. [20]. The performance of the electrolysis under two situations, namely with or without Cu deposition at anode was examined. With Cu deposition, the anode and cathode current efficiencies reach 95 and 80%, respectively, and oxidation rate reaches 98% when H_2SO_4 concentration in catholyte is 3–4 mol L^{-1} and current density is appropriate. The powdering rate is 93% and the corresponding current efficiency is 85%.

Electrolysis with simultaneous Cu precipitation Ce oxidation is more energy efficient and can reduce the cell voltage. For 1 kg of CeO_2 produced, 0.2118 kg Cu is generated by deposition.

7.3.6 Other Metals

7.3.6.1 Lutetium

In this part, ELM and SLM are exclusively concerned. Actually, LM dominates the membrane technologies associated with separation and purification of scattered and rare earth metals.

Lu is the hardest and most compact rare earth metal. Its density is 9.8404 g cm^{-3}. Lu is relatively stable when exposed to atmosphere and dissolves in dilute acid solution. The naturally occurred Lu isotopes are Lu 175 and Lu 176. Metal Lu is produced by reduction of hydrated lutetium fluoride ($LuF_3 \cdot 2H_2O$). Its application is currently limited to nuclear power industry and research.

Kumric et al. used SLM technique for recovery of Lu(III) from lutetium chloride ($LuCl_3$) aqueous donor or feed solution [21]. Two different SLM modes (schematically shown in Fig. 7.6) investigated are (1) static or nonrecirculating mode with the acceptor phase being immobilized in the lumen and (2) dynamic or recirculating mode, with a U-shaped module

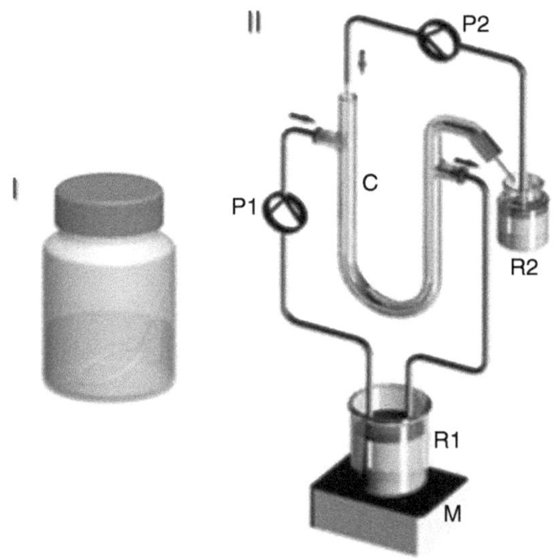

Figure 7.6: Schematic of the experimental setup used for the SLM extraction of Lu(III). (I) The static, HF-LPME system and (II) the dynamic, HF-LPE system. (C, contactor; R1, feed reservoir; R2, stripping reservoir; P1, peristaltic pump for the donor phase; P2, peristaltic pump for the acceptor phase; and M, magnetic stirrer.)

containing a single hollow fiber membrane with a closed loop recirculation of aqueous phases through the module. Both of them were tested using miniaturized module containing a single hollow fiber. The donor phase has a pH value of 3.5 and the stripping phase is HCl solution (2 mol L^{-1}). The organic phase that was immobilized within the polypropylene hollow fiber structure was 5% (v/v) DEHPA in di-n-hexyl ether (DHE).

Operation using dynamic mode shows that the removal and the extraction efficiencies of Lu(III) increased as the donor flow rate increased, while the memory effect was independent of the donor flow rate. The memory effect (M) is a fraction of Lu(III) captured in the immobilized organic phase, with respect to Lu(III) initially present in the donor phase. Theoretically, it is the difference between the removal efficiency (E_R) and the extraction efficiency (E_E).

Fig. 7.7 demonstrates the influence of the extraction time on the process efficiency at the highest donor flow rate of 5.3 cm^3 min^{-1} and the acceptor flow rate of 0.4 cm^3 min^{-1}. Both the E_R and E_E of Lu(III) were featured with a rising trend before being leveled off at around 88 and 84%, respectively, after ~80 min of operation. The memory effect (M) recorded at 5 min was 25%, but gradually decrease. After 60 min, the content of Lu(III) remained in the organic phase was <5% of that extracted from the feed. This result indicates that the stripping reaction was fast enough to avoid Lu(III) accumulation in the membrane. The overall mass-transfer coefficient of Lu(III) was proportional to the donor flow rate. Using the Wilson plot,

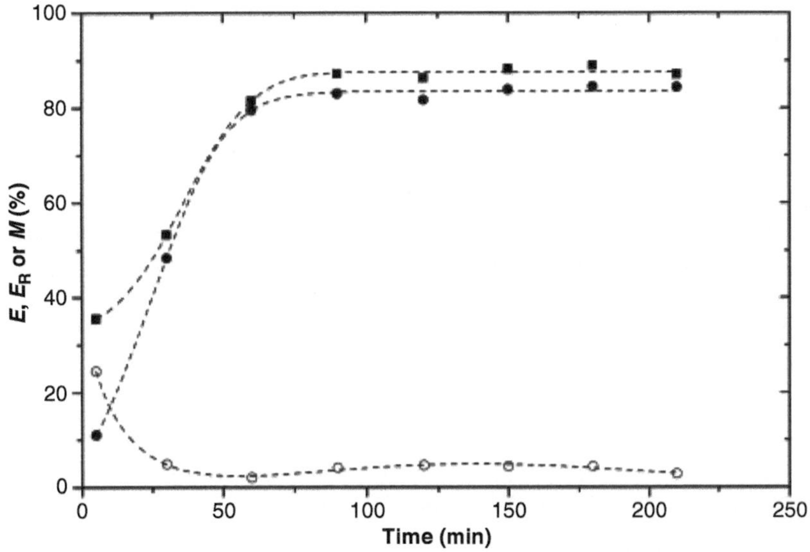

Figure 7.7: Time dependence of the efficiency parameters (●) *E*, (■) E_R, and (○) *M* in the SLM extraction system with the recirculation of aqueous phases. Donor phase: 20 cm^3, 11 μmol dm^{-3} Lu(III), pH 3.5. Acceptor phase: 3 cm^3, 2 mol dm^{-3} HCl. Organic phase: 0.16 mol dm^{-3} DEHPA in DHE. The donor flow rate was 5.3 cm^3 min^{-1}, and the acceptor flow rate was 0.4 cm^3 min^{-1}.

it was confirmed that the diffusion through the boundary layer of the donor phase was a rate-controlling step in the overall process.

The removal of Lu(III) achieved in the dynamic module reported in this research was insufficient for commercial applications. The higher removal efficiencies (above 95%) may be achieved using either a hollow fiber contactor with a larger membrane area or a two-stage process.

7.3.6.2 Yttrium [22]

The idea of recovering Y(III) ions by hollow fiber SLM involving two steps of separation (ie transfer from feed to membrane and membrane to receiving phase) has been proposed. Y has identified applications in manufacturing special glass and alloys. The idea of recovering Y(III)) by hollow fibers SLM involving two steps of separations (ie. feed to membrane and membrane to receiving phase) has been proposed.

The two-channel fibers SLM was used for investigating this idea. The fibers SLM was impregnated with different concentrations carrier. The plain woven cotton fibers used were with diameter 0.027 cm (270 μm) and porosity 30–50% (Century India Ltd.). The experimental variables explored were concentration of Y(III), pH of source phase, PC-88A concentration in membrane phase, acid concentration in receiving phase, and stirring speed. The preconcentration of Y(III) investigated showed that the enrichment factor was reduced when the initial Y(III) ions concentration in the feed solution became higher.

7.4 Applications Common for Rare Earth Metallurgy

7.4.1 Concentration by Pressure Driven Membrane Separation

With low grade ion-type rare earth ores, the concentration of rare earth metal ions in the leaching solution is low. If precipitation is used to separate the rare earth metal in the leaching solution, the amount of leachate to be tackled with is huge, leading to large sized equipment, high consumption of chemicals for impurity removal by precipitation, and costly leachate transport.

Fig. 7.8 is the flowsheet for solvent extraction of rare earth leachate with membrane separation technology as major innovation; it is reported by Wang [23]. After in situ leaching of low-grade rare earth ore, the leachate goes through a ceramic membrane module for purification and clarification. Thereafter, it is concentrated by nanofiltration (NF). The resultant leaching solution was reduced in volume by at least 10 times as compared with the original leachate and correspondingly higher metal ions concentration. The follow-up steps include removal of impurities, precipitation, etc. Rare earth metals were obtained as precipitated carbonate salts. Undoubtedly, the shrunken leachate volume eases the burden on transport. The tail liquid or supernatant from precipitation step may be used in the leaching step after some further treatment (eg, NF, neutralization).

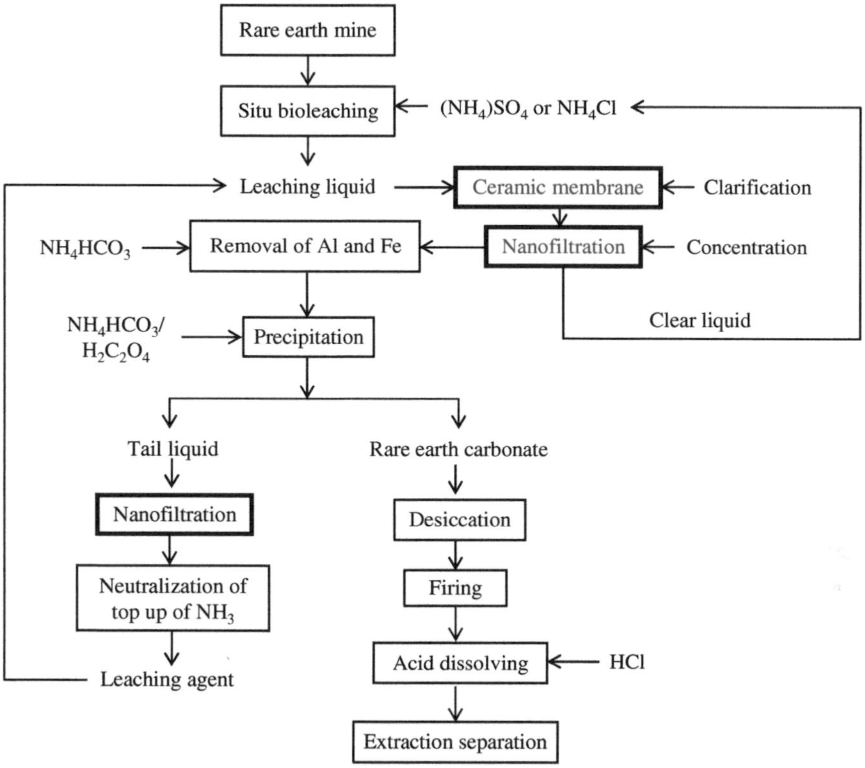

Figure 7.8: Flowsheet involving membrane separation technology for enrichment of rare earth metals in leaching liquid.

Economic efficiency estimation based on 2000 ton d^{-1} ammonium sulfate ($(NH_4)_2SO_4$) leaching solution indicates that the total investment for the system (no detailed breakdown) is 3,500,000RMB and payback period for equipment investment is 2 years. The benefits due the system operation are the recovery of rare earth metals and save in energy and chemicals.

7.5 Wastewater Treatment

7.5.1 Recovery of Ammonia-Nitrogen

The wastewater released from the ion-type rare-earth leaching and saponification in smelting has high ammonia-nitrogen (NH_3-N) concentration. This type of wastewater has also some other common properties. Generally, the average concentrations of $(NH_4)_2SO_4$ and ammonium chloride (NH_4Cl) are around 50,000 and 600,000 mg L^{-1}, respectively. Around 1–2% acid and a certain level of Ca, Mg, manganese (Mn), etc. are also present. The technologies available for removing NH_3-N are mainly air stripping, folding point chlorination, dialysis, chemical precipitation, and ion-exchange.

Table 7.4: NH3-N wastewater released from ion-type rare earth metal extraction

Resource	pH	NH_3-N conc. (mg L^{-1})	Other Major Component (mg L^{-1})	Amount Per Ton of Ore (ton ton^{-1})
Leaching by $(NH_4)_2SO_4$	6.5–7	6500–8500	SO_4^{2-}: 30,00050,000 Mg^{2+}: 5000–8000 Ca^{2+}: 400–800	13–15

Wang reviewed the usage of multistage reverse osmosis (RO) to capture NH_3-N [23]. The retentate with elevated NH_3-N concentration is then sent to the step of aqueous ammonia recovery using air stripping. Table 7.4 shows the feed water composition.

The experimental results for single-stage concentration of wastewater discharged from the process using NH_4Cl are demonstrated in Table 7.5. The rejection for both wastewater was initially >90%. As concentration process goes on, the rejection gradually decreases, sometimes to the level of 82–86%. The final effluent water has an NH_3-N concentration being 1/3 of that in the original feed. A two-stage RO was applied with the concentration factor of the first-stage RO being 3, and the second-stage 4.5. The NH_3-N in purified water in the first-stage RO was only 400 mg L^{-1}; this parameter was further reduced to $<5 \times 10^{-6}$ (mass fraction).

The removal of NH_3-N in metallurgical wastewater by membrane absorption was investigated by Hong and Zhang [24]. By regulating the pH condition of the wastewater, NH_4(I) were converted to NH_3. According to the Henry's law, a partial pressure difference of NH_3 will be established across the membrane. Therefore, NH_3 will move from the feed/membrane interface to the absorption liquid/membrane interface, followed by NH_3 absorption.

Xue et al.'s work displayed that electrodialysis (ED) can remove ammonium salt more efficiently than other inorganic salt [25]. Under certain conditions, almost complete removal

Table 7.5: The efficiency of one-stage RO in treating NH3-N wastewater

Items	HN_4^+-N (mg L^{-1})	NH_4Cl (mg L^{-1})	Rejection (%)
Original feed	2,543	9,719	
Concentrate	3,915	14,963	
30 min permeate	267	1,012	89.5
60 min permeate	169	647	93.3
Final permeate (4 h)	357	1,364	85.9
Mixed permeate	408	1,559	83.9
Original feed	4,998	19,099	
Concentrate	7,334	28,026	
10 min permeate	534	2,040	92.7
30 min permeate	847	3,237	88.5
Final permeate (6.5 h)	765	2,924	89.6
Mixed permeate	630	2,411	91.4

of NH_3-N was actualized. The desalinated water can be discharged directly; the highest concentration of ammonium salt in the concentrated solution is 12.6 wt.%, which satisfies the concentration requirement of being used as a liquid fertilizer.

Theoretically, the maximum concentration (C_{max}) that could be achieved with IEM based ED is estimated using following equation:

$$c_{max} = \frac{(t_C + t_A - 1)\frac{i}{F} - K_S \Delta c}{\beta_i + K_w \Delta c}$$

where t_C, t_A is the transfer number for cation in CEM and anion in AEM, respectively, i is the operation current, K_s is the diffusion coefficient of electrolyte, mmol $(cm^2 \cdot h \cdot mol \cdot L^{-1})^{-1}$, β is the electro-osmotic coefficient of water, mL mA^{-1} h^{-1}, K_w is the permeability coefficient of water, mL $(cm^2 \cdot h \cdot mol \cdot L^{-1})^{-1}$, Δc is the concentration change of solution in concentrate chamber, and F is the Faraday constant.

According to the equation, there exist two approaches for enhancing the final concentration: increase of the transfer numbers and decrease of the water electro-osmotic coefficient and permeability. The scaling phenomenon due to $CaSO_4$ can be solved by utilization of scale inhibitor.

7.5.2 Acid Recovery

Chen et al. used membrane diffusion dialysis (DD) to recover acid from solution with rare earth metal compounds [26]. The membrane was a kind of homogeneous AEM. It used epoxy propyl methacrylate (EPMA) as a raw material and formed by coating. The dialysis membrane has good resistance toward acid and alkali attacks.

Tang et al. studied the treatment of the waste acid released from the process of rare earth leaching process by vacuum membrane distillation [27]. The concentration by VMD was followed by the separation of acid from rare earth metal using DD. The H_2SO_4 recovery rate reached 70–80% in the actual operation. The experimental results also showed that H_2SO_4 recovery from rare earth sulfate solutions by the combination of VMD and DD has no influence on the rejection rate of rare earth metal ions. The benefit of the process is that the concentration of H_2SO_4 by VMD greatly reduced the volume of the solution to be treated by DD.

Based on the vapor/liquid equilibrium of the HCl-R_ECl_3-H_2O (note: R_E indicates rare earth metal) as illustrated in Fig. 7.9, separation of rare earth metal and hydrochloric acid (HCl) in highly acidic back extraction solution in one single step by VMD was attempted [28]. Eighty percent of the HCl in the waste solution could be extracted into the distillate. The membrane rejection toward cations of rare earth metal was generally higher than 98%, and relatively pure HCl solution was produced. On condition that the initial concentration of HCl or rare

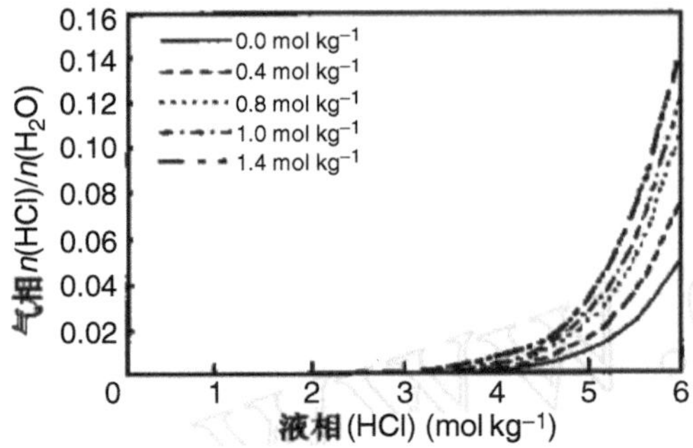

Figure 7.9: HCl-$R_E$$Cl_3$-$H_2O$ tertiary ternary system vapor/liquid equilibrium for different $R_E$$Cl_3$ concentrations (R_E indicates rare earth metal).

earth ions in the feed liquid is low, VMD could only lead to concentration of feed solution. HCl recovery is closely related to CF; the higher the concentration of rare earth metals in the feed, the higher the recovery rate. In addition, the HCl content in the distillate solution may be larger than that in the original feed.

7.6 Conclusions

Membrane technologies that have been involved in scattered and rare earth metal metallurgy include hydraulic pressure driven membrane separation, electrolysis, ED, liquid membrane, and VMD. For scattered metals, the focus is the recovery of metal ions from low concentration aqueous solutions and product generation via redox reaction. For rare earth metals, the application of membrane mainly covers the solvent extraction of leaching solution and wastewater treatment. One point that catches the attention is that many works have attempted SLM for these two categories of metals, probably due to the simultaneous extraction and enrichment of metal ions, and the fine discrimination capability intrinsic of extractant. Nevertheless, its stability requires further research effort. Due to disordered mining, pollution featured manufacturing technologies, and difficulty in treatment of solid and liquid wastes, rare earth metal metallurgy have caused serious environment problems. "Green Process" essentially characterized of extractant recycling, energy efficiency, zero discharge will become the major component for future metallurgy.

Glossary and Symbols

AEM	Anion exchange membrane
Al	Aluminum
BaF_2	Barium chloride
$CaSO_4$	Calcium sulfate
CEM	Cation exchange membranes
Ce	Cerium

$CeCl_3$	Cerium chloride
CeO_2	Cerium oxide
CF	Concentration factor
Cl_2	Chlorine
Cu	Copper
DEHPA	di-(2-ethylhexyl) phosphoric acid
DHE	di-n-hexyl ether
Dy	Dysprosium
ED	Electrodialysis
EDL	Electric double layer
EDTA	Ethylene diamine tetraacetic acid
E_E	Extraction efficiency
ELM	Emulsion liquid membrane
EPMA	Epoxy propyl methacrylate
E_R	Removal efficiency
Er	Erbium
Eu	Europium
Eu_2O_3	Europium oxide
F	Fluoride
Ga	Gallium
Gd	Gadolinium
Ge	Germanium
HCl	Hydrochloric acid
HEDTA	2-hydroxyethylethylenediaminetriacetic acid
Ho	Holmium
H_2	Hydrogen
$H_2C_2O_4$	Oxalic acid
H_2SO_4	Sulfuric acid
IEM	Ion exchange membrane
In	Indium
In_2O_3	Indium oxide
KCl	Potassium chloride
La	Lanthanum
$LaCl_3$	Lanthanum chloride
LaF_3	Lanthanum fluoride
$La(OH)_3$	Lanthanum hydroxide
LED	Light emitting diode
Li	Lithium
LiCl	Lithium chloride
LiF	Lithium fluoride
Lu	Lutetium
$LuCl_3$	Lutetium chloride
$LuF_3 \cdot 2H_2O$	Hydrated lutetium fluoride
M	Memory effect
Mo	Molybdenum
Mn	Manganese
NaCl	Sodium chloride
Na_2SO_3	Sodium sulfite
Nb	Niobium
Nd	Neodymium
NF	Nanofiltration

NH_3-N	Ammonia-nitrogen
NH_4Cl	Ammonium chloride
NH_4HCO_3	Ammonium bicarbonate
$(NH_4)_2SO_4$	Ammonium sulfate
Pb	Lead
Pm	Promitnium
Pr	Praseodymium
Pt	Platinum
Pu	Plutonium
PVDF	Polyvinyl difluoride
Re	Rhenium
RO	Reverse osmosis
Sc	Scandium
Se	Selenium
SLM	Supported liquid membrane-
Sm	Samarium
Tb	Terbium
Te	Tellurium
Th	Thorium
Tl	Thanllium
Tm	Denebium
U	Uranium
UF	Ultrafiltration
VMD	Vacuum membrane distillation
Y	Yttrium
Yb	Ytterbium
t_C, t_A	Transfer number for cation in CEM and anion in AEM, respectively.
i	Operation current
K_s	Diffusion coefficient of electrolyte, mmol/(cm^2·h·mol·L^{-1})
β	Electro-osmotic coefficient of water, mL mA^{-1}·h^{-1})
K_w	Permeability coefficient of water, mL/(cm^2·h·mol·L^{-1})
Δc	Concentration change of solution in concentrate chamber
F	Faraday constant

References

[1] Editorial Committee, Handbook of Rare Metals, Metallurgical Industry Press, Beijing, 2012.
[2] G.X. Xu (Ed.), Rare Earth, Metallurgical Industry Press, Beijing, 2002.
[3] J.Z. Cheng, L.P. Che, Current mining situation and potential development of rare earth in China, Chin. Rare Earths 31 (2010) 65–69.
[4] M.G. Zheng, Y.H. Chen, Supply and demand of global rare earth resources and China's rare earth industry policy, Nonferrous Metals Sci. Eng. 4 (2012) 70–74.
[5] Y.F. Jiao, X.L. He, C.F. Lia, P.G. Jiang, Progress of research on separation and purification of heavy rare earth elements in recent decade, Chin. Rare Earths 34 (2013) 74–79.
[6] Q.Y. Han, The present situation analysis of rare earth extraction separation technology, Hunan Nonferrous Metals 26 (2010) 24–27.
[7] C.H. Yan, S. Wu, C.S. Liao, J.T. Jia, F.X. Cheng, S.L. Wang, B. Zhang, Development of the theory of counter-current extraction and its industrial applications, Chin. J. Inorg. Chem. 24 (2008) 1200–1205.
[8] Q.W. Qin, Z.G. Deng, New extractant Cyanex272 as carrier in emulsion liquid membrane enrichment of rare earth research, J. South. Inst. Metall. 22 (2001) 304–309.

[9] Y. Jia, X.L. Lv, W.J. Zhou, W. Liu, C.R. Wu, X. Wang, H.Y. Chen, Q.J. Gao, Study of membrane degassing technology for HCl removal, The Fourth Symposium (China) on Application of Membrane Separation Technology in Metallurgical Industry, 2014, pp. 181–185.
[10] J. Xiao, W.Z. Chen, G.Y. Yin, W.R. Liu, The application of inorganic ceramic membrane in nano powder materials production, The Third Symposium (China) on Application of Membrane Separation Technology in Metallurgical Industry, 2013, pp. 158–160.
[11] Q.X. Zhang, L.S. Xiao, Application of membrane technology in metallurgy of rare metals, Chin. J. Rare Metals 27 (2003) 1–7.
[12] X.H. Dou, Rare earth elements lanthanum and its application, Rare Earth Inf. 251 (2005) 33–34.
[13] W. Cheng, M.S. Huang, Z.J. Wang, S.G. Jia, L.H. Yang, The latest research on the preparation and application of lanthanum metal, Rare Metals Cement Carbides 42 (2014) 62–65.
[14] S.M. Pang, S.H. Yan, Z.A. Li, D.H. Chen, L.H. Xu, B. Zhao, Development on molten salt electrolytic methods and technology for preparing rare earth metals and alloys in China, Chin. J. Rare Earths 35 (2011) 440–450.
[15] W. Cheng, Z.G. Li, D.H. Chen, S.M. Pang, Z.P. Wang, X.S. Wang, Preparation of high purity lanthanum by combined method of lithium-thermal reduction and vacuum distillation, Chin. J. Rare Earths 35 (2011) 782–785.
[16] F.S. Xu, F. Zhang, W.H. Xing, Y.Q. Fan, Recovering lanthanum (III) by ceramic ultrafiltration membrane with pore size of 5 nm, The Fourth Symposium (China) on Application of Membrane Separation Technology in Metallurgical Industry, 2014, pp. 48–53.
[17] M. Tomaszewska, Concentration of the extraction fluid from sulfuric acid treatment of phosphogypsum by membrane distillation, J. Membr. Sci. 78 (1993) 277–282.
[18] Q.S. Yang, Y.Q. Cao, L.J. Gao, Z.G. Liu, Y.Q. Lu, An environment-friendly technology for the production of high purity europia, Chin. Rare Earths 29 (2008) 64–67.
[19] Z.Q. Long, W.M. Huang, X.W. Huang, Electrolytic reduction technology for preparation of bivalent europium, Chinese Patent 02117300.1 (App. No.), 2003.
[20] Q.F. Wei, Q.X. Zhang, W.T. Xia, Oxidation of cerium (III) to cerium (IV) and simultaneous deposition of copper powder by anion—exchange membrane coupling electrolysis, J. Cent. South Univ. 36 (2005) 396–401.
[21] R.K. Kumrić, G.T. Vladisavljević, T.M. TrtićPetrović, Membrane-assisted liquid-phase extraction of Lu (III) in a U-shaped contactor with a single hollow fiber membrane, Ind. Eng. Chem. Res. 51 (2012) 14199–14208.
[22] A.G. Gaikwad, A.M. Rajput, Transport of yttrium metal ions through fibers supported liquid membrane solvent extraction, J. Rare Earths 28 (2010) 1–6.
[23] Z.G. Wang, Application of membrane separation technology in hydrometallurgy, The Fourth Symposium (China) on Application of Membrane Separation Technology in Metallurgical Industry, 2014, pp. 54–61.
[24] H.Y. Hong, X.X. Zhang, Membrane absorption of ammonia-nitrogen removal technology and its application of metallurgical wastewater treatment, The Fourth Symposium (China) on the Application of Membrane Separation Technology in Metallurgical Industry, 2014, pp. 133–135.
[25] D.M. Xue, G.W. Hong, G.F. Wu, X.Y. Xiao, Concentration and recovery of ammonium salts in the waste from rare-earth industry by electrodialysis, Membr. Sci. Technol. 20 (2000) 61–65.
[26] R.Q. Chen, D.C. Ge, J.W. Jiang, J. Hu, Preparation of KS-1 diffusion dialyzing membrane and its application in treatment of acid solution containing rare earth element, Technol. Water Treat. 17 (1991) 231–238.
[27] J.J. Tang, W. Zhang, K.G. Zhou, Q.X. Zhang, Feasibility study of the waste acids recovery by membrane techniques in rare earth metallurgy, Mining Metall. Eng. 27 (2008) 54–57.
[28] J.J. Tang, K.G. Zhou, Q.X. Zhang, Study on hydrochloric acid recovery from rechloride solution by vacuum membrane distillation, Membr. Sci. Technol. 22 (2002) 38–42.

CHAPTER 8

Radioactive Metals

L.Y. Jiang
School of Metallurgy and Environment, Central South University, Changsha, Hunan Province, China

8.1 Introduction

Radioactive metals refer to natural and synthetic metals that can release alpha (α), beta (β), and gamma (γ) rays. α-Ray is composed of particles that are positively charged helium (He) nuclei obtained from natural radioactive decay; β-ray is high speed electron flow; γ-ray is electromagnetic wave–photon flow, of which the wavelength is shorter than that of x-Ray. In the periodic table of element, the positions of the majority of the radioactive metals are located at the bottom.

Naturally occurring radioactive metals include polonium (Po), francium (Fr), radium (Ra), actinium (Ac), thorium (Th), protactinium (Pa), uranium (U), neptunium (Np), and plutonium (Pu) [1].

U is one of the most important radioactive metals for modern world. It is the principal raw material for nuclear bombs and nuclear power plants. The three natural occurring isotopes of U are U-234, U-235, and U-238. U (99.28%) is U238; about 0.715% of total U is U-235 and the ratio for U-234 is only 0.00006%. U-235 fission is the key reaction utilized for energy and heat generation. Its natural occurrence in the deposit is mainly in the form of oxides or silicates. Fig. 8.1 shows the simplified schematic of U fuel cycle [1,2]. Currently, Australia ranks first in U reserve and the largest U mine is located at River McArthur, Canada, from which the U throughput occupies 13% of world total. Those mines that rank second include Olympic Dam of Australia and Niger Ahil Mineral Company. The ratios of their output to the world total production of U are 6 and 5%, respectively. According to the reports in recent years to International Atomic Energy Agency (IAEA), the proved total reserve of U in China ranks behind tenth place. Table 8.1 displays the information about the major U mines worldwide [3,4].

Th-232 changes to fissionable Th-233 after capture of a neutron and follow-up reaction [1]. Behaving similarly to U-235, Th-233 can cause chain reaction that releases energy. Due to the much higher abundance of Th as compared with that of U (0.0012% vs 0.0004%), Th is another important resource for nuclear fuel. Addition of a few percentage of Th in

Membrane-Based Separations in Metallurgy
http://dx.doi.org/10.1016/B978-0-12-803410-1.00008-6

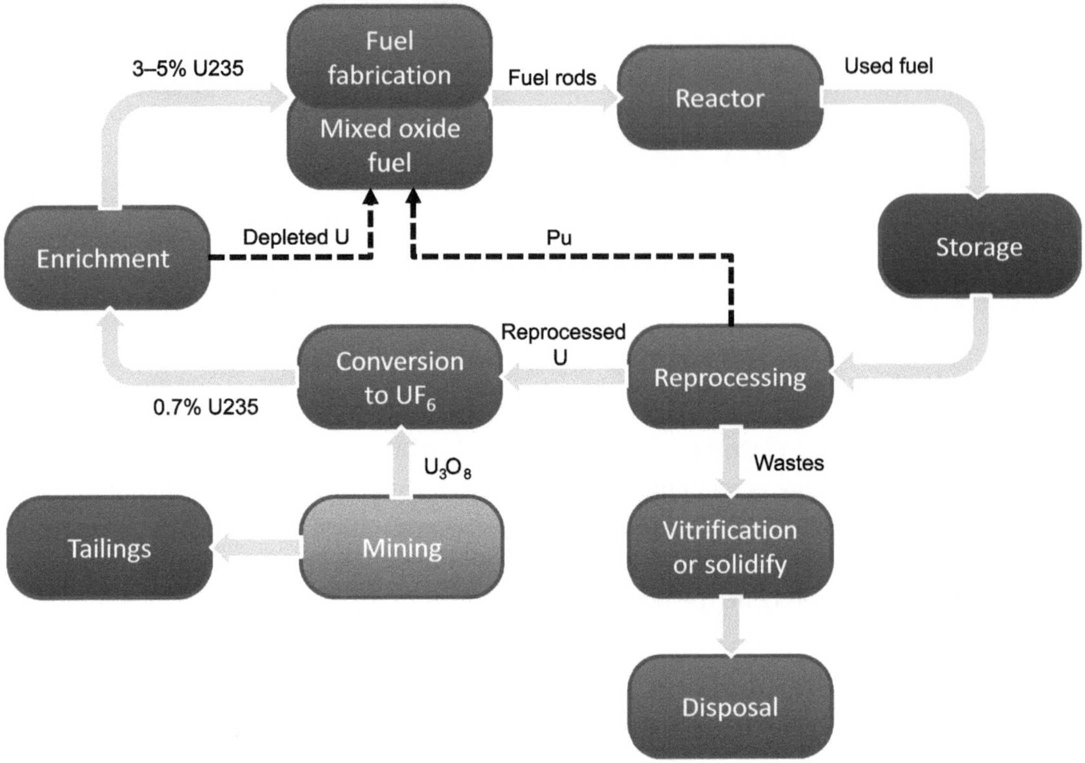

Figure 8.1: The nuclear fuel cycle.
Redrawn from http://www.world-nuclear.org/information-library/nuclear-fuel-cycle/introduction/nuclear-fuel-cycle-overview.aspx [2].

Table 8.1: Information about the major uranium mines in the world

Name	Country	Mining Type	Annual Production(tU)	Percentage
McArthur River	Canada	Underground mining	7744	13
Olympic Dam	Australia	By-product method/ underground mining	3400	6
SOMAIR	Niger	Surface mining	2730	5
Tortkuduk (est)	Kazakhstan	In situ leaching method	2550	4
Ranger	Australia	Surface mining	2510	4
Priargunsky	Russia	Underground mining	2133	4
Budenovskoye-2	Kazakhstan	Leaching method	2115	3
Langer Heinrich	Namibia	Surface mining	2094	4
Inkai	Kazakhstan	In situ leaching method	2047	3
Rossing	Namibia	Surface mining	2043	3
South Inkai	Kazakhstan	In situ leaching method	2030	3
Central Mynkuduk	Kazakhstan	In situ leaching method	1800	3
Rabbit Lake	Canada	Underground mining	1587	3
COMINAK	Niger	Underground mining	1508	3
Budenovskoye 1, 3, and 4	Kazakhstan	In situ leaching	1499	2

magnesium (Mg) delivers low density alloy with excellent property, which brought about applications of Th outside nuclear power industry.

Pu is another radioactive metal that has been identified with a certain range of applications. There are more than 20 isotopes of Pu, but only Pu-239 and Pu-244 are natural ones; the most important one is Pu-239. With its propensity toward nuclear fission, Pu-239 can be used for nuclear power generation. The large-scale production of Pu-239 counts on the reaction in U nuclear reactor, where U-238 captures a neutron and releases two electrons via β-decay to generate Pu-239. The product from the reaction is purified using solvent extraction and ion exchange. The Pu isotopes are also used in manufacturing isotope battery, alloys and as power sources for spaceships, man-made satellites, beacons, etc.

The major application of Ra is associated with the field of medical science, where the radioactivity of Ra is capitalized on to treat cancer and intractable dermatomycosis. It is also used in industry for flaw detection, thickness measurement, etc. [1]. Ac-227 has a half-life of 21.77 years. It exists in Pitchblende and other U bearing minerals. Presently, the commercial or scientific application of Ac is focused on thermal resource in spacecraft [1]. There exist more than 20 kinds of synthetic radioactive metals, which are listed in Table 1.2 (Chapter 1). They are produced by following methods: neutron irradiation of reactor, bombarding using atom smasher, extraction from irradiated nuclear fuel, and thermonuclear explosion. The synthetic metals are mainly applied in computer, rocket, thermonuclear reaction device, nuclear battery fuel, etc.

As mentioned previously, radioactive metals can be put to good use in many ways. One of the most prominent applications is power or electricity generation. Even though regarded as the most terrible source of power, nuclear energy has never been really abandoned. Many are quite optimistic about the survival of nuclear power generation for solving the energy crisis worldwide [5]. Einstein's formula relating the energy and mass means that if complete conversion of 1 lb of mass into energy was initiated, the total heat that would be produced is the same as would be generated from burning 2,800,000,000 lbs. of coal. On conditions that safety concerning operating a nuclear reactor, reprocessing spent fuel, and disposing of radioactive waste and obsolete facilities can be guaranteed, obviously, nuclear energy will be one of the most efficient methods in tackling the global energy crisis in the foreseeable future.

None of the resources on the earth is infinite or absolute renewable. This rule is also applicable to the deposit of U and other radioactive metals. Therefore, constitution of pragmatic plans regarding atomic energy development, exploration, mining, and extraction are highly necessary.

8.2 Treatment of Leaching Solution (or Metal Extraction)

8.2.1 Emulsion Liquid Membrane

Among the secondary resources of U, natural phosphates are found to be an important one [6]. According to the report of an Advisory Group Meeting organized by the IAEA in

Vienna in March 1987, research and development activities on U recovery from phosphoric acid are continuing along three main lines:

- improve the di(2-ethylhexyl) phosphoric acid (D2EHPA)/tri-*n*-octyl phosphine oxide (TOPO) process to achieve higher flexibility and economy benefit;
- develop new generation of liquid–liquid extractants;
- employ other separation technologies, such as ion exchange, liquid membranes, and processes compatible with the hemi-hydrate process for the production of phosphoric acid.

The potentiality of applying emulsion liquid membrane (ELM) technique for U recovery from locally produced phosphoric acid produced by Abu Zaabal Fertilizers and Chemical Co., Egypt using D2EHPA/TOPO system was investigated by Elsayed et al. [6].

A comprehensive summary about the researches contributed to U extraction by ELM prior to this report was reviewed. In Elsayed et al.'s work [5], the acid or commercial wet process phosphoric acid (WPPA), the raw material containing U supplied by the company aforementioned was first converted into clear green acid solution suitable for ELM treatment by removing most of the impurities. The steps for the purification are briefed as follows:

1. pretreatment to take away most of the organic and humic substances
2. fluoride (F) and iron (Fe) separation:
   The precipitation of sodium fluorosilicate (Na_2SiF_6) according to the following reactions:
   $$SiO_2 + 6HF \rightarrow H_2SiF_6\downarrow + 2H_2O$$
   Fe would indeed precipitate according to the following reaction:
   $$16H_3PO_4 + K_2SO_4 + 3Fe_2O_3 \rightarrow 2Fe_3KH_{14}(PO_4)_8{\cdot}4H_2O\downarrow + H_2SO_4 + 5H_2O$$
3. acid polishing and dilution: the filtered acid was finally treated by passing through a carbon column (80 cm × 10 cm diameter) several times to more thorough removal of residual humic matter

The final feedstock for U recovery by ELM had the following compositions: 28% P_2O_5, 0.12% F, 0.9% Fe, and 42 ppm U. The organic phase of the ELM was either D2EHPA/TOPO mixture or tridodecyl amine (TDDA) dissolved in kerosene. A certain amount of Span 80 was then added as a surfactant. Orthophosphoric acid (H_3PO_4) with 40% P_2O_5 was added into the internal aqueous phase, while 2 M sodium carbonate (Na_2CO_3) was used in case that TDDA was an extractant. In addition, the effects of two reducing agents, ascorbic acid (better known as Vitamin C) and Fe were examined.

The experimental results showed that the presence of 300 ppm Fe in the internal phase significantly enhanced the extraction efficiency, which is about 85% after 10 min. The ascorbic acid is a good additive for extraction using H_3PO_4 solution (40% P_2O_5) in the internal phase. According to the authors' explanation, the two inorganic additives can convert the extracted U from hexavalent U(VI) to nontransferrable or nonextractable tetravalent U(IV), which plays critical role in raising the extraction efficiency when used in internal phase. Concerning the treatment of Abu Zaabal phosphoric acid, U extraction efficiency reached 54% using the D2EHPA/TOPO system with 1% ascorbic acid added to the internal H_3PO_4 solution (40% P_2O_5) and 1% Span 80 at 800 ppm, while 34% by the addition of 300 ppm Fe to the internal phase.

In Kulkarni et al.'s work, the leach liquor of U concentrates delivered at Uranium Corporation of India Limited (UCIL), Jaduguda, India was processed by ELM technique for the recovery of U in the presence of impurity metal ions, such as Fe, calcium (Ca), Mg, and manganese (Mn) [7]. The entire process involving ELM for U recovery is shown in Fig. 8.2.

The feed is sulfate leaching solution with U being present as uranyl sulfate (UO_2 SO_4). The liquid membrane employed is composed of a diluent (light and heavy paraffin), a surfactant (Span 80), and an extractant (Alamine 336). Na_2CO_3 was used in the stripping solution.

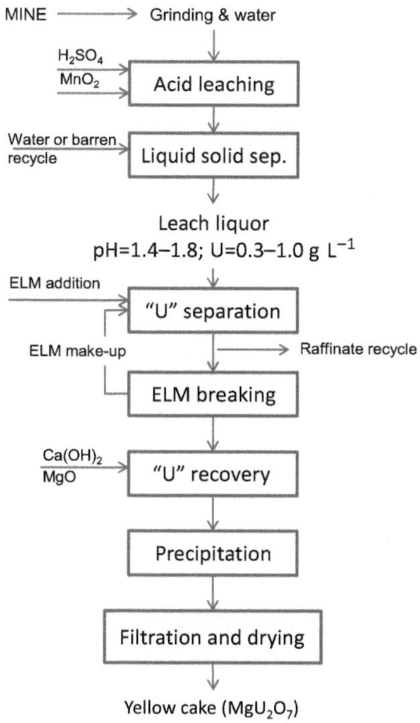

Figure 8.2: Proposed ELM-based process scheme for the separation and concentration of U from the leach liquor.

The extraction and stripping reactions involved in the ELM process are described as follows:

$$\text{Extraction}: UO_2(SO_4)_3^{4-} + 4H^+ + 4R_3N(org) \Leftrightarrow (R_3NH)_4UO_2(SO_4)_3(org)$$

$$\text{Stripping}: (R_3NH)_4UO_2(SO_4)_3(org) + 7Na_2CO_3 \Leftrightarrow Na_4UO_2(CO_3)_3 + 3Na_2SO_4 + 4NaHCO_3 + 4R_3N(org)$$

The first part of the work is optimization of the ELM process parameters by using simulated leach liquor as the feed. The role of the feed pH was found to be significant by helping via cotransport through liquid membrane in achieving the complete extraction and stripping of U. H^+ ions transport along with uranyl trisulfate ($UO_2(SO_4)_34^-$) through the membrane, facilitating the formation of sodium uranyl tricarbonate ($Na_4UO_2(CO_3)_3$, which is a complex molecule rather than a simple ion. Therefore, the stripping reaction can achieve a zero concentration of uranyl (UO_2^{2+}) ions in the internal phase, favoring the shift of reaction equilibrium to the stripping side. Therefore, the complete extraction and recovery of uranium should be possible in this specific ELM system.

The optimized parameters were then employed for the separation and concentration of U from industrial leachates. The results showed that >90% of UO_2^{2-} passed through the liquid membrane, in the presence of various metal ions in one step. The concentration of U in the internal phase was nearly 7 times higher than that of original feed. Separation factors (ratio of enrichment factors) of U with respect to Fe, Mg, Ca, and Mn were found to be 129, 781, 43, and 1462, respectively, in a 12 min experiment.

It was also confirmed that the usage of 30% heavy paraffin in composing the liquid membrane markedly enhanced the stability of the emulsion and the recovery of U.

The stripping phase with concentrate U existing as $Na_4UO_2(CO_3)_3$ is treated using calcium hydroxide [$Ca(OH)_2$] and magnesium oxide (MgO). The precipitates containing U are filtered and dried to form a yellow cake (MgU_2O_7).

8.3 Metal Compound Production

8.3.1 Application of Membrane Electrolysis [8]

Uranium tetrafluoride (UF_4) is the most important compound of U. It is the reactant for preparing elementary U or uranium hexafluoride (UF_6). The latter, due to its low boiling point, is generally used for U purification and concentration by gas diffusion method.

The flowsheet of conventional hydrometallurgical process for UF_4 production is given in Fig. 8.3 (left). The major steps of the process are leaching, solvent extraction, reduction,

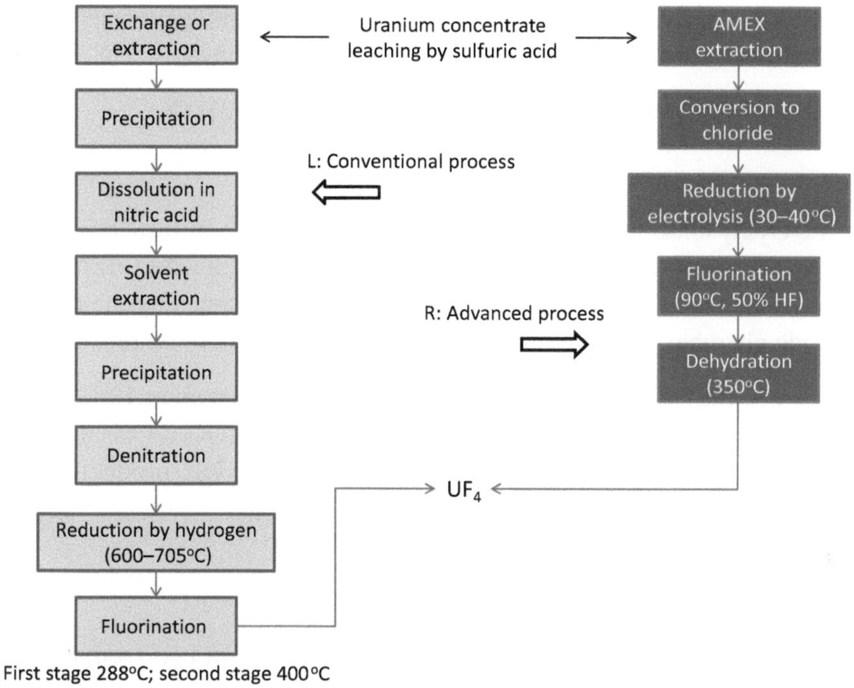

Figure 8.3: Conventional process and advanced process involving electrolysis for production of UF_4.

precipitation, and dehydration. The overall process is complicated and involves temperature higher than 600°C.

Japan Asahi-KASEI and Power Reactor and Nuclear fuel Company (PNC) developed together a membrane-based process for generating uranous chloride (UCl_4) from uranyl chloride (UO_2Cl_2), the raw material for producing UF_4 [8]. The schematic for the conversion mechanism is shown in Fig. 8.4. The reduction reaction takes place at the anode, whereas the

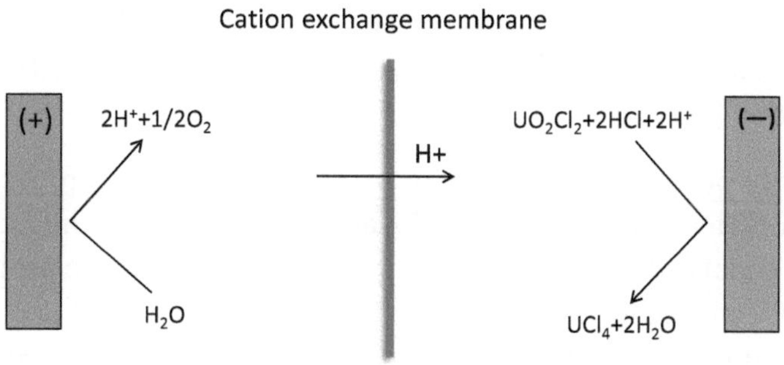

Figure 8.4: Schematic for the reduction of uranyl chloride by electrolysis [8].

oxidation reaction at the cathode. Water electrolysis at the cathode generates oxygen (O_2) and proton (H^+). The reactions at the two electrodes are given as follows:

$$UO_2Cl_2 + 2HCl + 2H^+ + 2e \rightarrow UCl_4 + 2H_2O$$

$$H_2O \rightarrow 2H^+ + 1/2O_2 + 2e$$

The analyte is UO_2Cl_2 solution and the catholyte is 0.25 mol L^{-1} sulfuric acid (H_2SO_4). Due to the selectivity of the cation exchange membrane (CEM), H^+ transport from the cathode chamber to the anode chamber, while the chloridion (Cl^-) is rejected by the membrane and stays in the anode chamber, which ensure the progress of the reduction reaction at the anode.

This technology was placed in service at Ningyo Pass Plant in Japan in 1980. The transfer efficiency of H^+ is up to 99% and the reduction percentage of UO_2^{2-} ions is as high as 99.5%. Application of membrane electrolysis has greatly simplified the UF_4 production process flow, as indicated by the comparison in Fig. 8.3.

8.4 Processing of Radioactive Wastes

High-level radioactive waste (HLW) mainly comes from (1) spent fuel from commercial or research reactors, (2) the reprocessing of spent fuel (eg, PUREX reprocessing), and (3) nuclear weapons and propulsions industry [9]. Low-level radioactive waste (LLW) comes from hospitals and industry, as well as the nuclear fuel cycle. The carriers comprise paper, rags, tools, clothing, filters, liquid, etc., which generally features small amounts of mostly short-lived radioactivity. Intermediate-level radioactive waste (ILW) is characterized of relatively higher levels of radioactivity. It typically comprises resins, chemical sludges and metal fuel cladding, as well as contaminated materials from reactor and facility decommissioning [2].

As shown in Fig. 8.1, the nuclear fuel cycle starts with the mining and ends with the disposal of nuclear waste. The spent fuel, viewed as HLW, contains not only short-lived fission products, but also a significant amount of minor actinides [ie, Pu, Np, americium (Am), and curium (Cm)] and fission products [ie, strontium (Sr)-90, cesium (Cs)-137, technetium (Tc)-99, and iodine (I)-129] with high toxicity and very long half-lives. The latter two groups constitute the long-term radiotoxic inventory which remains as a hazard beyond human perception or detection [10].

The public acceptance of the nuclear power as an alternative to the conventional energy sources takes it as a prerequisite the safe handle and management of the radioactive waste [11]. Reprocessing enriches U and Pu, the major actinides, from waste products and fuel assembly cladding by cutting up the fuel rods, dissolving them in acid and extracting with solvent. Meanwhile, it leads to a significantly reduced amount of waste, compared with treating all spent fuel as waste [2].

The remaining 3% of HLW may be directly stored in liquid and subsequently vitrified. Nevertheless, the decay of some components has an extremely long half-life and will release heat and radiation [12]. Particularly, some nuclides [eg, Tc-99, selenium (Se)-79, and I-129] are mobile in water. Therefore, storage approaches that can efficiently limit their impact on surroundings are highly needed [13]. In most cases, the partition of different nuclides is conducted for possible classified storage, say according to the half-life. For example, Am-241, Am-243, Cm-244, Np-237, etc. are long-lived minor particles. The wastes could then be disposed in a cost-efficient and environment-friendly manner. In addition, some materials may have value for reusing. For example, Cs-137 can be used as a potential source of γ- radiation in place of cobalt (Co)-60 for purpose, such as sewage sludge treatment, sterilization of foods and medical accessories, etc. [14].

Processes for actinide partitioning developed and tested at various laboratories include TRUEX, TRPO, DIDPA, and DIAMEX. They utilize the reagents, such as CMPO, trialkyl phosphine oxide (TRPO), di-isodecyl phosphoric acid (DIDPA), and *N,N,N,N*-dimethyl dibutyl tetradecyl malonamide (DMDBTDMA) and have been tested in the pilot plant [13].

8.4.1 Fuel Recovery

In the research conducted by Kulkarni [15], it was demonstrated that U can be successfully recovered from nitric acid (HNO_3) wastes of U processing plants by ELM. Acidic wastes containing low concentrations of U are generated during U purification and processing. Selective separation and recovery of U from this wastewater was accomplished using TOPO in paraffin as a carrier and Na_2CO_3 as a stripping chemical. The waste, having a composition of nearly 600 ppm U(VI), 360 ppm Fe(III), 325 ppm Ca(II), 390 ppm Mg(II) at an acidity of 1.2 M HNO_3, was used as the feed phase. Experimental findings with a two-stage counter extraction suggest that the U(VI) concentration in the final raffinate can be lower than 50 ppm. In the presence of various metal ions, selective permeation of UO_2^{2-} through liquid membrane was observed to be more than 70%.

Singh et al. investigated the application of supported liquid membrane (SLM) for the separation of U(VI) from H_3PO_4 acid medium [16]. Experiments have also been carried out to see the transport behavior of different fission products from a diluted HLW solution. The carrier is binary mixture of 2-ethyl hexyl phosphoric acid-mono-2-ethyl hexyl ester (PC88A) and neutral donor which is a mixture of four tri-alkyl phosphine oxide better known as Cyanex 923 in *n*-dodecane. Ammonium carbonate [$(NH_4)_2CO_3$] is used in the stripping phase.

More than 95% U(VI) is recovered from the 0.5 M H_3PO_4 feed in 6 h when using a binary mixture of 0.60 M PC88A and 0.15 M Cyanex 923 in *n*-dodecane as carrier and 0.5 M $(NH_4)_2CO_3$ as stripping phase. Lower concentration of H_3PO_4 (0.5 M) and carrier composed of 0.60 M PC88A and 0.15 M Cyanex is most suitable for transport of U(VI). The optimum conditions obtained from this study were also applied to recover U from analytical waste in H_3PO_4 medium generated in the laboratory. It has also been tried to model the physicochemical transport of U(VI) in SLM as well as establishing the mechanism (diffusion controlled) of transport.

Extraction and stripping of U from nitrate media using a hollow fiber liquid membrane contactor was studied [17]. Try-butyl phosphate (TBP) diluted in kerosene was used as an extractant and sodium hydroxide (NaOH) was applied as a stripping solution. U components were extracted using TBP 5% (v/v), while Th components were rejected into raffinate. The maximum percentage of extraction for U was 67%. When the concentration of HNO_3 increased, more U was extracted; however, when the concentration of U and Th in the feed solution was increased, the percentage of extraction and stripping slightly decreased. The permeability decreases when the concentration of the feed solution increases due to membrane fouling and concentration polarization.

8.4.2 Partitioning of Minor Actinides and Fission Products

8.4.2.1 Electrodeionization

Electrodeionization (EDI) combines the advantages of electrodialysis (ED) and ion exchange. Compared with single ion exchange, the ion-exchange resin in EDI can be continuously regenerated by the H^+ and OH^- produced in the vicinity of the surface of the resin and the membrane. Therefore, it is unnecessary to apply extra chemical reagent to regenerate the resin. As compared with single ED, EDI obviously strengthens the transfer of ions toward the ion exchange membrane (IEM), which facilitates the achievement of a very low concentration of targeted ions.

Walters et al. first designed the setup of packed bed ED, otherwise stated, EDI for treating radioactive wastewater [18,19]. The granular iron exchange resin was filled between the anion exchange membranes (AEMs) and CEMs of desalting chambers. In a semiindustrial experiment, the radiation strength of the treated fresh water was reduced to 1.53×10^{-10} Ci L^{-1}, and the saltiness to 2.67 mg L^{-1}. Therefore, the efficiencies for removing salt and radioactivity both reached 99.46%. Similarly, Luo et al. and Li et al. employed EDI in their research to deal with the radioactive waste water [18,20]. The average radiation intensity of original wastewater was 2×10^{-7} Ci L^{-1} and salt content was 550 mg L^{-1}. By EDI treatment, the values for the two parameters became 1.65×10^{-9} Ci L^{-1} and 3.33 mg L^{-1}, respectively. A further treatment by EDI generated an average radiation intensity of 3.1×10^{-10} Ci L^{-1}.

Zhang investigated the removal of Cs(I) ions in simulated radioactive wastewater using a four-chamber EDI system [21]. The schematic illustrating EDI principle is demonstrated in Fig. 8.5. The solutions flowing through the Chambers 1–4 are 0.5 mol L^{-1} sodium nitrate ($NaNO_3$), cesium nitrate ($CsNO_3$) of specific concentration, $CsNO_3$ of specific concentration, and 0.5 mol L^{-1} $NaNO_3$, respectively. Chamber 2 is packed with ion-exchange resin. In

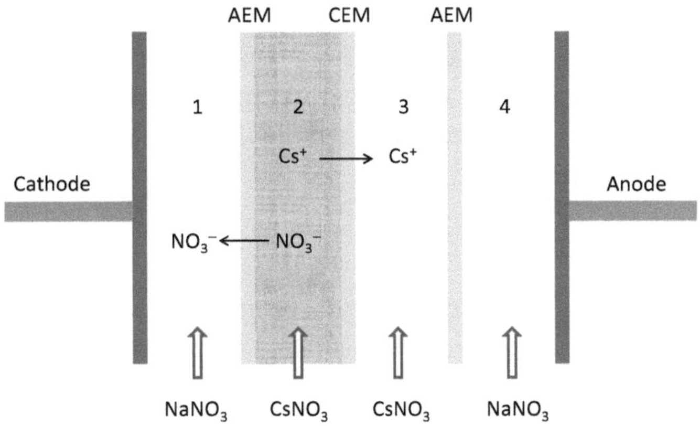

Figure 8.5: Schematic of laboratory EDI setup.

this study, various operation parameters including electric current, feed velocity, initial concentration, pH value, and volume reduction factor (VRF) were adjusted.

The separation process in the EDI system includes following steps: (1) the Cs(I) will be absorbed by the ion-exchange resins while passing through Chamber 2 with feed solution; (2) driven by the electric force, the Cs(I) ions in Chamber 2 will transport through the CEM and move into Chamber 3, whereas nitrate radicals (NO_3^-) will move into Chamber 1; (3) water molecules in Chamber 2 will split on the surfaces of the resin and the membrane to form H^+ that help regenerate the cation exchange resins. The resins used have been pretreated by saturated $CsNO_3$ aqueous solution to prevent the low efficiency of concentrating Cs(I). The optimum conditions involve: electric current of 0.14 A, feed flow rate of 6.0 L h^{-1}, initial Cs(I) concentration lower than 50 mg L^{-1}, initial pH value of 7.0, and VRF lower than 40. The Cs concentration in effluent from Chamber 2 is reduced to be lower than 0.05 mg L^{-1} and the removal efficiency is up to 99.9%.

8.4.2.2 Membrane extraction

Kathios et al. investigated the feasibility of microporous polypropylene (PP) hollow fiber membrane contactors for processing radioactive wastes [22]. They chose neodymium (Nd) as a model solute; this element is not radioactive but behaves similarly to Am(III). The organic extractants di-hexyl-*N*,*N*-diethylcarbamoylmethyl phosphonate (DHDECMP) and *n*-octyl(phenyl)-*N*,*N*-di-isobutylcarbamoylmethyl phosphine oxide (CMPO) were chosen because of their selectivity for actinides over fission products at moderate acid concentrations; diisopropylbenzene (DIPB) was used as a diluent. The extraction proceeded in the presence of 2 M HNO_3 via the formation of an Nd complex:

$$Nd^{3+} + 3NO_3^- + 3CMPO \Leftrightarrow Nd(NO_3)_3 \cdot 3CMPO$$

The reverse reaction describes the subsequent recovery of Nd from the organic phase by back extraction into dilute (0.01 M) HNO_3 solution. Aqueous and organic phases flowed through the tube and shell side, respectively, in countercurrent mode. The experiment also shows that reasonable separation can only be actualized when the flow rate is lower than that acceptable for practical nuclear processing.

8.4.2.3 Supported liquid membrane

Actinide partitioning is viewed as one of the key back end steps of the nuclear fuel cycle (interim storage, reprocessing, 2nd interim storage, land disposal). It involves selective extraction of minor actinides, such as Am, Cm, and Np and residual Pu from the HLW.

Am has the second highest contribution to the radiotoxicity in spent fuel. The one ranking first is Pu. Processes for actinide partitioning developed and tested in pilot plants include TRUEX, TRPO, DIDPA, and DIAMEX [23]. They utilize the reagents, such as CMPO, trialkyl phosphine oxide (TRPO), di-isodecyl phosphoric acid (DIDPA), and *N,N,N,N*-dimethyl dibutyl tetradecyl malonamide (DMDBTDMA) [23]. Ansari et al. [23] investigated the transport of trivalent lanthanides, namely La(III), Eu(III), Lu(III) and several fission products, such as Tc, molybdenum (Mo), zirconium (Zr), palladium (Pd), I, Cs, and ruthenium (Ru) across a flat sheet SLM. The membrane was made of polytetrafluoroethylene (PTFE) impregnated with *N,N,N′,N′*-tetraoctyl diglycolamide (TODGA), CMPO, TRPO, and DMDBTDMA in *n*-dodecane. The membranes were supplied by Sartorius, Germany.

As shown in Table 8.2, the transport rates of the Am increased with HNO_3 concentration in a manner similar to its distribution ratios. The relative magnitude of metal ion permeability after 75 min followed the order: La > Eu > Lu > Zr > Sr > I > Pd ~ Tc > Ru > Mo ~ Cs. This trend was maintained even after 5 h of operation. For HLW, the main concern is the decontamination of trivalent actinides (eg, Am) from fission products elements, such as Zr. The decontamination factor (DF) of La(III), which has similar transport behavior as Am(III), with respect to fission products is displayed in Table 8.3. The DF values with respect to Zr-97, Sr-89, and I-131 were not encouraging. In an attempt to improve the aforementioned DF, a

Table 8.2: Permeability coefficient (*P*) of Am(III) at different HNO_3 concentrations

$[HNO_3]$ (M)	D_{Am}	$P \times 10_3$ (cm s^{-1})	Transfer (%) (time: 3 h)
0.75	6.95	1.2 ± 0.01	85.68
1.0	30.12	2.1 ± 0.03	95.9
2.0	170.8	3.8 ± 0.06	99.9
3.0	203.1	3.7 ± 0.04	99.5
4.5	341.39	3.1 ± 0.16	99.2
6.0	348.59	2.2 ± 0.12	91.8

Carrier: 0.1 M TODGA in *n*-dodecane; stripping phase: distilled water.
Distribution ratio of Am: D_{am} %*T* is the cumulation of metal ion in the feed side.

Table 8.3: Decontamination factors (DF) of La(III) with respect to various fission products used in the present study

DF (La/FP)	DF at Time Intervals	
	75 min	300 min
Tc-95	8.7	3.2
Ru-103	36.3	11.8
Pd-107	8.0	1.8
Sr-90	3.17	1.15
Zr-97	1.55	1.0
I-131	4.7	3.8
Mo-99	1238	133
Cs-137	1651	585

Carrier: 0.1 M TODGA in *n*-dodecane; feed acidity: 3 M HNO_3; stripping phase: distilled water.
Note: As La(III) transport rates are comparable to that of Am(III), the table compares the DF values with respect to La-140 tracer.

modified feed solution was prepared by adding 0.4 M oxalic acid to the original feed solution of 3 M HNO_3. The transport rate of Zr(IV) decreased to <0.2% even after 5 h of operation, in contrast with >99.9% transport in the absence of oxalic acid. Consequently, the DF value improved from 1 to >700.

Another challenge in radioactive waste processing is separation of heat generating fission products, such as Cs-137 and Sr-90. SLM containing a solution of four calix-crown-6 ligands, namely, calix[4]arene-bis-crown-6 (CC), calyx arene-benzo-bis-crown-6 (CBC), calix[4] arene-naphtho-bis-crown-6 (CNC), and bis(octyloxy)calix[4]arene-mono-crown-6 (CMC) in a fluorinated diluent, phenyltrifluoromethyl sulfone (PTMS) along with small fraction of Alamine 336, were employed for the recovery of Cs from HNO_3 feed solutions [24]. It was found that the transport efficiency of the membranes followed the trend indicated by the extractant: CNC = CBC > CC = CMC. CBC, CC and CMC were used with a 4.0×10^{-3} M concentration in respective solution, while CNC at 1.0×10^{-3} M due to solubility limitations. Near quantitative Cs(I) transport was observed at tracer scale after 4 h using CBC. Acid cotransport was observed with all the calix-crown-6 ligands which increased sharply if the membranes were used for longer time. When the CBC based SLM was used, >60% transport of the metal ion and <3% acid cotransport were observed after 5 h in treating simulated high level waste (S-HLW) spiked with Cs-137 as the feed.

Facilitated transport of Cs(I) from aqueous nitrate feed solutions to a receiver solution containing distilled water was investigated by Raut et al. [14]. SLM relying on CNC was employed. The carrier solution consisted of CNC dissolved in a mixture of 80% 2-nitrophenyl octyl ether (NPOE) and 20% *n*-dodecane, and the support was PTFE membrane with average pore size of 0.45 μm. The transport rates were found to be influenced by the mobile carrier concentration, Cs(I) concentration, as well as the feed pH. The ratio of extracted metal to ligand conformed to a stoichiometry of 1:1, suggesting the extraction of an organophilic mono-Cs-crown complex. Fission products obtained from an irradiated natural U target were poorly transported whereas Cs-137 got enriched. Durability/chemical stability of the membrane was maintained at a high level during a more than 20 days of continuous operation.

8.4.3 Membrane Process for LLW

Contaminant removal and concentration or thickening are also essential for LLW management and critical for environmental protection and human health. For this target, complexation facilitated membrane filtration is more intensively investigated for more efficient rejection of low concentration radioactive nuclides.

The feasibility of an ultrafiltration (UF)/complexation hybrid system for reduction of long-lived radioisotope concentration in the effluent (permeate) was studied by Zakrzewska-Trznadel and Harasimowicz [25]. Such complexing agents as soluble chetating polymers and cyanoferrates of transient metals were used to bind radioactive ions. The preliminary pilot plant experiments are conducted using system equipped with a ceramic module of 23-channel and molecular weigh cut-off (MWCO) of 8 kDa. Inorganic membranes may have better resistance toward the radioactive environment. Model solutions and original radioactive wastes were used as feed. The radioactive substances concentrated in a small volume as retentate can be directly fossilized or vitrified.

Application of inorganic membranes was considered for radioactive wastes processing in Zakrzewska-Trznadel's work [26]. The experiments were performed with radioactive model solutions and original radioactive waste samples. To achieve high DFs the process was arranged as a "seeded UF". Complexing agents, such as poly(acrylic) acid and poly(acrylic) acid salts of different crosslinking, polyethylenimine (PEI), and cyanoferrates of transient metals were employed. It was proved that DFs of the hybrid process were higher than those of UF solely. The effectiveness of complexation by each ligand was markedly pH-dependent. According to the introduction by Chen et al. [27], the so-called "seeding" means that some types of additives are spiked into feed solution to facilitate flocculation. The flocculants develop into greater sizes when metal ions gradually adsorb to the seeding additives. Pollutants are retained by membrane and the rejection can be greatly improved if the flocculants' size is greater than the membrane pore size.

In another work about ceramic membrane-based UF combined with complexation for radioactive wastes processing [28], the ceramic UF membranes MEMBRALOX and CéRAM INSIDE were used. To enhance the UF separation effect, different soluble polymers and ferrocyanides that can bind Co(II) and Cs(I) ions were selected as complexing agents. The experiments showed that a significant improvement of DFs was achieved upon using the macromolecular compounds. Another study investigated the potential of a hybrid adsorption/microfiltration (MF) system for the selective removal of hazardous Co(II) and Sr(II) ions from nuclear power plant laundry wastewater [29]. This separation task is very critical for managing LLW. A crystalline silicotitanate (CST) and four different iron oxides were synthesized for adsorption. These oxides had different crystal structures and adsorptive mechanisms, such as ion exchange and surface coordination. CST showed the greatest affinity for Co and Sr ions. The Freundlich isotherm constant, K_F, was 23.5 and 33.9 mg g^{-1}, respectively, at pH 7. In addition, its adsorption capacity was independent of solution pH when it is in the range of 5–9). The iron oxides removed substantial amounts of Co ions at neutral and alkaline pH, but only trivial amounts of Sr. The impurity species, for example potassium (K)(I), Na(I), Ca(II), and Mg(II) had no significant impact on the interactions between the adsorbents and Co and Sr ions. The selectivity coefficients of CST for both Co(II) and Sr(II) ions ranged from 3 to 130. The addition of adsorbents to the MF system enhanced the membrane permeability (>30%). Very likely the Co species were removed before it precipitates at the membrane surface. Substantial efficient metal removals (>90%) were obtained for both Co and Sr during 50 continuous system operations. CST and iron oxides adsorbents were regenerated using NaCl and HCl solutions, respectively. Their postregeneration adsorption capacities were almost maintained, as confirmed by preservation of the activity after multiple adsorption and regeneration cycles.

The synergy achieved by dosing a mixture of CST and ferrihydrite (FH) into a submerged membrane adsorber for the removal of metal species [ie, Co(II), Sr(II), and Cs(I) ions] in LLW were investigated [30]. Maintaining the concentration of CST in the separation system at levels as high as 0.3 gL^{-1} generated substantial and stable removal of metal ions (>90%) This was attributed to the ion exchange-based sorption reaction in the CST cavities. The effect of FH varied significantly depending on the target metals. Interestingly, the mixed CST and FH (even small amount) efficiently eliminated fouling, which was improbable when only CST is used. It was found that FH exhibited high capacity for sorption of a key foulant in the feed solution (ie, the nonionic surfactant Tween 80). Possible binding of multivalent metal ions to the surfactant anchored on FH could induce further repulsive forces between the FH particles, thereby making the cake layer looser. In Chen et al.'s work [27], laser ablation inductively coupled plasma mass spectroscopy (LA-ICP-MS) was used to visualize the fouling induced by seeded reverse osmosis (RO) for removing liquid state LLW. Coincident with obvious permeate flux drop, significantly improved rejection (from 68% to 84%) for Co, Sr (from 55% to 71%), and Cs (from 20% to 34%) were observed with usage of APDC additives. The surface mapping by LA-ICP-MS showed that the hot zones of Co, Sr, and Cs associated with the APDC–Co, APDC–Sr, and APDC–Cs systems were superimposed with the locations of the S moiety of APDC. These characteristics indicated that the aforementioned rejection enhancement is partially due to the formation of APDC–metal complexes, and the possible coprecipitation of these elements with APDC–metal complexes.

It is also proved that APDC additives facilitate the formation of complete blocking, intermediate blocking, and standard blocking, and as a consequence enhance Co rejection. By contrast, cake formation induced by APDC additives was mainly responsible for the better rejection of Sr and Cs. Depth profile analyses revealed that the fouling mainly occurred in the first 2–6 μm of the RO membrane.

In other words, a large particle/colloid is able to penetrate deeply into the membrane even though the pore size of the RO membrane is much smaller than those particles/colloids.

8.4.4 Caustic Recovery

Sodium super ion conducting ceramic membranes—NaSICON were investigated for partitioning of Na(I) ions from radioactive wastes [31]. The schematic showing the electrochemical process to split the salt using the NaSICON is illustrated in Fig. 8.6. The waste is fed to the anode compartment, and an electrical potential is applied to the cell. The ceramic membrane only allows the transport of Na(I) ions to the cathode compartment while reject most other cations [eg, K(I), Cs(I), and anions]. The charge balance in the anode compartment is maintained by H^+ generated, whereas that in the cathode cell is preserved by OH^- produced. O_2 and hydrogen (H_2) are gaseous product at the anode and the cathode, respectively. The acidity of anode chamber strengthens with H^+ generation. OH^- produced in the cathode compartment lead to delivering NaOH.

The membrane was selective to Na(I) ions. An average DF of 2000 was observed with respect to Cs-137. In treating actual waste, the final average Na(I) transport rate was 10.3 $kg day^{-1} m^{-2}$ with an average Na(I) transport efficiencies of ~99%. The NaSICON ceramics are polycrystalline materials that contain channels within the crystal structures for Na(I) ion transfer.

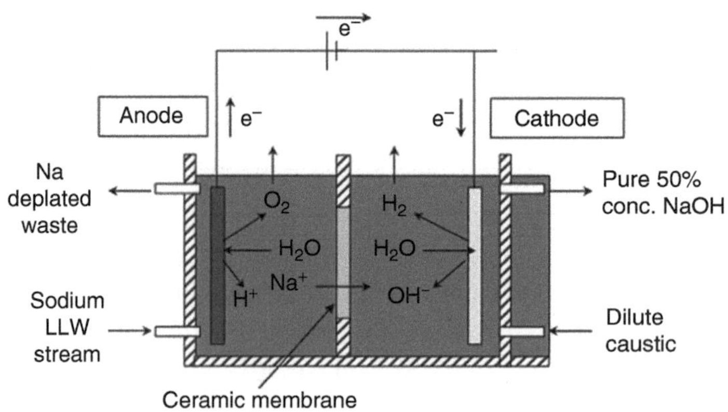

Figure 8.6: Schematic of an electrochemical process using the NaSICON membrane for caustic recycling.
Copied from Ref. [31].

8.5 More Researches and Examples

More researches before 2013 were reviewed by Zakrzewska-Trznadel [32]. Some examples showing the membrane processes untouched in Section 8.1 are briefed later. In Zakrezewska-Tranadel's series works, membrane distillation (MD), a thermally driven process, was developed and tested for liquid radioactive waste concentration [33]. With just one-stage separation all radionuclides were rejected, indicating a very high DF, and pure water, which could be safely discharged into the environment or reused, was obtained as distillate. ED is used to split salts [eg, sodium sulfate (Na_2SO_4)] in saline wastewater for recovery of acid and base [32]. ED was also investigated for tackling waste stream released from the U mining process, for which taking out Mo ions, carbonate (CO_3^{2-}), and bicarbonate (HCO_3^-) can bring some economy value [34]. ED assisted by diffusion dialysis (DD) was utilized for removing heat generating Cs-137 and Sr-90 from HNO_3 solutions [35]. ED exhibited ~90% transfer and DD actualized the separation of metal ions and acid. IEMs were tested for the recovery of boric acid (H_3BO_3) from radioactive wastewater [36]. Some application and laboratory examples are summarized in Table 8.4 with information about location and type of radioactive waste targeted being introduced simply [32].

Table 8.4: Some examples of applications or laboratory testing in treating radioactive wastewater using membrane separation

Major Process	Examples (Location and Some Description)
RO	• In Wolf Creek Nuclear Power Plant, RO with UF pretreatment for floor drains and other wastes was used. • Nine Mile Point Nuclear Power Plant (United States) were equipped with a full-scale RO systems based on Thermex to treat floor drains and other low level wastes in the nuclear plant.
NF	• NF membranes were employed in three-stage RO installation in Chalk River Laboratories (Atomic Energy of Canada Ltd., Canada) for cleaning the reactor coolant and boric acid recovery. • Bugey Nuclear Power Plant in France employed NF for separation of ionized silica and boric acid.
UF	• UF was applied in two stages in the Enhanced Actinide Removal Plant at Sellafield (United Kingdom). • Paks Nuclear Power Plant use UF for cleaning and recycling contaminated boric acid solutions.
MF	• MF was tested as a pretreatment step in RO installation constructed at AECL Chalk River Laboratory (Canada) for mixed low-level radioactive wastes processing. • Crossflow MF for radioactive waste dewatering was evaluated in Berkeley Nuclear Laboratories.
LM	• In Oak Ridge National Laboratory (USA) the emulsion phase contactor was built, where emulsion particles with size 1–5 μm were produced by electric pulses.

8.6 Conclusions

Membrane-based separation has covered the treatment of leaching solution enrichment, product conversion, and waste liquid in relation with radioactive metals. The research progress has been summarized by introducing some most recent typical examples. Literature survey would reveal that there actually exist many other publications, with focus being ELM, SLM, and membrane extraction.

A serious problem with harnessing radioactive metal is wastes. Once released into the air, water, or soil, they will result in possibly catastrophic consequence that may last quite long time and cause malignant diseases.

Evolving concepts lean toward the reprocessing and repository aided by partition of actinides and fission products. *"Nuclear power is the only large-scale energy-producing technology which takes full responsibility for all its wastes and fully costs this into the product"*. Membrane-based separation, particularly membrane contactor, has shown reasonable capability in decontamination of nuclides from HLW streams. For LLW discharge, membrane filtration (eg, RO, UF) coupled with complexation or adsorption is able to provide effective retention of the particles of radioactive metals and is the new research direction.

A critical challenge in radioactive waste treatment by membrane methods is a selection of membrane that can survive the radioactive environment for a long time. The membrane exposed to the ionizing radiation may experience variation in the chemical, physical, and separation properties. In the researches in the past few years, inorganic membranes are getting more interest due to their higher resistivity toward harsh conditions. The stability problem of the membrane contactors has also been reported by several researches.

Glossary and Symbols

Ac	Actinium
AEM	Anion exchange membrane
Am	Americium
Ca	Calcium
$Ca(OH)_2$	Calcium hydroxide
CBC	Calyx arene-benzo-bis-crown-6
CC	Calix[4]arene-bis-crown-6
CEM	Cation exchange membrane
Cl^-	Chloridion
Cm	Curium
CMC	Bis-(octyloxy)calix[4]arene-mono-crown-6
CMPO	*N*-Octyl(phenyl)-*N,N*-di-isobutyl carbamoyl methyl phosphine oxide
CNC	Calix[4]arene-naphtho-bis-crown-6
Co	Cobalt
CO_3^{2-}	Carbonate
Cs	Cesium

$CsNO_3$	Cesium nitrate
CST	Crystalline silicotitanate
Cyanex 923	Tri-alkyl phosphine oxide
D2EHPA	Di(2-ethylhexyl) phosphoric acid
DD	Diffusion dialysis
DF	Decontamination factor
DHDECMP	Di-hexyl-*N,N*-diethylcarbamoylmethylphosphonate
DIDPA	Di-isodecyl phosphoric acid
DIPB	Di-isopropylbenzene
DMDBTDMA	*N,N,N,N*-dimethyl dibutyl tetradecyl malonamide
ED	Electrodialysis
EDI	Electrodeionization
ELM	Emulsion liquid membrane
Eu	Europium
F	Fluoride
Fe	Iron
FH	Ferrihydrite
Fr	Francium
H^+	Proton
H_2	Hydrogen
H_3BO_3	Boric acid
HCO_3^-	Bicarbonate
He	Helium
HLW	High-level radioactive waste
HNO_3	Nitric acid
H_3PO_4	Orthophosphoric acid
H_2SO_4	Sulfuric acid
I	Iodine
IAEA	International Atomic Energy Agency
IEM	Ion exchange membrane
ILW	Intermediate-level radioactive waste
K	Potassium
La	Lanthanum
LA-ICP-MS	Laser ablation inductively coupled plasma mass spectroscopy
LLW	Low-level radioactive waste
LM	Liquid membrane
Lu	Lutetium
MF	Microfiltration
Mg	Magnesium
MgO	Magnesium oxide
Mn	Manganese
Mo	Molybdenum
MWCO	Molecular weight cut-off
Na_2CO_3	Sodium carbonate
$NaNO_3$	Sodium nitrate
NaOH	Sodium hydroxide
NaSICON	Sodium super ion conducting ceramic membranes
Na_2SiF_6	Sodium fluorosilicate
$Na_4UO_2(CO_3)_3$	Sodium uranyl tricarbonate
Nd	Neodymium
$(NH_4)_2CO_3$	Ammonium carbonate
Np	Neptunium

NPOE	2-nitrophenyl octyl ether
O_2	Oxygen
Pa	Protactinium
PC88A	2-ethyl hexyl phosphoric acid-mono-2-ethyl hexyl ester
Pd	Palladium
PEI	Polyethylenimine
Po	Polonium
PP	Polypropylene
PTFE	Polytetrafluoroethylene
PTMS	Phenyltrifluoromethyl sulfone
Pu	Plutonium
Ra	Radium
RO	Reverse osmosis
Ru	Ruthenium
Se	Selenium
Sr	Strontium
TBP	Tri-butyl phosphate
Th	Thorium
Tc	Technetium
TDDA	Tri-dodecyl amine
TODGA	*N,N,N′,N′*-tetraoctyl diglycolamide
TOPO	Tri-*n*-octyl phosphine oxide
TRPO	Trialkyl phosphine oxide
U	Uranium
UCl_4	Uranous chloride
UCIL	Uranium Corporation of India Limited
UF	Ultrafiltration
UF_4	Uranium tetrafluoride
UF_6	Uranium hexafluoride
$UO_2Cl)$	Uranyl chloride
UO_2^{2+}	Uranyl
UO_2SO_4	Uranyl sulfate
$UO_2(SO_4)_3^{4-}$	Uranyl trisulfate
VRF	Volume reduction factor
WPPA	Wet process phosphoric acid
Zr	Zirconium

References

[1] Y.X. Hua, Overview of Nonferrous Metallurgy, Metallurgical Industry Press, Beijing, 2014.
[2] Available from: http://www.world-nuclear.org/information-library/nuclear-fuel-cycle/introduction/nuclear-fuel-cycle-overview.aspx
[3] Q. Yan, A.J. Wang, G.S. Wang, Q.S. Chen, W.J. Yu, R.P. Li, Uranium resources survey and prediction for 2030, China Mining 2 (2011) 1–6.
[4] C.L. Tan, Basic situation of world production of uranium at the beginning of the century, World Nucl. Geosci. 22 (2005) 178–183.
[5] W.L. Yu, W.S. Ge W. Zhang, H. Liao, Q. Yuan, Development situation of global uranium resources exploration and China's "going overseas" strategy, Resources & Industries 3 (2015) 45–50.
[6] H.M. Elsayed, E.A. Fouad, N.M.T. El-Hazek, A.K. Khoniem, Uranium extraction enhancement form phosphoric acid by emulsion liquid membrane, J. Radioanal. Nucl. Chem. 298 (2013) 1763–1775.

[7] P.S. Kulkarni, S. Mukhopadhyay, S.K. Ghosh, Liquid membrane process for the selective recovery of uranium from leach solutions, Ind. Eng. Chem. Res. 48 (2009) 3118–3125.
[8] D.G. Winter, D.S. Flett, M.R. Adam, Ion-exchange Membranes in Hydrometallurgy, Warren Spring Lab, UK, (1987).
[9] R.C. Ewing, W.J. Weber, F.W. Clinard Jr., Radiation effects in nuclear waste forms for high-level radioactive waste, Prog. Nucl. Energy 29 (1995) 63.
[10] Available from: http://www.oecd-nea.org
[11] Available from: http://large.stanford.edu/courses/2010/ph240/thompson2/
[12] S.E. Hasan, International practice in high-level nuclear waste management, Dev. Environ. Sci. 5 (2007) 57.
[13] B. Grambow, Mobile fission and activation products in nuclear waste disposal, J. Contam. Hydrol. 102 (2008) 180.
[14] D.R. Raut, P.K. Mohapatra, S.A. Ansari, V.K. Manchanda, Evaluation of a calix[4]-bis-crown-6 ionophore-based supported liquid membrane system for selective ^{137}Cs transport from acidic solutions, J. Membr. Sci. 310 (2008) 229–236.
[15] P.S. Kulkarni, Recovery of uranium(VI) from acidic wastes using tri-n-octylphosphine oxide and sodium carbonate based liquid membranes, Chem. Eng. J. 92 (2003) 209–214.
[16] S.K. Singh, S.K. Misra, S.C. Tripathi, D.K. Singh, Studies on permeation of uranium (VI) from phosphoric acid medium through supported liquid membrane comprising a binary mixture of PC88A and Cyanex 923 in *n*-dodecane as carrier, Desalination 250 (2010) 19–25.
[17] P. Ramakul, T. Prapasawad, U. Pancharoen, W. Pattaveekongka, Separation of radioactive metal ions by hollow fiber-supported liquid membrane and permeability analysis, J. Chin. Inst. Chem. Eng. 38 (5-6) (2007) 489–494.
[18] H.Q. Li, P.F. Xue, D.X. Luo, C.C. Li Cheng, S.Q. Wang, H.D. Pang, G.P. Fan, Scale-up of low level radioactive wastewater treatment using packed bed filled with ion-exchange fiber, Technol. Water Treat. 1 (1982) 26–30.
[19] D.C. Ge, X.J. Guo. Investigation of application of anion and cation exchange membranes and bipolar membrane in metallurgy, The Fourth Symposium on Application of Membrane-Based Separation in Metallurgical Industry, 2014, pp. 100–102.
[20] D.X. Luo, H.Q. Li, P.F. Xue, Research of high purity ion exchange electrodialysis membrane, Atom. Energy Sci. Technol. 6 (1980) 660–669.
[21] Y.P. Zhang, L. Wang, S.S. Xuan, X.Y. Lin, X.G. Luo, Treatment of Cs containing wastewater using electrodeionization technology, The Fourth Symposium on Application of Membrane Technology in Metallurgical Industry, 2014, pp. 122–124.
[22] D.J. Kathios, G.D. Jarvinen, S.L. Yarbro, B.F. Smith, A preliminary evaluation of microporous hollow membrane modules for the liquid–liquid extraction of actinides, J. Membr. Sci. 97 (1994) 251–261.
[23] S.A. Ansari, P.K. Mohapatra, D.R. Prabhu, et al. Transport of lanthanides and fission products through supported liquid membranes containing *N,N,N′,N′*-tetraoctyl diglycolamide (TODGA) as the carrier, Desalination 232 (2008) 254–261.
[24] P. Jagasia, P.K. Mohapatra, D.R. Raut, P.S. Dhami, V.C. Adya, A. Sengupta, P.M. Gandhi, P.K. Watta, Pertraction of radio-cesium from acidic feeds across supported liquid membranes containing calix-crown-6 ligands in a fluorinated diluent, J. Membr. Sci. 487 (2015) 127–134.
[25] G. Zakrzewska-Trznadel, M. Harasimowicz, Application of ceramic membranes for hazardous wastes processing: pilot plant experiments with radioactive solutions, Desalination 162 (2004) 191–199.
[26] G. Zakrzewska-Trznadel, Radioactive solutions treatment by hybrid complexation–UF/NF process, J. Membr. Sci. 225 (2003) 25–39.
[27] L.-S. Chen, T.-H. Wang, Y.-K. Hsieh, C.-H. Hsu, Justin C.-T. Lin, C.-F. Wang, Visualization of clogs developed from interaction between APDC and low-level radwaste relevant nuclides on RO membranes: a LA-ICP-MS study, J. Membr. Sci. 456 (2014) 202–208.
[28] G. Zakrzewska-Trznadel, M. Harasimowicz, Removal of radionuclides by membrane permeation combined with complexation, Desalination 144 (1–3) (2002) 207–212.
[29] N.A. Weerasekara, K.-H. Choo, S.-J. Choi, Metal oxide enhanced microfiltration for the selective removal of Co and Sr ions from nuclear laundry wastewater, J. Membr. Sci. 447 (2013) 87–95.

[30] N.A. Weerasekara, K.-H. Choo, Synergistic combination of metal oxide adsorbents for enhanced fouling control and metal removal in a submerged membrane adsorber, J. Membr. Sci. 490 (2015) 9–17.
[31] M.S. Fountain, D.E. Kurath, G.J. Sevigny, A.P. Poloski, J. Pendleton, S. Balagopal, M. Quist, D. Clay, Caustic recycle from Hanford tank waste using NaSICON ceramic membranes, Sep. Sci. Technol. 43 (2008) 2321–2342.
[32] G. Zakrzewska-Trznadel, Advances in membrane technologies for the treatment of liquid radioactive waste, Desalination 321 (2013) 119–130.
[33] G. Zakrzewska-Trznadel, M. Harasimowicz, A.G. Chmielewski, Concentration of radioactive components in liquid low-level radioactive waste by membrane distillation, J. Membr. Sci. 163 (1999) 257–264.
[34] A. Lounis, C. Gavach, Treatment of uranium leach solutions by electrodialysis for anion impurities removal, Hydrometallurgy 44 (1997) 83–96.
[35] J.N. Mathur, M.S. Murali, M.V. Balarama Krishna, V. Ramahandhran, M.S. Hanra, B.M. Misra, Diffusion dialysis aided electrodialysis process for concentration of radionuclides in acid medium, J. Radioanal. Nucl. Chem. 232 (1-2) (1998) 237–240.
[36] J.K. Park, K.J. Lee, Separation of boric acid in liquid waste with anion exchange membrane contactor, Waste Manage 15 (4) (1995) 283–291.

CHAPTER 9

Noble Metals

H. Wang
Beijing Liquid Cube Technology Ltd., China

9.1 Introduction

Noble metals refer to ruthenium (Ru), rhodium (Rh), palladium (Pd), silver (Ag), osmium (Os), iridium (Ir), platinum (Pt), and gold [1,2]. Except Ag and Au, the others are collectively called platinum metals. Noble metals are widely used in petroleum, chemical, electronic, electrical, marine and aviation industry, known as the industrial "vitamin."

The minerals for Au are mainly natural Au, kustelite, telluride compounds. Ag has more mineral species, which are natural Ag, compounds of sulfur (S), arsenic (As) and Te, and kerargyrite. Huge amount of Ag also occurs in the iso-morphism manner within heavy metal minerals, such as tetrahedrite, galena, enargite, chalcopyrite, pyrite, arsenopyrite, etc. The platinum metals have multitudinous mineral species; say around 200, with great differences in compositions. The most common ones are stillwaterite, As-Rh-Pa-Pt mineral, millerite, etc. Noble metals have not only individual, but also symbiotic deposits.

Regarding naturally occurring noble metals, their abundance in the earth's crust is low and the distribution is scattered. Large-scale deposits of platinum metal deposits are mainly located in high altitude areas. Ninety percent of the world reserves of platinum group metals distributes mainly in South Africa, followed by Russia, which account for about 10% of the world's total storage. The rest of the world shares the balance of 1–2%. South Africa and Russia also possess most of the Au reserves, around 75% of the total. For Ag, the majority of the reserves are held by countries in America and Russia. China is the world's biggest Au producer.

Currently, noble metals recovered from secondary resources have taken a substantial proportion in the total output [1,2]. For example, the regenerated Ag is more than 300 tons annually in developed countries. The regeneration rate of Pt from catalyst in the petrochemical and chemical industries of United States of America (USA) reaches 85–95%. The major secondary resources for noble metals are as follows: (1) waste catalysts, (2) waste photosensitive materials, (3) waste electrodes, crucibles, thermocouples, furnace lining, firebricks, (4) wastewater of electroplating, (5) battery, conductive materials, welding materials, (6) jewellery and coins.

Membrane-Based Separations in Metallurgy
http://dx.doi.org/10.1016/B978-0-12-803410-1.00009-8

Au and Ag are soft and have excellent ductility. Au is soluble in aqua regia [1:3 nitric acid (HNO_3): hydrochloric acid (HCl) (v/v)], but not in common acids; Ag can dissolve in HNO_3. The specific density of Au is 19.32, and that for Ag is 10.49. The melting point for Au is 1063°C and the boiling point 2807°C, and those for Ag are 962°C and 2212°C, respectively [1,3].

The block flowsheet for a typical mining and metallurgical process for Au production is shown in Fig. 9.1 [4]. The general steps are grinding, dressing, pretreatment, leaching, extraction, refining and recovery, and "three wastes" treatment. The specific process and technologies depend on the nature of the deposits and minerals [4]. There exist several categories of methods for extracting Au: amalgamation, cyanidation, and noncyanide hydrometallurgy [eg, carbon in pulp (CIP) method, sulfourea method]. Cyanidation is still the major approach currently due to its simplicity and efficiency. Ag generally shares the same extractive methods with Au [1].

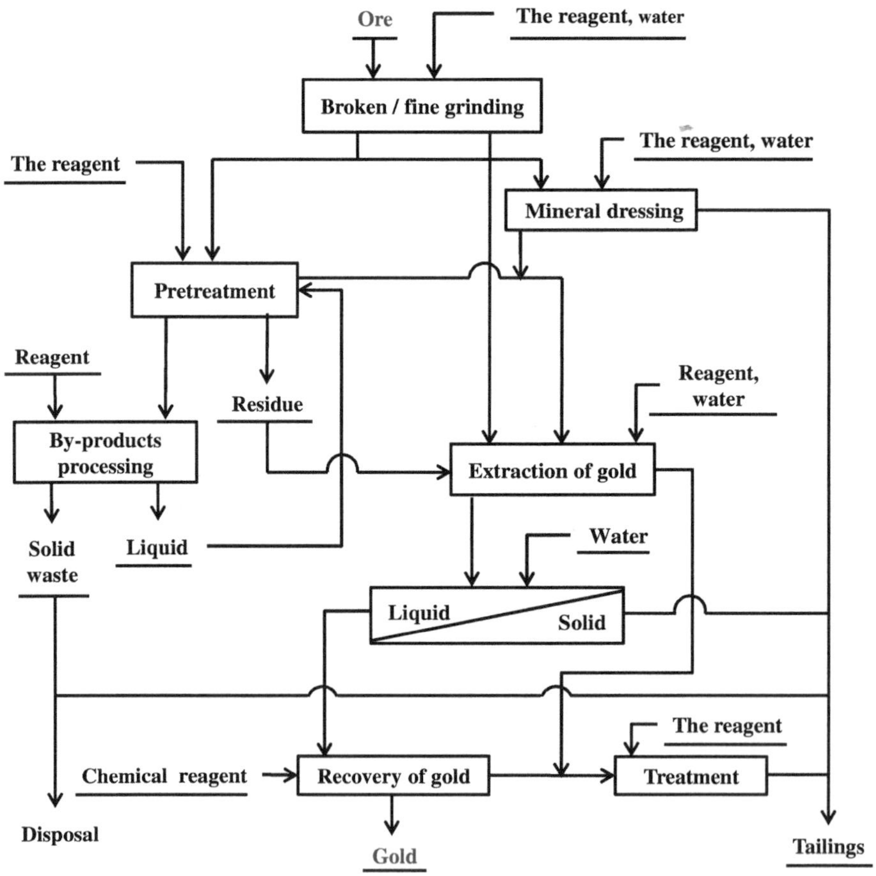

Figure 9.1: Simplified flow chart for gold production process.

Extraction of gold by cyanide process is based on the theory of oxidation by dissolved oxygen and corrosion. The overall reaction is as follows:

$$4Au + 8CN^- + O_2 + 2H_2O \Leftrightarrow 4Au(CN)_2^- + 4OH^-$$

The dissolution rate is described using the following equation:

$$\text{Dissolution rate} = \frac{2A \bullet D_{CN^-} - D_{O_2}[CN^-][O_2]}{\delta\{D_{CN^-}[CN^-] + 4D_{O_2}[O_2]\}}.$$

where D is the diffusion coefficient (mol s^{-1}), A is the liquid/solid contact area (cm^2), and δ is the Nernst boundary layer thickness.

Accordingly, in the lower concentration range of CN^-, the dissolution rate of Au is positively related to $[CN^-]$, whereas in the higher range, the rate is determined by dissolved O_2 concentration [1,3]. Therefore, controlling of both $[CN^-]$ and O_2 concentration is necessary for obtaining optimum dissolution rate.

9.2 Acid Mine Drainage Wastewater

Acid mine drainage (AMD) is one of the most tough environmental problems challenging mining industry. It is the result of oxidation of sulfide compounds in mining waste, tailings, and components of mines. It is characterized of low pH and high concentration of sulfate, metals, and metalloids. It has already been evidenced that membrane separation processes have the potential for treating AMD from mining processes of other metals [5,6].

AMD was collected at an Au mining company site in the state of Minas Gerais, Brazil [6]. The company has two underground gold mines and an industrial processing plant. Aguiar et al. conducted a research to evaluate the main operational conditions of the Au AMD treatment by membrane separation process, and a preliminary capital and operational cost evaluation [6].

In Aguiar et al.'s work [6], a series of commercial nanofiltration (NF) and reverse osmosis (RO) membranes were employed to treat AMD from Au mining. The AMD of an underground Au mining site owned by a company in the state of Minas Gerais, Brazil was tested [7]. The information regarding the membrane types, pH value variation, and maximum recovery rate (RR) in relation with permeate quality were accrued. RR is defined as the ratio of accumulated permeate volume (V_p) to total effluent feed volume (V_f).

Thereafter a preliminary capital and operational costs (CapEx and OpEx) evaluation were made [7].

The AMD composition and concentration fluctuated throughout the year. The components of particular interest to this research are presented in Table 9.1. Five different commercial

Table 9.1: AMD main characteristics

Characteristics	First Collection	Second Collection
pH	2.74	3.76
Color (μH)	31.1	268.4
Turbidity (UNT)	199	120
Total organic carbon (mg L^{-1})	2.3	—
Conductivity (μS cm^{-1})	2744	2573
Total solids (mg L^{-1})	2926	3102
Total suspended solids (mg L^{-1})	87	77
Sulfate (mg L^{-1})	1406.1	1959.5
Chloride (mg L^{-1})	54.5	151.5
Calcium (mg L^{-1})	323	284
Magnesium (mg L^{-1})	97	226

Table 9.2: Membrane characteristics, as provided by the suppliers

	Reverse Osmosis		Nanofiltration		
Characteristic	TFC-HR	BW30	MPF-34	NF90	NF270
Supplier	Koch Membrane	Dow Filmtec	Koch Membrane	Dow Filmtec	Dow Filmtec
Membrane material	Proprietary TFC polyamide	Polyamide thin-film composite	Polyamide SelRO COMPOSITE	Polyamide thin-film composite	Polyamide thin-film composite
Maximum operating pressure (bar)	41	41	35	41	41
Maximum operating temperature (°C)	45	45	70	40	40
pH range, continuous operation	4–11	2–11	0–14	2–11	2–11
NaCl retention	99.55%[a]	99.5%[b]	35%[c]	85–95%[d]	—
$CaCl_2$ retention	—	—	—	--	40–60%[e]
$MgSO_4$ retention	—	—	—	>97%[f]	>97%[f]

[a]Test conditions: 2000 mg L^{-1} NaCl at 15.5 bar applied pressure, 15% recovery, 25°C, and pH 7.5.
[b]Test conditions: 2000 ppm NaCl at 15.5 bar applied pressure, 15% recovery, and 25°C.
[c]Test conditions: 5% NaCl at 30 bar, 30°C.
[d]Test conditions: 2000 ppm NaCl at 4.8 bar applied pressure, 15% recovery, and 25°C.
[e]Test conditions: 500 ppm CaCl2 at 4.8 bar applied pressure, 25°C, 15% recovery.
[f]Test conditions: 2000 ppm $MgSO_4$ at 4.8 bar applied pressure, 25°C, and 15% recovery.

membranes were evaluated in the work. The basic information as provided by the suppliers about these membranes is summarized in Table 9.2.

The AMD was pretreated by ultrafiltration (UF) prior to NF and RO experiments, to prevent damaging the RO and NF membranes due to suspended solids in the raw AMD. It was found that the RO membranes tested exhibited considerably lower permeate fluxes than the NF

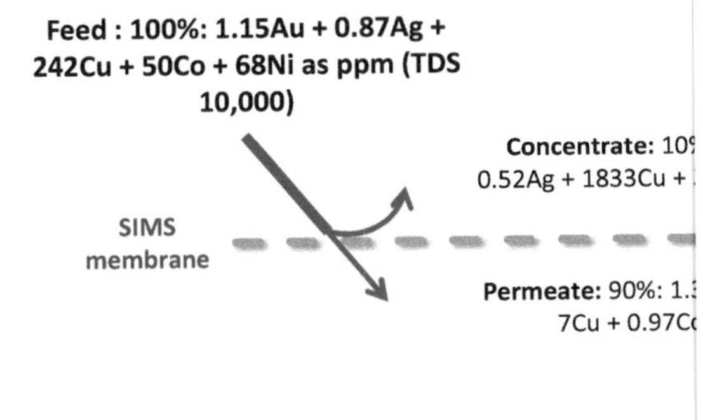

*SIMS: Special Industrial Membrane Separati

Figure 9.2: Removal of the interference from bivalen

Au was mixed into the leaching system. In 2003, the SIMS
solution treatment was installed and put into operation by ar
comprehensive processing capacity reached 2500 $m^3\ h^{-1}$.

A noble metal smelting factory produces 150 m^3 cyanide bar
comprising free CN^- of 800 mg L^{-1}, Cu about 600 mg L^{-1},
[11]. The overall process adopted for tackling this barren sol
pretreatment, cyanide barren solution flows into SIMS syste
25 $m^3\ d^{-1}$ with holding back bivalent ions. The liquid collect
125 $m^3\ d^{-1}$, which is used directly in the cyanidation process
Au, and Ag are immediately recovered, bringing about econ
solution was acidified, followed with recovery of Cu and Zn
of NaCN by using gas stripping and sodium hydroxide (NaC

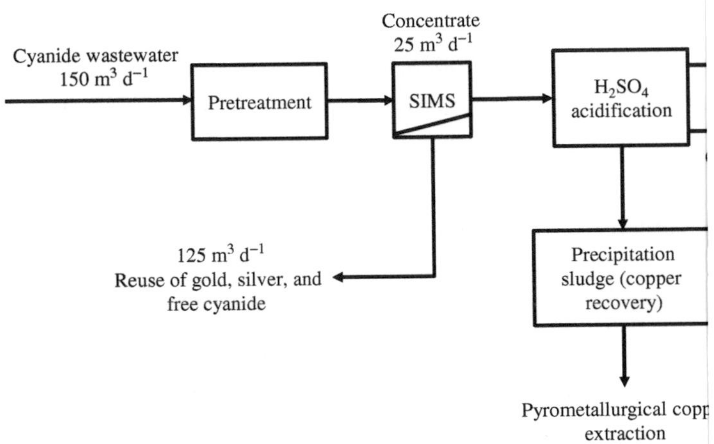

Figure 9.3: The process for recycle cyanide

membranes during the filtration time. The highest average permeate flux was obtained with the NF270, which was 88.6 L hm^{-2}. In addition, RO membranes showed more severe fouling. This was due to the influence of contact angle and higher salt retention of RO membranes.

The retention efficiency of all the membranes is high. The pollutant content in the permeate is lowest with process using TFC-HR RO membrane; the BW30 RO membrane displayed even lower rejection than all the NF membranes.

Based on the aforementioned information, NF90 and NF270 were more suitable for AMD treatment than the other membranes. The extremely low permeate fluxes of the RO membranes enhanced the initial investment cost of the membrane unit.

When using NF270 membrane at pH 5.5, the maximum RR for a single step was 60%. The permeate quality achieved at a recovery rate of 60% was superior to that of the process water generally utilized by this company (ie, 13.3 mg L^{-1} Ca and 262.1 mg L^{-1} sulfate). Above this RR, membrane fouling would be more severe and permeate quality would deteriorate.

In CapEx and OpEx estimation, the variables considered are membrane unit cost, membrane replacement, alkalizing agents, chemical cleaning agents, energy consumption, and maintenance. The capital cost for the UF-NF unit was evaluated according to the price of US\$8750.00$(m^3/h)^{-1}$ (effluent flow). Estimation of the capital cost per cubic meter of effluent involves the annualization by amortization factor shown in Eq. (9.1):

$$\mathrm{A/P} = \frac{i_c \cdot (1+i_c)^{DL}}{(1+i_c)^{DL} - 1} \tag{9.1}$$

where (A/P) is the amortization factor; i_c the investment rate; DL the design life of the plant (ie, 15 years). The capital cost per cubic meter was then calculated from Eq. (9.2):

$$C_{cap/m^3} = \frac{C_{cap} \cdot A/P}{Q_{des}} \tag{9.2}$$

where is the capital cost per cubic meter of effluent; C_{cap} is the system capital cost and Q_{des} is the capacity of the designed system.

Membrane replacement was based on a life span of 5 years. The alkalizing agent used was NaOH and the cleaning agent was HCl solution at 0.2% w/w.

The energy cost evaluation comprised the requirements on UF power and NF feed pump in a once-through operation process. The UF power requirement was around 0.2 kW h m^{-3} [8], whereas the cost for NF feed pump was analyzed using following equations:

$$Q_f = \frac{J \cdot A}{RR} \tag{9.3}$$

$$E_f = \frac{\Delta P \cdot Q_f}{\eta}$$

where Q_f is the NF feed flow; E_f the energy for feed pump; J
A the total NF membrane area; RR the water recovery rate (6
(10 bar); and η the feed pump efficiency (70%). Finally, the s
assessed at 5% of the initial investment cost for each year.

Based on the aforementioned conditions, the CapEx of the N
estimated at US$131,250.00, and the OpEx was 0.263 US$/n

9.3 Cyanide Barren Solution Treatment

The liquid separated from Au and Ag solid precipitated by z
solution [1]. It usually contains (resulted from Zn powder re
extracting elementary Au), copper (Cu) (resulted from cyani
noble metals. The process adopted by Au smelting plant for
treatment relies on steps of acidification and oxidization. As
higher content of Cu, precipitation by sulfide [sodium sulfid
(NaHS)] is employed. The most popular processes based ma
technologies are sulphidization, acidification, recycling and
and acidification volatilization and regeneration (AVR) proc
SART is sent to cyanidation step, whereas that in AVR to th
(CCD) [9]. These two technologies have already been practi
worldwide. The problems associated with the barren solutio
are summarized as follows [7]. The RR of noble metal(s) is
precipitation; the efficiency in eliminating or shielding the i
low. In many industrial level operations, with open circuit a
barren solution can be recycled and most of the bivalent ion
process. According to the practice of a large-scale Au smelt
leaching using purified water has an extraction efficiency 5-
water with bivalent ions is concerned. Therefore, more effic
barren solution is desired.

Wang employed special industrial membrane (SIMS) in sep
$[Cu(CN)_2]^-$, $[Cu(CN)_3]^{2-}$, $[Cu(CN)_4]^{3-}$ and gold cyanide co
membrane is NF and the separation principle of the membra
shown in Fig. 9.2. As revealed, SIMS allows the permeation
complex with cyanide. Cu can be concentrated by 5–10 tim
only 25–20% of the original. The retentate was then treated
and oxidization to recover Cu and sodium cyanide (NaCN),

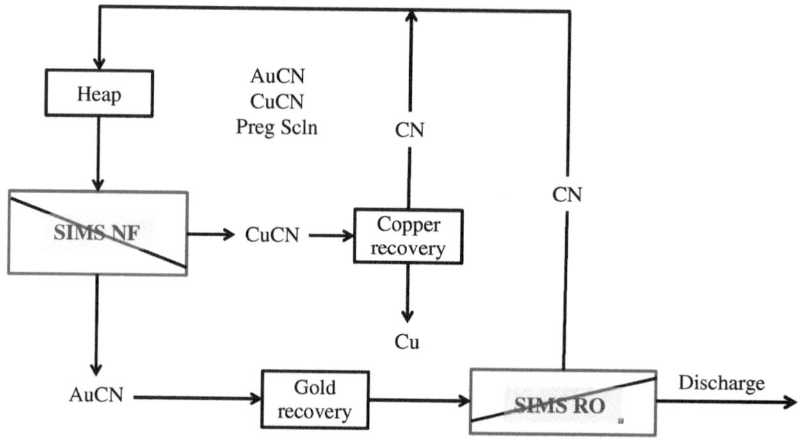

Figure 9.4: The flow chart of the fractionation of gold and copper–cyanide complex.

has an efficiency of only 98–99%, leaving a certain amount of CN^- in wastewater. For minimizing its hazardous impact, oxidization by sodium hypochlorite (NaClO) is used in most situation, in combination with high density slurry (HDS) precipitation. The slag is stored in the tailing dam, and the supernatant is used in the floatation step.

Previously in fact, conventional NF membranes were also utilized to separate the cyanide complexes of Cu and Au, but with low efficiency [12]. Considering the complicacy in valence state of Cu complexes, according to the producer, SIMS are developed with special materials, suitable pore size and charge on membrane surface. The scientific and/or engineering information about the membranes is proprietary. For some process, as shown schematically in Fig. 9.4, an additional SIMS system (ie, RO) may be employed to concentrate and recover Au in the permeate from NF unit; the purified water from this second membrane system may be used to solve the water balance problem. Several industrial-level applications of SIMS are summarized in Table 9.3.

Fig. 9.5 is the simplified flow diagram for the process centering on membrane separation for treating cyanide barren solution in an Au mine in Shandong province, China. Table 9.4 summarizes the water quality after treatment by the process. The concentration of Cu and Zn

Table 9.3: Some SIMS applications for Au cyanide barren solution

Year	Company	Application	Scale (m^3 h^{-1})
2003	Yanacocha, Peru 1	Au recovery, CN barren solution	250
2005	Yanacocha, Peru 2	Au recovery, CN barren solution	1000
2008	Yanacocha, Peru 3	Au recovery, CN barren solution	500
2008	Waibi Gold, New Zealand	CN barren solution process	250
2010	Yanacocha, Peru 4	Au recovery, CN barren solution	500
2012	Barrick Gold Co.	Au/Cu recovery	500

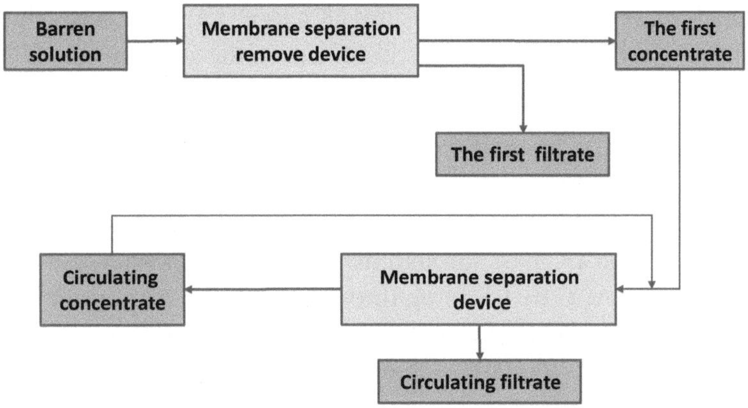

Figure 9.5: The process flow diagram of the two section membrane concentration process for the treatment of cyanide barren solution process in a gold mine in Shandong.

Table 9.4: The experimental results of treating Au cyanide barren solution using membrane separation

Item	Cu^{2+}(mg·L^{-1})	Zn^{2+}(mg·L^{-1})	CN^{-}(mg·L^{-1})
Barren solution	2023.8	702.0	14.5
First filtrate	180.8	23.5	7.0
Circulating filtrate (individual sampling)	544.8	107.0	8.8
Circulating filtrate (final sampling)	1359.5	326.3	13.2
Final concentrate	8775.0	3932.5	32.0

in the barren solution was reduced by half. The volume reduction of the feed is around 90%, which makes it possible to perform flocculant precipitation and pressurized filtration.

As implied by the aforementioned researches, an indivisible aspect associated with barren solution from Au extraction is regarding cyanide treatment. Due to its potential hazards, many countries and environmental protection agencies have imposed limiting standards for the discharge of cyanide-contaminated effluent [13].

There exist a variety of technologies for controlling cyanide discharge and some of them have been briefly touched upon in aforementioned investigations. The more traditional one is destruction by chemical reagents with high oxidation potential, which are very efficient but expensive [14]. Undoubtedly, the recovery of cyanide offers potentially both economic and environmental benefits. The most popular methods for cyanide recovery involve acidification such as AVR and SART [15].

In Zheng et al.'s work [15], a novel approach based on electrodialysis (ED) was designed for treating CN^{-} and other components in the effluent discharged from Au cyanidation process. The feed was provided by Zhaoyuan Gold Smelter Plant, China. The metals present in the solution include sodium (Na), Cu, Zn, potassium (K), iron (Fe), lead (Pb), aluminum (Al). As for Cu, Zn, and CN^-, their dominant existing forms in the solutions are $[Cu(CN)_4]^{3-}$, $[Cu(CN)_3]^{2-}$, $[Zn(CN)_4]^{3-}$, and CN^-.

Fig. 9.6 demonstrates the experimental setup schematic (A) and two operation modes (B,C). *Mode 1* is a five-compartment ED process with homogeneous and heterogeneous anion exchange membranes (AEMs) and cation exchange membranes (CEMs). As shown in Fig. 9.6(B), Na^+ and K^+ in DC pass through the CEM to CCC, whereas metal complexes, CN^- and OH^- in DC enter ACC by traveling through the AEM. As a result of aforementioned mass transfer, the cyanide in various forms was removed from the feed solution in DS, while the alkaline [sodium hydroxide (NaOH) and potassium hydroxide (KOH)] was recovered

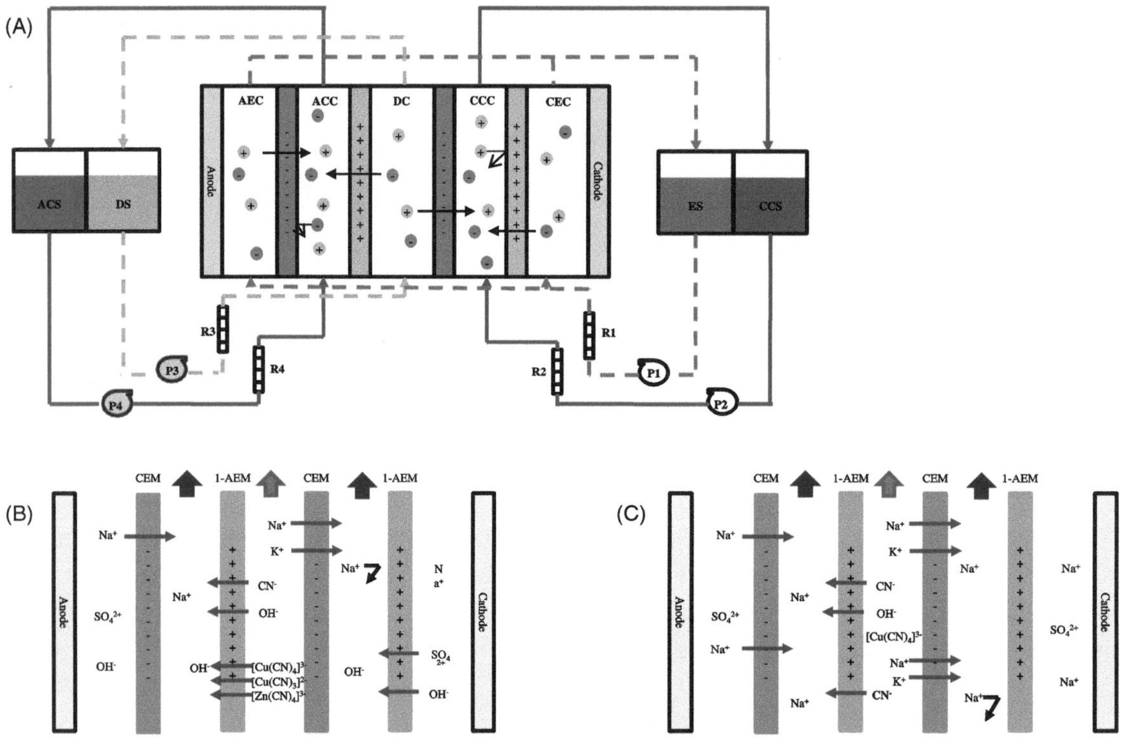

Figure 9.6: Recovery of CN^- using membrane electrodialysis (ED): (A) schematic of the experimental setup; ionic species transport in the ED stack with (B) AEM & CEM and (C) 1-AEM & CEM. Note: *AEC*, anodic electrode compartment; *ACC*, anodic concentrated compartment; *DC*, diluted compartment; *CCC*, cathodic concentrated compartment; *CEC*, cathodic electrode compartment. *CEM*, cation exchange membrane; *AEM*, anion exchange membrane; *1-AEM*, monovalent anion exchange membrane; *ES*, electrode solution; *ACS*, anodic concentrated solution; *DS*, diluted solution; *CCS*, cathodic concentrated solution. P1-pump for ES, P2-pump for CCS, P3-pump for DS, P4-pump for ACS; R1-rotameter for ES, R2-rotameter for CCS, R3-rotameter for DS, R4-rotameter for ACS.

in CCC. With *Mode 2,* the recovery of free cyanide, CN^-, was achieved with homogeneous monovalent AEMs (1-AEMs) and CEMs. According to Fig. 9.6(C), the NaOH and KOH pass through the CEM to enter CCC. However, only CN^- and OH^- in DC could transport through 1-AEM from DC to join ACC, without cotransport of metal complexes. Therefore, the free cyanide, CN^-, could be recovered in ACC.

The experimental results indicate that an extraction or RR of 87.1% for CN^- was achieved at an applied voltage of 30 V with the energy consumption of 8.33 $kWhm^{-3}$ in *Mode 2*.This value was lower than 17.88 $kWhm^{-3}$ obtained with traditional three-compartment ED [16]. The concentration of free CN-recovered in Mode 2 reached 101.95 mg L^{-1}. Reference to the operation practice of cyanidation process indicates that this concentration is suitable for tank and heap leaching. The removal rate of cyanide reached up to 97.9% with homogeneous ion exchange membranes (IEM) at an applied voltage of 38 V in *Mode 1.* This value was much higher than that when heterogeneous IEM was used, which was only 74.0%.

A study developed by Estay et al. focused on the cyanide recovery from Au mining solutions using the gas-filled membrane absorption (GFMA) process [17]. As shown in Fig. 9.7, the dissolved hydrocyanic acid (HCN) is volatilized and transferred through the membrane pores from feed solution to the absorption phase containing NaOH. The conversion of HCN into nonvolatile cyanide (ie, NaCN) in the receiving solution is instantaneous. A transport model was developed to explain the critical steps of the cyanide transfer in the GFMA process. The phenomenological model combines chemical equilibrium of various cyanide species in solution, phase equilibrium, HCN mass transfer, and cyanide balance as a function of time. It was solved by an iterative program based on the Regula Falsi method. The GFMA

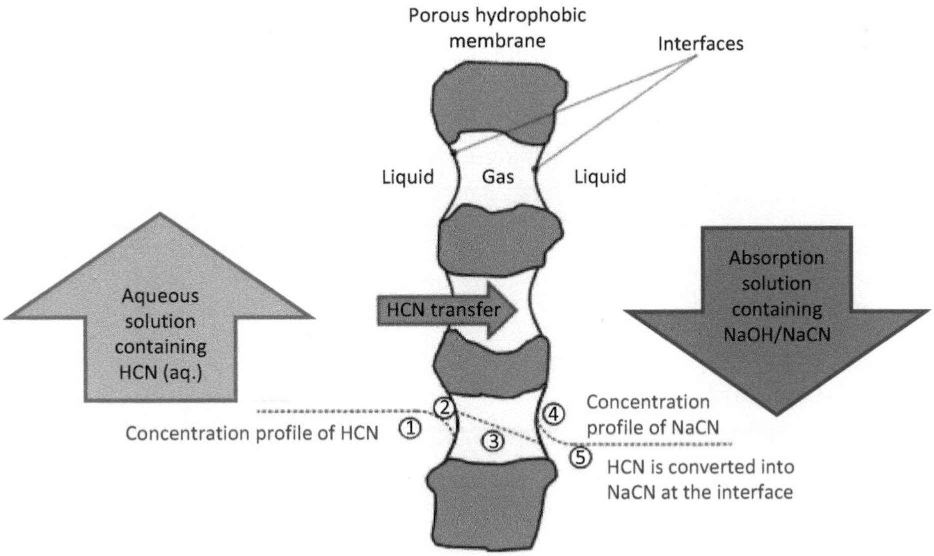

Figure 9.7: Outline of the gas-filled membrane absorption (GFMA) process.

Table 9.5: Characterization of the cyanide solution used for scaling-up study

Description	Value	Unit
[CN^-]	1200	mg L^{-1}
[Zn]	500	mg L^{-1}
[Cu]	80	mg L^{-1}
pH inlet	10.5	—
Temperature	10	°C
Atmospheric pressure	600	mmHg
Feed flow	120–300	m^3 h^{-1}

system obtained a recovery close to 90% after 10 min of starting the extraction; the average cyanide transfer flux was 0.01 kg m^2 h^{-1}. The model was validated by comparison with the experimental results of recovering cyanide in the presence of Cu and Zn ions.

Another work investigating the GFMA in recovering free cyanide from Au mining solutions [18] focused on identification of the main technical variables that control GMFA's economic feasibility and its economical competitiveness with the cyanide recovery technologies currently practiced in Au mining (ie, AVR and SART). Table 9.5 summarizes the characteristics of the simulated or hypothetical feed used in the scale-up estimation. Scaling-up of the GFMA process relied on experimental tests conducted with a Celgard Liqui-Cels 1.7 × 5.5 Mini-module containing 7400 fibers and the theoretical predictions obtained with a mass transfer model validated by a previous research [17]. The membrane module characteristics are summarized in Table 9.6. The schematic flow diagrams for the batch and continuous GFMA processes are illustrated in Fig. 9.8. In both processes, cyanide solution is stored in tanks. It is then delivered into the parallel n-circuits composed of the acidification tank connected with membrane modules in-series. As for continuous operation, Fig. 9.8(B) illustrated one parallel circuit with 39 membrane modules installed in-series.

Table 9.6: Technical characteristics of the industrial membrane module chosen in this scaling-up study

Item	Description
Membrane material	Polypropylene
Number of fibers	584, 299
Mass transfer area (m^2)[a]	560
External diameter of fibers (μm)	300
Internal diameter of fibers (μm)	200
Internal diameter of shell (m)	0.3556
Module length (m)	1.016
Fiber porosity (%)	25
Contact specific area (m^2 m^{-3})[a]	5545
Manufacturer	Liqui-Cel
Model number	14 in. × 40 in.

[a]Based on external diameter of fibers.

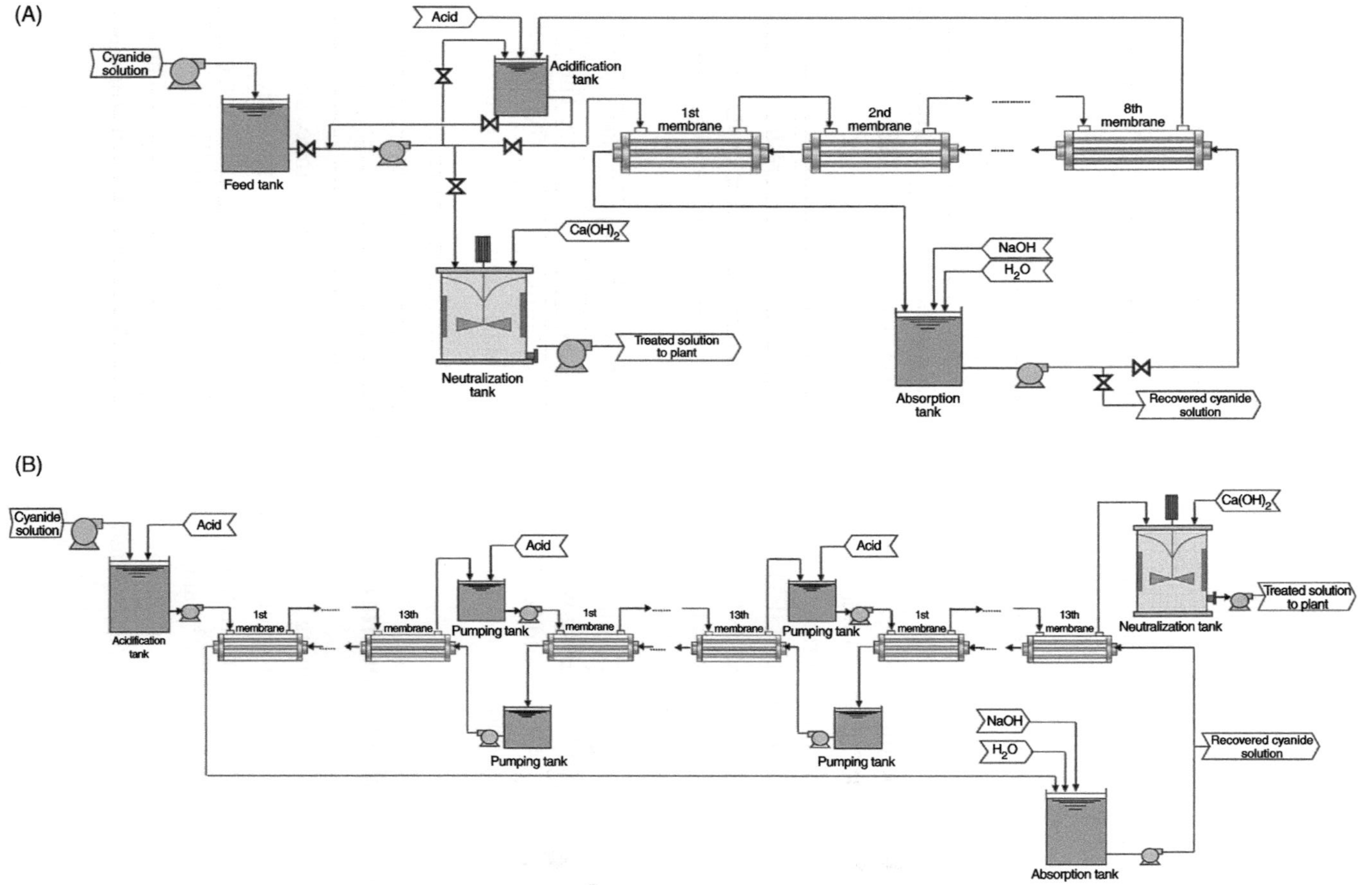

Figure 9.8: (A) Schematic flow diagram of batch configuration in the GFMA process. (B) Schematic flow diagram of continuous configuration for the GFMA process.

The OpEx for the GFMA and the other two conventional processes (ie, SART, AVR) was estimated considering the following aspects: labor, reagents, water, energy, metallurgical control, maintenance, contractors, membrane modules replacement in GFMA case, packing replacement in AVR case, and contingencies (corresponding to 5% of the total cost). The CapEx estimation was broken down into equipment acquisition, construction, and indirect (engineering costs, general expenses, contractor profits, spare parts, commissioning, owner costs, and start-up).

The major indicator for the economic assessment was net present value (NPV) based on incomes and OpEx and CapEx. The effect of the cyanide/Cu molar ratio on the economic evaluation was evaluated using a sensitivity analysis approach for the feed solution composition selected in the experimental part.

The experimental results showed that a high footprint area is required to treat 240 $m^3\ h^{-1}$ of solution containing 1200 mg L^{-1} of total cyanide using either batch or continuous mode of GFMA. Notwithstanding, the area range [2.5–4.2 $m^2(m^3/h)^{-1}$] for both GFMA modes was below that in SART. GFMA under batch configuration is more complex in design and operation and needs more membrane modules as compared with the continuous GFMA configuration. Additionally, batch GFMA, AVR, and SART processes were more energy-intensive than continuous GFMA configuration. The SART presented the best NPV (US$4.24 k(ton NaCN/y)$^{-1}$), while the continuous GFMA exhibited the lowest CapEx of US$56.5 k($m^3$/h)$^{-1}$ and the second best NPV (US$2.47–3.65 k(ton NaCN/y)$^{-1}$).

In view of the aforementioned findings, it was concluded by the work that the new GFMA process was not suitable for dealing with solutions with low cyanide/Cu ratio. The SART or hybrid of GFMA and SART would be the best alternative. Unlike SART, GFMA retains HCN gas inside the membrane pores during its operation, and hence depressing the amount of gas leakages.

9.4 Metallurgical Wastewater

In Lin et al.'s work [19], softening, sedimentation, filtration, ion-exchange, and RO were investigated for processing Au smelting wastewater. The solution was the effluent from the "sulfide-pressure filter-neutralizing treatment" step with main components being Na^+, and Ca^{2+}, and Mg^{2+}, Cu^{2+}, and Pb^{2+}, and As^{2+}, Cl^-, and SO_4^{2-}. The two treatment processes are shown in Fig. 9.9 (with ion-exchange) and Fig. 9.10 (w/o ion-exchange). The water leaving the second stage RO was qualified for recycling. More than 85% water can be reclaimed.

An Au smelting enterprise in Henan Province, China produces a large volume (approximately 120 tons day^{-1}) of wastewater containing heavy metals. Wang used a bipolar membrane (BPM) ED device to electrolyze sodium sulfide (Na_2S), producing hydrogen sulfide (H_2S) and

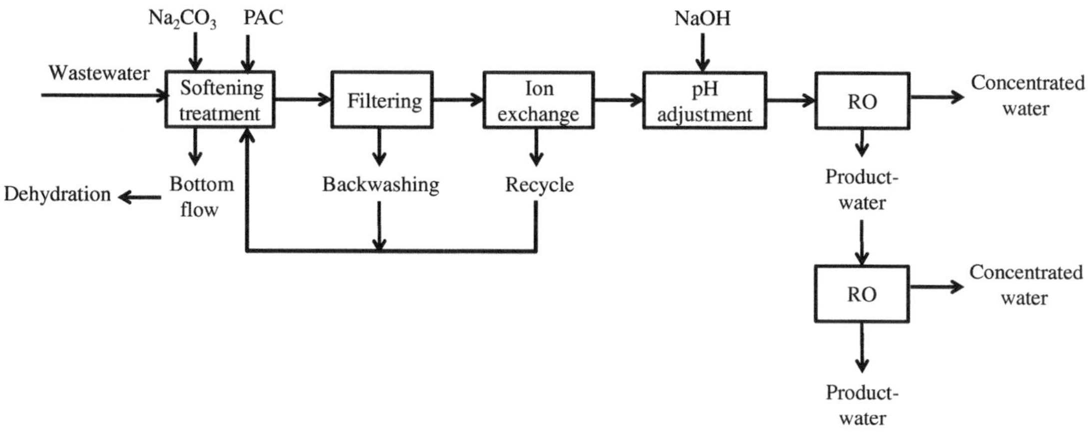

Figure 9.9: Flow chart of the process with ion-exchange.

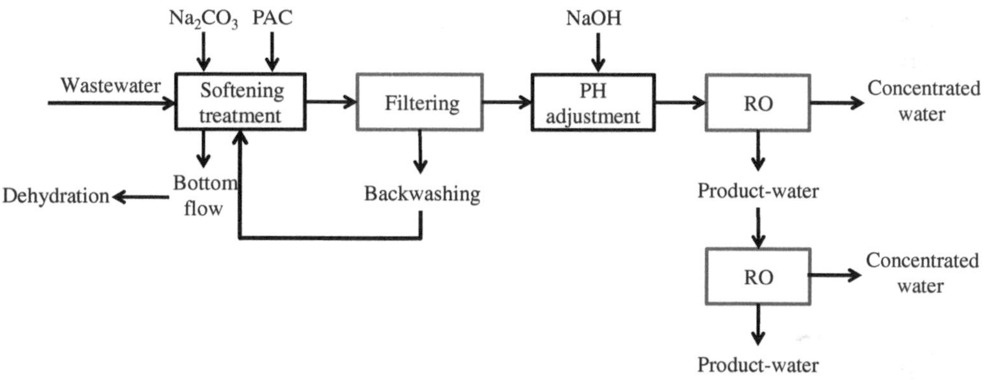

Figure 9.10: Flow chart of the process without ion-exchange.

NaOH [20]. The device schematic is shown in Fig. 9.11. H_2S recovered is useful in online precipitation of metal ions in the Au smelting wastewater.

A demonstration project was constructed with the membrane stack being composed of BPM, CEM, protective CEM, and AEM; the area of each membrane is 7 m^2. The major auxiliary components of the membrane stack and module include baffle plates, spacers, and electrodes. The membrane stack follows the BPM–AEM–CEM configuration. Due to precipitation occurring in the acid room, the channel of the compartment is expanded, the wheat clapboard is employed, and the export diversion hole is enlarged, so as to reduce dead corner as much as possible. To prevent osmosis caused by pressure difference between two compartments, all the solutions are circulated using the pump with identical design (flow $Q = 7\ m^3\ h^{-1}$, head $H = 15$ m). Additionally, the inlet and outlet of the membrane stack are equipped with diaphragm to adjust the pressure and flow rate. In the actual application, the BPM ED device

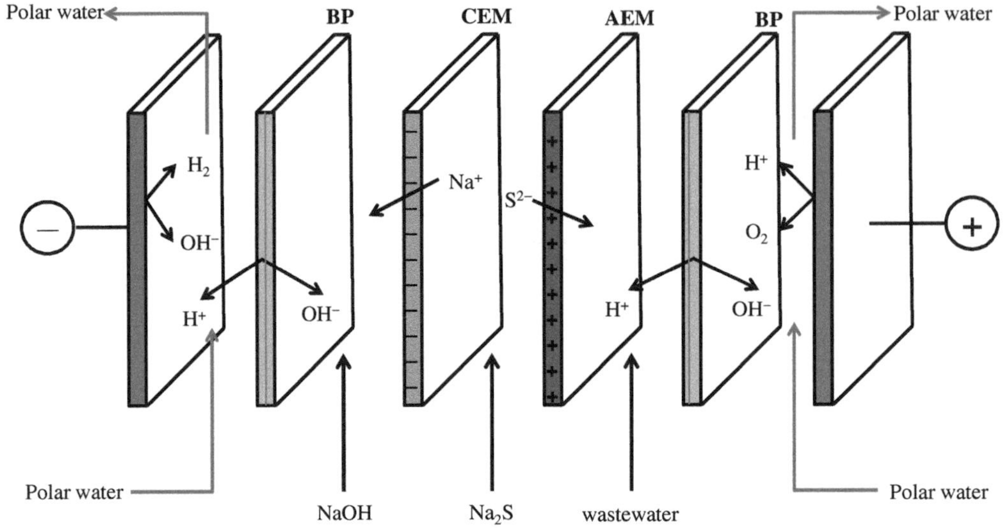

BP: Bipolar membrane; A: anion membrane; C: cation membrane

Figure 9.11: Principle diagram of bipolar membrane (BPM) ED removing gold smelting wastewater.

exhibited >90% heavy metal removal efficiency; the current efficiency of acid chamber is 60%, and that for the alkali chamber is 70%.

9.5 New Developments

As mentioned in the introduction part, secondary resources, such as electronic scraps, anode mud, spent catalyst, metal concentrates, take progressively more important position in noble metal production. At present, the recovery mainly relies on electrolysis, ion exchange, displacement, and precipitation. These methods have their respective advantages and disadvantages. Precipitation is a traditional method, which could be further divided into sodium sulfide (Na_2S) method, organic reducing sugar method, thiosulfate method, boron compounds reduction, and metal reduction [21]. Membrane contactor and membrane adsorber are the most intensively investigated membrane technologies for recovering noble metals from this category of resources.

Recovery of Au and other precious metals from secondary resources is generally actualized by an initial dissolution using aqua regia. Under such conditions, Au is oxidized to Au(III) and forms chloro-complexes, predominantly $AuCl_4^-$; meanwhile base metals such as Cu and Zn can also form chloro-complexes [22]. The other metals, for example, Pt, Pd, iron (Fe), may also be present in the feed stock.

Selective separation of Au(III) ions from acidic aqueous solutions, using methyl iso-butyl ketone (MIBK) as a carrier and LK-80 as an emulsifier, has been investigated [23]. The mechanisms for extraction of Au(III) complex by MIBK and stripping were reported to be as follows:

Ionization:

$$AuHCl_4 \cdot 4H_2O{+}H_2O \rightarrow AuCl_4^-{+}5H_2O{+}H^+$$

Complexation and extraction:

$$4(MIBK) + (H^+)_{aq} \rightarrow ((MIBK)_4H)^+_{aq}$$

$$[(MIBK)4H]^+_{org} + (AuCl_4^-)_{aq} \Leftrightarrow ([(MIBK)_4H]^+ AuCl_4^-)_{org}$$

Stripping:

$$2Na_2SO_3 + ([(MIKE)_4H]^+ AuCl_4^-)_{org} \rightarrow NaAuSO_3 + Na_2SO_4 + 4(MIBK)_{org} + 2HCl + 2NaCl$$

Au(III) was reduced to Au(I) in the sulfite solution. Au(I) ions are not able to form complex with MIBK.

The extraction results from modeled solution containing Au(III), Pd(II), Pt(IV), Cu(II), and Fe(III), showed that only the Au(III) ions transported across the liquid membrane and nearly all of other ions remained in the feed without variation in their concentration. There exist several models for description of emulsion liquid membrane (ELM) systems, and the one that considers diffusion and simultaneous chemical reaction has better compatibility with the system under investigation. MIBK in effect is a carrier that can combine with many metal ions such as those of Pd and Pt. Therefore, the selective separation of Au(III) from other ions relies on selection of reagent used in the stripping phase as shown in aforementioned reaction and the adjustment of extraction condition.

Two macrocyclic compounds based on thiacalix[4] arene bearing three and four amide substituents, respectively, are investigated for the extraction of Au and Pd in chloride media [24]. Results obtained from liquid–liquid extraction experiments show that the new compound effectively extracts both Au and Pd. Based on these results, a SLM system has been designed and used for Pd and Au extraction the new ligands dissolved in 2-nitrophenyl octyl ether (NPOE) as organic phase, and using 0.5 M sodium sulfocyanate (NaSCN) at pH 2 as stripping agent [24]. For the SLM using the thiacalix[4]arene derivative with three amides dissolved in NPOE and thiocyanate in stripping phase, both Pd(II) and Au(III) in a chloride feed transported to the other side of the membrane. The flux of both metals were almost constants even when the chloride content in the feed was enhanced, which suggests that the extraction was based on the mechanism of ion pair proposed by Fontas et al. [25]. Under the competitive conditions, Au obtained more favorable transfer than Pd.

The work has also investigated the influence of different solid supports [ie., Accurel polypropylene (PP), Durapore polyvinylidene difluoride (PVDF), and Fluoropore polytetrafluoroethylene (PTFE)]. Fluoropore with lowest ε(porosity)/d_0(thickness)τ(tortu osity) ratio presents the lowest experimental flux.

In past 10 years, several new processes emerged attracting increasing attention from the researchers on rare metal extraction are membrane absorber or affinity membrane, and solid–liquid extraction or polymer inclusion membrane (PIM).

The schematic illustrating the steps involved in the separation by affinity membrane, or membrane adsorber is given in Fig. 2.16. It was originally developed as an alternative to traditional packed bed for improved mass transfer. In Li et al.'s work, PVDF resin was grafted with thiourea groups and made into porous membranes for recovering trace amount of Au(III) [26]. Grafting was actualized by nucleophilic addition reaction first and the membranes were formed by means of nonsolvent induced phase separation. Au was recovered from aqueous solutions by filtration with a stack of the membranes prepared. The maximum Au uptake by the membrane was 17.2 mg g^{-1} when the Au concentration in feed solution was 2 mg L^{-1}. It was found that the breakthrough point of the membrane was not significantly influenced by pH and flow rate. Au can even be selectively adsorbed by the membrane from a solution containing Au and, 10 times concentrated Cu and Fe irons and was proven to be a promising alternative for selective capture of Au(III) from a real industrial wastewater. The adsorbed Au can be easily eluted from the membrane with a solution containing 0.1 M thiourea and 1 M HCl. When the Au(III) concentration in effluent became the same as that in feed, the loaded membrane was rinsed with DI water followed by elution with acidic thiourea solution. The elution peak for Au(III) was sharp, and the maximum Au(III) concentration was 37 mg L^{-1}. The elution yielded 98% recovery of Au(III). The regenerated membrane could be used for three times without obvious loss of Au uptake capacity. Based on X-ray photoelectron spectrophotometry analyses, the mechanism for Au adsorption was proposed.

A novel membrane adsorber with a high density of adsorption sites that can selectively capture Au(III) ions is fabricated using poly-thiosemicarbazide (PTSC) [27]. PTSC is a polymer that contains one chelate site per monomeric unit, as shown in Fig. 9.12. This approach avoids the necessity to add an external agent into the membrane matrix.

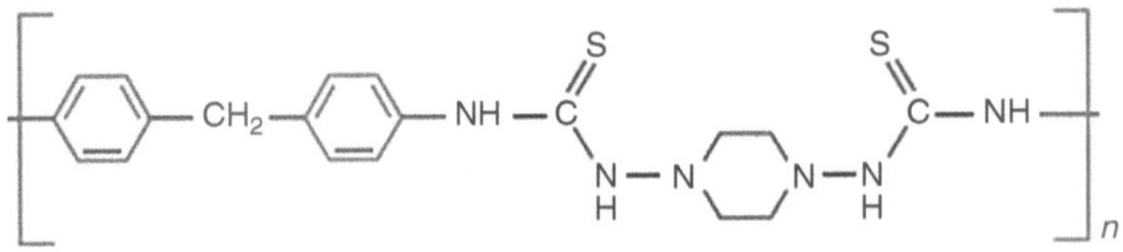

Figure 9.12: Poly-thiosemicarbazide (PTSC) structure.

The resulting PTSC membranes show mechanical stability and can be used in a pressure-driven permeation process, which can eliminate the diffusion limitations commonly required in packed column adsorption. Ninety percent of the Au in a solution which has also 9-fold higher Cu concentration was achieved. operating at a flux as high as 1868 L m^{-2} h^{-1}. The maximum Au uptake of the membrane was 5.4 mmol Au g^{-1}.

The extraction of low concentration (100 mg L^{-1}) of Au(III) from hydrochloric acid (HCl) solutions using PIM made of poly(vinyl chloride) (PVC) is described in Na et al.'s work [28]. The PIM was embedded with commercial extractant Aliquat 336 as a carrier and NPOE, 1-hexanol, 1-octanol, 1-decanol, 1-dodecanol, or 1-tetradecanol as a modifier The PIM composed of 10% (m m^{-1}) 1-dodecanol, 20% (m m^{-1}) Aliquat 336 and 70% (m m^{-1}) PVC offered the highest extraction rate and exhibited sound stability. Upon application of sonication increase of the extraction rate of Au(III) was resulted, was attributed to the acoustic cavitation. It not only disturbed aqueous stagnant diffusion layer situated at the membrane/feed interface, but also weakens the dependence of the initial extraction rate on the original Au[III] concentration.

In the work by Bonggotgetsakul et al. [29], PIM investigated was fabricated using 70 wt.% poly(vinylidene fluoride-co-hexafluoropropene) (PVDF-HFP) embedded with 30 wt.% Cyphoss ionic liquid (IL) 104. The unit structures of the polymer and IL constructing the membrane are shown in Fig. 9.13. The membrane was used for the separation of Au(III) from its HCl solutions and from a solution obtained from the digestion of electronic scrap by aqua regia [29]. The ions in the feed are $[AuCl_4]^-$, NO^{3-}, Cl^-, and H^+. The most probable mechanism for the extraction is described as follows:

$$[P]^+ + [PO_2]^-_m + H^+[AuCl_4]^-_{aq} - \Leftrightarrow [P]+[AuCl_4]^- H^+[PO_2]^-_m$$

where $[P]^+$ and $[PO_2]^-$ represent the trihexyl(tetradecyl)phosphonium cation and the bis(2,4,4-trimethylpentyl)phosphinate anion, respectively; and subscripts m and aq indicate membrane and aqueous phases, respectively.

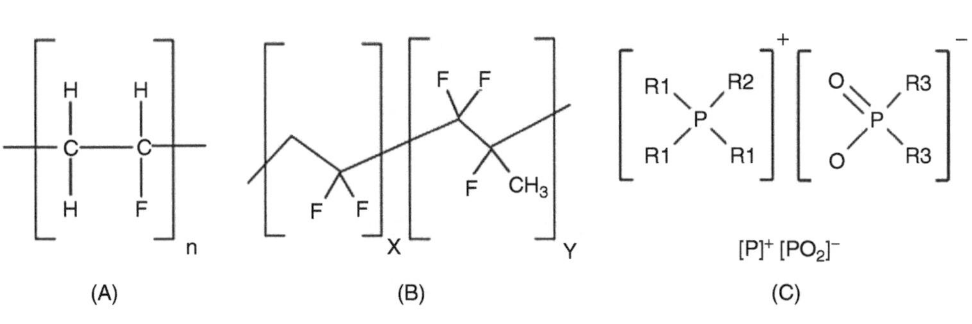

Figure 9.13: Chemical structures of (A) PVDF, (B) PVDF-HFP, and (C) Cyphoss IL 104 ($[P]^+$ $[PO_2]^-$) where R1 = hexyl, R2 = tetradecyl, and R3 = trimethylpentyl.

The resultant PIM is very stable in the aqua regia solution. Au(III) is back-extracted from the PIM by Na_2SO_3. A possible mechanism for the back-extraction is described by following equation:

$$[P]+[AuCl_4]^- \cdot H^+[PO_2]_m^- + 3SO_{2\,aq}^{3-} + H_2O \Leftrightarrow$$
$$[P]^+[PO_2]_m^- + [Au(SO_3)_2]_{aq}^{3-} + 3H_{aq}^+ + 4Cl_{aq}^- + 4Cl_{aq}^- + SO_{4\,aq}^{2-}$$

Research using digested solution as the feed and 0.50 mol L1 Na_2SO_3 solution for back-extraction achieved a complete transport of Au(III) from the feed to the receiver. The content of the other metals in the solution was much higher, for example Cu(II) had a concentration more than 1500-fold higher than that of Au(III).

Yet another new type of PIM membrane for Au(III) transport from concentrated HCl medium was based on organic–inorganic hybrid material [30]. In San Miguel et al.'s work, such a kind of membrane was synthesized by crosslinking of hydroxy terminated poly(dimethylsiloxane) with (3-aminopropyl) triethoxysilane in the presence of the commercial extractant Kelex 100. Au(III) with a relatively low concentration (10ppm) could be selectively separated from Cu(II) of 50ppm and Zn(II) of 100ppm, with a Au(III) recovery of 70%. Nevertheless, the membrane could only used for three cycles. Additionally, Au accumulated inside the membrane. A latest review given by Du et al. shows to us a potential membrane material for ions separation [31]. As shown in Fig. 9.14, electroactive ion exchange materials (EIXMs) are composed of a base structure with fixed negative or positive charge and doped free counterions with opposite charge for maintaining electroneutrality. Applying external electrochemical potential leads to the oxidization and reduction of EIXMs reversibly. The transfer of electronic charges of EIXMs based on the redox reaction can be compensated by ion exchange in electrolyte solution. According to this mechanism, EIXMs may be a good choice for membranes separation of high value-added or toxic ions.

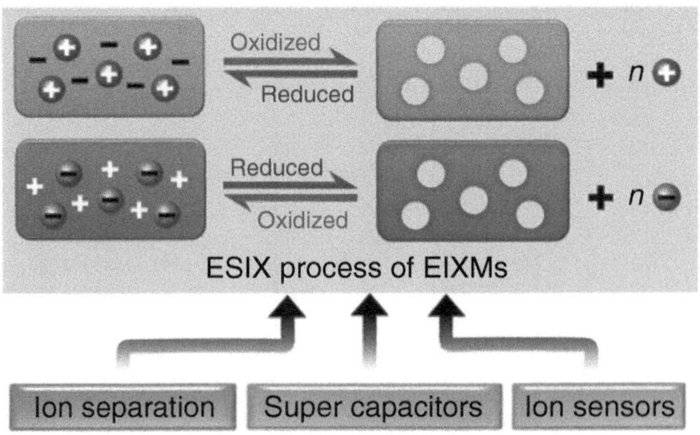

Figure 9.14: Schematic of electrochemically switched ion exchange process of electroactive ion exchange materials.

9.6 Conclusions

With the development of noble metal smelting technology, new technologies and processes emerge in large numbers, greatly enhancing the utilization of this group of materials. In the leaching unit, cyanide method is still the major approach. The high cyanide consumptions in leaching operations have increased the cyanide contents in leach tailings. For this reason, the inclusion of processes to recover or eliminate cyanide from the discharge should be the focus to reduce pollution risk. ED and GFMD were confirmed to be feasible in recovering cyanide. Application of SIMS special filtration membrane in Au metallurgy enables the recycling of cyanide, and in turn enhances the noble metal recovery efficiency. As secondary resources, particularly electronic wastes, take more and more critical position in noble metal production, a series of membrane technologies were investigated to recover noble metals from acid solutions. Liquid membrane is undoubtedly the major research target, but not given most space in this review. The others include affinity membrane, PIM, organic–inorganic hybrid membrane. Most of them demonstrate reasonable performance.

Nomenclature

ACC	Anodic concentrated compartment
ACS	Anodic concentrated solution
AEC	Anodic electrode compartment
AEMs	Anion exchange membranes
Ag	Silver
Al	Aluminum
AMD	Acid mine drainage
As	Arsenic
Au	Gold
AVR	Acidification volatilization and regeneration
BPM	Bipolar membrane
CapEx	(Total) capital cost
CCC	Cathodic concentrated compartment
CCD	Counter-current decantation
CCS	Cathodic concentrated solution
CEC	Cathodic electrode compartment
CEMs	Cation exchange membranes
Cu	Copper
DS	Diluted solution
ED	Electrodialysis
ELM	Emulsion liquid membrane
ES	Electrode solution
Fe	Iron
GFMA	Gas-filled membrane absorption
HCl	Hydrochloric acid
HCN	Hydrocyanic acid
HDS	High density slurry
HNO_3	Nitric acid
H_2S	Hydrogen sulfide

IEM	Ion exchange membranes
IL	Ionic liquid
Ir	Iridium
K	Potassium
Mg	Magnesium
MIBK	Methyl iso-butyl ketone
Na	Sodium
NaClO	Sodium hypochlorite
NaCN	Sodium cyanide
NaHS	Sodium bisulfine
NaOH	Sodium hydroxide
NaSCN	Sodium sulfocyanate
Na_2S	Sodium sulfide
NF	Nanofiltration
NPOE	2-nitrophenyl octyl ether
NPV	Net present value
OpEx	(Total) operational cost
Os	Osmium
Pb	Lead
Pd	Palladium
PIM	Polymer inclusion membrane
PP	Polypropylene
Pt	Platinum
PTFE	Polytetrafluoroethylene
PTSC	Poly-thiosemicarbazide
PVC	Poly(vinyl chloride)
PVDF	Polyvinylidene difluoride
PVDF-HFP	Poly(vinylidene fluoride-co-hexafluoropropene)
Rh	Rhodium
RO	Reverse osmosis
RR	Recovery rate
Ru	Ruthenium
S	Sulfur
SARY	Sulphidization, acidification, recycling, and thickening
SIMs	Special industrial membrane
SLM	Supported liquid membrane
Te	Telluride
UF	Ultrafiltration
Zn	Zinc

References

[1] L.Y. Wang, Noble metal, in: D.L. Wang (Ed.), Rare Metals Handbook, Metallurgical Industry Press, Beijing, 2008.

[2] Editorial Board, The Guidebook for Gold Production Technology, Geological Publishing House, Beijing, (2000).

[3] Z.M. Wu, J.C. Yao, Gold extraction using cyanide method and high purity gold refining, Gold Sci. Technol. 17 (2009) 60–63.

[4] F.F. Chen, Y.F. Zhang, G. Xue, Status and development of production technology for gold smelting, China Nonfer. Metall. A 1 (2011) 11–18.

[5] B.A.M. Al-Rashdi, D.J. Johnson, N. Hilal, Removal of heavy metal ions bynanofiltration, Desalination 315 (2013) 2–17.
[6] M. Mullett, R. Fornarelli, D. Ralph, Nanofiltration of mine water: impact of feed pH and membrane charge on resource recovery and water discharge, Membranes 4 (2014) 163–180.
[7] A.O. Aguiar, L.H. Andrade, B.C. Ricci, W.L. Pires, G.A. Miranda, M.C.S. Amaral, Review—gold acid mine drainage treatment by membrane separation processes: an evaluation of the main operational conditions, Sep. Purif. Technol. 170 (2016) 360–369.
[8] G.K. Pearce, UF/MF pre-treatment to RO in seawater and wastewater reuse applications: a comparison of energy costs, Desalination 222 (2008) 66–73.
[9] M.M. Botz, J.C. Scola, R. Fueyo, W. de Moura, Cyanide recovery practice at cerro vanguardia, SME Annual Meeting, Feb. 23–25, Denver, Colorado. Available from: http://www.botz.com/recovery.html
[10] H. Wang, Separation of copper cyanide and gold cyanide complexes using special industrial membrane, The Fourth Conference on the Application of Membrane Separation Technology in Metallurgical Industry, 2014.
[11] H. Wang, Membrane treatment of cyanide barren solution in noble metals smelting reduction technologies, The Third Conference on the Application of Membrane Separation Technology in Metallurgical Industry, 2009.
[12] Z.G. Wang, Application of membrane separation technology in hydrometallurgy, The Fourth Symposium on the Application of Membrane Separation Technologies in Metallurgical Industry, 2014.
[13] R.R. Dash, A. Gaur, C. Balomajumder, Cyanide in industrial wastewaters and its removal: a review on biotreatment, J. Hazard. Mater. 163 (2009) 1–11.
[14] G. Demopoulos, T. Cheng, A case study of CIP tails slurry treatment: comparison of cyanide recovery to cyanide destruction, Eur. J. Miner. Process. Environ. Prot. 4 (2004) 1–9.
[15] Y. Zheng, Z. Li, X. Wang, X. Gao, C. Gao, The treatment of cyanide from gold mine effluent by a novel five-compartment electrodialysis, Electrochim. Acta 169 (2015) 150–158.
[16] Y. Zheng, Z. Li, X. Wang, Y. Wang, X. Gao, C. Gao, Application of electrodialysis to remove copper and cyanide from simulated and real gold mine effluents, Electrochim. Acta 169 (2015) 150–158.
[17] H. Estay, M. Ortiz, J. Romero, A novel process based on gas filled membrane absorption to recover cyanide in gold mining, Hydrometallurgy 134-135 (2013) 166–176.
[18] H. Estay, E. Troncoso, J. Romero, Design and cost estimation of a gas-filled membrane absorption (GFMA) process as alternative for cyanide recovery in gold mining, J. Membr. Sci. 466 (2014) 253–264.
[19] L.H. Lin, X.Y. Zheng, G.C. Lin, X.S. Lin, Process research of zero emissions of gold smelting wastewater, The Fourth Symposium on the Application of Membrane Separation Technologies in Metallurgical Industry, 2014.
[20] Y.M. Wan, L. Wu, C.R. Li, Demonstration project construction for gold smelting wastewater treatment using bipolar membrane, The Fourth Conference on the Application of Membrane Separation Technology in Metallurgical Industry, 2014.
[21] Y. Su, Study on the retrieval of silver from waste liquid, J. Southwest Nationalities College (Natural Sci. ed.) 25 (1999) 173–175.
[22] I. Villaescusa, V. Salvado, J. de Pablo, Liquid–liquid and solid–liquid extraction of gold by trioctylammonium chloride (TOMACl) dissolved in toluene and impregnated on amberlite XAD-2 resin, Hydrometallurgy 41 (1996) 303.
[23] A. Kargari, T. Kaghazchi, B. Mardangahi, M. Soleimani, Experimental and modeling of selective separation of gold(III) ions from aqueous solutions by emulsion liquid membrane system, J. Membr. Sci. 279 (2006) 389–393.
[24] A. Zaghbani, R. Tayeb, M. Dhahbi, M. Hidalgo, F. Vocanson, Isabelle Bonnamour, Patrick Seta, Clàudia Fontàs, Selective thiacalix[4]arene bearing three amide groups as ionophore of binary Pd(II) and Au(III) extraction by a supported liquid membrane system, Sep. Purif. Technol. 57 (2007) 374–379.
[25] C. Fontas, E. Antic o, F. Vocanson, R. Lamartine, P. Seta, Efficient thiacalix[4]arenes for the extraction and separation of Au(III), Pd(II) and Pt(IV) metal ions from acidic media incorporated in membranes and solid phases, Sep. Purif. Technol. 54 (2007) 322–328.
[26] H.G. Li, X.M. Wang, L.X. Cao, X. Zhang, C.F. Yang, Gold-recovery PVDF membrane functionalized with thiosemicarbazide, Chem. Eng. J. 280 (2015) 399–408.

[27] L.F. Villalobos, T. Yapici, K.-V. Peinemann, Poly-thiosemicarbazide membrane for gold recovery, Sep. Purif. Technol. 136 (2014) 94–104.
[28] Ya Ya N. Bonggotgetsakul, M. Ashokkumar, R.W. Cattrall, S.D. Kolev, The use of sonication to increase extraction rate in polymer inclusion membranes. An application to the extraction of gold(III), J. Membr. Sci. 365 (2010) 242–247.
[29] Ya Ya N. Bonggotgetsakul, R.W. Cattrall, S.D. Kolev, Recovery of gold from aqua regia (digested electronic scrap using a poly (vinylidene fluoride-co-hexafluoropropene) (PVDF-HFP) based polymer inclusion membrane (PIM) containing Cyphoss IL 104, J. Membr. Sci. 514 (2016) 274–281.
[30] E.R. de San Miguel, A.V. Garduño-García, M.E. Núñez-Gaytán, J.C. Aguilar, J. de Gyves, Application of an organic–inorganic hybrid membrane for selective gold(III) permeation, J. Membr. Sci. 307 (2008) 1–9.
[31] X. Du, X. Hao, Z. Wang, G. Guan, Electroactive ion exchange materials: current status in synthesis, applications and future prospects, J. Mater. Chem. A, 4 (2016) 6236–6258.

CHAPTER 10

More Works on Waste Treatment and Process Improvement

C.-J. Tang

School of Metallurgy and Environment, Central South University, Changsha, Hunan Province, China

From Chapter 3 to Chapter 9, the investigations and applications of membrane-based separation in metallurgy have been reviewed according to metal categories. There are still some researches that are not specifically addressing any kind of metallic material's production or processing, or they may be essentially applicable to more than one metal or process. It is also found that some novel ideas have been proposed without presenting comprehensive experimental or application information at present. In response to these situations, this chapter is composed as a supplement to previous chapters so that the book covers almost all relevant researches.

10.1 Waste Gas Treatment

Pyrometallurgy is still a very important branch for current metallurgical industry. As mentioned in previous chapters, it encompasses the high-temperature processes of roasting, smelting, converting, and refining. Due to the existence of myriads of compounds inside the metallic ores and the fuels, as well as the adjustment in process operation for meeting certain quality requirements (eg, addition of flux), the flue gas or exhaust gas generated in a series of high temperature procedures always contains huge amount of particles entrapping various components and a wide range of gaseous chemicals. The particles in exhaust emission constitute a major source of fog and haze, and sometimes lead to resource dissipation if valuable metals or other components are embedded. As for the gaseous part, in most cases it compromises sulfur dioxide (SO_2), nitrogen oxide (NO_x), carbon monoxide (CO), carbon dioxide (CO_2), chlorine (Cl_2), some of which can cause negative impact on biological system and human health. SO_x, including SO_2 and a small amount of SO_3, is the main cause of acid rain. In some specific situations, methane (CH_4), hydrogen (H_2), and CO are dominant.

Taking China's metallurgical industry as an example, there exist thousands of various furnaces, power boilers, etc., constituting a strong gas pollution source. Processes such as

Membrane-Based Separations in Metallurgy
http://dx.doi.org/10.1016/B978-0-12-803410-1.00010-4

blasting furnace casting, secondary flue gas treatment in oxygen blasting open-hearth furnace, and dust removal in coke oven, coke quenching, refractory furnace, and metal mixer are in urgent need of technology innovations. Gas pollution is also associated with hydrometallurgy. A typical example is chlorine-containing waste gas discharged from nickel electrolysis which has been introduced in Chapter 4. Yet another common characteristic of metallurgical flue gas is the high temperature, particularly for pyrometallurgy. The temperatures of flue gas from flash smelting, refining reverberatory, and fuming furnace are around 1300–1350, 1250–1300, and 1200–1250°C, respectively.

How will this industry change its smokestack image to meet the environmental restrictions for the benefit of societal sustainability? Owing to the properties of flue gas mentioned previously, the following aspects are undoubtedly considered for green development: (1) removal of dust, (2) capture of toxic components, (3) recovery of valuable material, and (4) utilization of waste heat. In fact, pyrometallurgical processes themselves have been the subject of considerable development in response to the gas emission problem arising from the traditional operations. Intense smelting techniques developed by INCO, Noranda, and Mitsubishis have made significant inroads in copper pyrometallurgy. In this part, the potential of membrane separation in tackling the waste gas is introduced and analyzed.

10.1.1 Gaseous Components Removal

Membrane gas absorption (MGA) utilizes microporous hollow fiber to actualize gas–liquid contact [1,2]. Their separation mechanisms have been introduced in Chapter 2. As compared with conventional absorption, it has the advantages of compact and light equipment, modularized integration of devices, and easy operation and control. In the MGA process, the gas and liquid phases can be adjusted independently, with a wide range of flow rate. There also occurs no flooding, channelling, foam entrainment, etc. [3,4]. The indirect contact of bulk gas and liquid prevents the gases other than the components from going into liquid, and avoid corrupting the equipment by eliminating substantial amount of humidity carried by flue gas. MGA has been investigated for treating CO, CO_2, SO_2, hydrogen sulfide (H_2S), ammonia (NH_3), volatile organic compounds (VOCs), and mercury (Hg) vapors [5–10].

SO_2 is probably the most serious problem among all the gaseous emissions from metallurgical industry, due to its strong toxicity and high content. Therefore, it has been the major reference for technological design for flue gas treatment. A variety of methods for removing SO_2, can be generally classed into precombustion, combustion, and postcombustion desulfurization. The last one is subcategorized into wet, semidry, and dry methods. Only few of these methods have made for commercial applications. Hereinafter some relatively new investigations about MGA are presented. In most investigations actually, synthetic waste gas composed of only SO_2 and some inert components (eg, nitrogen (N_2)) is used as feed, instead of true waste gas.

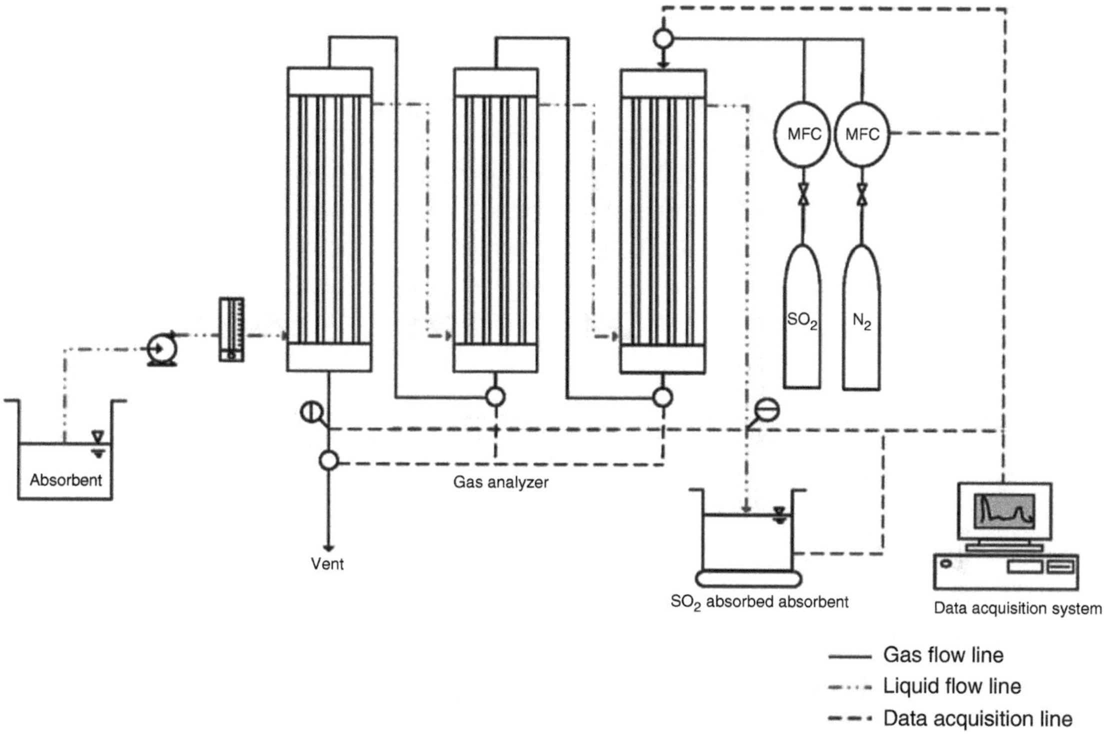

Figure 10.1: A schematic of the gas–liquid contactor separation process [10].

Park et al. investigated the application of polyvinyldenefluoride (PVDF) hydrophobic membranes for absorption of SO_2 from flue gas [10]. The experimental set-up for the MGA process is shown in Fig. 10.1. The SO_2 removal efficiency of ~85% has been achieved with 2M sodium hydroxide (NaOH) solution. The membrane's coefficient evaluated from the structure parameters, that is, average pore radius and effective porosity, was found smaller than that obtained from absorption experiments; however membrane's coefficient evaluated from the mathematical model agreed well with the latter. Hollow fiber membrane contactors have shown quite high SO_2 removal efficiency and may be proved to be a better alternative in future for acid gas removal considering the capital investment, operating cost, and reactor size. Gas–liquid contactor equipped with polypropylene (PP) micro-porous hydrophobic membranes was applied by Han for removal of SO_2 from flue gas [11]. Thirty percent methyldiethanolamine (MDEA) solution were adopted as the absorbent. The performance of the hollow fiber membrane for absorption of SO_2 was evaluated in terms of the gas and liquid flow rate, the absorbent temperature, and the gas residence time in PP hollow fiber membranes. It was found that the desulfurization efficiency can be up to 86% by this approach. In addition, theoretical work regarding the desulfurization efficiency, η, by MGA has been investigated based on the empirical and differential equations of mass transfer. Additionally, vacuum membrane distillation (VMD) was employed in the work to regenerate loaded MDEA aqueous solution that has absorbed SO_2 from the flue gas. It is identified that the optimum temperature for VMD is 70°C, and the regeneration efficiency reaches 98% under the selected experimental conditions.

In Sun et al.'s work, seawater was used as an absorbent in dealing with SO_2 from fluing a hollow fiber membrane contactor [12]. To verify the pros and cons of SO_2 removal by using seawater, and tap water as absorbents, aqueous NaOH solution with a pH value of 8.35 were introduced for comparison. The experimental results indicated that the mass transfer coefficient in the system using seawater was about twice as high as that of aqueous NaOH solution with a pH value of 8.35. This is probably associated with the complex $CO_2 - H_2O -$ bicarbonate $(HCO_3^-) -$ carbonate (CO_3^{2-}) equilibrium, which provides a higher absorption capacity. It has also been found that seawater could withstand the more intensive fluctuation of SO_2 loading. When the pressure gradient between gas and liquid was maintained below the penetration pressure, the membrane contactor could obtain high overall mass transfer coefficient, especially when seawater was used as an absorbent. The height of transfer unit (HTU) of the membrane contactor was relatively small, and HTU increased hardly with increasing flow rate of flue gas when seawater was used as an absorbent.

Chen et al. developed a novel batch-type absorption system using sea water for flue gas desulfurization [13]. Actual flue gas from coal firing was used in the research; the compositions are listed in Table 10.1. The key component of the absorption system is hollow fiber MGA module with either parallel- or crossflow mode, of which the schematic is illustrated in Fig. 10.2. For crossflow mode, the absorption liquid flows in the lumen channel along the axis direction of the hollow fiber, whereas the gas at the other side of the membrane moves in the direction perpendicular to the hollow fiber axis or surface. The advantage of this operation mode is that it can efficiently enhance local perturbation to form turbulence, and improve the mass transfer coefficient at the shell side. In addition, the regular arrangement and even distribution of the fibers can not only avoid the development of the shell channelling, but also reduce the spoiler and back mix flow. As a result, the effective membrane area and transfer efficiency have been raised.

The flow diagram of the overall process is demonstrated in Fig. 10.3. The high temperature flue gas was cooled down after flowing through the heat exchanger. Thereafter, it was led into the absorber by a booster fan. The seawater after being pretreated by disc filter entered the lumen channel of the hollow fibre membrane installed inside absorber, where the seawater took away the sulfur content in the flue gas. In the next procedure, the seawater was discharged from the MGA into the aeration tank and mixed fresh seawater. The treated seawater was pumped into the sea upon its quality reaching the standard.

The seawater inflow adopted a bi-directional design, which made it possible to conduct both forward and reverse flush to prevent inlet pollution. In addition, a spray cleaning system was

Table 10.1: Composition of coal-fired flue gas [13]

Parameters	SO_2	NO_x	CO	CO_2	O_2	Moisture Content	Gas Temp (°C)		Relative Humidity (%)		Smoke Conc.
	(mg Nm^{-3})		(%)			(g kg^{-1})	Before Heat Exchange	After Heat Exchange	Before Heat Exchange	After Heat Exchange	(mg N m^{-3})
Value	1000–6000	600–1000	4–10	6.5	7	61	100–120	30–38	4.5	100	1553

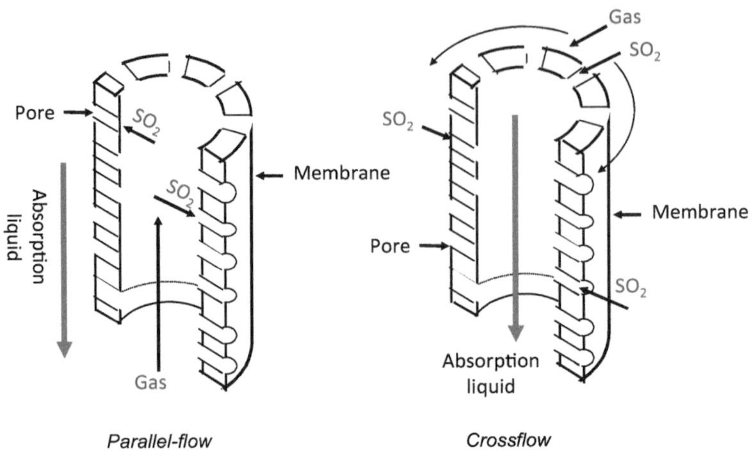

Figure 10.2: Parallel- and crossflow modes of the membrane gas absorption [13].

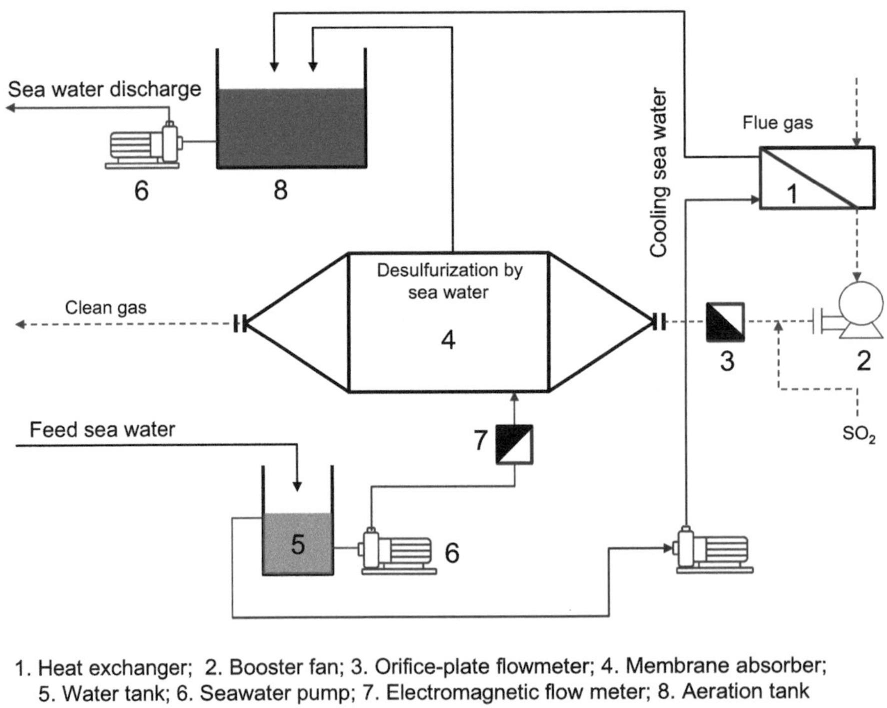

Figure 10.3: Flow chart for membrane-based flue gas desulfurization [13].

Table 10.2: Comparison of membrane-based desulfurization by seawater and conventional technologies [13]

	Subject of Entry	Lime–Gypsum Method	Desulfurization by Seawater	Membrane-Based Desulfurization by Seawater
Technical performance	Unit capacity/MW	300	300	300
	Technology system	Direct contact Absorbing tower	Direct contact Absorbing tower	Nondirect contact Membrane absorption
	Maturity level	Commercialized	Commercialized	Pilot scale
	Reheat requirement	Need	Need	Unnecessary
	Sulfur content in fire coal (%)	1–3	0.85	1–3
	Size of desulfurization system (m^3)	11.9 × 7.9 × 34.05	Fi12 × 37.1	10.0 × 10.0 × 18.0
	Mass content of SO_2 in flue gas ($mg\ m^{-3}$)	1430–2860	1800	1000–4300
Economic performance	Total investment (million CNY)	80–90	146	78
	Investment/(CNY/kW)	267–300	218	260
	Running cost/ (10,000 CNY/a)	2792.6–2907.8	1939.31	2087
	Cost of desulfurization per ton of SO_2/CNY	2531.63	984.72	1384
	Rising cost per kW/CNY	0.02	0.0125	0.0126
Safety	System corrosion and blockage	Serious	Serious	Avoidance
	Environmental impact	Secondary pollution by gypsum	Heavy metal and smoke pollution	Avoidance of heavy metal and smoke pollution

equipped to remove flue gas pollution. The comparison among three different desulfurization methods including lime–gypsum method, seawater desulfurization, and membrane-based seawater desulfurization method is displayed in Table 10.2.

The experimental results led to the following conclusions:

1. The pilot scale system exhibited stable S components removal efficiency while treating coal-fired flue gas released from real working conditions; it could also flexibly deal with the impact of SO_2 concentration fluctuation, satisfying the requirements for treating flue gas generated from coal with high S content.
2. The heavy metal concentration in the seawater before and after desulfurization did not change. This was because the gas and the liquid were not in direct contact.
3. As far as technology performance is concerned, membrane-based desulfurization was suitable more for medium- or high-S coal. The desulfurized flue gas could be discharged directly without reheat. Analysis on the economy cost indicated that the investment

associated with the membrane-involved process was lower than those of conventional approaches. The operation cost was lower than lime–gypsum method, and slightly higher than seawater desulfurization method. Nevertheless, the environmental benefit related to MGA is apparent.

In spite of the various advantages, the high price, poor membrane pollution resistance, difficulty in cleaning, low stability, etc. become the impediments to the actual application of MGA technology at industrial level for the SO_2 and many other gases treatment.

10.1.2 Dust Elimination

Dust capturing is another critical aspect for flue gas processing. Due to the high temperature of the flue gas, as introduced previously, the conventional technology ordinarily involves cooling as the first step for adapting the gas to the requirements of dust collection equipments [14]. Wet collection is seldom employed due to the loss of the heat by water evaporation and yield of wastewater. The primary technologies for dry collection include cyclone method, electrostatic method, bag filter, and high performance resistance filtration. In flue gas, particles smaller than 10 μm are most challenging. Cyclone approach has low operation cost, but is not efficient in getting rid of dust with particle size smaller than 7 μm. Therefore, it is usually applied as a pretreatment technology. The efficiency of electrostatic approach can be up to 99%–99.6%. Nevertheless, it has poor corona stability in high pressure and high temperature environment commonly associated with metallurgical gaseous waste. In addition, material stability and thermal expansion are also encountered. The bag filter is relatively simple, but is liable to burning and has to be replaced every 3 months or even shorter period.

In order to achieve higher efficiency in reducing toxic emission and recovering metals, high resistance filtration has been investigated or applied in treating the flue gas. High resistance filtration generally refers to the application of ceramic, metal, and solid bed filters to retain the particles. Several industrialized applications using metal membranes are introduced in Chapters 3, 4, and 6. The case in Chapter 6 is regarding the flue gas generated in molybdenum (Mo) smelting using iron (Fe)/aluminum (Al) metal membrane [14]. Yet another example is the utilization of Fe_2 chromium (Cr) Al silicate (Si) copper (Cu) membrane in removal of arsenic (As) from the gas discharged from the roasting of high-grade tin concentrates, which has been introduced in Chapter 4 [15]. The series of Fe/Al, Fe/Al/Si, and Fe/Al/Cr/Si porous materials have already been successfully applied in dealing with high-temperature gaseous waste generated during the production of ferroalloy in Qinghai Jihua Jiangyuan Company and alumina (800 kt a^{-1}) in Yunnan Wenshan Company.

Ceramic membranes have also been used to remove dust from flue gas [16–18]. Table 10.3 tabulated the efficiency of regular dedusters and ceramic membrane filters. Obviously, ceramic filter is superior to all the others. In addition to reducing the dust discharged into atmosphere, more efficient dust retention prevents the development of severe blockage of

Table 10.3: Efficiency comparison of conventional dust separator and ceramic membranes

Type	Applied Range			
	Grain Size (μm)	Temp. Tolerance	Resistance (Pa)	Dust Removal Efficiency (%)
Inertial	>15	<400	200–1000	96
Cyclone	>15	<400	400–2000	96
Electrostatic	>10	<400	100–500	99
Bag	>5	<300	800–2000	99
Ceramic membrane	>5	<900	800–200	99.5

heat exchange pipeline. Conventional bag filtration or electrostatic dust collector necessitated precooling. The presence of the dust in the flue gas depresses significantly the heat-exchange efficiency and blocks the pipeline. Ceramic or metal membranes can sustain high temperature, and therefore can directly remove the dust without precooling. Heat recovery after dust removal can greatly enhance the rate of heat recovery.

Industrial application of ceramic membranes was initiated in 1990s. The ceramic membranes produced by British Glosfume Company have been applied in smelting of waste Al and burning of nuclear waste. The ceramic membranes Fabriclean from British Birmingham Company have been successfully applied in metal recovery industry, especially high-value dust from metal processing. The dust usually contains platinum (Pt), rhodium (Rh), nickel (Ni), tin (Sn), lead (Pb), Cu, titanium (Ti), Al etc.

10.2 Water Treatment

10.2.1 Electroplating Wastewater

Electroplating is a process that uses chemical or electrochemical method to form a coherent metallic coating layer on the surface of base materials. To improve the quality of the plating pieces, a variety of additives that can form relatively strong complexes with heavy metals are used, including tartaric acid, ethylene diamine tetra-acetic acid, pyrophosphate, citric acid, NH_3, etc. As a result, a lot of polluters that are toxic or even carcinogenic will be present in electroplating wastewater. In addition, the amount of valuable metals cannot be ignored.

One of the membrane-based methods for electroplating wastewater treatment is dual-membrane method. In Wu and Liu's work, the dual-membrane process based on ultrafiltration (UF) and reverse osmosis (RO) was applied [19]. Before treatment, the ion conductivity of the wastewater is as high as 6000 $\mu s\ cm^{-1}$. The value became 100 $\mu s\ cm^{-1}$ after processed by the system which makes it possible to reuse the water in rinsing step of electroplating plant. Continuous membrane filtration (CMF) using microfiltration (MF) has been used by Zhang as the pretreatment for the step using ion-exchange resin in treating electroplating wastewater [20]. The system exhibited quite stable performance in a three-month continuous operation, dramatically improving the quality of source water.

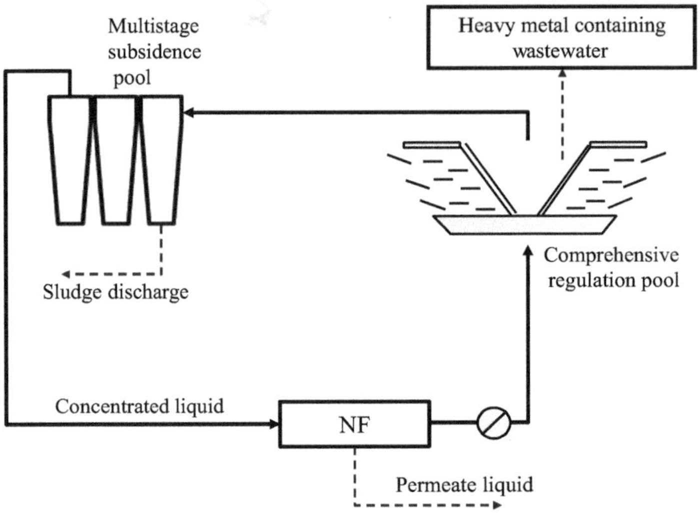

Figure 10.4: The schematic of IBMS-CP process [21].

Interaction-balanced membrane separation-chemical precipitation process (IBMS-CP) and process combining IBMS-CP and other technologies have been proven highly efficient in tackling electroplating wastewater [21]. The flow diagram for IBMS-CP process is shown in Fig. 10.4. As shown, wastewater treated by chemical agent is sent to multistage settling tank to remove precipitates and heavy metal. Thereafter, the supernatant is delivered to nanofiltration (NF) system for concentration. The permeate may be directly discharged; the retentate from the membrane system will flow to regulation pool to enter the next cycle of treatment. Due to the concentration by nanofiltration, the volume of wastewater is reduced and hence the precipitation tank is downsized. In addition, the precipitation reaction is rendered quicker and more thorough.

The engineering application of IBMS-CP process in dealing with the electroplating wastewater at a plant was reported. The quality and quantity of a series of wastewaters generated in the electroplating plant are summarized in Table 10.4. Self-circle process is adopted for gold (Au)

Table 10.4: Water quality and quantity of electroplating waste water [21]

No.	Project	Type and Concentration of Pollutants ($mg \cdot L^{-1}$)	Water ($t \cdot d^{-1}$)
1	Wastewater containing chromium (Cr)	Cr^{6+}26, pH < 3	25
2	Cyanide wastewater	CN^-50, Cu^{2+}50, Ni^{2+}50, pH = 8–10	160
3	Continuous plating wastewater	Cu^{2+}50, Ni^{2+}50, pH > 3	80
4	Wastewater containing copper and nickel	Cu^{2+}200, Ni^{2+}50, pH > 3	90
5	Acidic wastewater	HCl, pH = l	35
Total			390

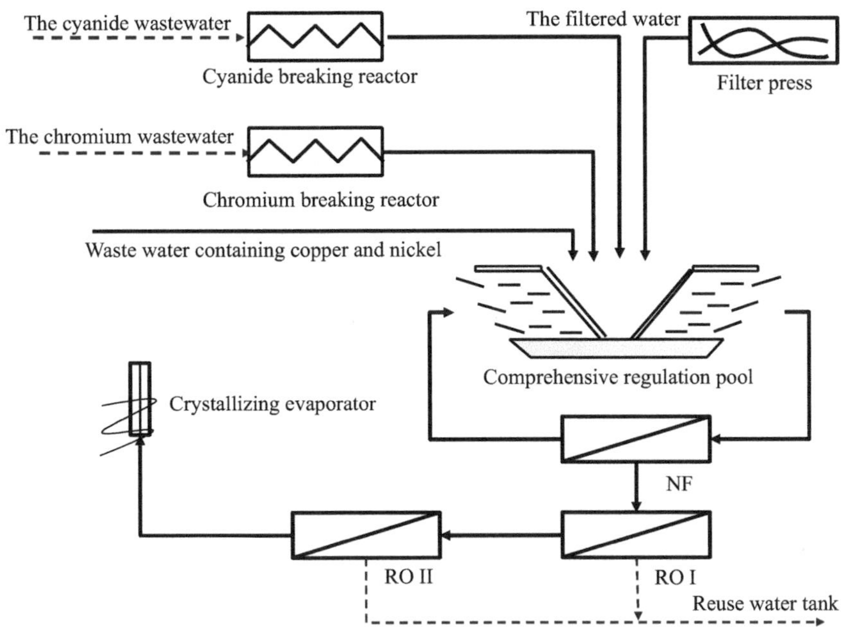

Figure 10.5: The diagram of the industrialized process for treatment and reuse of electroplating waste water [21].

electroplating. Cyanide and Cr wastewaters have respective collection systems. All the other wastewaters, collectively referred to as Cu-Ni wastewater, are mixed and stored in one tank. The flow diagram for processing these wastewaters is given in Fig. 10.5. After being treated separately, the series of wastewaters merged inside the IBMS-CP system. The permeate from the NF procedure is further condensed by RO.

The IBMS-CP system presented relatively stable performance during a 17-month operation [21]. Membrane cleaning was conducted every 3 months and can recover the separation performance of membrane efficiently. Table 10.5 summarizes the rejection efficiencies of heavy metals of the system obtained at three different sampling dates. The rejection toward heavy

Table 10.5: Rejection efficiency of heavy metals in electroplating wastewater by IBMS-CP combined with RO [21]

Date	Sample	Cu	Ni	Cr
2008-02-18	Feed	76.3	43.7	5.1
	Permeate	1.17	1.67	N/A
2008-03-01	Feed	54.6	45.1	5.5
	Permeate	0.89	1.62	N/A
2008-11-30	Feed	61.1	39.4	4.3
	Permeate	1.04	0.84	N/A

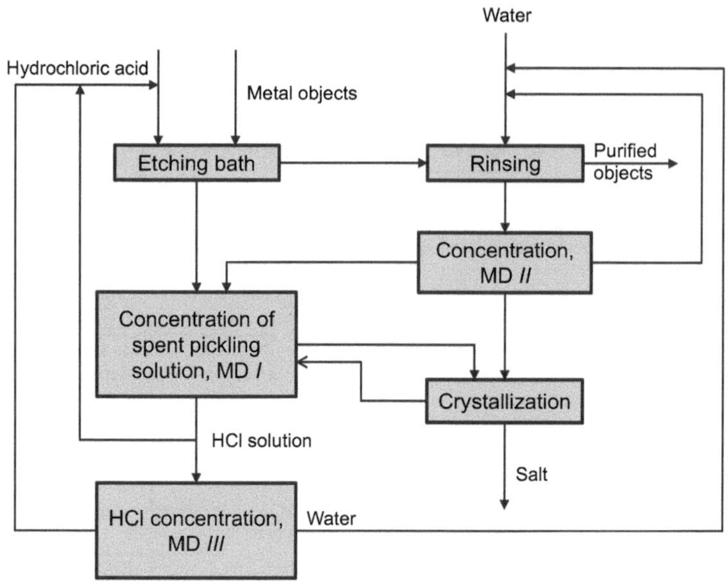

Figure 10.6: The flow diagram of MD treatment of an electroplating effluent [22].

metals has been maintained at a high level. Nevertheless, the concentration of residual ions does not reach the discharge requirements, which necessitates the treatment of reverse osmosis.

10.2.2 Combination of Etching With Pickling

An example of a solution of the etching metal objects process integrated with the treatment of spent pickling solution by MD is presented in Fig. 10.6 [22]. The MD process was proposed for the separation of hydrochloric acid (HCl) from a spent etching bath and for the concentration of solutions from rinsing objects after etching. The retentate from the first stage of MD, after concentration of the feed to supersaturated state could be discharged to a crystallizer. The separated salts may be utilized or restored. The liquor after crystallization could be recycled to the first stage of MD, where a high salt concentration will favor the HCl recovery. The distillate from the first stage contains about 100 g HCl dm^{-3} and can be recycled to fresh etching bath. When a higher acid concentration is required, the distillate can be concentrated in the third stage. The distillate from the latter stage can be recycled for rinsing. The solutions after rinse can be concentrated in stage 2. Due to low solute concentrations, the distillate is pure water, which can be reused for rinsing. The retentate of the second stage will be concentrated together with the spent etching bath. This method permits the occasional closure of the cycles of hydrochloric acid, water, and salt.

10.2.3 Recycling of Metal-Working Fluids

In Hilal's work, membrane separation consists of an UF followed by NF was employed for separating oil from water and treating waste metal-working fluids (MWF) [23]. This work presented two different filtration systems and investigated the possibility of reusing the permeates. In this case, UF membranes of 200 kD and 100 kD molecular weight cut-off (MWCO), respectively are used. Another membrane system consists of solely NF membranes. The treatment are carried out in a large-scale, semi-batch process with a volume 200L feed. Additionally, simulating the reuse of water in the practical manufacturing process, the permeate water produced by each system was used to prepare fresh MWF. This recycled water was tested by foaming, emulsion stability etc. and the results are compared with those obtained similarly with tap water. The tests confirm the feasibility of reusing the permeate water.

10.2.4 Advanced Wastewater Treatment

Metallurgical wastewater has very complex components which include many toxic chemicals and has high salt content. In current industrial-wastewater treatment, the concentrated water from the reverse osmosis process is generally sent to the industrial wastewater collection point to be processed along with other wastewater, followed with discharging. This virtually imposes extra burden on the water treatment system. With tightening the requirements on environmental protection, the problems associated with RO processing arouse intensified concern.

The RO retentate mainly comprises dissolved organic matters (DOM) and total dissolved solid (TDS). Particularly, DOM include endocrine disrupt chemical, medicine and pathogenic viruses that have toxic effects on the human health and hazardous effect on environment. Generally, the RO concentrate is discharged into surface water or wastewater treatment system, evaporation pond or deep well, directly or indirectly. The other treatment approaches include Fenton method and membrane distillation. Integrated utilization (eg, thermo-storage media) has also been investigated.

Zhang et al. carried out research on recondensing RO concentrate using tubular UF membranes (commercial product) [24]. The flowchart for the process combining tubular membrane UF is shown in Fig. 10.7. The effluent from the chemical dosing system was

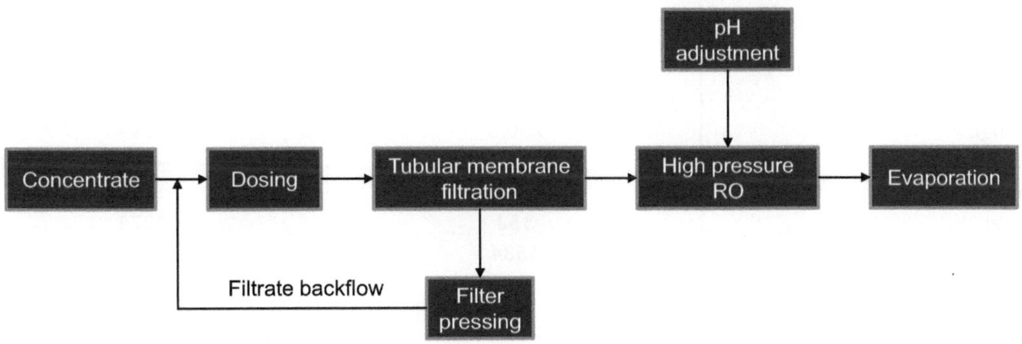

Figure 10.7: The flowchart for the process combining tubular membrane ultrafiltration [23].

sent to the tubular membrane module to form a counter current recycling. Part of the water in the concentrate transported through the membrane as yielding water, while the retentate was repressurized and delivered back to the membrane module. During the peak time of the feed, the volume and pressure for the recycling were enhanced to increase the system yielding. Back wash was conducted to delay the membrane fouling; chemical cleaning was necessitated when the fouling reached a critical extent. There also exist some other technologies that employ MF or hollow fiber UF [24]. As far as the pretreatment of high pressure RO is concerned, the membrane filtration can be a simpler replacement for the complicated multistage procedure involving settling tank, sand filtering, activated carbon adsorption, etc. As for hollow fiber UF, the small diameter of the fiber tend to cause channel blockage. In comparison, the tubular membrane is more advantageous. Additionally, its high mechanical strength endows it with long operation life span, and its space efficiency is comparable to hollow fiber membrane, having high space efficiency. evertheless, the practice using MF cannot reject chemicals such as calcium carbonate ($CaCO_3$), magnesium carbonate ($MgCO_3$), which might cause serious scaling on RO membrane surface [25].

As the slightly soluble chemicals including carbonate of calcium, sulfate of Mg, Ca, barium Ba, stronium (Sr), and silicate may experience supersaturation part of the chemicals may precipitate on membrane surface, and thus causing scaling. Scaling is a complicated phenomenon that is closely related to crystallization dynamics and transfer mechanism. There are two approaches for scaling: surface devitrification and block crystallization. The conventional methods for controlling scaling include acidification, adding scale inhibitor, resin adsorption, and chemical softening. Li et al. investigated the softening of wastewater (constituents shown in Table 10.6) by using NaOH and (Na_2CO_3) [26]. It was found that the formation of $CaCO_3$ was at the control step in softening process. When alkalinity was higher than Ca hardness, addition of only NaOH could obtain satisfactory softening effect. When the alkalinity was lower than Ca hardness, addition of Na_2CO_3 was necessitated. The dosage

Table 10.6: Composition of the feed to RO system [26]

Item	Value	
	Feed I	Feed II
pH	7.4	8.05
Electroconductivity ($\mu s\ cm^{-1}$)	7650	13221
Mg^{2+} mgL^{-1}	55.1	68.6
Ca^{2+} mgL^{-1}	250.3	154.4
Hardness mgL^{-1}	855	672
Alkalinity mgL^{-1}	227	1110
Cl^- mgL^{-1}	332	4273
SO_4^{2-} mgL^{-1}	3684	3584
TOC mgL^{-1}	9.8	61.4
COD mgL^{-1}	27	205

of softening agents depended on the concentrations of alkalinity, Ca^{2+}, and Mg^{2+} ions. The concentration factor (CF) (volume ratio of feed/retentate) for the brine waste water by NaOH and Na_2CO_3 was up to 4.

10.3 Potential Applications in Hydrometallurgical Processes [27]

In hydrometallurgy, the mineral slurry generated from grinding and flotation process has a concentration range of 10–15%, giving rise to high cost and high processing capacity upon direct leaching. Customarily, thickener is used to increase the concentration. But thickener has the weakness of bulky system, large footprint, low efficiency, long operation time, and inconvenient maintenance.

Ceramic membrane filtration is used to replace thickener for solid/liquid separation. The handling capacity requirements can be satisfied by adjusting membrane area or operation pressure. By choosing crossflow mode, the flux can be maintained at a high level. Most importantly, its separation is based on aperture screening. Through selecting the suitable pore size, therefore, high quality permeant with good clarity can be achieved; along with a slurry that is more concentrated than that obtained with thickener. Membrane filtration also saves the cost on applying coagulation and flocculation; due to the abolish of settlement process. Combination of ceramic membrane with hydrocyclone will actualize an even higher concentration factor and lower cost.

Solvent extraction is the major precise separation technology for metal enrichment in hydrometallurgy. The problem in the process results from the formation of emulsification and flocs, for which the main cause is the presence of solid particles smaller than 1 μm and colloidal silica. As is often the case, the flocs will be taken out periodically for recovering the organic phase. Nevertheless, there still exists partial loss of the extractant. In comparison, employments of MF and UF prove more efficient and mature. Ceramic membranes are particularly well adapted for the situation, owing to their resistance toward acid, alkali, and oxidizing agents.

10.4 Conclusions

In this chapter, the researches or applications that have not been specifically targeted in any metallurgical processes are summarized. Gas waste is very common in metallurgical industry. For SO_2 removal, MGA based on sea water showed a higher absorption capacity due to the complex chemical reaction equilibrium. The advantages of ceramic membranes in dust capturing have been verified. The combination of IBMS-CP with reverse osmosis led to zero or significantly reduced discharge in treating electroplating wastewater. Some problems necessitating our attention are the stability of membranes in MGA, fouling in nanofiltration, etc. Reported also in this chapter are some new ideas for hydrometallurgy, including membrane involvements in slurry treatment and solvent extraction.

Glossary and Symbols

Al	Aluminum
As	Arsenic
Au	Gold
Ba	Barium
$CaCO_3$	Calcium carbonate
CF	Concentration factor
CH_4	Methane
Cl_2	Chlorine
CMF	Continuous membrane filtration
CO	Carbon monoxide
CO_2	Carbon dioxide
CO_3^{2-}	Carbonate
Cr	Chromium
Cu	Copper
DOM	Dissolved organic matters
Fe	Iron
H_2	Hydrogen
HCO_3^-	Bicarbonate
HCl	Hydrochloric acid
Hg	Mercury
H_2S	Hydrogen sulfide
HTU	Height of transfer unit
IBMS–CP	Interaction balanced membrane separation–chemical precipitation process
SO_2	Sulfur dioxide
MDEA	Methyldiethanolamine
MF	Microfiltration
MGA	Membrane gas absorption
$MgCO_3$	Magnesium carbonate
Mo	Molybdenum
MWCO	Molecular weight cut-off
MWF	Metal-working fluids
N_2	Nitrogen
Na_2CO_3	Sodium carbonate
NaOH	Sodium hydroxide
NF	Nanofiltration
NH_3	Ammonia
Ni	Nickel
NO_x	Nitrogen oxide
Pb	Lead
PP	Polypropylene
Pt	Platinum
PVDF	Polyvinyldene fluoride
Rh	Rhodium
RO	Reverse osmosis
Si	Silicon
Sn	Tin
Sr	Stronium
TDS	Total dissolved solid
Ti	Titanium

UF	Ultrafiltration
VMD	Vacuum membrane distillation
VOCs	Volatile organic compounds
η	Desulfurization efficiency

References

[1] Z. Qi, E.L. Cussler, Microporous hollow fibers for gas absorption. I. Mass transfer in the liquid, J. Membr. Sci. 23 (1985) 321–332.

[2] B. Ozturk, H.B. Al-Saffar, R. Hughes, Sulphur dioxide absorption in hollow fibre membrane modules, Chem. Eng. Commun. 177 (2000) 157–175.

[3] B.B. Li, Y.X. Zhang, S.L. Wang, Present situation and development of membrane distillation and membrane absorption technology, Chem. Eng. Sci. Technol. 15 (5) (2007) 61–65 (in Chinese).

[4] Y.X. Zhang, B.B. Li, S.L. Wang, Modeling building and software development for gas membrane separation, Comput. Appl. Chem. 25 (2) (2008) 137–140 (in Chinese).

[5] A.C. Ghosh, S. Borthakur, N.N. Dutta, Absorption of carbon monoxide in hollow fiber membranes, J. Membr. Sci. 96 (1994) 183–192.

[6] D.O. Cooney, C.C. Jackson, Gas absorption in a hollow fiber device, Chem. Eng. Commun. 79 (1989) 153–163.

[7] K. Li, D. Wang, C.C. Koe, W.K. Teo, Use of asymmetric hollow fiber modules for elimination of H_2S from gas streams via a membrane absorption method, Chem. Eng. Sci. 53 (1998) 1111–1119.

[8] R. van der Vaart, J. Akkerhuis, P. Feron, B. Jansen, Removal of mercury from gas streams by oxidative membrane gas absorption, J. Membr. Sci. 187 (2001) 151–157.

[9] B. Ozturk, H.B. Al-Saffar, R. Hughes, Sulphur dioxide absorption in hollow fibre membrane modules, Chem. Eng. Commun. 177 (2000) 157–175.

[10] H.H. Park, B.R. Deshwal, I.W. Kim, H.K. Lee, Absorption of SO_2 from flue gas using PVDF hollow fiber membranes in a gas–liquid contactor, J. Membr. Sci. 319 (1) (2008) 29–37.

[11] Y.J. Han, Study on flue gas desulfurization using membrane absorption and membrane distillation technology, Master's Thesis, Changzhou University, China, 2011 (in Chinese).

[12] X. Sun, F. Meng, F. Yang, Application of seawater to enhance SO_2 removal from simulated flue gas through hollow fiber membrane contactor, J. Membr. Sci. 312 (1–2) (2008) 6–14.

[13] Y. Chen, Y.P. Guan, C.G. Guo, G.C. Liu, X.M. Li, Z. Liu, H. Li, Application of box-type membrane absorber in coal-fire flue gas treatment, The 4th National (China) Symposium for Application of Membrane Technologies in Metallurgical Industry, 2014.

[14] P.Z. Shen, L. Gao, H.Y. Gao, Y.H. He, The sulfidation resistance of Fe/Al intermetallic porous material and its application in purification of high temperature flue gas containing sulphur, The Third National (China) Symposium for Application of Membrane Technologies in Metallurgical Industry, 2014.

[15] Z.H. Wang, X.J. Zhang, W.X. Zeng, M. Jiang, L. Gao, YT membrane separation technology for arsenic removal application in the non-ferrous metallurgical industry, The Fourth National (China) Symposium for Application of Membrane Technologies in Metallurgical Industry, 2014, pp. 34–37.

[16] T.Z. Li, L.H. Lin, X.S. Lin, Y.Z. Lu, Ceramic membranes technology and its application in ferrous metallurgical flue gas, The Fourth National (China) Symposium for Application of Membrane Technologies in Metallurgical Industry, 2014.

[17] X. Lv, High temperature ceramic filter system and structure design, Master's Thesis, Shangdong University of Technology, 2004.

[18] Y. Lu, Research on industrial tests device of ceramic membranous canal dust catcher, Master's Thesis, Guizhou University, 2009.

[19] J. Wu, C. Liu, Application and research of membrane separation technology in electroplating wastewater treatment and the metal recovery, Proceedings of the Seminar of Fourth National (China) Symposium for Application of Membrane Technologies in Metallurgical Industry, 2014.

[20] H. Zhang, Treatment of electroplating wastewater with continuous membrane filtration, Chin. J. Rare Metals 31 (Suppl.) (2007) 17–20.
[21] X.P. Bai, W.C. Hao, Z.L. Xu, Interaction balanced membrane separation-chemical precipitation process for recycling electroplating wastewater, The Third National (China) Symposium for Application of Membrane Technologies in Metallurgical Industry, 2014.
[22] M. Tomaszewska, M. Gryta, A.W. Morawski, Recovery of hydrochloric acid from metal pickling solutions by membrane distillation, Sep. Purif. Technol. 22-23 (2001) 591–600.
[23] N. Hilal, G. Busca, N. Hankins, A. Wahab, Mohammad, The use of ultrafiltration and nanofiltration membranes in the treatment of metal-working fluids, Desalination 167 (1–2) (2004) 227–238.
[24] J. Zhang, X.Y. Guan, D.M. Zhang, B.C. Zhang, Application of I5 low energy consumption tubular membrane in treating RO concentrate, The Fourth National (China) Symposium for Application of Membrane Technologies in Metallurgical Industry, 2014, pp. 165–168.
[25] D. Hasson, A. Drak, R. Semiat, Induction times induced in an RO system by antiscalants delaying $CaSO_4$ precipitation, Desalination 157 (2003) 193–207.
[26] H. Li, Y.P. Wu, W.X. Li, W.H. Xing, Scaling control in application of RO membrane in treating brine waste with extreme concentration, The Fourth National (China) Symposium for Application of Membrane Technologies in Metallurgical Industry, 2014, pp.136–139.
[27] Z.G. Wang, Application of membrane technology in hydrometallurgy, The Fourth National (China) Symposium for Application of Membrane Technologies in Metallurgical Industry, 2014, pp. 54–61.

PART 3

Development of Special Industrial Membranes for Metallurgy

CHAPTER 11

Overview

L.Y. Jiang

School of Metallurgy and Environment, Central South University, Changsha, Hunan Province, China

Environment and energy are crucial for global quality of life and sustainable socio-economic development. With intensified attention toward environmental protection and energy efficiency, reasonable exploitation of natural resource and employment of new mining and metallurgical methods undoubtedly will become focuses. Membrane-based separation, a versatile technology, among which partial categories can be operated physically without secondary pollution and some can be combined with biological, chemical, or other techniques to bring about higher efficiency has already been attracting world-wide attention [1]. As obviously demonstrated in previous chapters, they have been successfully applied in many metallurgical processes [2–4].

The processes that have entailed membrane separation include but are not necessarily limited to mining wastewater treatment, metal ions separation, solution concentration in hydrometallurgy, exhaust gas treatment, etc. The in-service conditions for these applications specific to metallurgical processes are different from those often existing in conventional water or gas treatment by exhibiting complexity and following characteristics [4,5]:

1. Very common acidic or alkaline environment is present.
2. Foreign ions' concentration is high with a common presence of high concentration of ferrous metal ions.
3. The concentrations of Ca^{2+} and Mg^{2+} are high. In the common sulfuric acid system the ion concentration products of $CaSO_4$ and $MgSO_4$ approximate corresponding K_{sp} value, which can easily cause scaling during membrane concentration process.
4. Very often there exists the necessity for selective separation of ions with different charges, which requires the cooperation from experts familiar with both membrane separation and metallurgy.
5. Strong oxidizing environment exist in situations involving nitric acid, sexavalent chromate, etc.
6. Separation is demanded sometimes in removing oils and metal ions in high temperature solutions.
7. High temperature gaseous or liquid discharge laden with particles, colloids.

Membrane-Based Separations in Metallurgy
http://dx.doi.org/10.1016/B978-0-12-803410-1.00011-6

The aforementioned points raise the concern about matching the properties of membrane with the requirements of the myriads of processes. An apparent aspect of this concern is that conventional membranes may not sustain the harsh environment. Problems such as serious membrane fouling, finite processing capacity per unit area of membrane are also the consequences of this mismatching. Therefore, the industrial adoption of this technology is still limited. Another reason behind this situation is that the cost of the membranes is so high that many applications become economically unfavorable, despite their excellent technical feasibility.

On the basis of the findings by many researches, a consensus has been reached that to overcome these challenges, the most critical aspect is to construct material design and membrane development systems with fully taking into account the industrial in-service conditions. This special market requires combined effort among scientists from different domains including membrane fabrication, membrane application, and metallurgy.

As for ceramic membrane development, the strategy addressing the aforementioned concern over application (not necessarily metallurgical industry)-orientated membrane design has been proposed by Xing et al. [6]. The essential steps are illustrated by the frame diagram shown in Fig. 11.1. As shown, there are three principal aspects constituting the building blocks of the strategy. The first one is new separation processes developed based on new materials, which capitalizes on the special properties (microstructure and surface property) of the membrane materials to achieve separation of specific components, and unveil the transport and reaction mechanisms inside the membrane microstructure via exploring the property–structure relationship. The second one is regarding solving the pivotal problems

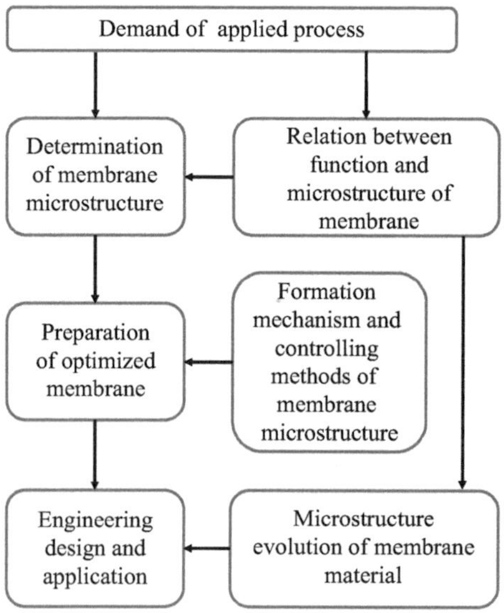

Figure 11.1: Theoretical framework of application oriented ceramic membrane design.

occurring in the process of membrane formation by using theories and methods of chemical engineering, which targets conversion from empirically directed fabrication to quantified and customer-orientated fabrication via controlling material micro-structure and property. The last one focuses on understanding microstructure evolution in application process. Its nature is actually the interaction between material and environment. The investigation about the relationship between materials' physicochemical properties and application system will help disclose the mechanism for membrane fouling, solve membrane dysfunction in some relevant major projects, develop new approaches for fouling control, and finally actualize stable operation.

The aforementioned basic frame may be transferred to help establish the systems for other types of membranes, such as polymeric membranes, metal membranes, etc. To fully elaborate the overall progress that has been achieved based on the framework mentioned previously is beyond the scope of this book. Attention in the forthcoming sections will be given to some developments in material design and membrane fabrication.

References

[1] N.P. Xu, Process-Oriented Design, Preparation and Application of Ceramic Membranes, Science Press, Beijing, (2005).
[2] Q.X. Zhang, Status quo of applying membrane for liquid separation in metallurgy, The third Symposium on Application of Membrane-based Separation in Metallurgical Industry, 2009, pp.1–8.
[3] Y.M. Cao, Research and application of gas separation membranes, The third Symposium on Application of Membrane-based Separation in Metallurgical Industry, 2009, pp. 9–10.
[4] H. Wang, Application of Special Industrial Membranes (SIMS) in Mining Industry, The third Symposium on Application of Membrane-based Separation in Metallurgical Industry, 2009, pp. 64–65.
[5] L. Gao, Report on industrial application frontiers of novel porous membrane and membrane separation technologies, The third Symposium on Application of Membrane-based Separation in Metallurgical Industry, 2009, pp.14–24.
[6] W.H. Xing, Y.Q. Fan, Z.X. Zhong, N.P. Xu, Recent advances in process-engineering oriented preparation and application of ceramic membranes, CIESC J. 60 (2009) 2679–2668.

CHAPTER 12

Polymeric Membranes

L. Wu*, H. Wang, T.-W. Xu*, Z.-L. Xu†**

**School of Chemistry and Materials Science, University of Science and Technology of China, Hefei, Anhui Province, China; **Beijing Liquid Cube Technology Ltd., China; †State Key Laboratory of Chemical Engineering, Membrane Science and Engineering R&D Lab, Chemical Engineering Research Center, East China University of Science and Technology, Shanghai, China*

12.1 Nanofiltration

The key characteristics of nanofiltration (NF) separation are the low rejection of monovalent ions, high rejection of divalent or higher-valent ions, and relatively higher flux compared to reverse osmosis (RO) separation [1–5]. For more efficient applications of NF in metallurgical industry as reported in previous chapters, properties such as accurate rejection, high temperature resistance, acid and alkali resistance, oxidation resistance, low cost, antifouling, etc. are the critical requirements. To achieve these targets, both material and membrane formation are essential. Polymeric materials for NF membranes mainly include the following categories: cellulose, polysulfone (PSF), polyethersulfone (PES), polyamide (PA), polyvinyl alcohol (PVA), polyvinylidene fluoride (PVDF), chitosan or chitin, polyethersulfoneketone (PESK). Their unit structure and major physiochemical properties are summarized in Table 12.1. The primary approaches for NF membrane fabrication are interfacial polymerization (IP), phase inversion, and surface coating. Thin film composite membrane by interfacial polymerization (TFC-IP) attracts the effort of a large portion of the investigations worldwide. This should be attributed to its flexibility in membrane structural control and performance reproducibility, which also leads to its fast and steady commercialization. For these reasons, this part will focus on researches regarding interfacial polymerization. The process for producing the TFC-IP involves the reaction of two monomers at the interface of an aqueous and an oil phase dissolving the two monomers, respectively. Traditionally, the monomers for the reaction or polycondensation amine and acyl chloride, resulting in polyamide (PA) network as selective barrier [6]. Polyester is a relatively new direction [7].

12.1.1 General Strategies for Improving Nanofiltration Membrane Separation Efficiency

The traditional method for improving separation efficiency adopted by interfacial polymerization relies on molecular design of monomers, adjustment of reaction conditions, and surface tuning [8–12]. The researches that have been devoted to these directions

Membrane-Based Separations in Metallurgy
http://dx.doi.org/10.1016/B978-0-12-803410-1.00012-8

Table 12.1: Polymeric material for NF membranes

Name	Unit Structure
CTA	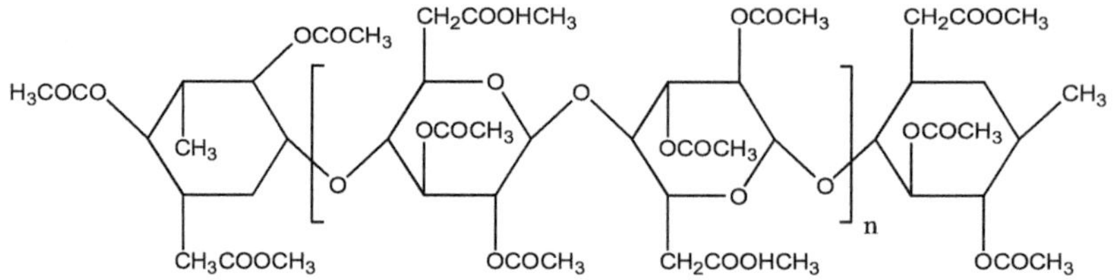
APA	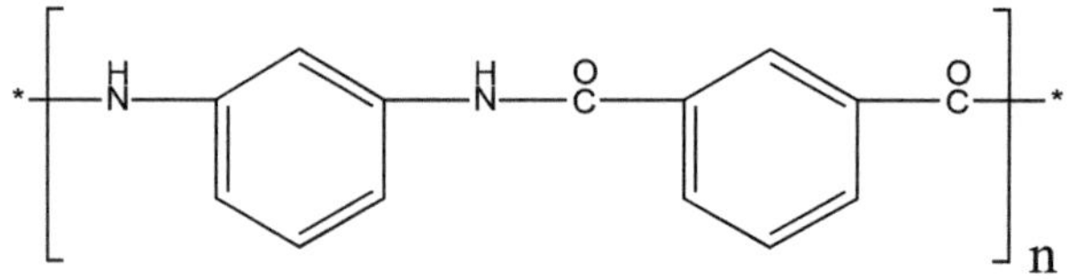
PPA	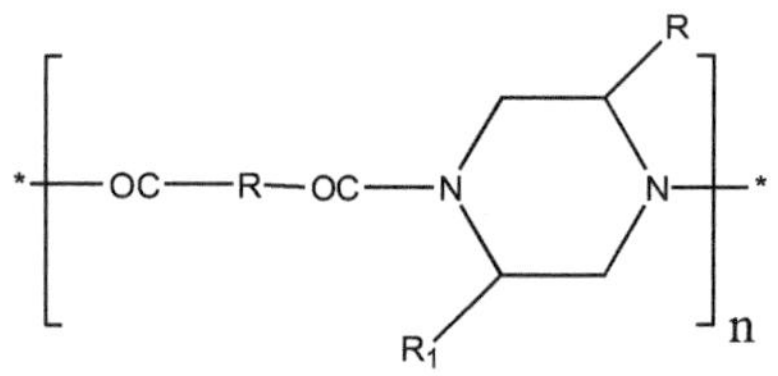
PVA	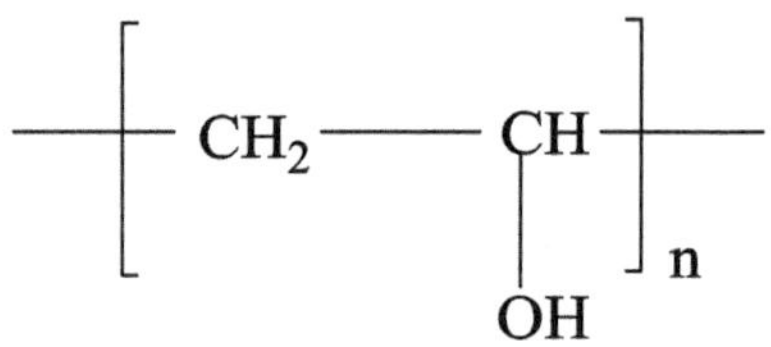
PVDF	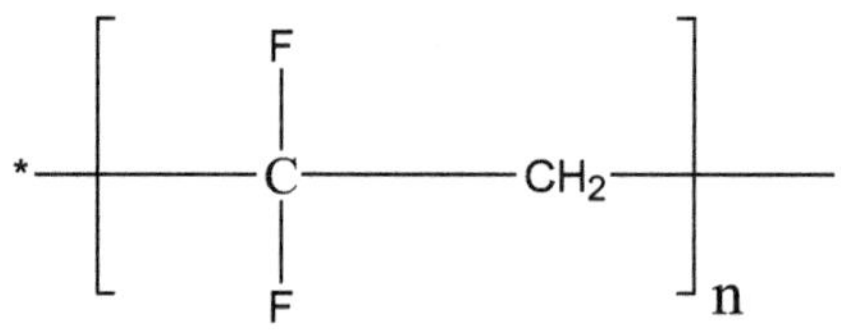
PS	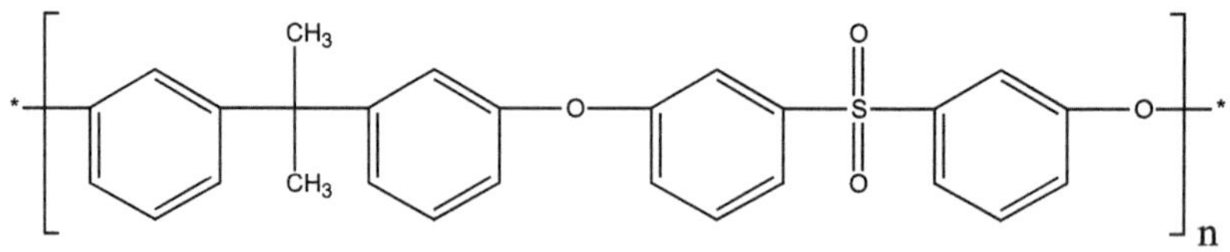
SPS	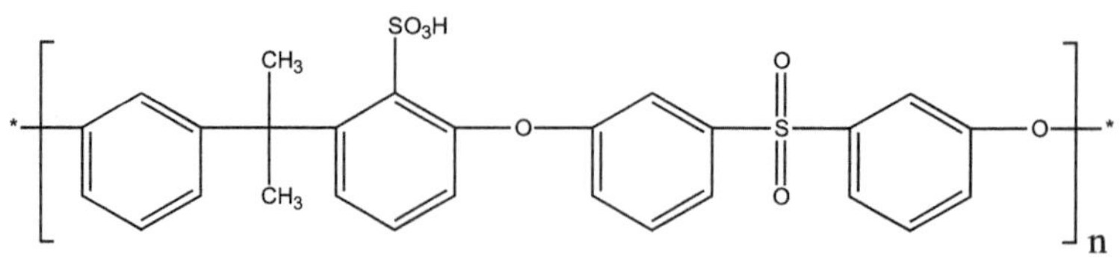

Table 12.1: Polymeric material for NF membranes (*cont.*)

Name	Unit Structure
SPES	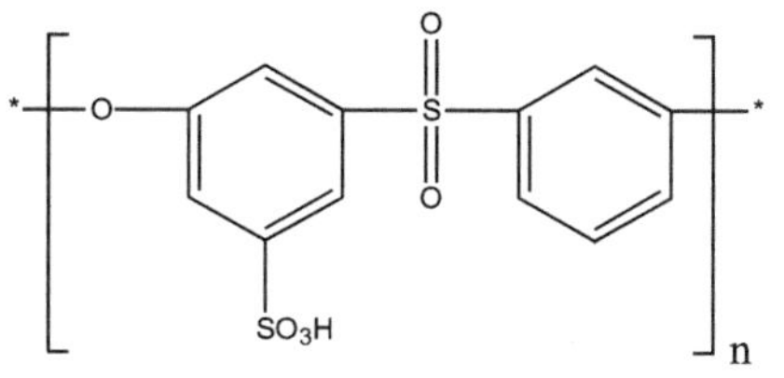
CS	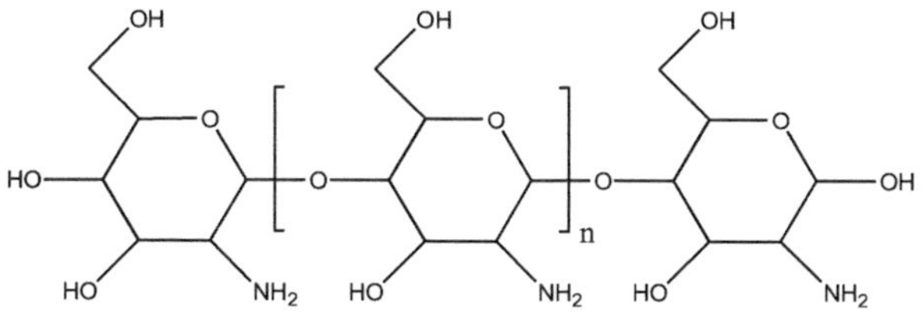
PPESK	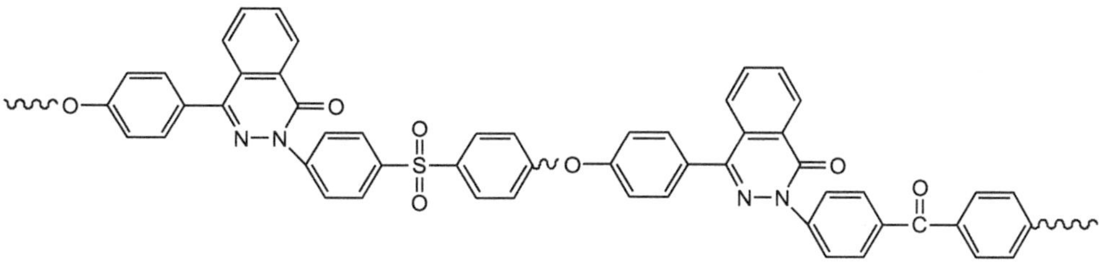

CTA, Cellulose triacetate; APA, Aromatic polyamide; PPA, Polypiperazine amide; PVA, Polyvinyl alcohol; PVDF, Polyvinylidene fluoride; PS, Polysulfone; SPS, Sulfonated polysulfone; SPES, Sulfonated polyethersulfone; CS, Chitosan; PPESK, Poly (phthalazione ether sulfone ketone).

are abundant and many reviews already exist to give systematic summarization. A latest direction is organic or inorganic filled mixed matrix membranes [13]. The fillers investigated principally include zeolite [14], carbon nanotube [15], mesoporous and nonporous material [16,17], aquaporin [18], graphene, etc.

Ratto proposed for the first time the introduction of carbon nanotube of 0.8 nm diameter into interfacial PA layer [19]. The modification led to almost doubled water flux and maintained the salt rejection. It was also found that surface modification of the tube could adjust the salt rejection properties of the membrane [20,21].

Zhao et al. suggested the heat pressing combined with dissection method [15], realizing micro-scale orientation of carbon nanotube on porous polymeric substrate via macro-scale physical operation. The mechanism is schematically shown in Fig. 12.1. Thereafter, coating was applied on the membrane surface to prepare mixed matrix membrane with well arranged tubes embedded inside the polymer matrix. In Wu et al.'s work [22], a novel β-cyclodextrin (β-CD)/polyester thin film NF composite membrane was prepared via interfacial

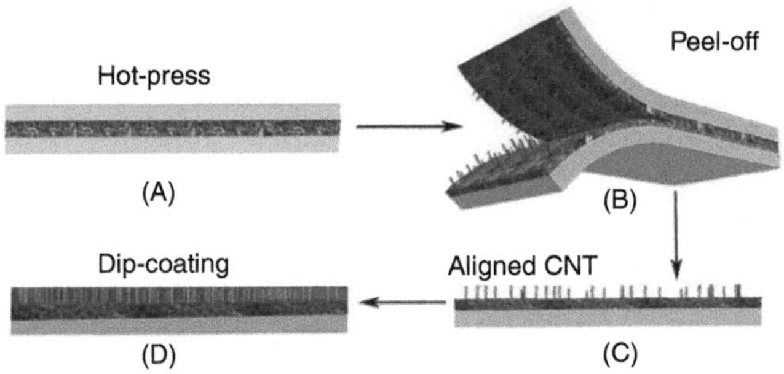

Figure 12.1: Schematic illustration of the method to align SWNT by hot-press and peel-off.

polymerization of trimesoyl chloride (TMC) and triethanolamine (TEOA) in the presence of β-CD. Because the hydroxyls of β-CD can react with the acid chloride group of TMC, β-CD molecules can be chemically introduced into the NF composite membrane in situ and thus a stable structure is formed to avoid the leaching of β-CD during long time operation. The membrane preparation procedure and reaction formula are illustrated in Fig. 12.2. The addition of sulfated β-CD can cause a remarkable increase in the negative charge density and salt rejection of membrane. Furthermore, NF composite membrane incorporated with

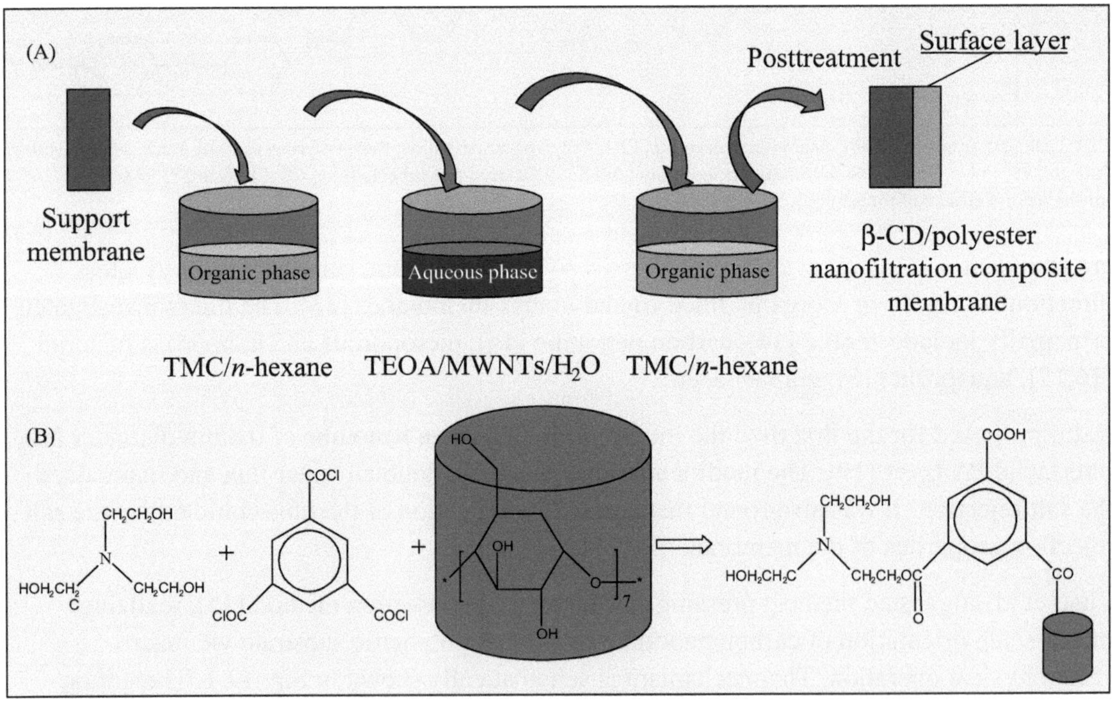

Figure 12.2: (A) Schematic of the preparation procedure of β-CD/polyester nanofiltration composite membrane; (B) the formulas of interfacial polymerization between TMC and TEOA in the presence of β-CD.

β-CD displays a significant improvement on antifouling performance. It was pointed out that the relationship of material/fabrication/structure/performance, mathematical modeling, and molecular dynamic simulation should be the focuses to get more insights about membrane separation mechanisms and help guide the future research. In An et al.'s work, the Age-momentum Correlation (AMOC) positron annihilation (PALS) Technique was used to probe the free volume and nanocavity characteristics of the NF membranes prepared [23]. Zwitterionic polymer thin film composite NF membranes with well-defined structures were prepared in a series of investigations [23,24]. The NF membranes showed an increase in the water flux from 32.6 to 46.3 L m^{-2} h^{-1} without sacrificing the salt selectivity when the zwitterionic monomer 3-dimethyl-(methacryloyloxyethyl) ammonium propane sulfonate (DMAPS) content was increased from 0 to 4.35 mol.%. The reactions for synthesizing piperazine propane sulfonate (PPS) and the interfacial polymerization of piperazine (PIP) and PPS with TMC are displayed in Fig. 12.3. According to the AMOC PALS data, the free volume was reduced as the molecular chains were arranged much closer to each other with

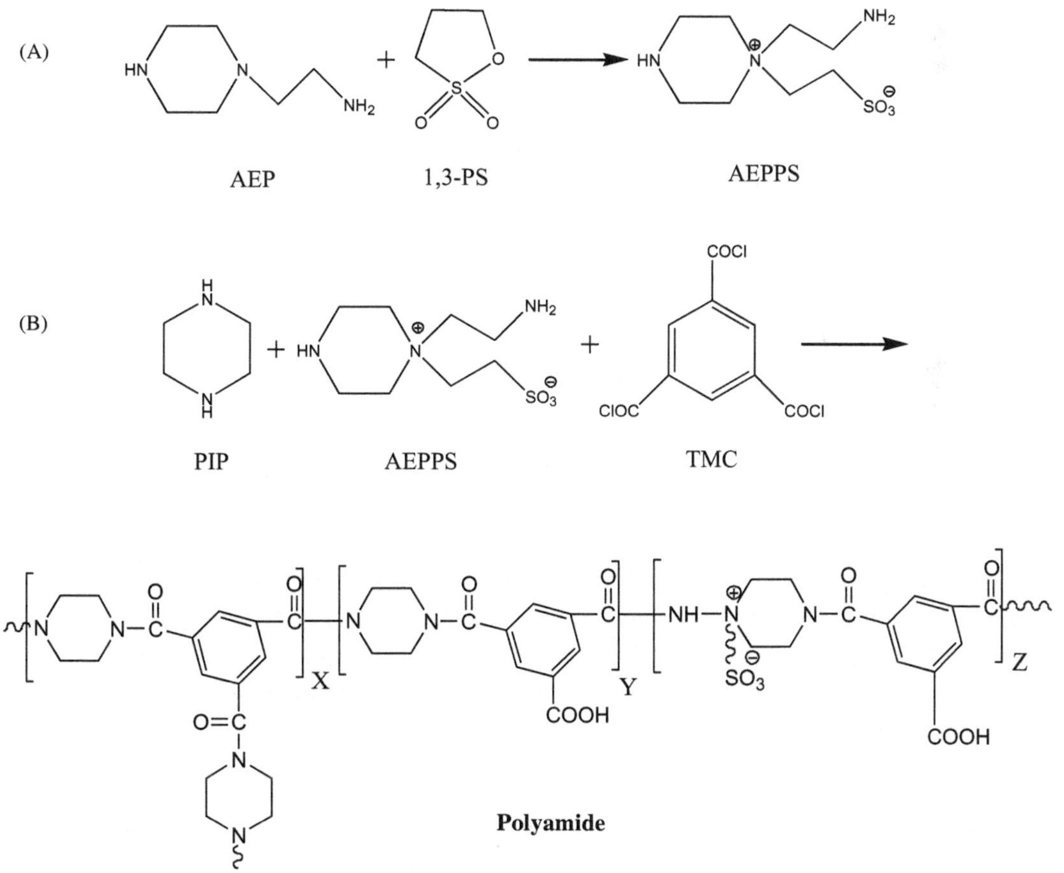

Figure 12.3: (A) Synthesis route of zwitterionic monomers AEPPS and (B) synthesis of NF membranes by interfacial polymerization.

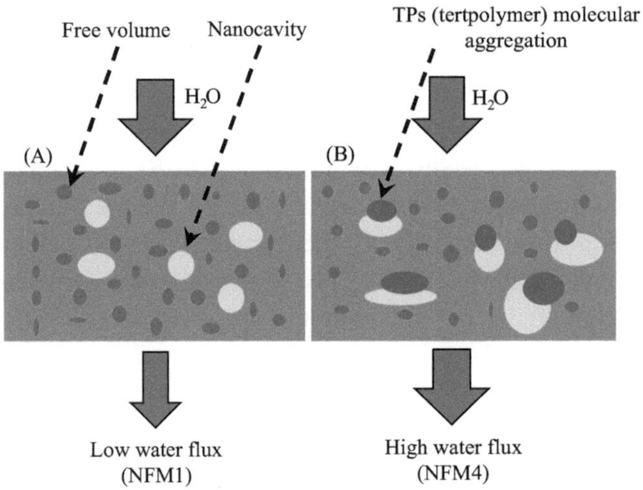

Figure 12.4: Schematic model of water transport through (A) NFM1 and (B) zwitterionic NFM4.

increasing electrostatic interaction, and the nanocavity size was enlarged as the interfacial space was increased as a result of the increased molecular association. The microstructure of NF membranes and the water transport through them are schematically represented in Fig. 12.4. In this model, it is assumed that there are two types of passage in NF membranes: free volume and nanocavity. Free volume is a small space between the molecular chains, whereas nanocavity is a large space between the molecular associations. NF membranes with larger nanocavities and smaller free volume have higher water permeability and higher salt (divalent/monovalent) selectivity. The size of the nanocavity in NF membranes increases as the concentration of zwitterionic DMAPS increases, which may contribute to the increase in the water flux.

12.1.2 Acid-Resistant Nanofiltration Membranes

The actual metallurgical process, acidic condition is very common under in-service environment. Huge amount of acidic effluent or waste demanding further treatment are generated. The major resources for acids include the following: (1) waste acid from extraction; (2) acid pickling solution; (3) electroplating waste acid; (4) waste acidic solution in mold spray. Strongly acidic solution is often used today to extract precious metals. A typical example is usage of phosphoric acid in liquid extraction of precious metals such as copper (Cu) or gold (Au). Pickling step is essential to ensure high quality surface of steel or other metals for further processing. Acid is used to remove scale from the surface containing oxides [25].

The common polymers such as PA and cellulose acetate (CA) fail to survive the strong acidic condition. The mechanism for PA degradation under extremely acidic condition is shown in Fig. 12.5 [26]. There exist a series of acid-resistant polymeric materials, such

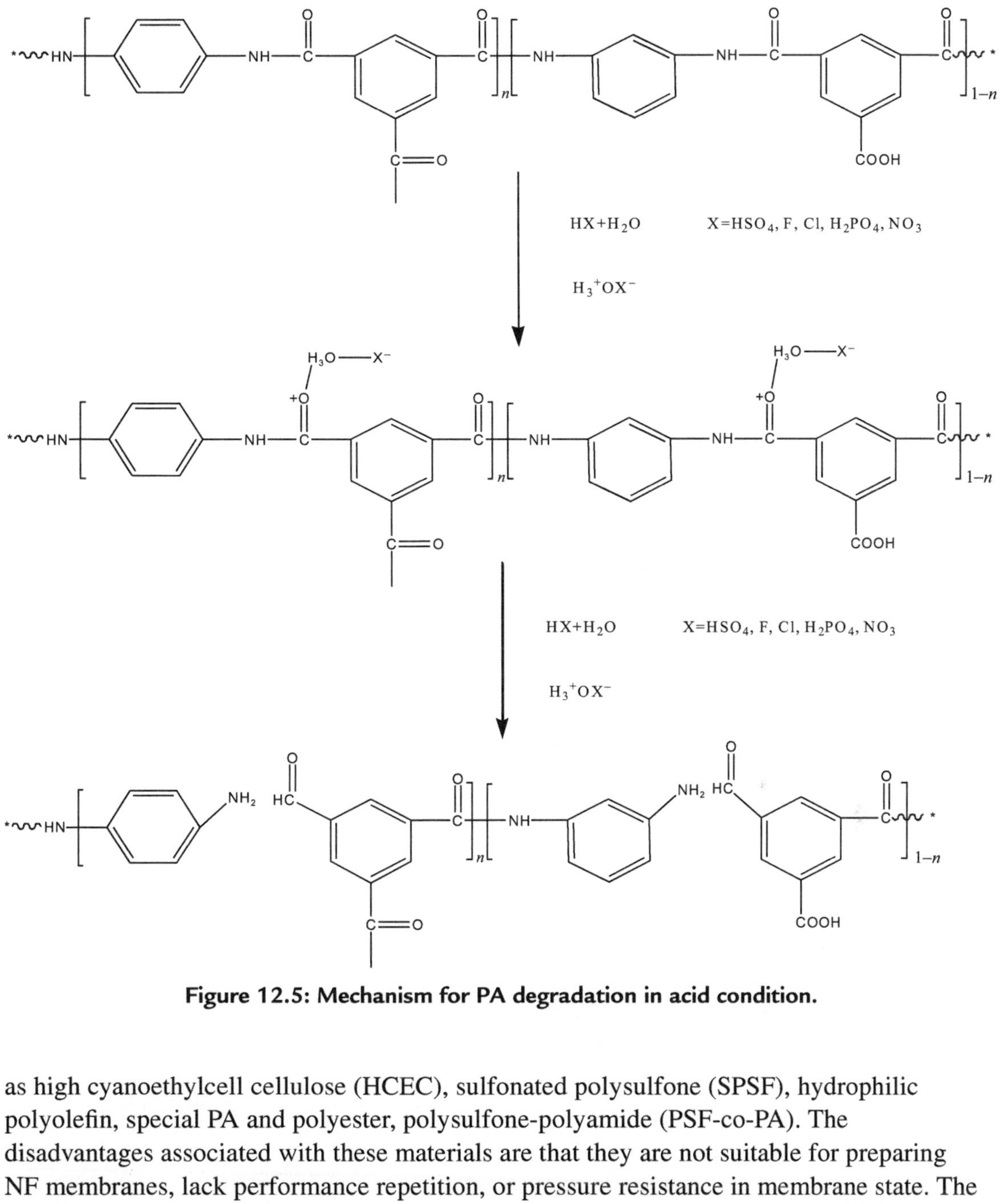

Figure 12.5: Mechanism for PA degradation in acid condition.

as high cyanoethylcell cellulose (HCEC), sulfonated polysulfone (SPSF), hydrophilic polyolefin, special PA and polyester, polysulfone-polyamide (PSF-co-PA). The disadvantages associated with these materials are that they are not suitable for preparing NF membranes, lack performance repetition, or pressure resistance in membrane state. The membrane chemical or physical changes after being attacked by acid or other aggressive chemicals involve the variations in film thickness, pore size, hydrophilicity, etc. Long-term process of degradation will inevitably bring about irreversible membrane structure and performance deterioration [27]. Therefore, searching acid resistance NF membranes will be of significance.

Some commercialized NF membranes have to a certain degree the acid resistance, such as the series of SelRO, Desal KH, NTR 729HG, NF-45, BPT-NF series of membranes. In Ma's work [28], the polysulfonamide (PSA) or Psf-co-PA was selected as the membrane material to prepare NF membrane possessing both good separation performance and superior acid resistance. Firstly, the PSA selective layer was formed on the PSf support membrane through the interfacial polymerization of PIP in aqueous phase and 1,3,6-naphthalenesuifonate chloride (NTSC) in organic phase. Acid soaking tests demonstrated that after soaking in 20 w/v% H_2SO_4 aqueous solution for 8 days, the PSA membrane showed no change in the surface morphological structure. The PSA membrane could maintain the chemical property and separation performance during the 60 days' acid treatment process. In Qin et al.'s work, a new piperazine–ethylenediamine (PIP–EDA) NF membrane to desalt and recover naphthalenesulfonic acids from wastewater with concentrated bivalent salt was designed [29]. The chemical structure of the PIP and influence of EDA bonding are shown in Fig. 12.6. The membrane was fabricated by interfacial polymerization and the chain structure and membrane property of PIP NF membrane was regulated using linear EDA. The optimized PIP–EDA NF membrane exhibited excellent rejection (>94%) for 1-naphthylamine-3,6,8-trisulfonic acid (T acid) and the rejection for ammonium sulfate ($(NH_4)_2SO_4$) is only 36%. NF experiment and characterization results indicated that the addition of a small amount of EDA can obviously improve the permeation flux and the difference between rejections for T acid and salt simultaneously. A novel TFC-IP NF membrane was successfully prepared from 3,3′,5,5′-biphenyl tetraacyl chloride (mm-BTEC) and PIP through interfacial polymerization [30]. The resulting NF membrane exhibited significantly enhanced water permeability while maintaining high rejection to salt. The flux and rejection of NF membrane to sodium sulfate

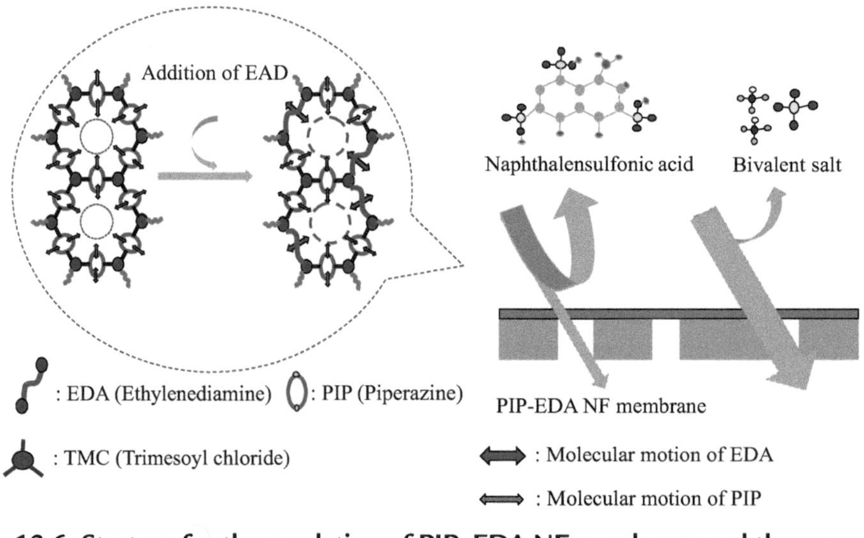

Figure 12.6: Strategy for the regulation of PIP–EDA NF membrane and the recovery of naphthalenesulfonic acid from bivalent salt solution.

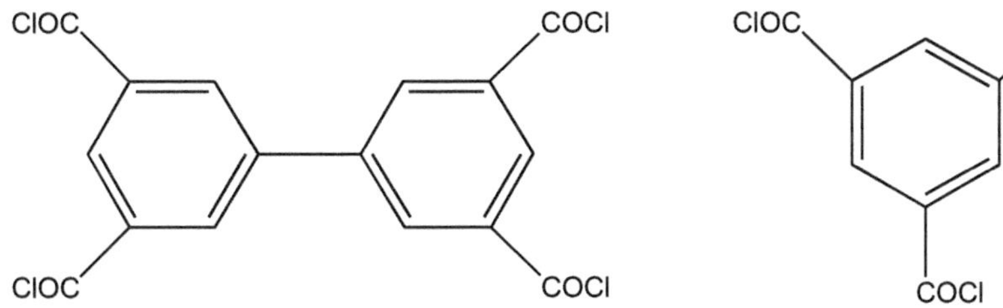

mm-BTEC TMC

Figure 12.7: Structures of mm-BTEC and TMC.

(Na_2SO_4) (500 ppm) reached to 51.5 L (m^{-2} h^{-1}) and 95% under 0.5 MPa. The streaming potential tests indicated that the TFC-IP membrane surface had high charge density and very low isoelectric point which was situated between pH 1 and 2. The chemical structure of the TMC and mm-BTEC is shown in Fig. 12.7.

12.1.3 Alkali-Resistant Nanofiltration Membrane

In metallurgical industry, sodium carbonate (Na_2CO_3) or caustic soda is often used to convert the active ingredient contained in the mineral into soluble sodium salts. In the process, the nonsoluble impurities are removed or separated from the useful components. In aluminum (Al) smelting, for example, sodium carbonate or caustic soda are employed in preparation of cryolite and treatment of bauxite. Yet another example is tungsten (W) smelting which involves roasting of the mixture comprising W concentrate and caustic soda to form sodium tungstate (Na_2WO_4). In addition, caustic solution is often used in metal cleaning process.

Recycle and reuse of the useful components in the alkaline wastewater can generate significant economic and environmental benefits [31]. Applications for alkali-resistant NF membranes included the separation of dissolved metal hydroxides from caustic pickling liquors. The commercially available membranes that can withstand the strong alkaline conditions include N30F, SelRO MPF-34, NTR-7470, G-5 (GE), and GR95PP [32]. Their suppliers are Microdyn-Nadir, Koch, Nitto-Denko, Osmonics, DSS, respectively. The researches in searching alkaline-resistant NF membrane are few.

In Lee et al.'s work [33], polyamine membranes were prepared by the interfacial polymerization of polyethylenimine (PEI) and cyanuric chloride on porous PES supports. The thin film composite membranes thus generated showed superior pH stability as compared to PA membranes that are produced from PEI and TMC. Exceptionally strong resistance toward nucleophilic attack induced by extreme pH conditions was verified by the stable salt rejection and molecular weight cut- off (MWCO) performance after 5 weeks of treatments at pH = 1 and pH = 13. Hydrolysis of the PA membranes under high pH conditions causes a significant in their salt rejection. The polyamine has a fivefold lower permeance as compared to the PA membrane, and a sodium chloride (NaCl) rejection of 65%, exceeding that of the PA by ~10%.

In Wu et al.'s work [34], multi-wall nanobutes (MWNTs)/polyester thin film nanocomposite (TFN) membrane was prepared. The membrane performance and morphologies are remarkably influenced by MWNTs concentration and surfactant in the aqueous phase as well as the reaction time of interfacial polymerization. The experiment of membrane separation properties suggests it is possible for the TFN membrane to be applied in the removal of sodium sulfate (Na_2SO_4) from the concentrated brine of chlorine-alkali industry due to high rejection selectivity of SO_4^{2-} from Cl^-. In addition, the effect of pH on the flux and salt rejection for Na_2SO_4 solution was investigated, indicating that the TFN membrane was relatively suitable for neutral and alkali feed. The TFN membrane also exhibited a good long-term stability.

12.1.4 Oxidation-Resistant NF Membranes

Fouling is a factor impeding the efficient and large-scale application of NF process. Organic and microorganism fouling often occurs in NF process, leading to flux decrease. Oxidizing agent can remove the bio-fouling to recover the membrane flux. Nevertheless, polymeric membrane materials are labile to the attack of active chlorine, resulting in performance degradation [35]. The possible mechanisms include amide chlorination, aromatic chlorination, and Orton Rearrangement [36–38]. Chlorination will lead to PA layer deformation and scission of amide bonding, resulting finally in membrane properties worsening.

The strategies for enhancing the chlorine oxidizing resistance are categorized into following five approaches [35]: (1) removal of N–H bonding of amide group; (2) introduction of functional groups (eg, –CH, −Cl) at o-position of aromatic ring to cause steric hindrance; (3) use of electron withdrawing groups (eg, nitroso (–NO_2), carboxylic acid (−COOH), sulfonic acids (−SO_3H)) to reduce electron density, lowering activity of aromatic ring and preventing o-position replacement; (4) replacing aromatic amide with alicyclic or aliphatic amides; (5) surface coating to protect the PA from chlorine attack.

12.2 Ion-Exchange Membrane

In metallurgical industry, the two important environmental problems that have involved utilization of ion-exchange membranes (IEMs) are [39,40] as follows: recovery and enrichment of valuable ions and removal of undesirable ions, particularly, the toxic metal ions, from wastewater. The operation modes for the above separations can be electrodialysis (ED), electrolysis or diffusion dialysis (DD) [41–43] based on the application of anion exchange membrane (AEMs), cation exchange membranes (CEMs), and/or bipolar membranes (BPMs).

Depending on the degree of heterogeneity [44], commercially available IEMs can be divided into the following groups: (1) homogeneous membranes, (2) interpolymer membranes, (3) micro heterogeneous graft and block-polymer membranes, (4) snake-in-the-cage membranes, and (5) heterogeneous IEM. For successful application of IEMs in metallurgical sector, the membranes should meet several requirements: high permeability and permselectivity, low electrical resistance, good mechanical stability, and sound chemical stability. It was the development of stable, highly selective IEMs of low electric resistance in 1950 by Juda and McRae of Ionics Inc. and Winger et al. at Rohm in 1953 that opened the age of industrialized application for IEM-based processes [41]. The literature body for IEMs synthesis is large including many high quality reviews [45–47]. The current frontiers are "green" synthesis (ie, methods free of carcinogenic effect) of AEMs, membrane stability improvement, and nanochannels construction and manipulation.

Among the possible anion-exchange groups displayed in Fig. 12.8 quaternary ammonium salts (strong anion-exchange groups) and tertiary amines (weak anion-exchange groups) are principal choices due to their higher chemical stability as compared with the sulfonium and oxinium [47–49] anion-exchange materials are conventionally based on polystyrene (PS) cross-linked with divinylbenzene (DVB) [50]. The ion-exchange sites are linked via benzylic methylene groups to the polymer backbone. The benzyl chloride group is introduced by the use of vinyl benzyl chloride as monomer. The benzyl chloride may also be introduced by chloromethylation of polymer using chloromethyl methyl ether (CMME) or formaldehyde and trimethylsilyl chloride (TMS-Cl) [47]. The reaction scheme for the above process is illustrated in Fig. 12.9. The latter method is applicable to polymers other than PS, such as epichlorohydrin or epoxy-based polymers, PVA, poly(acrylate)s or poly(olefine)s, perfluorinated polymers, chitosan, poly(ether ketone ketone) (PEKK), polybenzimidazoles (PBI) [47].

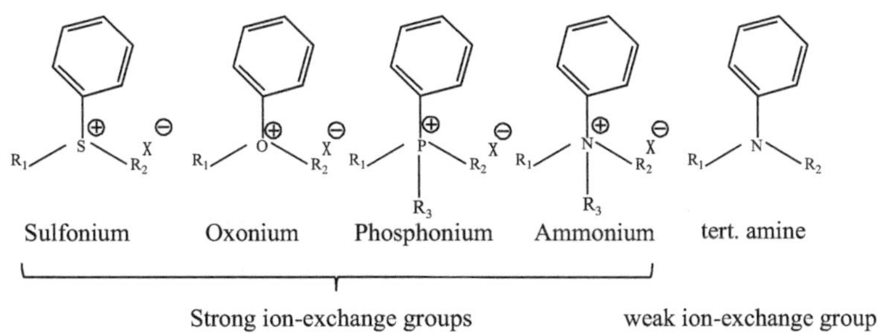

Figure 12.8: Possible functional groups for anion-exchange.

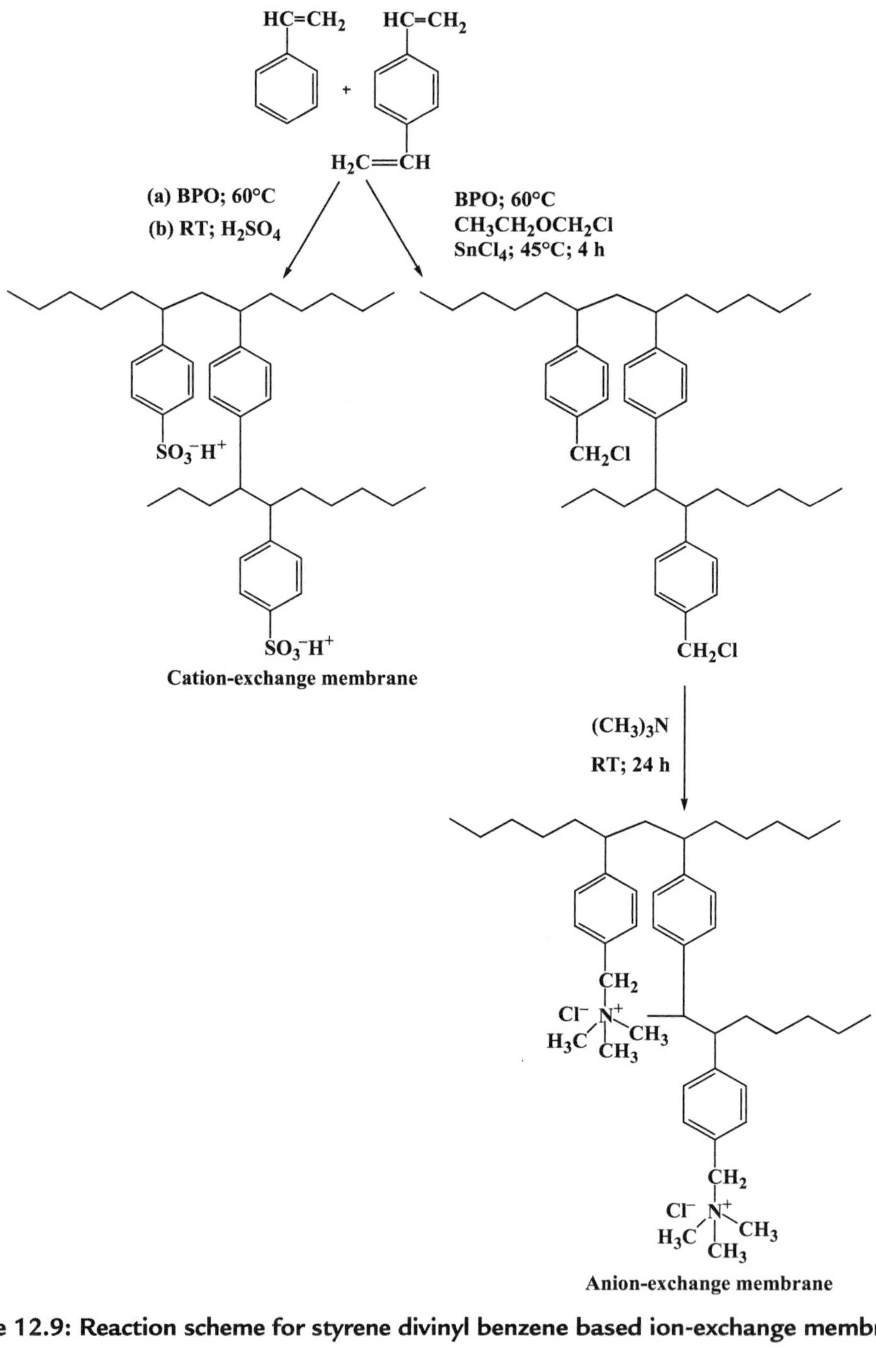

Figure 12.9: Reaction scheme for styrene divinyl benzene based ion-exchange membranes.

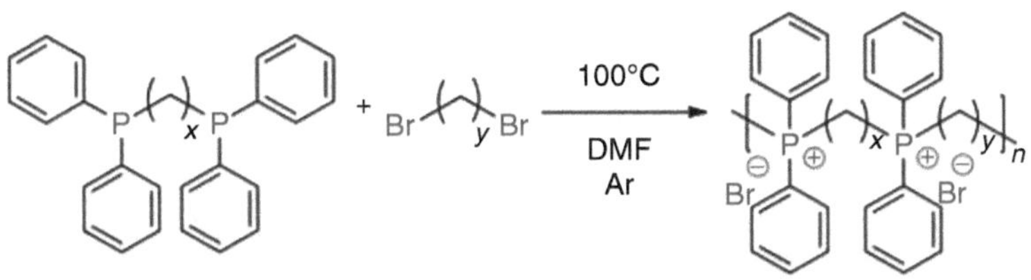

Figure 12.10: Step-growth polymerization of bis(diphenyl) phosphines and dibromides to synthesize phosphonium ionenes for anion exchange membranes.

In recent years, phosphonium groups receive attention to a certain degree [48,49,51,52]. According to Hemp et al.'s work, phosphonium ionenes were synthesized with bis(diphenyl) phosphinoalkanes and dibromoalkanes. The scheme for the reaction is shown in Fig. 12.10. The type of the alkyl group between the two phosphines or bromides generated a series of phosphonium ionenes with different spacing between the two cationic sites.

The most commonly used cation-exchange groups are $-SO_3H$, $-COOH$, phosphonic acids ($-PO_3H_2$) and phenolic hydroxide groups. Hereinto, $-SO_3H$ group is preferred due to the strong acidity with a high dissociation. Therefore, the materials relying on this functional group have good proton (ion) conductivities [53]. The $-PO_3H_2$ is thought to be more suitable for applications involving medium or high temperatures due to its amphoteric character and the ability to transfer protons under anhydrous conditions by the Grotthus mechanism [54].

Posttreatment of base membrane by concentrated sulfuric acid (H_2SO_4) is the most convenient way for preparing sulfonated polymers [55]. There are also other sulfonation agents and solvent system, for example, chlorosulfonic acid in halogenated solvents [56] and trimethylsilylsulfonyl chloride (TMSSC) in the corresponding solvent [57].

Perfluoro sulfonic acid (PFSA) polymers have excellent chemical, thermal, and mechanical stability. They have been commercialized for many years and were originally utilized in the chloralkali industry. Among PFSAs, Nafion (DuPont) is well known as one of the most promising state-of-the-art proton exchange membranes (PEMs). PFSAs are also well known under the trade name from other companies, such as Flemion, Selemion from Asahi Glass, Aciplex from Asahi Kasei, Fumion F from FuMa Tech, Aquivion from Solvay. Most of the PFSAs and perfluoroalkyl carboxylic acids (PFCA) were developed for the membrane based chlor-alkali electrolysis. A series of nonfluorinated polymers have also been investigated in the past decades as potential materials for CEMs. The most famous companies for producing CEMs free of fluorine are GE (USA), Aston (Japan), and FuMA-Tech (Germany).

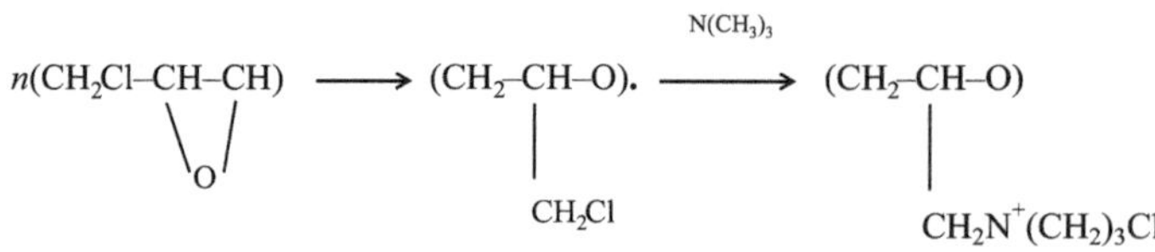

Figure 12.11: Preparation of chlorohydrin rubber anion-exchange membrane.

12.2.1 "Green" Synthesis of Anion-Exchange Membranes

With traditional routes of preparing AEMs containing ammonium salts and amines as mentioned earlier, chloromethylation process is generally applied, which involves highly toxic carcinogen, chlorine methyl ether. Therefore, new approaches free of the hazardous effects are intensively searched. In Zhang's work [58], several strategies are introduced, which are summarized as:

1. Novel reactions
   Reaction between epoxide and amine groups: Epoxide group can react directly with amine compound. In this regard, polymers containing epoxy groups can be first fabricated into basilemma or base membrane, which is then aminated to obtain AEM.
   Friedel—Crafts acylation: PS was functionalized by chloracetyl followed with amination/quaternization.
2. Functional groups on side chain
   Halogen methyl group: The halogen methyl groups on side chain can be quaternized to make AEM. One example is chlorohydrin rubber membrane. In the reaction, as shown in Fig. 12.11, the epichlorohydrin was used as monomer to form polymer with methyl chloride as side chain. Benzene (or toluene) was solvent and triisobutylaluminum (TIBA) was the catalyst.
   Methyl groups attached to side chain of benzene: One example relies on the use of a much less hazardous chemical, *N*-bromosuccinimid (NBS) [59]. The reactions for the conversion are shown in Fig. 12.12. With this method, in effect, ion exchange groups in terms of number and location can be conveniently adjusted by (1) the amount of

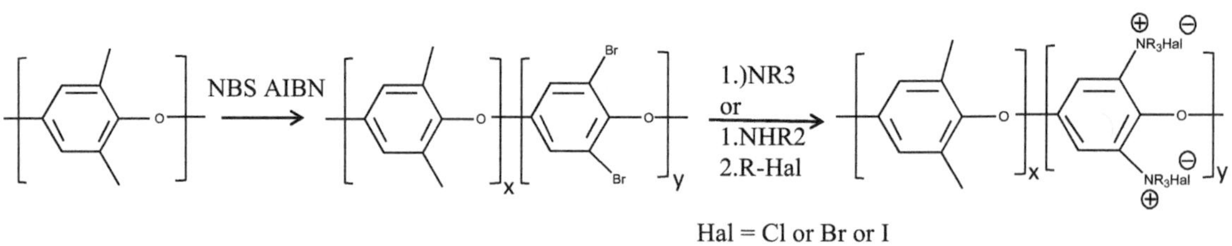

Figure 12.12: Synthesis of polymeric anion-exchange materials via bromination of methylated polymers.

methylated monomers used in the polymerization step, (2) the degree of bromination, which is controlled by the amount of NBS used in the first step, and (3) the polymer-type (random or block copolymer).

3. New chloromethylation reagents [60]
   Long chain halide methyl halide alkyl ether: The chlorine methyl ether with a relatively long chain has a high boiling point, low volatility. No proof has ever been collected reporting its carcinogenicity. In Lu et al.'s work [60], The chloromethylation is accomplished by using 1, 4-bis (chloromethoxy) butane (BCMB) as the chloromethylating reagent and concentrated sulfuric acid as the solvent and catalyst. BCMB is a kind of high-efficient, noncarcinogenic and inexpensive chloromethylating reagent, and concentrated sulfuric acid replaces toxic halogenated hydrocarbon.
4. Surface grafting [61]
   Porous membrane made from high density polyethylene (PE) was irradiated by the electron beam to generate free radicals for grafting glycidyl methacrylate (GMA) with epoxy group and DVB as crosslinker. Thereafter, ring opening reaction was initiated by introducing diethylamino (DEA) group to obtain AEMs.

12.2.2 Membrane Stability Control

During operation, as already mentioned, the IEMs should have good mechanical strength, which generally is reflected by the degree of swelling or shrinking phenomenally. Additionally, the membranes have to survive a wide pH-range and temperature in the presence of caustic components (eg, oxidizing agents). A commonly accepted impression is that the organic membranes are characterized with poor mechanical strength, and weak resistance toward attack from operation condition. This is assumed as one of the reasons for the limited applications of IEMs in industries for quite a long time, which also motivates the search for new materials or modifications in the past decades.

The foremost approach deals with molecular design or architecture of polymeric chains. As far as AEMs are concerned, one group of materials with higher stability is aforementioned phosphonium containing macromolecules. They exhibited improved thermal and base stabilities compared to ammonium analogs, which is attributed to the propensity of the ammonium cation toward Hofmann elimination [61]. For example, no degradation was observed when treating the polymers in 2M sodium hydroxide (NaOH) at room temperature for 24 h. As for CEMs, polymeric phosphonic acids have attracted much attention as during the past decade due to their higher thermal stability compared to polymeric sulfonic acid [62].

A series of poly (arylene ether sulfone) (PAES) with bulky imidazole groups (PSF-Im-x) have been synthesized using monomer 2,2′-bis-(2-ethyl-4-methyl-imidazole-1-ylmethyl)-biphenyl-4,4′-diol (EMIPO) [63]. Subsequently with quaternization by *n*-bromobutane, the AEMs were formed. It was found that with 2-ethyl-3-butyl-4-methyl-imidazolium as the functional group, structures around the imidazolium ring reduce the access of OH^- to imidazolium, which in turn enhances the alkaline stability of the membranes. The membrane exhibited ion exchange capacity (IEC) of 2.07 and ionic (OH^-) conductivity of 0.014 Scm^{-1} at 30°C. After being treated in 1M potassium hydroxide (KOH) at 60°C for 144 h, additionally, and 80% of the ionic conductivity was maintained.

It has been reported, that block copolymers having the same overall chemical composition as random copolymers exhibit a lower inplane swelling but a much higher swelling in *z*-axes (thickness) [64–67]. These properties were attributed to a much better phase separation between sulfonated and nonsulfonated domains in the block copolymers. Another finding regarding highly sulfonated IEMs with high dimensional stability was reported by Matsumoto et al. [68,69] and Tian et al. from Hays's group [70,71]. These works reported compact and highly sulfonated clusters based on tetrasubstituted hydroquinone or hexasubstituted 4,4′-dihydroxybiphenyl which is present either in the polymer main chain, or at the terminal end, or in both states. Whatever the situation, the clustered structure led to a distinct phase separation between sulfonated and nonsulfonated domains. Membranes with IECs in the range of 2 mmol g^{-1} showed low water uptake and good dimensional stabilities comparable to Nafion.

The most conventional approach to adjust membrane stability refers to cross-linking like covalent bonding [72] or ionic bonding [73]. Poly(α,β,β-ntrifluorostyrene sulfonic acid) was used as IEMs material [74], which is known as BAM-3 (Ballard Advanced Material, generation-3). Sulfonation cross-linking might occur by formation of sulfone bridges to enhance chemical stability of the membrane.

However, the IEC will decrease with ionically cross-linked membranes due to the reaction of sulfonic acid groups with basic polymer for neutralization, while covalent cross-linking will in most cases weaken the elasticity and flexibility of the membrane.

In addition to the earlier mentioned schemes involving molecular design in functional group and/or polymer chain, another relatively new direction is the combination of organic polymer and inorganic materials [75–77]. The organic–inorganic hybrid membrane has the potential to overcome the shortcomings of respective material to a certain extent. Wu et al. [78] made a review about the preparation of organic–inorganic hybrid membranes. According to this review, sol–gel method is one of the most important methods for preparingn this type of membrane. For example, a new type of anion exchange material synthesized via sol–gel method was reported. The precursors for the membranes were tetra methoxy orthosilane (TMOS) and *N*-trimethoxysilylpropyl-*N*, *N*, *N*-trimethylammonium chloride (TMSPTA). The reactions in the sol–gel process using TMOS and TMSPTA are shown in Fig. 12.13. Among the other inorganic materials, nanotube is the most widely used for preparing organic–inorganic hybrid IEMs.

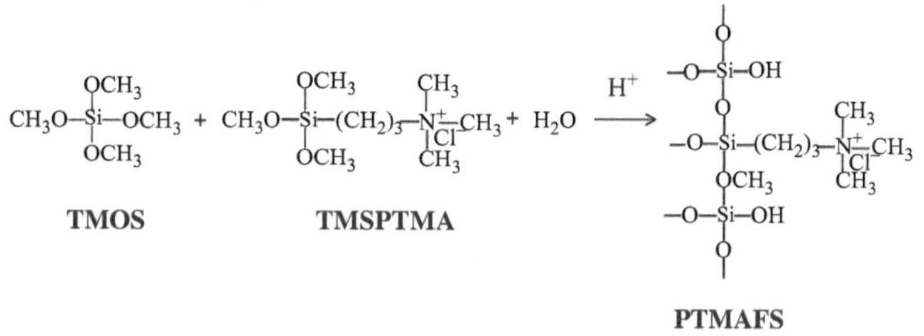

Figure 12.13: Anion-exchange materials with TMOS and TMSPTMA as precursor.

12.2.3 Nanochannels Architecture and Modification

Nanochannels containing ionic groups, through which the transport of "hydrated" ions occurs, are thought to be the most critical factor for IEMs to achieve sound ion transport properties relating to flux and selectivity. Several strategies have been proposed for constructing and modifying the channels.

As for separation factor, a strategy was proposed in some researches on tuning crystallinity by annealing treatment that enhances the density of PVA-based CEM and thus improves a pore-size sieving effect for ions [42]. Zinc ions (Zn^{2+}) with a larger Stokes radius is more difficult to transport through the dense membrane than smaller H^+. However, the reduction of H^+ flux is unavoidable due to the increase in the density of membranes.

Alternatively, cross-linking via acid–base interaction can strengthen the density of a membrane and thus balance its perm-selectivity and ionic flux. Constructing H^+ transfer channels via the formation of acid–base pairs was proposed by several researches [79]. The hydrogen bonding between basic 1-vinylimidazole monomers and acidified sulfonated poly(2, 6-dimethyl-1, 4-phenylene oxide) (SPPO-H) leads to a network making the membranes compact. Results showed that, the acid–base pair based membranes exhibit very low Zn^{2+} leakage and excellent perm-selectivity between H^+ and Zn^{2+} during the ED process. This research here demonstrates a versatile route that balances perm-selectivity against ionic flux of the PSMs, which pave a way for applications in separation of proton and a metal ions system.

The acid-base interaction was also used by Wu et al. to form a "barrier in proton conductive channel [80]. The poorly hydrophilic ionic acid–base interactions ($—SO_3^- \ldots NH^+—$) present blockage for methanol transport; whereas, more hydrophilic hydrogen bonded type acid–base interaction ($—SO_3H\ldots(H_2O)_n\ldots N—$, $n < 3$) facilitated the transport of protons via the switch between its two states: and breaking and connecting states. The membrane was featured with a quite low methanol permeability (30% of that with Nafions 112) and high proton conductivity of 90 mS cm^{-1} at ambient temperature.

As aforementioned, Nafions has been widely used in PEMs. This may be attributable to its favorable proton conductivity resulting from the nanoscale phase separation. Inspired by the molecular architecture of Nafions, research has been carried out to fabricate PEMs with sulfonic acid containing side chains. It is found that the characteristics of the side chains in terms of mobility, aggregation, and distribution are of primary importance for the morphology and proton conduction. Another idea enlightened by Nafions is to balance properties of PEMs by introducing perfluoroalkyle sulfonic acid groups into ionomers [81]. Compared with the acidity of conventional aryl or alkyl sulfonic acid groups (pKa ~ -1), perfluorosulfonic acids are better proton conductor; its pKa of 6 allows easier deprotonation under less humid conditions.

Application of block and graft copolymers composed of the covalent unions of two or more chemically distinct polymeric segments is also a way for manipulating the nanochannels [82]. Chemical incompatibility between constituent segments leads to a nanoscale phase separation. It was found that this type of copolymers containing hydrophilic sulfonated segments exhibit optimal combination of properties in IEC, proton conductivity and mechanical strength that are usually unlikely for statistical copolymers. Presumably, the better interconnected proton conductive channels (PCCs) derived from the higher degree of nanoscale phase separation between hydrophilic and hydrophobic domains via self assembly is the principal contributing factor.

Similar to block copolymers, graft copolymers with sulfonated pendant segments attached at various points to the polymer backbone can also generate well-connected PCCs.

Enlightenment from investigation regarding commercial Nafion PEM also covers the finding about alignment of the membrane structure. It was confirmed by Medsen and coworkers [83] that the extruded Nafions (Nafions 112 and Nafions 117) displayed a alignment of PCCs parallel to the extrusion direction, whereas the solution cast membrane (Nafions 212) possessed uniaxially aligned channels perpendicular to the plane of the membrane. Based on this and a series of other findings, attempts were made to obtain PPCs orientation in PEMs by applying solvent annealing or field forces (eg, electric field, pressure, flow field) during the membrane preparation process. For example, Park et al.'s research about poly(styrenesulfonate-b-methylbutylene) (PSS-b-PMB) copolymer PEMs used a "solvent-annealing" process [84].

The characterizations indicated that at low sulfonation levels, the PCCs take on orientation perpendicular to the film surface when the film was in humid air. Through-plane measurements revealed a 30% higher conductivity when the PCCs were perpendicular to the surface of the membrane as compared with samples without electric field alignment.

Further effort was devoted to fabricating artificial anisotropic PCCs to improve the through-plane conductivity [46]. Proton conductive nanofibers were produced by electrospinning. The proton channels due to the network of the sulfonic acid groups formed within the nanofiber facilitate the transport of the protons. PEMs composed of uniaxially aligned nanofibers will

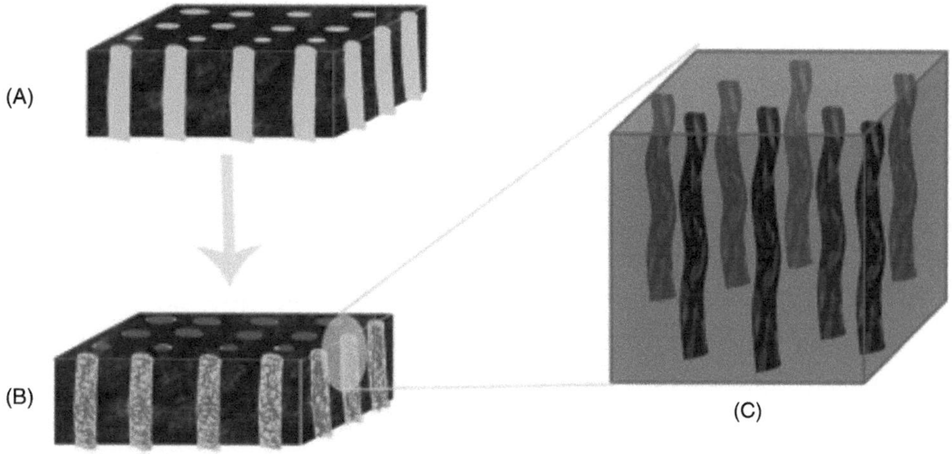

Figure 12.14: Schematically illustrated synthesis strategy for aligned mesoporous PEMs: starting from the preparation of a highly aligned mesoporous film (A), followed by functionalization of the pore wall (B); the obtained membrane exhibits a through-plan aligned morphology of PCCs (C).

exhibit excellent through-plane proton transport. However, the alignment of nanofibers in PEMs is disordered. Fabrication of porous scaffolds containing highly ordered and monodispersed PCCs is attracting increasing interest as an alternative route to control proton conduction. The schematic for the synthesis strategy for obtaining high degree of alignment is given in Fig. 12.14.

12.2.4 Novel Membrane Formation Technologies

A new type of IEMs fabrication technology attracting attention in recent years is electrostatic spinning [85]. Novel AEMs were fabricated from quaternized poly(2,6-dimethyl-1,4-phenylene oxide)/silicon dioxide hybrid material (QPPO–SiO_2) by electrospinning and posttreatment (solvent fumigation and hot-press). In treating a simulated Fe polishing waste solution containing 1 mol L^{-1} HCl, and 0.225 mol L^{-1} ferrous chloride ($FeCl_2$), both higher acid permeability (0.053 m h^{-1}) and selectivity (68.05) in comparison with the QPPO–SiO_2 hybrid membrane formed by direct casting and solvent evaporation. Its performance is even better than commercial DF-120 membrane. The macroscale dense morphology can guarantee the high selectivity of the membrane while the microscale loose morphology endows the membranes with high permeability [86].

12.2.5 Industrialization Status in China

High performance and low cost membranes are very critical for their penetration into utilization market. Currently the major suppliers of IEMs are DuPont, Asahi, Asahi Kasei, General Electric, Aston, MA-Tech etc.

Research on IEMS in China was initiated in late 1950s. The products in the preliminary stage were mainly heterogeneous membranes made from ion exchange resins via grinding and hot press molding in sequence.

From then on, many membrane companies and research institutes in China started the IEMs R&D for the applications in treatment of water and process solutions. At present the companies that provide commercialized IEMs products mainly are Hefei Chemjoy Polymer Materials Co. Ltd., Shandong Tianwei Mmebrane Technology Co. Ltd., Bejin Tingrun Membrane Technology Co. Ltd. Table 12.2 summarizes the properties of some IEMs by the companies earlier mentioned. They are already on sale in the Chinese market.

The industrialization of perfluorinated IEMs production in China was actualized by Dongyue Group around 2009. Their application in chlorine-alkali industry represents an important breakthrough for the chlor-alkali industry of China.

12.3 Materials and Membranes for Ultrafiltration

The major preparation methods of polymeric ultrafiltration (UF) membranes are nonsolvent induced phase separation (NIPS) [87] and thermally induced phase separation (TIPS) [88]. TIPS can be further attributed to two phase inversion mechanisms: liquid–liquid (L–L) separation and solid–liquid (S–L) separation [88,89].

The method based on electrospun nanofibers is a relatively new development. Zhao et al. prepared high flux composite UF membranes using electrospun nanofibrous PVDF scaffolds [90]. The membranes thus generated exhibited a sound flux, high rejection, and low fouling tendency. TFC UF membranes have also been marketed in recent years [91].

A variety of materials have been used for producing commercial UF membranes. These materials include bisphenol A type of cellulose derivatives, PSF, PES, PSA, PVDF, CA, and polyacrylonitrile (PAN). With abundant resources, nontoxicity, cheapness, good hydrophilicity, simple and convenient membrane formation at industrial-scale CA has long been used as the basic materials for preparing UF membrane and is still one of the research focuses. Another polymeric material that is worthy of mentioning is PVDF, which possesses excellent chemical resistance. The hydrophobic nature of PVDF, nevertheless, makes it susceptible to fouling when applied in water filtration [92–94]. Similar problem exists with materials such as polyvinyl chloride (PVC), PSF, PES, PP, and PC, which are generally viewed as hydrophobic polymers.

12.3.1 Molecular Design for Reinforced Ultrafiltration

12.3.1.1 Micellar-enchanced Ultrafiltration

In the 1980s micellar-enchanced ultrafiltration (MEUF) was initiated [95,96]. It is mainly used for removing almost all the toxic metal ions and/or soluble organic solutes

Table 12.2: The properties of commercialized ion exchange membranes currently in Chinese market

Company	Properties					Websites
Chemjoy	Cation exchange membrane		Anion exchange membrane		Bipolar membrane	http://www.cj-membrane.com/
	CJMCED	CJMCDD	CJMAED	CJMADD	CJBPM	
	Properties: Thickness: 0.14–0.16 mm; IEC: 0.9–1.5 mmol g^{-1}; WU: 30–40%; R_m:≤2 Ω. cm^2; t_m:≥0.97	Properties: Thickness: 0.18–0.2 mm; IEC: 1.0–1.5 mmol g^{-1}; WU: 60–80%; ≥ 1.5×10^{-7} m s^{-1} S^-_{OH}/AlO^+_2 ≥13	Properties: Thickness: 0.14–0.16 mm; IEC: 0.9–1.2 mmol g^{-1}; WU: 25–35%; R_m:≤4 Ω. cm^2; t_m:≥0.98	Thickness: 0.18–0.2 mm; IEC: 1.0–1.5 mmol g^{-1}; WU: 60–70%; U^+_H ≥4×10^{-7} m s^{-1} S^+_H/Fe^{2+} ≥15	Properties: Thickness: 0.2–0.25 mm; WU: 25–30%; V_{disso}:≤1.8v;	
	Application: Desalination (ED)	Application: Alkali recovery (DD).	Application: Desalination (ED)	Application: Acid recovery (DD)	Application: Organic acid production, resule of salty water, etc. (ED)	
Tianwei	Cation exchange membrane		Anion exchange membrane		Bipolar membrane	http://www.sdtianwei.com/
	TWEDC	TWDDC	TWEDA	TWDDA	TWBPI	
	Properties: Thickness: 0.10–0.13 mm; WU: 20–30%; R_m:≤4 Ω.cm^2; t_m:≥0.97	Properties: Thickness: 0.15–0.28 mm; WU: 40–80%; U_{OH} ≥2×10^{-7} m s^{-1}; S^-_{OH}/AlO^+_2≥15	Properties: Thickness: 0.13–0.16 mm; WU: 30–40%; R_m:≤4 Ω.cm^2; t_m:≥0.98	Properties: Thickness: 0.2–0.32 mm; WU: 20–70%; U^+_H ≥5×10^{-7} m s^{-1} S^+_H/Fe^{2+} ≥15	Properties: Thickness: 0.18–0.23 mm; WU: 20–30%; V_{disso}:≤1.8v; water dissociation efficient ≥0.97	
	Application: Desalination (ED)	Application: Alkali recovery (DD).	Application: Desalination (ED)	Application: Desalination (ED)	Application: Alkali recovery (DD).	
Tingrun	Cation exchange membrane		Anion exchange membrane		Bipolar membrane	http://tingrun.com/
	JCM-II		JAM-II			
	Properties: Thickness: 0.16–0.23 mm; IEC: 1.8–2.9 mmol g^{-1}; WU: 33–43%; R_m:1–7 Ω. cm^2; t_m:≥0.95		Properties: Thickness: 0.16–0.23 mm; IEC: 1.6–2.2 mmol g^{-1}; WU: 22–30%; R_m: 4–10 Ω.cm^2; t_m:≥0.90		Properties: Thickness: 0.16–0.23 mm; WU: 35–40%; V_{disso}:≤1.6v;	
	Application: Desalination (ED)		Application: Alkali recovery (DD).		Application: Desalination (ED)	

U H^+/OH^-, diffusion coefficient; IEC, Ion exchange capacity; R_m: membrane area resistance (0.5 M NaCl, 25°C); *S*: separation factor; t_m, transport number (0.5 M/0.1 M NaCl, 25°C); V_{disso}, water dissociation voltage (100 mA cm^{-2}, 25°C); WU: Water uptake.

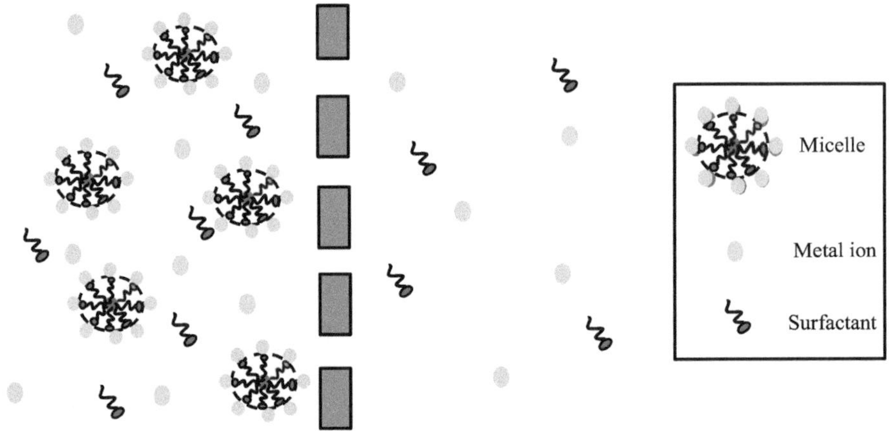

Figure 12.15: A schematic diagram illustrating micellar-enhanced ultrafiltration for the removal of metal ions.

from aqueous solutions [97]. As schematically shown in Fig. 12.15 [98], when the amount of surfactants present in aqueous solutions are above a certain concentration, ie, the critical micelle concentration (CMC), their aggregates are formed that may attract or solubilize a solute in the solution. The micelles and solutes composite with bigger hydrodynamic size can be rejected more efficiently by UF membranes than unbounded solute molecules, due to the size exclusion mechanism. In this manner the UF system obtain higher or enhanced retention of solutes. The surfactants used in MEUF are mainly synthesized organics. A typical surfactant is lecithin [99]. The surfactants commonly used in recent years are summarized in Table 12.3.

Sodium dodecyl sulfate (SDS) is the most widely used surfactant in MEUF. In Lee et al.'s work [100], the formation of SDS micelles, the characteristics for Zn^{2+} adsorption on SDS micelles in terms of the kinetics and two-parameter adsorption isotherm models had been studied. Investigation on adsorption isotherm revealed that Freundlich isotherm may better account for the adsorption characteristic of Zn^{2+} on SDS aggregates.

UF experiments were used to evaluate the loading of micelles with Zn(II) ions using following equation:

$$L_m(mM/mM) = \frac{[C]_R - [C]_P}{[S]_R - CMC}$$

where, $[C]_R$ and $[C]_P$ are Zn^{2+} ion concentration in the retentate and permeate, respectively; and $[S]_R$ represents the SDS concentration in retentate.

Table 12.3: Commonly used surface active agents

SAA	Molecular Structure	Example	Molecular Structure
Nonionic SAA	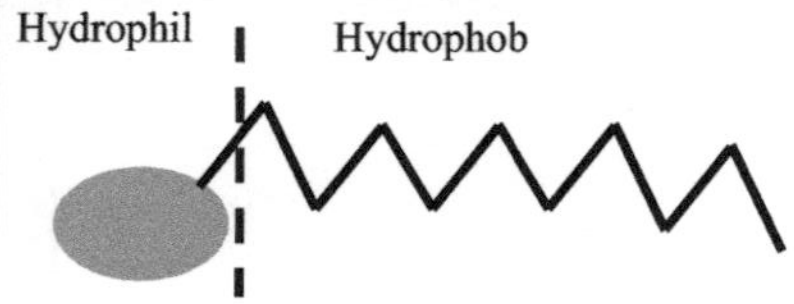	Polyoxyethylene octyl phenyl ether (TritonX-100) APEO; AEO; AE	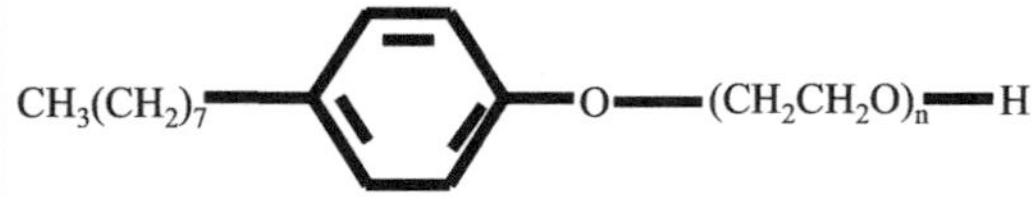 n can be 4, 6, 7
Anionic SAA	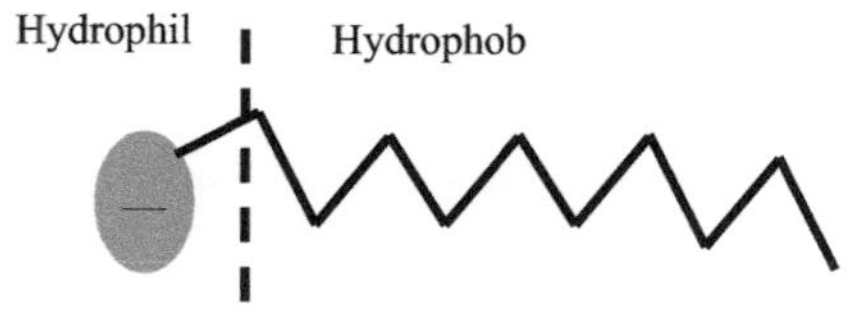	LAS	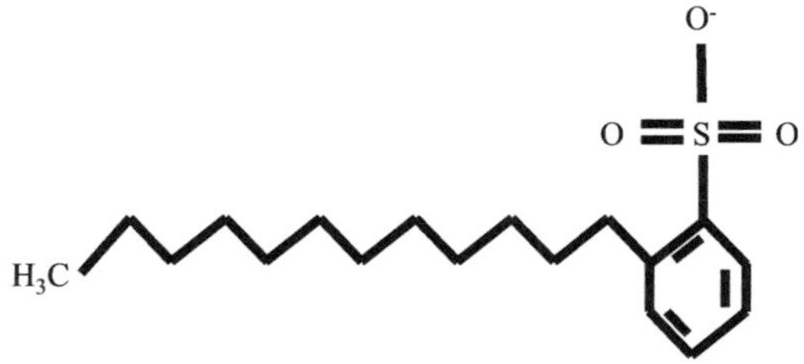
		SDS	
Cationic SAA	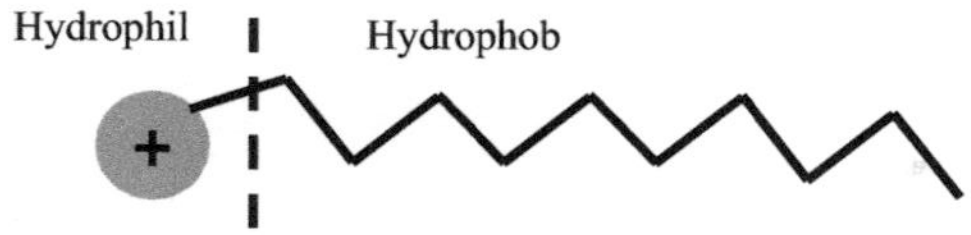	Quaternized SAA	

(*Continued*)

Table 12.3: Commonly used surface active agents (*cont.*)

Amphoteric SAA	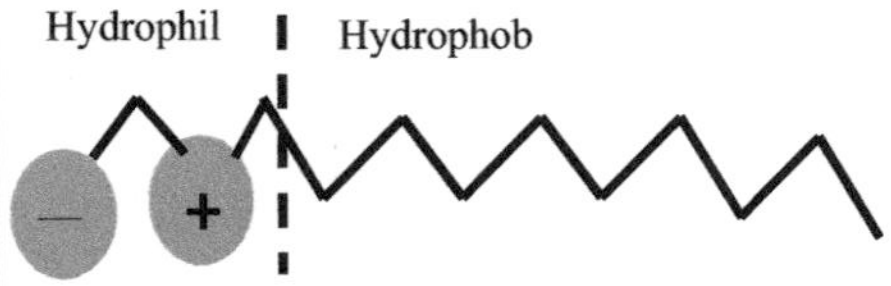	Lecithin Amino acid-based surfactants; betaine zwitterionic surfactant	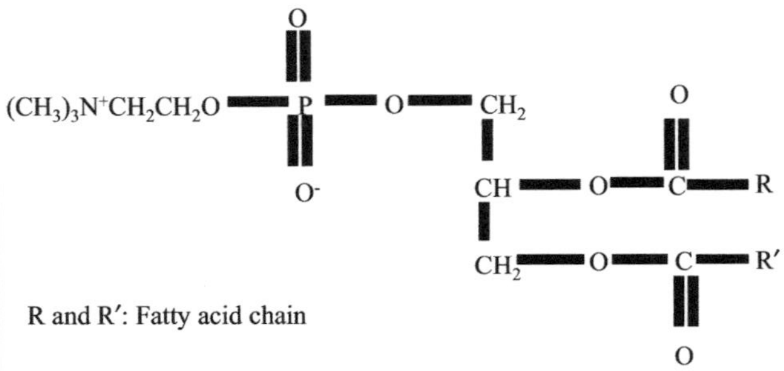R and R′: Fatty acid chain

SAA, Surface active agent; APEO, Alkylphenol ethoxylates; AEO, High fatty alcohol polyoxyethylene ether; AE, polyoxyethylene fatty acid; LAS, Linear alkylbenzene sulfonate; SDS, Sodium dodecyl sulfate.

Table 12.4: Average zinc(II) removal percentage, micellar binding constant (K_p), micelle loading (L_m), volume fraction (V_p/V_f), and distribution coefficient (D) values of zinc(II) ions in SDS micellar solution

Volume Fraction (V_p/V_f)	Initial Zinc(II) Concentration (C_0)(mM)	Retentate Zinc(II) Concentration $[C]_R$(mM)	Permeate Zinc(II) Concentration $[C]_P$(mM)	Distribution Coefficient (D)	Micelle Loading (L_m) (mM mM^{-1})	Micelle Binding Constant (K_p)(l mM^{-1})	Zinc(II) Removal Percentage
0.1003	0.42	0.1019	0.0924	1.1028	0.0452	0.4897	84.67
0.0823	0.84	0.1875	0.1844	1.0168	0.0148	0.0801	81.43
0.1772	1.68	0.3905	0.3614	1.0806	0.1386	0.3834	78.98
0.0775	2.10	0.4896	0.4483	1.0921	0.1967	0.4387	82.42

The loading of micelles (L_m) for various initial Zn^{2+} concentrations is shown in Table 12.4. The table also lists some other parameters, such as micellar binding constant (K_p), distribution coefficient (D). D is defined as the ratio of $[C]_R$ and $[C]_P$. The data in the Table 12.4 suggest the agreement between the removal efficiency and the D values. Higher the distribution of Zn^{2+} ions between the retentate and permeate, higher will be the removal percentage. The increment in the value of D indicates more surfactant molecules are present in the micellar phase, therefore more metal ions can be bound and removed [93]. No further explanation was given by the authors regarding the other relationships among the data in Table 12.4.

The results also indicated marginal increment and sharp reduction in Zn^{2+} removal with increasing SDS concentration below CMC and above CMC, respectively. Zn^{2+} removal percentage decreased from 84.67% to 82.42% when the initial Zn^{2+} concentration increases from 0.42 to 2.10 mM at constant SDS concentration. This indicated that the SDS micelles were able to trap Zn^{2+} ions to form more metal–micelle complexes with the increasing initial Zn^{2+} concentration.

Anionic surfactants usually have a relatively high CMC, and thus a large amount of surfactant is needed to achieve effective separation. Economic viability of MEUF process greatly relies on the ability to recover the surfactants from the retentate. Therefore, high concentration of surfactant will ultimately enhance the operating costs. In addition, unbound surfactant monomers in permeate may result secondary pollution as well. One approach to solve the problem is related with the concentration polarization (CP) [101].

Samper et al. studied the removal of metal ions including cadmium ion (Cd^{2+}), Cu^{2+}, nickel ion (Ni^{2+}), lead ion (Pb^{2+}), and Zn^{2+} of low concentration by MEUF [98]. Two anionic surfactants, SDS and linear alkylbenzene sulfonate (LAS), were employed. When the initial SDS concentration was below the CMC, metal retention higher than 90% was unexpectedly obtained, except for Ni^{2+}. In addition, complete removal of metal ions, except for Ni^{2+}, could be achieved when LAS concentration was below its CMC. LAS with concentration as low as 1 mM led to rejection higher than that obtained of SDS with 9 mM concentration.

Nonionic surfactants were also attempted to partially replace the anionic surfactant, which can reduce the process cost and [102]. As revealed by several researches [103,104], an increase in the mole fraction of nonionic surfactant (Triton series of X100, X114, and X405 containing different ethylene oxide groups) would cause a reduction in the CMC of a mixed surfactant system. The ethylene oxide groups were weakly charged due to the interaction between the ether oxide groups and the protons [104,105]. They may neutralize the negative charge carried by SDS. Therefore, there would be an overall reduction in the electrostatic repulsive force among the molecules of the mixed surfactants and the CMC tends to move toward a lower level of surfactant concentration.

Additionally, the presence of nonionic surfactants in the feed has shown a favorable effect on removal efficiency [98]. Huang et al. investigated removal of methylene blue (MB) and Cd^{2+} by MEUF using mixed surfactants of SDS and polyoxyethylene octyl phenyl ether (TritonX) [103]. The results regarding the rejection of MB and Cd^{2+} are revealed in Table 12.4. The addition of TritonX-100 within a specific range of concentration was in favor of MB and Cd^{2+} removal. Nevertheless, excessive amount of TritonX-100 of would reduce the MB retention [103]. The formation of more compact micelles with smaller apparent hydrodynamic diameter may be the cause [103]. Another example is Tanhaei et al.'s investigation, where nonionic surfactant (the Triton series of X100, X114, and X405 with glycol tertoctyl phenyl ether containing different ethylene oxide groups) was added to SDS solution. It was found that the Ni^{2+} rejection increased observably [104].

A very important parameter for mixed surfactant system is the interactions parameter (β) based on regular solution approach and theoretical methods. It can be evaluated by CMCs values [105]. Three cases for β interaction parameter are available: $\beta < 0$ indicates the synergistic interaction of mixed micelle formation, $\beta > 0$ is an indication of antagonism, and $\beta = 0$ represents the ideal combination of mixed micelles [102]. The value of β could be evaluated by the correlation presented by Clint and Rubingh [104].

The efficiency and characteristics of removing multipolluters simultaneously via MEUF are more practical problems for wastewater treatment. Huang et al. also reported the treatment of wastewater with coexistence of MB and Cd^{2+} [103]. It was found that the presence of Cd^{2+} could promote the rejection of MB by increasing the micelles' number in the retentate but without influencing the *unit binding capacity of micelles* and the relative affinity of MB for micelles. The presence of MB also increased Cd^{2+} rejection in a similar manner. Tanhaei et al. also found the mutual benefit effect [106].

12.3.1.2 Water soluble polymer enhanced ultrafiltration

As mentioned above, the disadvantages associated with MEUF are that most synthetic surfactants are toxic and soluble in water which can cause secondary pollution. Water soluble polymer enhanced ultrafiltration (PEUF) method is substitute for MEUF. In 1980, Michaels for the first time proposed the complex of water soluble polymer with heavy metal ions can be rejected by UF to concentrate targeted metals in aqueous solution [107]. Because the complexation of heavy metal ions and polymer is mostly reversible the metal ions may be released by the function of some chemical agents added to the solution, to fullfil the purpose of recovering metal ions. The complexation of heavy metal ions and polymer is usually selective. The possible examples are poly (sodium 4-styrene sulfonate) (PSS, Fig. 12.16(a)) with Cd^{2+} preferentially [108], poly(allylamine) (PALA, Fig. 12.16(b)) with Ni^{2+} [107], poly [acrylamide-co-1-(2-hydroxyethyl) aziridinyl] with Cu^{2+} [109], poly aziridine or PEI with mercury (Hg) [108], etc. Therefore, the polymer selected will determine the type of metal to be removed and corresponding removal efficiency. A more detailed theoretical description of the method is given in Spivakov et al.'s work [110]. The disadvantage associated with PEUF is high cost and poor effect in treating wastewater with multi heavy metals.

Cafiizares et al. studied the process for recovery heavy metals (ie, Cu^{2+}, Ni^{2+}, Pb^{2+}, and Cd^{2+}) from industrial wastewaters PEUF [111]. The polymers considered were polyethyleneimine (PEI), and poly(acrylic) acid (PAA). The effects due to membrane state, transmembrane pressure, feed stream velocity, polymer concentration, ionic strength, temperature, and pH on the separation performance (ie, permeate fluxes and rejection coefficients) were studied. It was revealed that the key variable is pH. As pH decreases, the mass transfer coefficient (k) rises, except for pH nearly 1, below which k decreases again.

Palencia et al. [112] studied PEUF using poly(vinyl sulfonic acid) (PVSA). The focus of the work was to investigate fouling during PEUF operation. ATR-FTIR was used for determining

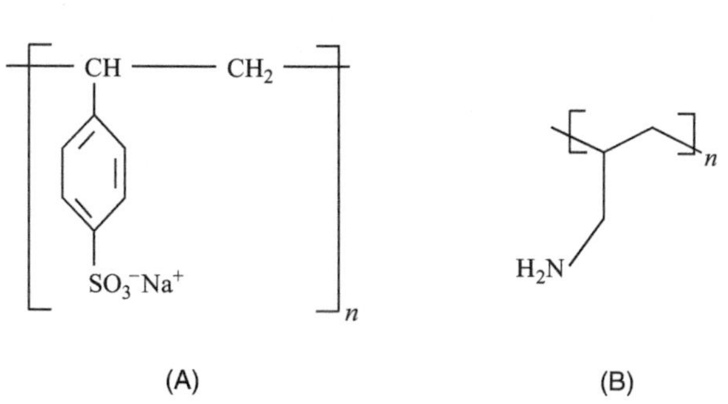

Figure 12.16: Unit structure for (A) Poly (sodium 4-styrene sulfonate) (PSS) and (B) poly (allylamine) (PALA).

the fouling layer thickness. The two bands selected investigation were 3340 cm^{-1} which was associated with polyvinylpyrrolidone PVP absorptions, and 1412 cm^{-1} which was associated with the arylether groups of PES.

It was found that the surface charge density and the sign of both the PES membrane and of polyelectrolyte PVSA, the ionic strength, and the pH of solution will influence the fouling. According to Hermian's model, the predominant mechanism of fouling was the cake filtration on the membrane surface.

Cojocaru [113] studied optimization of complexation conditions of removal of cobalt (Co) ions from aqueous solutions by PEUF. Experimental design was employed to obtain the optimum conditions which are summarized as follows: initial concentration of $[Co^{2+}]_0 = 65$ mg L^{-1}, polymer/metal ratio = 5.88, and pH = 6.84. The maximum rejection efficiency of 96.65% was obtained under these conditions.

Uzal et al. [114] studied the removal of Co^{2+} ions from water solutions using UF-complexation process. They used PVA polymer (M_w = 10,000 g/mol) and its sulfonated form as binding agents for Co ions. The experimental results indicated that sulfonated PVA has higher capability to remove Co^{2+} than its non-sulfonated precursor [114].

12.3.2 Membrane Surface Design

None of the neat membrane material can satisfy the entire requirement from in-service operation, such as chemical stability, membrane-forming property, thermal stability, acid and alkali resistance, microbial corrosion resistance, oxidation resistance and good mechanical strength. For UF membrane as aforementioned, the utmost problem encountered is the low flux and serious fouling, regarding the characteristics of its applications. Therefore, material design and membrane modification is generally adopted to obtain strengthened hydrophilicity, which is a property that can facilitate water transport and antifouling. Conventional approaches include use of hydrophilic additives [115] and modification of membrane surface and bulk structures [116]. Surface modification can be divided into chemical reaction, ultraviolet (UV) grafting, low temperature plasma grafting, irradiation grafting, electron beam modification etc. [117–122]. The latest development has been devoted to the application of amphiphilic polymers [123] and nanoparticles [124,125]. The latter two approaches may introduce better stability and durability in membrane hydrophilicity.

12.3.2.1 Polymer blending

PVDF arouses great attention in the past decade, due to its excellent chemical resistance, thermal stability, and mechanical properties. The industrial level PVDF membrane manufacturing has already been actualized. A wide range of amphiphilic copolymers have attracted much attention for preparing UF membranes prepared using PVDF and its derivatives. The typical examples of amphiphilic polymers are poly methyl methacrylate—*g*-(ethylene glycol) methyl ether methacrylate (P(MMA-r-PEGMA)), poly {4-[11-(acryloyloxy) undecoxy] benzoic acid}-b-polystyrene-b-poly {4-[11-acryloyloxy] undecoxy} benzoic acid) (PAUBA-b-PS-b-PAUBA)), poly (vinylidene fluoride-*co*-chlorotrifluoroethylene-g-poly (oxyethylene methacrylate) ([P(VDF-co-CTFE)]-g-POEM). Their hydrophobic backbone provides compatibility with PVDF, while the hydrophilic segment provides the hydrophilicity.

Liu et al. synthesized amphiphilic copolymer P(MMA-r-PEGMA) using a radical polymerization [126]. PVDF/P(MMA-r-PEGMA) hollow fiber membranes were then fabricated by the phase inversion. It was revealed that the usage of P(MMA-r-PEGMA) results in a delayed demixing and depressed precipitation as compared with neat PVDF membrane. The copolymer tended to aggregate at the membrane/coagulant (ie, H_2O) interface to form a hydrogel layer. FTIR and XPS characterizations confirms the above enrichment which probably due to the hydration of the copolymer with H_2O molecules. The contact angle of the inner skin of the blend membrane changed from 82 to 0 degree in 15 s. An excellent protein fouling resistance and membrane cleaning efficiency were obtained with the blend membrane. The pure water flux reached 693.0 L m^{-2} h and carbon ink particles above 128 nm were completely rejected by these membranes.

12.3.2.2 Nanoparticles

In Liang et al.'s work [126], a novel UF membrane was fabricated by irreversibly tethering superhydrophilic silica (SiO_2) NPs to the PVDF membrane surface. Fig. 12.17 illustrated the schematized protocol of PVDF membrane functionalization, SiO_2 groups, were used to modify their surface.

The functionalized membranes acquired surface energy almost as two times as pristine ones; hence, the wettability of the membrane was improved. The membranes with tethered silica NPs also exhibited strong antifouling property as demonstrated by fouling filtration tests using bovine serum albumin (BSA) as a model organic foulant.

In Zhu et al.'s work [127] poly(2-hydroxyethylmethacrylate) (PHEMA) grafted SiO_2 nanoparticles (PHEMA@SiO_2 NPs) was used to prepared mixed matrix UF membranes with PES) as base material. The PHEMA@SiO_2 NPs were grafted with PHEMA brushes by surface-initiated reversible addition fragmentation chain transfer (RAFT) polymerization. During the hybrid membrane formation by phase inversion, the hydrophilic NPs tended to migrate toward the membrane/water interface. The water permeability, solute rejection, and antifouling ability of the hybrid membranes were enhanced in comparison with neat PES membrane.

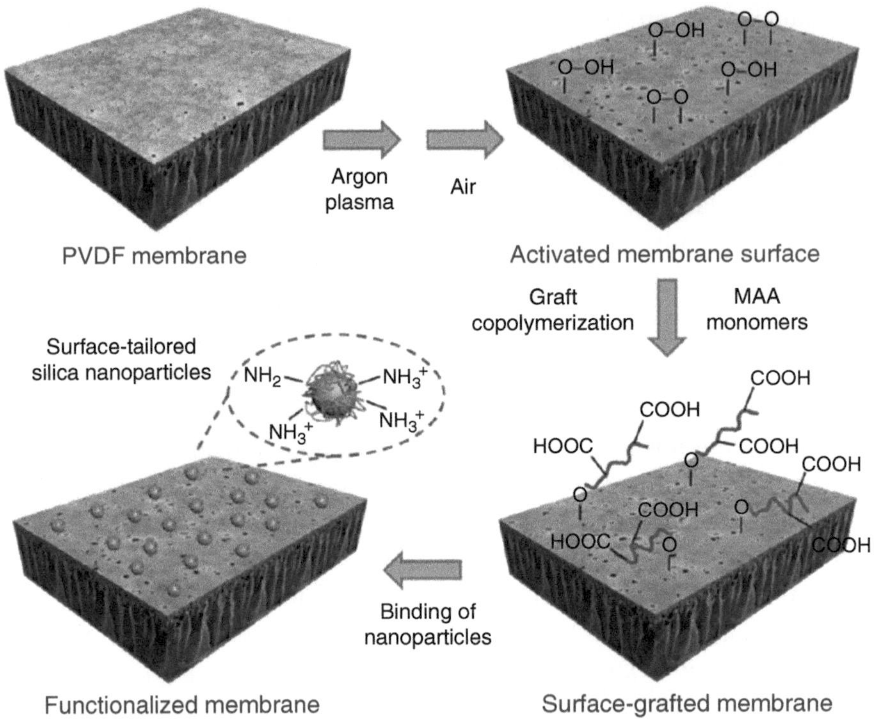

Figure 12.17: Schematic protocol of PVDF membrane functionalization, including in series a plasma treatment, graft copolymerization, and nanoparticle (NP) binding.

Glossary and Symbols

β	Interaction parameter
AEM	Anion exchange membrane
Al	Aluminum
AMOC	Age-momentum correlation
BCMB	1,4-Bis (chloromethoxy) butane
BPM	Bipolar membranes
$[C]_P$	Permeate concentration
$[C]_R$	Concentrations in retentate
CA	Cellulose acetate
CD	Cyclodextrin
Cd	Cadmium
CEM	Cation exchange membranes
CMC	Critical micelle concentration
CMME	Chloromethyl methyl ether
-COOH	Carboxylic acids
Co	Cobalt
CP	Concentration polarization
Cu	Copper
D	Distribution coefficient

DD	Diffusion dialysis
DEA	DiethylaminoDMAPS 3-dimethyl-(methacryloyloxyethyl) ammonium propane sulfonate
DVB	Divinyl benzene
ED	Electrodialysis
EMIPO	2,2′-Bis-(2-ethyl-4-methyl-imidazole-1-ylmethyl)-biphenyl-4,4′-diol
$FeCl_2$	Ferrous chloride
GMA	Glycidyl methacrylate
HCEC	Cyanoethyl cellulose
Hg	Mercury
H_2SO_4	Sulfuric acid
IEC	ion Exchange capacity
IEM	ion-Exchange membrane
k	Mass transfer coefficients
K_p	Micellar binding constant
KOH	Potassium hydroxide
L_m	Loading of micelles
LAS	Linear alkylbenzene sulfonate
L-L separation	Liquid-liquid separation
MB	Methylene blue
MEUF	Micellar-enchanced ultrafiltration
mm-BTEC	3,3′,5,5′-Biphenyl tetraacyl chloride
MWNTs	Multiwall nanotubes
NaCl	Sodium chloride
Na_2CO_3	Sodium carbonate
NaOH	Sodium hydroxide
Na_2SO_4	Sodium sulfate
Na_2WO_4	Sodium tungstate
NBS	*N*-bromosuccinimid
NF	Nanofiltration
$(NH_4)_2SO_4$	Ammonium sulfate
Ni	Nickel
NIPS	Nonsolvent induced phase separation
$-NO_2$	Nitroso group
NP	Nanoparticle
NTSC	1,3,6-Naphthalenesuifonate chloride
PA	Polyamide
PAA	Poly(acrylic) acid
PAES	Poly(arylene ether sulfone)
PALA	Poly(allylamine)
PALS	Positron annihilation
PAN	Polyacrylonitrile
PAUBA-b-PS-b-PAUBA	Poly{4-[11-(acryloyloxy)undecoxy] benzoic acid}-b-polystyrene-b-poly {4-[11-acryloyloxy] undecoxy} benzoic acid)
Pb	Lead
PBI	Polybenzimidazoles
PCCs	Proton conductive channels
PE	Polyethylene
PEI	Polyethylenimine; poly aziridine; polyethyleneimine
PEKK	Poly(ether ketone ketone)
PEMs	Proton exchange membranes

PES	Polyethersulfone
PESK	Polyethersulfoneketone
PEUF	Polymer complexation or enhanced ultrafiltration
PFCA	Perfluoroalkyl carboxylic acids
PFSA	Perfluoro sulfonic acid
PIP	Piperazine
PIP-EDA	Piperazine-ethylenediamine
P(MMA-r-PEGMA)	Polymethyl methacrylate-*g*-(ethylene glycol) methyl ether methacrylate acid}-b-polystyrene-b-poly {4-[11-acryloyloxy] undecoxy} benzoic acid)
PPS	Piperazine propane sulfonate
PS	Polystyrene
PSA or PSF-*co*-PA	Polysulfonamide or polysulfone-polyamide
PSF	Polysulfone
PSS	Poly(sodium 4-styrene sulfonate)
PSS-b-PMB	Poly(styrenesulfonate-b-methylbutylene)
PVA	Polyvinyl alcohol
PVDF	Polyvinylidene fluoride
[P(VDF-*co*-CTFE)]-g-POEM	Poly(vinylidene fluoride-*co*-chlorotrifluoroethylene-g-poly (oxyethylene methacrylate)
PVP	Polyvinylpyrrolidone
PVSA	Poly(vinyl sulfonic acid)
QPPO-SiO_2	Quaternized poly(2,6-dimethyl-1,4-phenylene oxide)/Silicon dioxide hybrid material
RAFT	Reversible addition fragmentation chain transfer
RO	Reverse osmosis
SDS	Sodium dodecyl sulfate
SiO_2	Silica, silicon dioxide
$-SO_3H$	Sulfonic acids
SPPO-H	Poly(2, 6-dimethyl-1, 4-phenylene oxide)
S-S separation	Solid-liquid separation
T acid	1-Naphthylamine-3,6,8-trisulfonic acid
TFC-IP	Thin film composite membrane by interfacial polymerization
TFN	Thin film nanocomposite
TIBA	Triisobutylaluminum
TIPS	Thermally induced phase separation
TMC	Trimesoyl chloride
TMOS	Tetramethoxy orthosilane
TMS-Cl	Trimethylsilyl chloride
TMSPTA	*N*-trimethoxysilylpropyl-*N,N,N*-trimethylammonium chloride
TMSSC	Trimethylsilylsulfonyl chloride
TritonX	Polyoxyethylene octyl phenyl ether
UF	Ultrafiltration
UV	Ultra violet
W	Tungsten
Zn	Zinc

References

[1] D.-X. Wang, M. Su, Z.-Y. Yu, X.-L. Wang, M. Ando, T. Shintani, Separation performance of a nanofiltration membrane influenced by species and concentration of ions, Desalination 175 (2005) 219–225.

[2] W.M. Deen, Hindered transport of large molecules in liquid-filled pores, AIChE J. 33 (1987) 1409–1425.

[3] F.G. Donnan, Theory of membrane equilibria and membrane potentials in the presence of non-dialysing electrolytes: a contribution to physical-chemical physiology, J. Membr. Sci. 100 (1995) 45–55.
[4] A.E. Yaroshchuk, Rejection mechanisms of NF membranes, Membr. Technol. 100 (1998) 9–12.
[5] D.L. Oatley, L. Llenas, R. Pérez, Review of the dielectric properties of nanofiltration membranes and verification of the single oriented layer approximation, Adv. Colloid Interface Sci. 173 (2012) 1–11.
[6] H.-C. Yang, M.-B. Wu, Y.-J. Li, Y.-F. Chen, L.-S. Wan, Z.-K. Xu, Effects of polyethyleneimine molecular weight and proportion on themembrane hydrophilization by codepositing with dopamine, J. Appl. Polym. Sci. (2016) DOI:10.1002/app.43792.
[7] H. Wu, B. Tang, P. Wu, Preparation and characterization of anti-fouling β-cyclodextrin/polyester thin film nanofiltration composite membrane, J. Membr. Sci. 428 (2013) 301–308.
[8] L.-G. Wu, Y. Zhou, L. Zhang, H.-L. Chen, K.-C. Gao, The research progress in functional materials of reverse osmosis composite membrane, Chem. Prog. 20 (2008) 1216–1221.
[9] B.-B. Tang, T.-W. Xu, P.-Y. Wu, The composite membrane prepared by interfacial polymerization, Chem. Prog. 19 (2007) 1428–1435.
[10] S. Qiu, L. Wu, L. Zhang, et al. Preparation of reverse osmosis composite membrane with high flux by interfacial polymerization of MPD and TMC, J. Appl. Polym. Sci. 112 (2009) 2066–2072.
[11] G.-R. Xu, J.-N. Wang, C.-J. Li, Strategies for improving the performance of the polyamide thin film composite (PA-TFC) reverse osmosis (RO) membranes: surface modifications and nanoparticles incorporations, Desalination 328 (2013) 83–100.
[12] J.-X. Qin, S.-S. Lin, S.-Q. Song, L. Zhang, H.-L. Chen, 4-Dimethylaminopyridine promoted interfacial polymerization between hyperbranched polyesteramide and trimesoyl chloride for preparing ultralow-pressure reverse osmosis composite membrane, ACS Appl. Mater. Interfaces 5 (2013) 6649–6656.
[13] H. Dong, L. Zhang, H.-L. Chen, C.-J. Gao, Synthesis and application of mixed-matrix membranes for water treatment, Chem. Eng. Prog. 12 (2014) 2007–2018.
[14] H. Huang, X.-Y. Qu, X.-S. Ji, X. Gao, L. Zhang, H.-L. Chen, L. Hou, Acid and multivalent ion resistance of thin film nanocomposite RO membranes loaded with silicalite-1 nanozeolites, J. Mater. Chem. A (1) (2013) 11343–11349.
[15] H. Zhao, Z. Zhou, H. Dong, L. Zhang, H. Chen, L. Hou, A facile method to align carbon nanotubes on polymeric membrane substrate, Sci. Rep. 3 (2013).
[16] E.S. Kim, B. Deng, Fabrication of polyamide thin-film nano-composite (PA-TFN) membrane with hydrophilized ordered mesoporous carbon (H-OMC) for water purifications, J. Membr. Sci. 375 (2011) 46–54.
[17] S.Y. Lee, H.J. Kim, R. Patel, R. Patel, S.J. Im, J.H. Kim, B.R. Min, Silver nanoparticles immobilized on thin film composite polyamide membrane: characterization, nanofiltration, antifouling properties, Polym. Adv. Technol. 18 (2007) 562–568.
[18] Y. Zhao, C. Qiu, X. Li, A. Vararattanavech, W. Shen, J. Torres, C. Hélix-Nielsene, R. Wang, X. Hu, A.G. Fane, C.Y. Tang, Synthesis of robust and high-performance aquaporin-based biomimetic membranes by interfacial polymerization-membrane preparation and RO performance characterization, J. Membr. Sci. 423 (2012) 422–428.
[19] Ratto, T.V. Holt, J.K. Szmodis, A.W., 2011. Membranes with embedded nanotubes for selective permeability. US Patent 7,993,524, 2011-8-9.
[20] B. Corry, Designing carbon nanotube membranes for efficient water desalination, J. Phys. Chem. B 112 (2008) 1427–1434.
[21] C. Song, B. Corry, Intrinsic ion selectivity of narrow hydrophobic pores, J. Phys. Chem. B 113 (2009) 7642–7649.
[22] H. Wu, B. Tang, P. Wu, Preparation and characterization of anti-fouling β-cyclodextrin/polyester thin film nanofiltration composite membrane, J. Membr. Sci. 428 (2013) 301–308.
[23] Q.-F. An, Y.-L. Ji, W.-S. Hung, K.R. Lee, C.-J. Gao, AMOC positron annihilation study of zwitterionic nanofiltration membranes: correlation between fine structure and ultrahigh permeability, Macromolecules 46 (2013) 2228–2234.
[24] Q.-F. An, W.-D. Sun, Q. Zhao, Y.-L. Jia, C.-J. Gao, Study on a novel nanofiltration membrane prepared by interfacial polymerization with zwitterionic amine monomers, J. Membr. Sci. 431 (2013) 171–179.
[25] M. Regel-Rosocka, A review on methods of regeneration of spent pickling solutions from steel processing, J. Hazard. Mater. 177 (2010) 57–69.

[26] M. Dalwani, N.E. Benes, G. Bargeman, D. Stamatialisa, M. Wessling, A method for characterizing membranes during nanofiltration at extreme Ph, J. Membr. Sci. 363 (2010) 188–194.
[27] S. Platt, M. Nyström, A. Bottino, G. Capannelli, Stability of NF membranes under extreme acidic conditions, J. Membr. Sci. 239 (2004) 91–103.
[28] Ma, M., 2011, Fabrication of novel acid-resistant composite NF membrane and performance investigation. Master Thesis, Zhejiang University of Science and Technology.
[29] J. Qin, X. Dai, Y. Zhou, L. Zhang, H. Chen, C. Gao, Desalting and recovering naphthalenesulfonic acid from wastewater with concentrated bivalent salt by nanofiltration process, J. Membr. Sci. 468 (2014) 242–249.
[30] L. Li, S. Zhang, X. Zhang, Preparation and characterization of poly (piperazineamide) composite nanofiltration membrane by interfacial polymerization of 3, 3', 5, 5'-biphenyl tetraacyl chloride and piperazine, J. Membr. Sci. 335 (2009) 133–139.
[31] Xu, P., Bi, F.-Y., Campbell, J., Zhao, L., 2006. Acid and alkali thermostable HYDRACoRe70pHT nanofiltration membranes for sugar industrial waste alkali recovery. The 2nd National Symposium on the Application of Membrane Separation Technology in Food Industry p. 5.
[32] R. Schlesinger, G. Götzinger, H. Sixta, A. Friedl, M. Harasek, Evaluation of alkali resistant nanofiltration membranes for the separation of hemicellulose from concentrated alkaline process liquors, Desalination 192 (2006) 303–314.
[33] K.P. Lee, J.M. Zheng, G. Bargeman, A.J.B. Kemperman, N.E. Benes, pH stable thin film composite polyamine nanofiltration membranes by interfacial polymerisation, J. Membr. Sci. 478 (2015) 75–84.
[34] H. Wu, B.-B. Tang, P.-Y. Wu, Optimization, characterization and nanofiltration properties test of MWNTs/polyester thin film nanocomposite membrane, J. Membr. Sci. 428 (2013) 425–433.
[35] Z.-L. Zhang, S. Wang, Z.-J. Guo, J. Wang, Development of the oxidation resistance of aromatic polyamide thin-film composite nanofiltration membranes, Chem. Ind. Eng. Prog. 30 (2011) 2240–2245.
[36] Y.N. Kwon, J.O. Leckie, Hypochlorite degradation of crosslinked polyamide membranes I. Changes in chemical/morphological properties, J. Membr. Sci. 283 (2006) 21–26.
[37] Y.N. Kwon, J.O. Leckie, Hypochlorite degradation of crosslinked polyamide membranes II. Changes in hydrogen bonding behavior and performance, J. Membr. Sci. 282 (2006) 456–464.
[38] G.-D. Kang, C.-J. Gao, W.-D. Chen, Study on hypochlorite degradation of aromatic polyamide reverse osmosis membrane, J. Membr. Sci. 300 (2007) 165–171.
[39] H. Strathmann, Electrodialysis and related processes, in: R.D. Noble, S.A. Stern (Eds.), Membrane Separation Technology—Principles and Applications, Elsevier Science B.V, Amsterdam, 1995, pp. 214–278.
[40] T.W. Xu, Ion exchange membranes: state of their development and perspective, J. Membr. Sci. 263 (2005) 1–29.
[41] T.W. Xu, The research progress of diffusion dialysis method for recycling of industrial acidic liquid waste, Water Treat. Technol. 30 (2) (2004) 63–66.
[42] D.C. Ge, The application of homogeneous ion exchange membrane in several industries in China, Membr. Sci. Technol. 23 (4) (2003) 202–208.
[43] G.P. Zhou, J.N. Yang, S.P. Luo, Recovery of high concentration of sulfuric acid waste water containing sodium sulfate by anion exchange membrane dialysis, Membr. Sci. Technol. 22 (6) (2002) 24–27.
[44] G.E. Molau, Heterogeneous ion-exchange membranes, J. Membr. Sci. 8 (1981) 309–330.
[45] N. Li, M.D. Guiver, Ion transport by nanochannels in ion-containing aromatic copolymers, Macromolecules 47 (2014) 2175–2198.
[46] Z. Yang, J. Ran, B. Wu, L. Wu, T. Xu, Stability challenge in anion exchange membrane for fuel cells, Curr. Opin. Chem. Eng. 12 (2016) 22–30.
[47] C. Vogel, M-H. Jochen, Preparation of ion-exchange materials and membranes, Desalination 342 (2014) 156–174.
[48] S.T. Hemp, M. Zhan g, M. Tamami, T.E. Long, Phosphonium ionenes from well-defined step-growth polymerization: thermal and melt rheological properties, Polym. Chem. 4 (2013) 3582–3590.
[49] L.D. Chen, D. Mandal, J.A. Gladysz, P. Bühlmann, Chemical stability and application of a fluorophilic tetraalkylphosphonium salt influorous membrane anion selective electrodes, New J. Chem. 34 (2010) 1867–1874.

[50] M.Y. Kariduraganavar, R.K. Nagarale, A.A. Kittur, S.S. Kulkarni, Ion-exchange membranes preparative methods for electrodialysis and fuel cell applications, Desalination 197 (2006) 225–246.
[51] K.J.T. Noonan, K.M. Hugar, H.A. Kostalik, E.B. Lobkovsky, H.D. Abruna, G.W. Coates, J. Am. Chem. Soc. 134 (2012) 18161.
[52] G. Merle, M. Wessling, K. Nijmeijer, Anion exchange membranes for alkaline fuel cells: a review, J. Membr. Sci. 377 (2011) 1–35.
[53] M. Yamabe, K. Akiyama, Y. Akatsuka, M. Kato, Novel phosphonated perfluoro-carbon polymers, Eur. Polym. J. 36 (2000) 1035–1041.
[54] K. Kreuer, S.J. Paddison, E. Spohr, M. Schuster, Transport in proton conductors for fuel-cell applications: simulations, elementary reactions, and phenomenology, Chem. Rev. 104 (2004) 4637–4678.
[55] C. Vogel, H. Komber, A. Quetschke, W. Butwilowski, A. Pötschke, K. Schlenstedt, J. Meier-Haack, Side-chain sulfonated random and multiblock poly(ether sulfone)s for PEM applications, React. Funct. Polym. 71 (2011) 824–842.
[56] P.C. Deb, L.D. Rajput, V.R. Hande, S. Sasane, A. Kumar, Modification of sulfonated poly(etheretherketone) with phenolicresin, Polym. Adv. Technol. 18 (2007) 419–426.
[57] J. Meier-Haack, C. Vogel, W. Butwilowski, D. Lehmann, Sulfonated poly(ether sul-fone)s for fuel cells by solvent-free polymerization, Pure Appl. Chem. 79 (2007) 2083–2093.
[58] S.L. Zhang, Preparation of organic-positively charged inorganic hybrid material and hybrid membranes, Master Thesis, University of Science and Technology of China, 2005.
[59] A.H.N. Rao, R.L. Thankamony, H. Kim, S. Nam, T. Kim, Imidazolium-functionalized poly(arylene ether sulfone) block copolymer as an anion exchange membrane for alkaline fuel cell, Polymer 54 (2013) 111–119.
[60] W. Lu, Z.-G. Shao, G. Zhang, J. Li, Y. Zhao, B. Yi, Preparation of anion exchange membranes by an efficient chloromethylation method and homogeneous quaternization/crosslinking strategy, Solid State Ion. 245–246 (2013) 8–18.
[61] K. Sunaga, M. Kim, A. Kyoichi Saito, K. Sugita, T. Sugo, Characteristics of porous anion exchange membranes prepared by co-grafting of glycidyl methacrylate with divinylbenzene, Chem. Mater. 11 (1999) 1986–1989.
[62] T. Rager, M. Schuster, H. Steiniger, K. Kreuer, Poly(1,3-phenylene-5-phosphonic acid), a fully aromatic polyelectrolyte with high ion exchange capacity, Adv. Mater. 19 (2007) 3317–3321.
[63] Y. Yang, J. Wang, J. Zheng, S. Li, S. Zhang, A stable anion exchange membrane based on imidazolium salt for alkaline fuel cell, J. Membr. Sci. 467 (2014) 48–55.
[64] M. Rikukawa, K. Sanui, Proton-conducting polymer electrolyte membranes based on hydrocarbon polymers, Prog. Polym. Sci. 25 (2000) 1463–1502.
[65] F. Wang, M. Hickner, Y.S. Kim, T.A. Zawodzinski, J.I.E. McGrath, Direct polymerization of sulfonated poly(arylene ether sulfone) random (statistical) copolymers: candidates for new proton exchange membranes, J. Membr. Sci. 197 (2002) 231–242.
[66] K. Nakabayashi, K. Matsumoto, M. Ueda, Synthesis and properties of sulfonated multiblock copoly(ether sulfone)s by a chain extender, J. Polym. Sci. Part A Polym. Chem. 46 (2008) 3947–3957.
[67] K. Nakabayashi, K. Matsumoto, T. Higashihara, M. Ueda, Influence of adjusted hydrophilic–hydrophobic lengths in sulfonated multiblock copoly(ether sulfone) membranes for fuel cell application, J. Polym. Sci. A 46 (2008) 7332–7341.
[68] K. Matsumoto, T. Higashihara, M. Ueda, Locally and densely sulfonated poly(ether sulfone)s as proton exchange membrane, Macromolecules 42 (2009) 1161–1166.
[69] K. Matsumoto, T. Higashihara, M. Ueda, Locally sulfonated poly(ether sulfone)s with highly sulfonated units as proton exchange membranes, J. Polym. Sci. A 47 (2009) 3444–3453.
[70] S. Tian, Y. Meng, A.S. Hay, Membranes from poly(aryl ether)-based ionomers containing multiblock segments of randomly distributed nanoclusters of 18 sulfonic acid groups, J. Polym. Sci. A 47 (2009) 4762–4773.
[71] S. Tian, Y. Meng, A.S. Hay, Membranes from poly(aryl ether)-based ionomers containing randomly distributed nanoclusters of 6 or 12 sulfonic acid groups, Macromolecules 42 (2009) 1153–1160.

[72] W. Zhang, V. Gogel, A.K. Friedrich, J. Kerres, Novel covalently cross-linked poly(ether ether ketone) ionomer membranes, J. Power Sources 155 (2006) 3–12.
[73] S. Feng, Y. Shang, S. Wang, X. Xie, Y. Wang, Y. Wang, J. Xu, Novel method for the preparation of ionically crosslinked sulfonated poly(arylene ether sulfone)/polybenzimidazole composite membranes via in situ polymerization, J. Membr. Sci. 346 (2010) 105–112.
[74] J. Wei, C. Stone, A.E. Steck, Trifluorostyrene and substituted trifluorostyrene copolymeric compositions and ion-exchange membranes formed therefrom, US Patent 5,422,411, 1995.
[75] L.H. Lee, W.C. Chen, High-refractive-index thin films prepared from trialkoxysilane-capped poly(methyl methacrylate)–titania materials, Chem. Mater. 13 (3) (2001) 1137–1142.
[76] C.M. Wu, T.W. Xu, W.H. Yang, A new inorganic–organic negatively charged membrane: membrane preparation and characterizations, J. Membr. Sci. 224 (1∼2) (2003) 117–125.
[77] C.M. Wu, T.W. Xu, W.H. Yang, Fundamental studies of a new hybrid (inorganic–organic) positively charged membrane: membrane preparation and characterizations, J. Membr. Sci. 216 (1–2) (2003) 269–278.
[78] C.M. Wu, T.W. Xu, W.H. Yang, The progress in preparation of inorganic and organic composite membrane technology, J. Inorg. Mater. 17 (4) (2002) 641–648.
[79] A. Roy, X. Yu, S. Dunn, J.E. McGrath, Influence of microstructure and chemical composition on proton exchange membrane properties of sulfonated-fluorinated, hydrophilic-hydrophobic multiblock copolymers, J. Membr. Sci. 327 (2009) 118–124.
[80] L. Wu, C.H. Huang, J.J. Woo, D. Wu, S.H. Yun, S.J. Seo, T.W. Xu, S.H. Moon, Modifying a proton conductive membrane by embedding a barrier, J. Phys. Chem. B 114 (2010) 13121–13127.
[81] K.D. Kreuer, On the development of proton conducting polymer membranes for hydrogen and methanol fuel cells, J. Membr. Sci. 185 (2001) 29–39.
[82] S. Foerster, T. Plantemberg, Fromself-organizing polymers biomaterials, Angew. Chem. Int. 41 (2002) 688–714.
[83] J. Li, K.G. Wilmsmeyer, L.A. Madsen, Hydrophilic channel alignment modes in perfluorosulfonate ionomers: implications for proton transport, Macromolecules 41 (2008) 4555–4557.
[84] M.J. Park, S. Kim, A.M. Minor, A. Hexemer, N.P. Balsara, Control of domain orientation in block copolymer electrolyte membranes at the interface with humid air, Adv. Mater. 21 (2009) 203–213.
[85] J. Pan, Y. He, L. Wu, C. Jiang, B. Wu, A.N. Mondal, C. Cheng, T. Xu, Anion exchange membranes from hot-pressed electrospun QPPO–SiO_2 hybrid nanofibers for acid recovery, J. Membr. Sci. 480 (2015) 115–121.
[86] J. Pan, L. Ge, X. Lin, L. Wu, B. Wu, T. Xu, Cation exchange membranes from hotpressed electrospun sulfonated poly(phenylene oxide) nanofibers for alkali recovery, J. Membr. Sci. 470 (2014) 479–485.
[87] T. Ishigami, K. Nakatsuka, Y. Ohmukai, E. Kamio, T. Maruyama, H. Matsuyama, Solidification characteristics of polymer solution during polyvinylidene fluoride membrane preparation by nonsolvent-induced phase separation, J. Membr. Sci. 438 (2013) 77–82.
[88] M.H. Gu, J. Zhang, X.L. Wang, H.J. Tao, L. Ge, Formation of poly(vinylidene fluoride) (PVDF) membranes via thermally induced phase separation, Desalination 192 (2006) 160–167.
[89] D.R. Lloyd, S. Kim, K.E. Kinzer, Microporous membrane formation via thermally induced phase separation. II. Liquid–liquid phase separation, J. Membr. Sci. 64 (1991) 1–11.
[90] Z.G. Zhao, J.F. Zheng, M.J. Wang, H.Y. Zhang, C.C. Han, High performance ultrafiltration membrane based on modified chitosan coating and electrospun nanofibrous PVDF scaffolds, J. Membr. Sci. 394-395 (2012) 209–217.
[91] Y.X. Yang, W.B. Wu, M. Zhang, Progress and development trend in studying on the materials of preparing ultrafiltration membrane, J. Cereals Oils 5 (2005) 15–18.
[92] S.R. Chae, H. Yamamura, K. Ikeda, Y. Watanabe, Comparison of fouling characteristics of two different poly-vinylidene fluoride microfiltration membranes in a pilot-scale drinking water treatment system using precoagulation/sedimentation, sand filtration, and chlorination, Water Res. 42 (2008) 2029–2042.
[93] R. Du, S. Peldszus, P.M. Huck, X. Feng, Modification of poly(vinylidene fluoride) ultrafiltration membranes with poly(vinyl alcohol) for fouling control in drinking water treatment, Water Res. 43 (2009) 4559–4568.
[94] M.R. Moghareh Abed, S.C. Kumbharkar, A.M. Groth, K. Li, Economical production of PVDF-g-POEM for use as a blend in preparation of PVDF based hydrophilic hollow fibre membranes, Sep. Purif. Technol. 106 (2013) 47–55.

[95] C.M. Zhong, X.H. Fang, Z.L. Xu, Membrane technology and its application in heavy metal wastewater treatment, Environ. Sci. Technol. 31 (2008) 44–48.
[96] Z.L. Xu, H.M. Xu, X.D. Zhai, Treatment of waste streams containing Pb^{2+} and Cd^{2+} by micellar enhanced ultrafiltration, Membr. Sci. Technol. 22 (2002) 15–20.
[97] Z. Zhang, G.-M. Zeng, J.-H. Huang, Y.-Y. Fang, K. Xu, Y.-H. Qu, C.-P. Yang, J.-B. Li, Removal of zinc ions from aqueous solution using micellar enhanced ultrafiltration at low surfactant concentrations, Water SA 33 (2007) 129–136.
[98] E. Samper, M. Rodríguez, M.A. De la Rubia, Removal of metal ions at low concentration by micellar-enhanced ultrafiltration (MEUF) using sodium dodecyl sulfate (SDS) and linear alkylbenzene sulfonate (LAS), Sep. Purif. Technol. 65 (2009) 337–342.
[99] S. Ahmadi, I.K. Tseng, B. Bactchelor, Micellar enchanced ultrafiltration of heavy metal using lecithin, Sep. Sci. Technol. 29 (1994) 2435–2450.
[100] S.H. Lee, S. Shrestha, Application of micellar enhanced ultrafiltration (MEUF) process for zinc (II) removal in synthetic wastewater: kinetics and two-parameter isotherm models, Int. Biodeter. Biodegr. 95 (2014) 241–250.
[101] Z. Zhang, G.M. Zeng, J.H. Huang, Y.Y. Fang, K. Xu, K.Y.H. Qu, C.P. Yang, J. Li, Removal of zinc ions from aqueous solution using micellar enhanced ultrafiltration at low surfactant concentrations, Water SA 33 (2007) 129–136.
[102] T. Joshi, J. Mata, P. Bahadur, Micellization and interaction of anionic and nonionic mixed surfactant systems in water, Colloid Surf. A 260 (2005) 209–215.
[103] J.H. Huang, L. Peng, G.M. Zeng, X. Li, Y. Zhao, L.X. Liu, F. Li, Q. Chai, Evaluation of micellar enhanced ultrafiltration for removing methylene blue and cadmium ion simultaneously with mixed surfactants, Sep. Purif. Technol. 125 (2014) 83–89.
[104] B. Tanhaei, M. Pourafshari Chenar, N. Saghatoleslami, M. Hesampour, M. Kallioinen, M. Sillanpaa, Mika Manttari, Removal of nickel ions from aqueous solution by micellar-enhanced ultrafiltration, using mixed anionic–non-ionic surfactants, Sep. Purif. Technol. 138 (2014) 169–176.
[105] K. Xu, G.M. Zeng, J.H. Huang, J.Y. Wu, Y.Y. Fang, G. Huang, J. Li, B. Xi, Removal of Cd^{2+} from synthetic wastewater using micellar enhanced ultrafiltration with hollow fiber membrane, Colloid Surf. A 294 (2007) 140–146.
[106] B. Tanhaei, M.P. Chenar, N. Saghatoleslami, M. Hesampour, T. Laakso, M. Kallioinen, M. Sillanpää, M. Mänttäri, Simultaneous removal of aniline and nickel from water by micellar-enhanced ultrafiltration with different molecular weight cut-off membranes, Sep. Purif. Technol. 124 (2014) 26–35.
[107] H. Strathmann, Selective removal of heavy metal ions from aqueous solutions by ultrafiltration of macromolecular complexes, Sep. Sci. Technol. 15 (1980) 1135–1152.
[108] I. Moreno-Villoslada, Metal ion enrichment of water soluble chelating polymer studied by ultrafiltration, J. Membr. Sci. Technol. 208 (2002) 69.
[109] I. Moreno-Villoslada, B.L. Rivas, Retention of metal ions in ultrafiltration of mixture of divalent metal ions and water soluble polymers at constant ionic strength based on Freundlich and Langmuir isotherms, Membr. Sci. Technol. 215 (2003) 195.
[110] B. Spivakov, V. Shkinev, Theory and mathematical models of ion retention by water soluble-polymers during membrane filtration, Macromol. Theory Simul. 5 (1996) 357–365.
[111] P. Cafiizares, A. P6rez, R. Camarillo, Recovery of heavy metals by means of ultrafiltration with water-soluble polymers: calculation of design parameters, Desalination 144 (2002) 279–285.
[112] M. Palencia, B.L. Rivas, E. Pereira, Metal ion recovery by polymer-enhanced ultrafiltration using poly(vinyl sulfonic acid): fouling description and membrane–metal ion interaction, J. Membr. Sci. 345 (2009) 191–200.
[113] C. Cojocaru, G. Zakrzewska-Trznadel, A. Jaworska, Removal of cobalt ions from aqueous solutions by polymer assisted ultrafiltration using experimental design approach. Part 1: optimization of complexation conditions, J. Hazard. Mater. 169 (2009) 599–609.
[114] N. Uzal, A. Jaworska, A. Miśkiewicz, G. Zakrzewska-Trznadel, C. Cojocaru, Optimization of Co^{2+} ions removal from water solutions via polymer enhanced ultrafiltration with application of PVA and sulfonated PVA as complexing agents, J. Colloid Interface Sci. 362 (2011) 615–624.
[115] E. Yuliwati, A.F. Ismail, Effect of additives concentration on the surface properties and performance of PVDF ultrafiltration membranes for refinery produced wastewater treatment, Desalination 273 (2011) 226–234.

[116] J.F. Hester, P. Banerjee, Y.Y. Won, A. Akthakul, M.H. Acar, A.M. Mayes, ATRP of amphiphilic graft copolymers based on PVDF and their use as membrane additives, Macromolecules 35 (2002) 7652–7661.
[117] K.B. Hvid, P.S. Nielsen, F.F. Stengaarg, Preparation and characterization of a new UF membrane, J. Membr. Sci. 53 (1990) 189–202.
[118] J. Garcia-Ivars, M. Iborra-Clar, M. Alcaina-Miranda, J. Mendoza-Roca, L. Pastor-Alcañiz, Development of fouling-resistant polyethersulfone ultrafiltration membranes via surface UV photo grafting with polyethylene glycol/aluminum oxide nanoparticles, J. Sep. Sci. Technol. 135 (2014) 88–99.
[119] M. Ulbricht, G. Belfort, Surface modification of UF Membranes by low temperature eplasma. II. Graft polymerization onto polyacrylonitrile and polysulphone, J. Membr. Sci. 111 (1996) 193–215.
[120] S. Mok, D.J. Worsfold, A. Fouda, T. Matsuura, Surface modification of polyethersulfone hollow-fiber membranes by γ-ray irradiation, J. Appl. Polym. Sci. 51 (1994) 193–199.
[121] F. Liu, C.H. Du, B.-K. Zhu, Y.-Y. Xu, Surface immobilization of polymer brushes onto proous poly (vinylidenefluoride) membrane by electron beam to improve the hydrophilicity and fouling resistance, Polymer 48 (2007) 2910–2918.
[122] L.-P. Zhu, Y.-Y. Xu, X.-Z. Wei, B.-K. Zhu, Hydrophilic modification of poly(phthalazine ether sulfone ketone) UF membranes by the surface immobilization of poly(ethylene glycol) acrylates, J. Desalination 242 (2009) 96–109.
[123] Y. Sui, Z.N. Wang, X.L. Gao, C.J. Gao, Antifouling PVDF ultrafiltration membranes incorporating PVDF-g-PHEMA additive via atom transfer radical graft polymerizations, J. Member. Sci. 413-414 (2012) 38–47.
[124] G.E. Chen, S.J. Xu, Z.L. Xu, W.W. Zhu, Q. Wu, W.G. Sun, Preparation and characterization of a novel hydrophilic PVDF/PVA UF membrane modified by carboxylated multi-walled carbon nanotubes, Polym. Eng. Sci. 56 (2016) 955–967.
[125] S. Liang, Y. Kang, A. Tiraferri, E.P. Giannelis, X. Huang, M. Elimelech, Highly hydrophilic polyvinylidenefluoride(PVDF) ultrafiltration membranes via post fabrication graftingof surface-tailored silica nanoparticles, ACS Appl. Mater. Interfaces 5 (2013) 6694–6703.
[126] F. Liu, Y.-Y. Xu, B.-K. Zhu, F. Zhang, L.-P. Zhu, Preparation of hydrophilic and fouling resistant poly(vinylidene fluoride) hollow fiber membranes, J. Membr. Sci. 345 (2009) 331–339.
[127] L.J. Zhu, L.-P. Zhu, J.-H. Jiang, Z. Yi, Y.-F. Zhao, B.-K. Zhu, Y.-Y. Xu, Hydrophilic and anti-fouling polyethersulfone UF membranes with poly(2-hydroxyethyl methacrylate) grafted silica nanoparticles as additive, J. Membr. Sci. 451 (2014) 157–168.

CHAPTER 13

Membrane Contactor

X.-L. Lu

State Key Laboratory of Separation Membranes and Membrane Processes, Tianjin Polytechnic University, Tianjin, China

13.1 Membrane Distillation

In the past decade, more and more attention has been given to apply membrane distillation (MD) in metallurgical processes [1–5]. The principal factor behind the consideration relates to the extreme concentration of some solutions that fall beyond the capability of traditional reverse osmosis (RO) and sometimes nanofiltration (NF) [6]. The existence of a huge amount of low quality waste heat available at industries is another booster for the direction.

In MD, in principle, water vapor diffuses through the membrane pores under a driving force given by the vapor pressure difference between both membrane sides, related to the water activities in the two streams. Chapter 2 has given more detailed introduction about MD transport mechanism. Depending on the way to generate the driving force, MD is further classed into direct contact membrane distillation (DCMD), air gap membrane distillation (AGMD), sweeping gas membrane distillation (SGMD), vacuum membrane distillation (VMD), and osmotic membrane distillation (OMD).

MD has not been implemented at industrial level till present. The problems to be tackled include: development of membrane with good separation performance and low cost, further understanding transport mechanism, and improvement of heat utilization efficiency.

13.1.1 Membrane Development

As a physical barrier in MD, membrane with high flux, good thermal stability, and suitable hydrophobicity is very critical for good separation efficiency [7]. Membrane structure in terms of fiber size, pore size, porosity, and thickness is closely related to the specific application of MD [5]. For inner contact MD, the heat carried by the fluid in lumen side is limited, and therefore, fiber diameter should not be too small to cause extremely low temperature at channel outlet. In order to reduce the heat loss from side to the cold side, leading to weakening of the driving force, thickness of DCMD membrane should not be thick

Membrane-Based Separations in Metallurgy
http://dx.doi.org/10.1016/B978-0-12-803410-1.00013-X

enough to insure good insulation effect. With guaranteeing membrane mechanical strength, membrane in VMD process should be as thin as possible. In addition, the cross-section with dense sponge structure and good thermal conductivity is preferred for preventing water vapor condensing within the pores and improving evaporation efficiency. According to heat and mass transfer models developed by Phattaranawika and coworkers [8] and Qtaishat and coworkers [4], the most relevant membrane properties related to the flux are the membrane pore size and pore size distribution, porosity, and tortuosity.

At present, there exist several principal approaches for preparing membrane for MD. They are nonsolvent induced phase inversion (NIPS), and thermally induced phase inversion (TIPS). The choice of the method will depend on the membrane materials. Currently, the materials for MD membranes are generally selected from a series of fluorine polymers, such as polyvinyldenefluoride (PVDF), polytetrafluoroethylene (PTFE), polypropylene (PP), Poly (ethylene chlorotrifluoroethylene) (ECTFE), Hyflon AD, Teflon AF, and Cytop. For preparing PVDF membrane, NIPS is the most frequently used method; whereas TIPS is a better choice for developing PP and PTFE membranes. Hyflon AD, Teflon AF, and Cytop have good solubility in perfluorinated solvents and low solution viscosity, which make them suitable for membrane coating.

In NIPS, adding additives into dope solution can generate different mechanisms with mutual interaction, thus achieving effective and precise control of pore structure and separation properties. A lot of investigations have been contributed to single-layer MD membranes. Porous hydrophobic/hydrophilic membranes have been proposed for desalination by DCMD (Fig. 13.1) [9,10]. The membranes were prepared by the simple phase inversion technique from polymer solutions containing a hydrophilic polymer, a solvent, a nonsolvent additive, and a surface modifying macromolecule (SMM) (Fig. 13.2). Membranes of different pore sizes have been prepared using polyetherimide (PEI) as a host polymer. The SMM actively migrated to the air surface and changed the surface properties of the PEI membranes. Contact angle (CA) measurements indicate that the PEI membrane surface becomes more

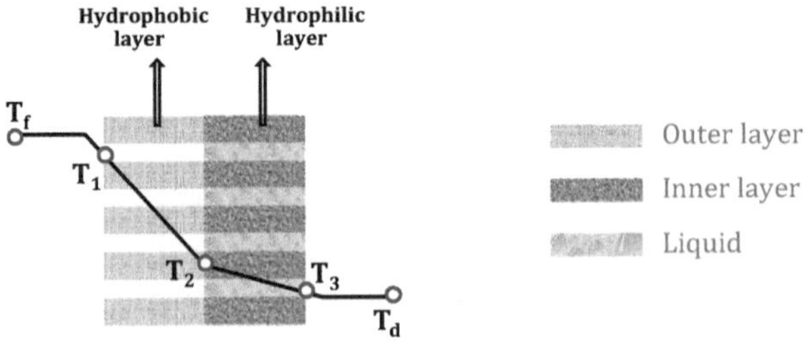

T_f: feed bulk temperature;
T_1: feed/membrane interface temperature;
T_2: two layer interface temperature;
T_3: distillate membrane interface temperature;
T_d: distillate bulk temperature

Figure 13.1: Hydrophobic/Hydrophilic Dual Layer Membrane for DCMD.

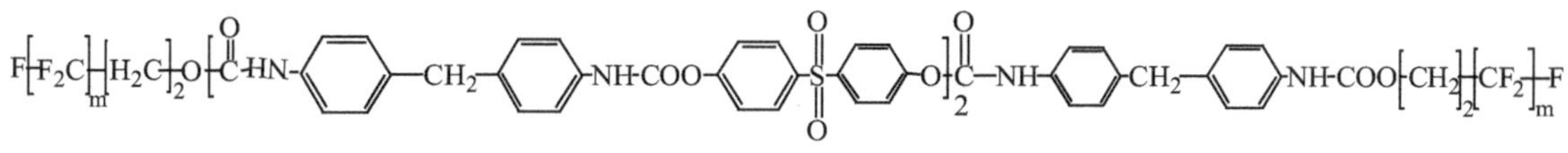

Figure 13.2: Chemical Structure of the SMM.

hydrophobic after adding the SMM to the PEI casting solutions, whereas X-ray photoelectron spectroscopy analysis shows enrichment of fluorine on the PEI membrane surfaces. The SMM modified and unmodified PEI membranes were also characterized by means of liquid entry pressure (LEP) of water measurements and gas permeation test using nitrogen as standard gas. SMM modified PEI membranes have potential to be used in MD. The new membranes seem to be promising in the field of MD as they combine a low resistance to mass flux achieved by the diminution of the water vapor transport path length through the hydrophobic thin top layer of the membrane and a low conductive heat loss through the membrane obtained by using thicker hydrophilic sublayer of the membrane [10].

A new strategy to enhance the desalination performance of PVDF hollow fiber membrane (HFM) for MD via architecture of morphological characteristics is explored in Wang and coworkers work (Fig. 13.3) [11]. It is proposed that a dual-layer hollow fiber consisting of a fully finger-like macrovoid inner layer and a sponge-like outer layer may effectively enhance the permeation flux while maintaining the wetting resistance. Dual layer fibers with the proposed morphology have been fabricated by the dry-jet wet spinning process via careful choice of dopes composition and coagulation conditions.

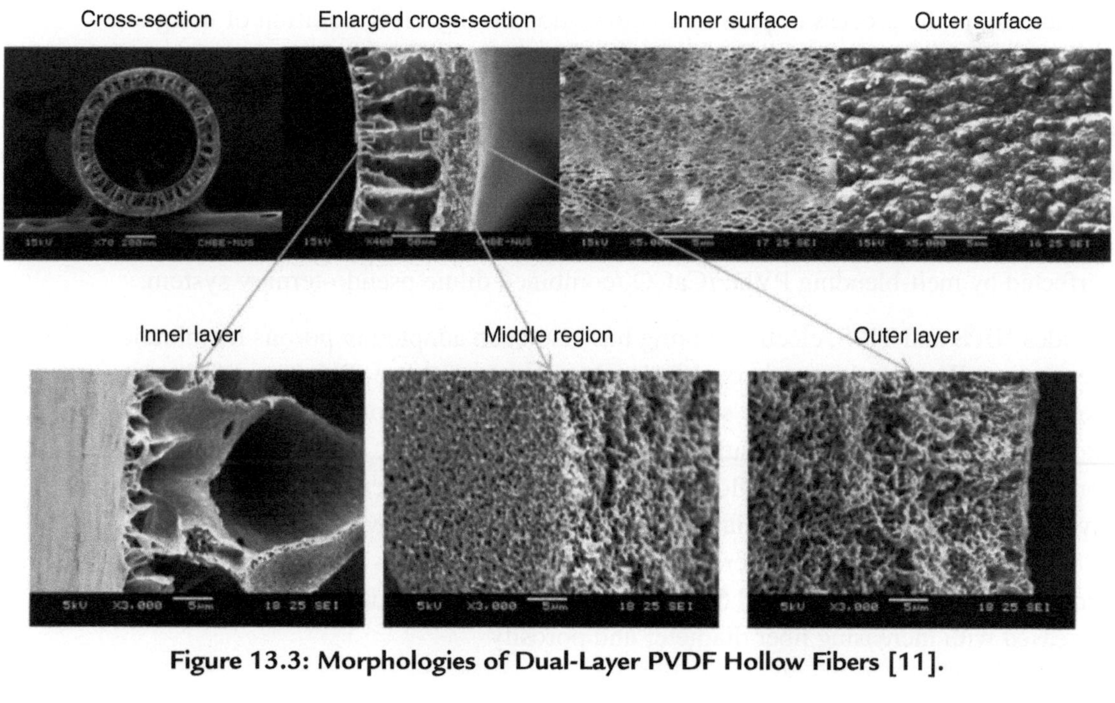

Figure 13.3: Morphologies of Dual-Layer PVDF Hollow Fibers [11].

Wavy PVDF HFM with regular diameter variation was fabricated by Wang et al. by controlling the extrusion frequency and intensity of dopant and bore fluid during NIPS process [12]. The results showed that decrease in undulation period and increase in diameter ratio led to the depression of initial bubble point pressure and improvement in pure water flux. When the wave period was 5.0 cm and the diameter ratio increased from 1.0 to 1.4, the pure water flux increases significantly from 180 to 234 L h^{-1} m^{-2}. The wavy structure was also found to have obviously enhanced the membrane antifouling property.

In recent years, there are a lot of researches on using thermally induced phase separation (TIPS) to prepare membrane for MD. The homogeneous solution consisting of hydrophobic polymer and chemical with small molecular weight and high boiling weight (or diluents) is prepared at high temperature. Solid–liquid or liquid–liquid phase separation is initiated with lowering temperature to generate porous structure [13]. Porous PVDF membrane was prepared using TIPS in Xu's research [14]. Three diluents were selected including diethylene glycol monobutyl ether, dimethyl phthalate, and dibutyl phthalate. The results showed that all the casting systems exhibited solid–liquid phase separation when a single diluent was used and the membrane was apt to form a spherulite structure disadvantageous for membrane mass transfer. For combined diluents with both diluents having poor compatibility and good compatibility with polymer, the liquid–liquid phase separation region broadened as the increasing content of poor solvent in combined diluent. Consequently, spherulite congeries structure in membrane was restrained due to the occurrence of liquid–liquid phase separation. Additionally, the addition of nucleating agent (e.g., NA-40, $CaCO_3$, TM-1, adipic acid) increased the nucleation density in some sense and improved the membrane uniformity consequently.

In addition, in the process of phase transformation during TIPS, addition of inorganic particles (e.g., $CaCO_3$) is a good way to suppress the occurrence of large spherical structures [15,16]. In Lv et al.'s work [17], it was found that $CaCO_3$ particles affected the crystallization dynamics of PVDF. Hence, it limited the spherulite congeries structure and a symmetrical bicontinuous morphology that dominated all the membrane bulky structure. The decreasing size of $CaCO_3$ favored strengthening this characteristic. Furthermore, the porosity of the membrane improved with acid treatment or etching. In all cases, membrane structure can be perfected by melt-blending PVDF/$CaCO_3$/combined dilute pseudo-ternary system.

Besides NIPS and TIPS, electrospinning has also been adopted in porous membrane preparation. Wang and coworkers [18] developed polyurethane (PU) based fibro-porous membranes and investigated the size effect of hierarchical porous structure on permeability and surface properties of the resultant electrospun membranes. Solvents, solution concentration, voltage applied, flow rate, and distance to collector were systematically investigated for tailoring fiber diameters. Membranes having average fiber diameters—347, 738, and 1102 nm were characterized, revealing average pore sizes of 800, 870, and 1060 nm and pore volumes of 44, 63, and 68, respectively. It was also found that hydrophobicity increased with increasing fiber diameter and porosity.

Polyvinylidene fluoride-*co*-hexafluoropropylene (PVDF-HFP) membranes containing different amounts of nanocrystalline cellulose (NCC) were fabricated by electrospinning technique for application in MD [19]. Incorporation of NCC in PVDF-HFP matrix improved the tensile strength and Young's modulus and narrowed down the pore size distribution of the membranes. LEP was improved from ~19 to ~27 psi with the addition of 2 wt.% NCC. Fabricated membranes were tested in DCMD, revealing water flux of 10.2–11.5 L h^{-1} m^{-2} with salt rejection of 99% for these NCC-incorporated membranes.

13.1.2 Molecular Design of Membrane Hydrophobicity

A critical challenge confronted by MD is wetting that will deteriorate MD performance in terms of rejection during the operation. The most serious consequence is the failure of the membrane, leading to termination of the process. A major underlying mechanism for wetting phenomenon is associated with membrane physiochemical properties. As illustrated in Laplace (Cantor) equation [20], the pressure difference at liquid–vapor interface ($\Delta p_{\text{interface}}$ in kPa) determined by the hydrostatic pressure and vapor pressure should be smaller than the entry pressure (Δp_{entry} in kPa) fixed by physiochemical properties of membrane surface and feed solution:

$$p_{\text{L}} - p_{\text{V}} = \Delta p_{\text{interface}} < \Delta p_{\text{entry}} = -\frac{2\Theta\sigma_{\text{L}}\cos\theta}{r_{\max}} \tag{13.1}$$

where, p_{L} and p_{V} are liquid and vapor pressure, respectively, (kPa); Θ is the pore geometric factor (1 for cylindrical pores), σ_{L} represents the liquid–vapor surface tension for water (N/m), θ is water-membrane CA, $r_{\max}$ indicates the maximum surface pore size (m).

Therefore, a very important research direction in recent years on MD membrane development improves surface hydrophobicity that can enhance the resistance toward liquid penetration under pressure.

The most straightforward approach for hydrophobicity enhancement is employment of new material. PVDF, PP, and polytetrafluoroethylene (PTFE) are the most widely investigated and applied polymers in preparing MD membrane. A relatively new polymer attracting attention is PVDF-HFP. Fluorine content of PVDF-HFP is higher than that of PVDF due to incorporation of amorphous HFP group, which makes it a more hydrophobic material than PVDF [21,22]. Preparation of PVDF-HFP (Fig. 13.4) flat-sheet asymmetric porous membrane has been studied for the first time by Feng and coworker [23]. The determination of CA to distilled water also reveals higher hydrophobic nature than that of PVDF membrane. PVDF-HFP asymmetric microporous HFMs were successfully fabricated using NIPS method by Shi and coworkers [24]. PDVF-HFP HFMs were more hydrophobic than those membranes made from pure PVDF; the former also exhibited pronounced ductile behavior when compared with the latter. Membranes made from the PVDF-HFP/NMP dopes were found to have small pure

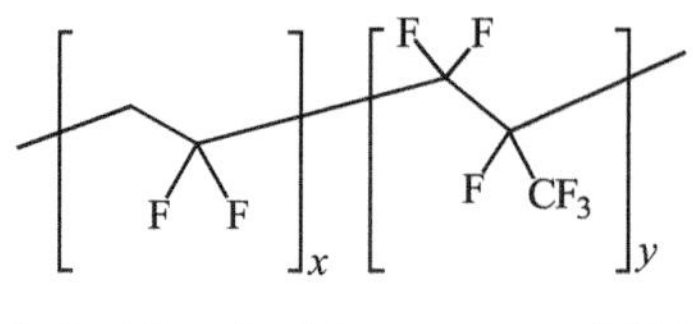

Figure 13.4: Chemical Structure of PVDF-HFP.
[Poly(vinylidene fluoride-*co*-hexafluoropropylene)].

water fluxes no matter what the polymer concentration, the bore fluid composition, and the air gap were used in the spinning processes. The pure water flux can be increased by adding PVP into the dope solution; up to 5 wt.% PVP in the dope can promote the formation of macrovoids in the membrane.

Manipulating the phase inversion characteristic is found to be effective in adjusting membrane surface properties. Lin and coworkers prepared superhydrophobic PVDF membranes by NIPS, investigating the influences of PVDF concentration and nonsolvent (low molecular weight diols) in the casting solution on the membrane hydrophobicity [25]. The results showed that with lower content of PVDF and a moderate PG loading in the casting solution, the static CA of PVDF membrane surface could be raised to 161.7°, and its scroll angle was reduced only by 15.8°. Observation by SEM revealed the presence of 100 nm protrusions, which induces the superhydrophobicity.

In Zhang and coworkers work, a facile methodology to fabricate a superhydrophobic–superoleophilic PVDF membrane via an inert solvent-induced phase-inversion process is reported [26]. A modified phase-inversion process was used in this work, as schematically shown in Fig. 13.5, where ammonia water was added as an inert solvent additive into the

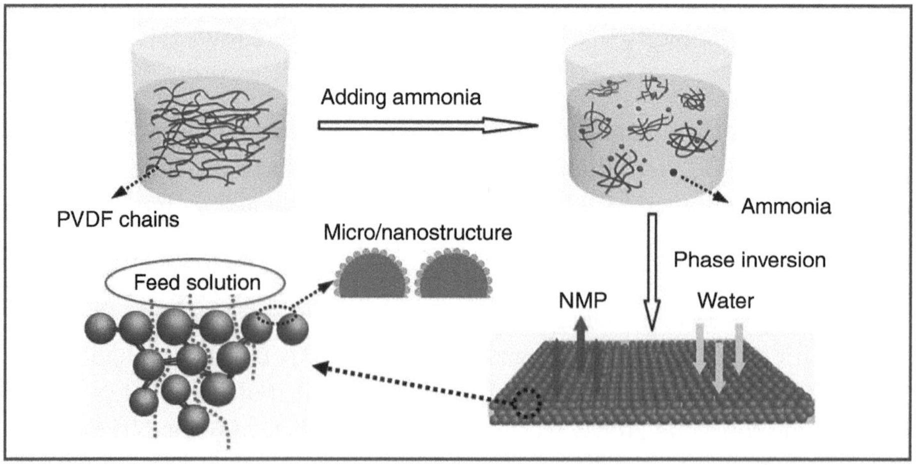

Figure 13.5: Schematic Illustration of the Formation of a Superhydrophobic–Superoleophilic PVDF Membrane via a Modified Phase-Inversion Process [26].

Figure 13.6: SEM Images of the Membrane With Different Magnification.
Scale bars: 50 μm (A), 5 μm (B), 1 μm (C). The membrane was prepared with an ammonia concentration of 2.94 mM.

polymer solution. Different from common-phase inversion, the addition of ammonia is prone to induce localized microphase separation and causes the formation of "incipient precipitation" in the form of PVDF clusters in the solution [27]. A water CA of 158° and oil CA of less than 1° are obtained in this case, showing both superhydrophobic and superoleophilic properties. SEM images (Fig. 13.6) indicate that the surface of the membrane is rather rough, composed of spherical microparticles. The surfaces of these microparticles are not smooth, but form microprotrusions that are densely and evenly distributed on the surface of each microsphere. These micro- and nanoscale structures are crucial for obtaining superwetting characteristics that are similar to what is observed in lotus leaves [28].

Surface modification for a more hydrophobic membrane can be conducted by using any of four major classes of methods: plasma treatment, chemical oxidation, organic chemical functionalization, and radiation induced surface grafting. In Wu and coworkers work [29], hydrophilic cellulose acetate and cellulose nitrate membranes were modified into hydrophobic membranes by radiation grafting polymerization and plasma polymerization for MD. The plasma polymerization, in which many particular kinds of monomer could be polymerized onto the surface of porous materials, has become an efficient method to prepare hydrophobic porous membrane with high performance from hydrophilic membranes.

An omniphobic microporous membrane for MD was prepared by modifying a hydrophilic glass fiber membrane with silica nanoparticles (NPs) followed by surface fluorination and polymer coating [30]. The modified glass fiber membrane exhibits an antiwetting property not only against water but also against low surface tension organic solvents that easily wet a hydrophobic PTFE membrane commonly used in MD applications. By comparing the performance of the PTFE and omniphobic membranes in direct contact MD experiments in the presence of a surfactant (sodium dodecyl sulfate, SDS), it was shown that SDS wets the hydrophobic PTFE membrane but not the omniphobic membrane. Our results suggest that omniphobic membranes developed in this work may be suitable MD applications with feed waters containing surface active species, such as oil and gas produced water, to prevent membrane pore wetting.

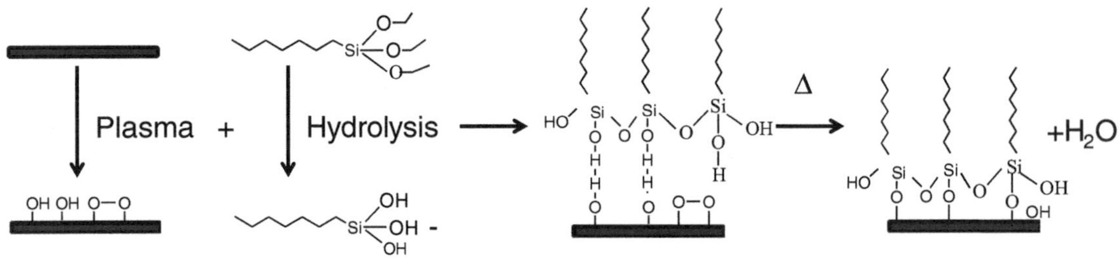

Figure 13.7: The Proposed Scheme for the Modification of the PVDF Membrane.

In recent work [31], a versatile modification tactic that focused on increasing the porosity/pore size of membrane by plasma flow irradiation and subsequent enhancing of the hydrophobicity by grafting silane coupling agent was studied. The proposed scheme for the modification of the PVDF membrane is exhibited in Fig. 13.7. The influences of plasma irradiation time and grafting conditions, including pH value, concentration of silane coupling agent and temperature of heat treatment, on the VMD performances for 5% (w/w) ethanol aqueous solution were investigated [32]. The analysis results revealed the changes in the surface pore size/porosity, chemical composition, and hydrophobicity of the membrane. A proper etching via plasma flow made the membrane surface more porous without damaging the membrane bulk structure. And the water CA increased greatly from 70.1° to 121.0° after modification.

13.2 Other Membrane Contactors

As mentioned earlier, there are several other membrane technologies that are categorized as membrane contactor. Those that are more meaningful to the specific requirements of metallurgical industry at present are membrane absorption and membrane extraction.

Membrane gas absorption (MGA) combines membrane with traditional physical or chemical absorption, and sometimes cryogenic distillation for gas separation [33,34]. It is characterized of advantages such as fast mass transfer, large gas–liquid contact area, mild operation conditions, no haze entrainment [35–37]. Both dense and porous membranes are used in MGA. Unlike porous membrane, dense membrane, also known as nonporous membrane, itself can present a certain level of selective permeability. The basics about the transport mechanisms in membrane absorption are discussed in Qi and Cussler's research works [38,39], Gabelman and Hwang's review [33], Cui and Qi's review [34].

Polymeric, inorganic, and composite materials can all be used in preparing membranes in membrane absorption. Organic silicon membrane materials combine the advantages of organic and inorganic polymers and have high flux, good resistance to oxidation, and photolysis. The methyl silicone rubber as a kind of mature membrane absorption material has a high permeation coefficient for CO_2 gas [40]. The other polymeric materials used include polydimethylsiloxane, polyolefins, polysulfone, cellulose acetate, and polycarbonate, PVDF, PTFE, PP, Polyetheretherketone etc. The inorganic membranes are generally made of metal, metal oxides, ceramics, porous glass, and zeolite that have good mechanical strength, excellent resistance toward temperature, corrosion, and biodegradation [41].

The original challenge in developing membrane contactor was the solid membrane framework; how to reduce the membrane thickness and enhance the permeabilities were the major targets [33]. With accumulation of the knowledge about membrane materials and the progress in manufacturing methods, it became possible to obtain membranes with much lower mass transfer resistance, mainly by improving the physical and chemical structure of the pore and the pore distribution. Owing to this advancement, membrane absorption using asymmetric HFMs have got a point of entry. They have been applied in several industrial processes [33]. Currently, the focus of membrane contactor is shifted to development of stable, cheap and refreshable absorbant, optimization of membrane module and process, the system scale-up, and the theoretical aspects.

For selecting a suitable absorbant for MGA, the following aspects should be given attention [42]:

1. The absorbents and their degradation compounds should not cause harmful emission.
2. Low vapor pressure or volatility and good thermal stability to minimize loss during operation.
3. Good compatibility, for example, damage to membrane physically or chemically, with the membrane materials.
4. Preferably high surface tension to provide a high critical LEP when using a porous membrane.
5. Low viscosity.
6. Commercially available at a low price.

The hazardous gaseous components released by metallurgical industries mainly include CO_2, H_2S, SO_2, SO_3, NO_2, etc. [43–49]. The common absorbant include water (e.g., seawater) and aqueous solutions dissolving NaOH, KOH, K_2CO_3, Na_2CO_3, Na_2SO_3, and $NaHCO_3$ [42]. NH_3 is toxic and inorganic salts will cause membrane scaling. In addition, vaporization of aqueous solvent is serious upon operating at high temperature. In order to solve the problems and further enhance the efficiency of membrane contactor, employment of a new category of absorbents, amines such as monoethanolamine, diethanolamine, and piperazine (PZ), also highly reactive with gaseous component was conducted. These chemicals in some cases are mixed with other active chemicals to promote absorption efficiency and facilitate lowering the operation temperature. In Lu and coworkers work [49], adenosine monophosphate and PZ were used as activating agents. They were added into in methyldiethanolamine aqueous solution to form activated absorbant solutions that had excellent influence in CO_2 removal. In the same work [49], it was found that mixed

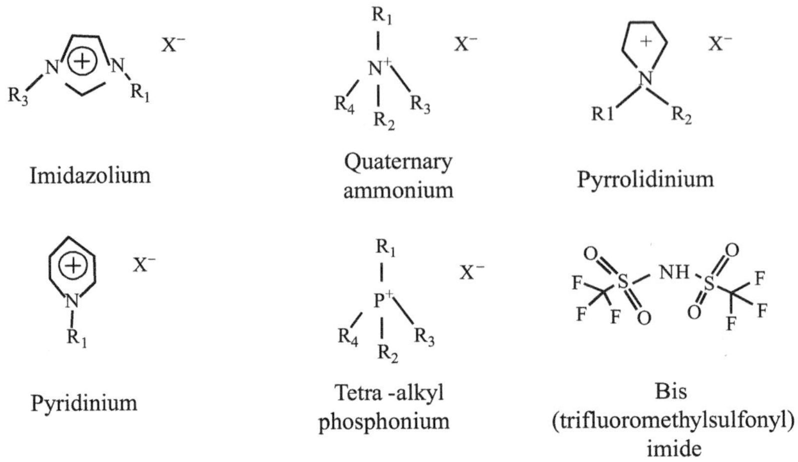

Figure 13.8: The Structures of Common Cations and Anions of Ionic Liquids [55].

absorbants brought about higher mass transfer coefficients and absorption efficiency. Thereinto, the mixture comprising glycol and PZ exhibited best effects. Molina and coworkers found that solution with a pH value of 11 made more sense considering the absorption effect and economic benefit [50].

A relatively new type of liquid media for absorption is ionic liquids (ILs). They normally consist of an organic cation, and a polyatomic inorganic anion, or more and more current, an organic anion [51]. The structures of some common cations and anions of ILs are shown in Fig. 13.8. Some types of ILs that have already been used in MGA include, but not limited to [52–55] 1-ethyl-3-methylimidazolium ethylsulfate [emim][EtSO4], 1-ethyl-3-methylimidazolium acetate [emim][Ac], 1-butyl-3-methylimidazolium dicyanamide ([bmim][DCA]), ILs, 1-butyl-3-methyl-imidazolium tetrafluoroborate ([bmim][BF4]), and 1-(3-aminopropyl)-3-methyl-imidazolium tetrafluoroborate ([apmim][BF4]), etc. In Chau and coworkers work [52], separation of the feed mixture containing 40.67% CO_2–He balance was done using pure IL [bmim][DCA] as the liquid absorbent. In Lu and coworkers work, mixtures of IL [bmim][BF4] or [apmim][BF4] with an alkanolamine of 2-amino-2-methyl-1-propanol (AMP) were attempted to form IL-based aqueous blends for CO_2 absorption [53].

Yet another aspect regarding membrane absorption requiring attention is the recycle of the absorbant [56]. Amine is good choice in terms of recycling, since its complex formed with acid gases is characterized of weak chemical bond that can be broken easily by heating or depressurizing. A good example is aminoethyl alcohol and triethanolamine in Yeon et al.'s work. A zero solvent emission process for SO_2 scrubbing by IL MGA has been developed [57]. The schematic of the process is shown in Fig. 13.9. Utilization of IL as absorption solvent avoids solvent losses due to its negligible vapor pressure. In addition, desorption may

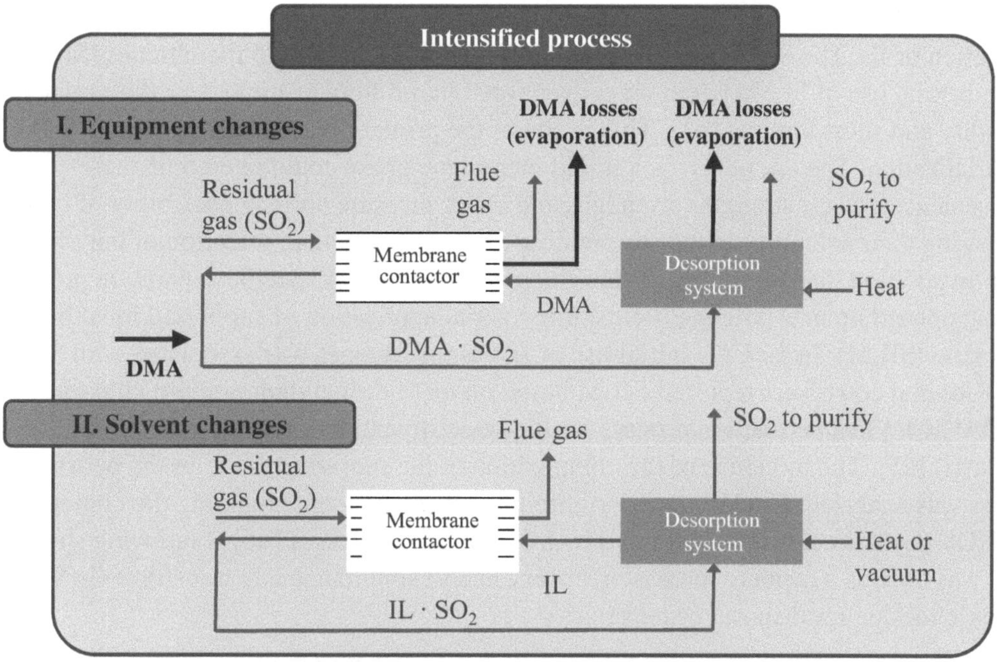

Figure 13.9: Intensification of SO_2 Absorption Process.
DMA: *N*,*N*-dimethylaniline; *IL*: Ionic Liquid.

be based on heating or vacuum. Simply stated, the thermal stability and low volatilization of IL construct the basis of "zero" discharge.

A problem that have been ignored in many researches regarding MGA is the counter diffusion of water vapor directed against the gaseous (e.g., CO_2) species transport when the water vapor is even higher than the partial pressure of targeted gas [58]. In Sirkar et al.'s work, the calculations based on multicomponent diffusion theory did not appear to indicate a substantial counter diffusion effect. However, there is a substantial gap in the value of the membrane permeance between what was obtained from the extrapolation to $\Delta P = 0$ and the experimentally observed permeances of CO_2 through the microporous membrane under $\Delta P = 0$ condition.

Selective separation or recovery of metals from industrial process and waste streams is another problem of paramount importance in hydrometallurgical industry. For these separations, microporous membrane liquid–liquid extraction (LLE) [59,60] and supported liquid membrane (SLM) [61] integrate extraction and stripping have been widely and intensively investigated. A rough estimation (Elsevier Science Database) shows that there exist more than 100 publications from 2010 to 2016 about using SLM for selective separation of metal ions. The various aspects about SLM such as solid membrane development and extractant selection are comprehensively summarized in some review [62].

Similar to the situation with MGA, a latest development for membrane extraction is application of IL. The SLM using IL is called supported ionic liquid membranes (SILM). The industrial use of SLMs based on conventional liquid films is limited by their relative instability and short lifetime [63]. This is one of the primary reasons hindering the SLM industrialization. The use of ILs as a liquid membrane phase could overcome these inconveniences. Their strengths are negligible vapor pressure and the possibility of minimizing their solubility in the surrounding phases by adequate selection of the cation and anion. In addition, the possibility of designing suitable ILs for specific separation problems has also opened up new potential fields of industrial application of supported ionic liquid membrane (SILM). In fact, the initial use of ILs as membranes was associated with SILMs [64]. Rios and coworkers reported SILM based on methyltrioctylammonium chloride [$MTOA^+$][Cl^-] in a pertraction process for the selective extraction of Fe(III), Zn(II), Cd(II), and Cu(II) [65]. The influence of the composition of the receiving phase in the pertraction process was analyzed. Cr (VI) transport through a SLM system containing the commercial IL CYPHOS IL101 as carrier was studied by Miguel and coworkers [66] A reducing stripping phase was used as a mean to increase recovery and to simultaneously transform Cr(VI) into a less toxic residue for disposal or reuse.

Three methods are commonly used for the preparation of SILMs: direct immersion, pressure, or vacuum [67–69]. In the former case, first method, IL immobilization is realized by contacting the supporting membrane with the IL, allowing it to soap up the liquid [67]. In the pressure method, the immobilization is achieved by placing the membrane in an ultrafiltration (UF) unit, filling IL and applying nitrogen pressure to force the IL to enter the pores of the membrane, thus displacing air inside the pores of the membrane [68]. In the vacuum method, the supporting membrane is submerged in a volume of IL and vacuum is applied to remove all the air inside the membranes [69]. After all procedures, the excess IL should be removed from the membrane surface either by leaving to drip overnight or by blotting with paper tissue.

13.3 Control of Pore-wetting State

As for physical stability problems during operation, they are critical for almost all kinds of membrane contactors. The review about physical stability or wetting of the hydrophobic pores in MD operation has already been introduced in Section 13.1. For membrane contactors involving liquid phase for selective separation, which are MGA, LLM, or SLM, there also exist challenging stability problems. For MGA, the nonwetting operation mode is desired for achieving higher absorption rate [42]. Three different wetting degrees of the pores during MGA operation occur: nonwetting, complete wetting, and partial wetting. The liability of the pores toward liquid wetting can be estimated by Eq. 13.1 for MD, as well, considering their same requirement on nonwetting. For LLE and SLM, pore wetting by liquid for extraction is needed in most cases. The liquid wetting the pores for selection may be displaced by the nonwetting fluid originally situated outside the membrane. As a result, the membrane lost its selective function, which is especially a serious problem for SLM. There have been long disputes about the causes for the stability of liquid membrane. According to the literature, there are several reason causing the physical instability of liquid contained in the membrane: (1) pressure difference over the membrane [70]; (2) mutual solubility of species from the aqueous phase and LM phase [71]; (3) progressive wetting of the pores in the membrane support by the aqueous phase [72,73]; (4) emulsion formation in the LM phase [74,75]. In the study by Zheng and coworkers [76], emulsion formation proved to be one of the main causes of SLM instability in the system. Conditions facilitating emulsion, such as higher membrane liquid concentrations, faster stirring speeds, lower salt concentrations, and higher hydrophile lipophile balance values, accelerated the membrane liquids degradation. Other mechanisms, including solubility and osmotic pressure, compound the membrane liquid loss.

For liquid/liquid system, the minimum transmembrane pressure required to push the impregnating phase out of the largest pores can be calculated using Laplace equation:

$$P_c = \frac{2\gamma\cos\theta}{r} \tag{13.2}$$

where γ is the interfacial tension between strip or feed solution and SLM phase, θ the CA between the membrane pores and the impregnating liquid, and r is the pore radius. This contact angle increases with increasing polarity difference between the membrane material and the fluid.

Prasad and Shirker [33] tabulated values of breakthrough pressure for a variety of membrane types and chemical systems; values range from about 100–500 kPa. Reed et al. [77] described the relative magnitude of breakthrough pressure for various configurations of gas/liquid and liquid/liquid systems; factors considered include membrane type (hydrophobic or hydrophilic), fluid polarity (e.g., aqueous, polar organic, nonpolar organic), and the wetting fluid (e.g., gas, aqueous, organic). A troublesome reduction in breakthrough pressure can occur in the presence of even trace levels of surfactant, which cause a reduction in interfacial or surface tension. Breakthrough can also be a problem with solvent systems where the interfacial tension is already low (e.g., water/*n*-butanol), or with a solvent that behaves like a surfactant (e.g., the pharmaceutical compound MK-819, mevinolinic acid) [70,71]. In addition, membrane swelling by the organic solvent may also intensify the tendency of wetting [42].

Based on the mechanisms introduced previously, several approaches to solve this stability problem have been suggested:

- Mix a compound (e.g., isopropanol) which increases the interfacial surface tension [78].
- Use a membrane with smaller pore size, which increases the breakthrough pressure according to Eq. 13.2 . Nevertheless, small pores will lead to higher resistance towards diffusion of large molecules, giving rise to a notable reduction in their mass transfer rate [79].

- Coat the microporous membrane surface with a thin, highly permeable polymer or gel. This layer covers the pores, effectively removing the influence from flow associated pressure [80,81]. It in fact also minimizes the possibility of dissolution, emulsification, and wetting.
- Continuous reimpregnation of the porous solid membrane with membrane liquid [82]. The regeneration works well, but feed and/or strip solutions will be polluted with the membrane liquid.
- The immobilization approach was also found to influence the membrane liquid stability [68]. The loss of IL was higher when immobilization was carried out under vacuum, even with the most viscous ILs, such as 1-butyl-3-methylimidazolium hexafluorophosphate ($[bmim^+][PF_6^-]$) and 1-butyl-3-methylimidazolium chloride ($[bmim^+][Cl^-]$).

13.4 Modules and Commercial Products

There exist three types of configurations for membrane contactor modules. They are plate, tubular, and hollow fiber configurations [33,42]. HFM contactors are well suited for use as an onsite treatment option. Hollow fiber modules are usually more expensive but they offer much higher surface area per unit of module volume up to 500 m^{-1}. Commercially available modules can be big enough and have up to 220 m^2 areas. In the case of hollow fiber SLM, there are several modes of operation [83]. In the simplest case the organic solvent with a carrier fills the microstructures of hollow tubes. Generally, the feed solution is circulated through the inner channel and the strip solution on the shell side of the hollow fibers. To avoid the problems due to low stability of organic liquid phase in the pores, in another mode only one unit operation, either extraction or stripping. Therefore, it is necessary to use two modules for the entire process of extraction and back-extraction.

As shown in Fig. 13.10, modules with three sets of hollow fibers were introduced to conduct simultaneous separation of cations and anions using two different carriers and two different stripping solutions [84]. In order to achieve extraction and stripping in a single module, Ho and Poddar [85] introduced the pseudo-emulsion-based hollow fiber strip dispersion technique (Fig. 13.11). This arrangement combines the advantages of emulsion liquid membranes for one-step extraction and stripping and nondispersive hollow fiber contactors. It reduces the leakage of the internal phase into the external aqueous phase in the absence of a high shear rate in the stirred contactor. It also makes it possible to minimize swelling of the internal phase.

Some new strategies for MGA process have been proposed in the past few years [86–88]. For example, a proof-of-concept experiments via a cyclic process and a novel device containing two commingled sets of hollow fibers applied by Mulukuta et al. for CO_2 absorption. Through the bore of one set of porous hydrophobic hollow fibers passed the flue gas; CO_2 diffuses to the shell side to get absorbed via chemical interaction in the stagnant polyamidoamine (PAMAM) dendrimer-IL on the shell side. When CO_2 breakthrough occurs at the other end

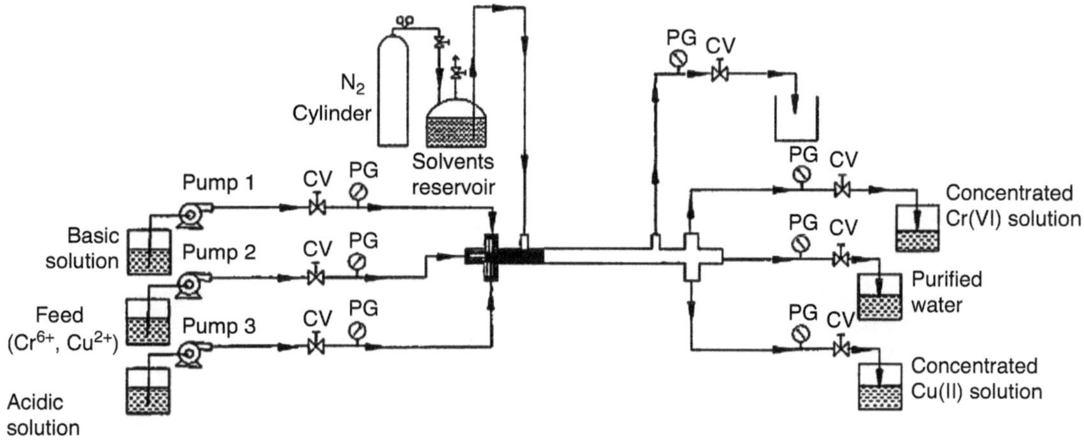

Figure 13.10: Setup of Novel Membrane-Based Simultaneous Extraction and Individual Stripping Device for Removal and Recovery of a Mixture of Two Cationic and Anionic Heavy Metals. *CV*, Check value; *PG*, pressure gauge.

of the porous hollow fibers, the feed gas supply is stopped. The second set of hollow fibers supply hot water to heat up the shell-side liquid absorbent rapidly for gas desorption; it is impermeable to moisture, CO_2, N_2, and O_2 under the conditions of use. For commercial products, the most well-known module for concentration-driven mass transfer in MGA is the Liqui-Cel Extra-Flow module offered by CELGARD LLC (*Charlotte, NC; formerly Hoechst Celanese*). Its design schematic is shown in Fig. 13.12 [89]. This module employs microporous PP fibers (Celgard). They are woven into a fabric and wrapped around a central tube feeder that supplies the shell side fluid. Design with fibers being woven into fabric delivers more uniform fiber spacing, which in turn contributes to higher mass transfer coefficients than those obtained with individual fibers. The more detailed information about the commercially available modules is summarized in Table 13.1.

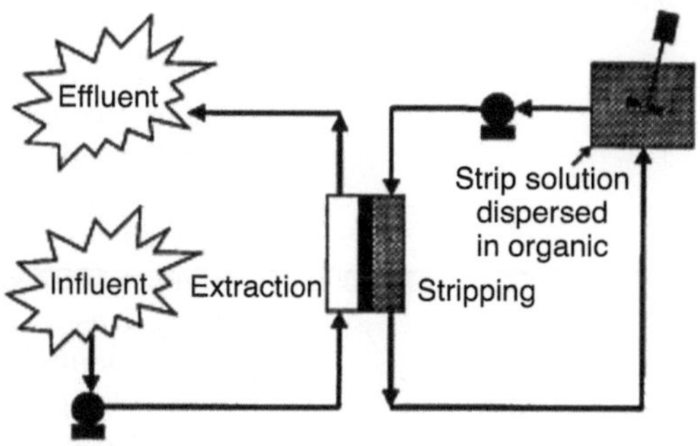

Figure 13.11: A Schematic Diagram of the SLM With Strip Dispersion.

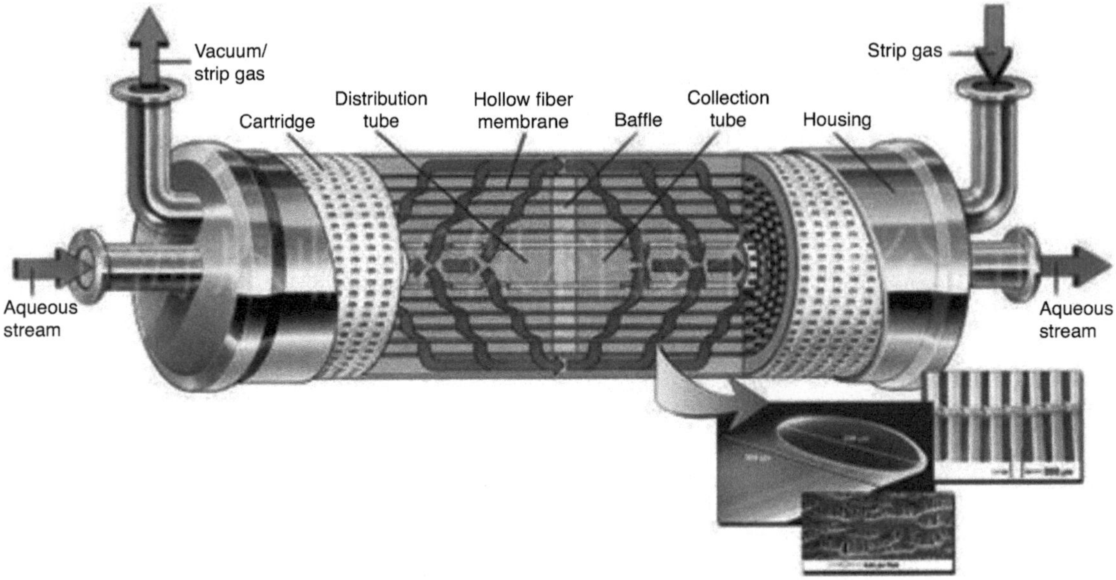

Figure 13.12: The Liquid-Cel Extra-Flow Membrane Contactor From CELGARD LLC.
The central baffle minimizes shell side bypassing and provides a component of velocity normal to the fibers.

Table 13.1: Information about commercially available membrane contactor modules [33]

Trademark	Company	Introduction
Liqui-Cel Extra-Flow	CELGARD LLC (Charlotte, NC; formerly Hoechst Celanese)	Membrane and configuration: Celgard microporous PP hollow fibers that are woven into a fabric and wrapped around a central tube that supplies feed solutions. Operation: Woven fabric allows more uniform fiber spacing, which in turn leads to higher mass transfer coefficients than those obtained with individual fibers.
/	Membrane Corporation (Minneapolis, MN)	Membrane and configuration: The fibers are composed of a 1 mm layer of polyurethane sandwiched between two layers of microporous polyethylene (PE). The modules fit within standard PVC pipes and contain multiple fiber bundles; the packing density is only 10%. Operation: All the gas components diffuse across the membrane into absorbing liquid, leading to 100% gas transfer efficiency. The low packing density offer sufficient space for liquid and membrane fluidization, which in turn leads to high gas/liquid mass transfer rate, low pressure drop, and free passage of solids.

(Continued)

Table 13.1: Information about commercially available membrane contactor modules [33] (*cont.*)

Trademark	Company	Introduction
DISSO3LVE	W.L. Gore & Associates (Elkton, MD)	Membrane and configuration: Expanded PTFE fibers that are resistant to highly corrosive chemicals such as ozone. The fibers are arranged with a helix arrangement. Operation: The expanded PTFE allows each fiber to serve as a point-of-use particle filter.
Separel (a trademark of Dainippon Ink and Chemicals) EFM-530 module	Pall Corporation (East Hills, NY)	Membrane and configuration: Nonporous polyolefin fibers that are woven into a fabric and wrapped around a central feed tube. Operation: Water introduced through the feeder flows to the fibers, which are either maintained under vacuum, flushed with nitrogen sweep gas, or both, for degassing.

Glossary and Symbols

AMP	Adenosine monophosphate
AMP	2-Amino-2-methyl-1-propanol
[bmim][DCA]	1-Butyl-3-methylimidazolium dicyanamide
[bmim][BF_4]	1-Butyl-3-methyl-imidazolium tetrafluoroborate
[apmim][BF_4]	1-(3-Aminopropyl)-3-methyl-imidazolium tetrafluoroborate
[$bmim^+$][PF_6^-]	1-Butyl-3-methylimidazolium hexafluorophosphate
[$bmim^+$][Cl^-]	1-Butyl-3-methylimidazolium chloride
CA	Cellulose acetate
CA	Contact angle
DCMD	Direct contact membrane distillation
DEA	Diethanolamine
ECTFE	Poly(ethylene chlorotrifluoroethylene)
[emim][$EtSO_4$]	1-Ethyl-3-methylimidazolium ethylsulfate
[emim][Ac]	1-Ethyl-3-methylimidazolium acetate
HFM	Hollow fiber membrane
IL	Ionic liquid
LEP	Liquid entry pressure
LLE	Liquid–liquid extraction
MD	Membrane distillation
MDEA	Methyldiethanolamine
MEA	Monoethanolamine
MGA	Membrane gas absorption
MSR	Methyl silicone rubber
[$MTOA^+$][Cl^-]	Methyltrioctylammonium chloride
NCC	Nanocrystalline cellulose
NIPS	Nonsolvent induced phase inversion
PAMAM	Polyamidoamine
NPs	Nanoparticles
PC	Polycarbonate
PDMS	Polyeimethylsiloxane

PEEK	Polyetheretherketone
PEI	Polyetherimide
PP	Polypropylene
PSF	Polysulfone
PTFE	Polytetrafluoroethylene
PU	Polyurethane
PVDF	Polyvinyldene fluoride
PVDF-HFP	Polyvinylidenefluoride-*co*-hexafluoropropylene
PZ	Piperazine
SDS	Sodium dodecyl sulfate
SILM	Supported ionic liquid membranes
SLM	Supported liquid membrane
SMM	Surface modifying macromolecule
TIPS	Thermally induced phase inversion
UF	Ultrafiltration
VMD	Vacuum membrane distillation
P_c	Minimum transmembrane pressure required to push the impregnating phase out
Δp_{entry}	Entry pressure
$\Delta p_{interface}$	Pressure difference at liquid–vapor interface
p_L	Liquid pressure
p_V	Vapor pressure
r	Membrane surface pore size
r_{max}	The maximum surface pore size of the largest pores
Θ	Pore geometric factor (1 for cylindrical pores)
σ_L	Liquid–vapor surface tension for water
θ	Water-membrane contact angle
γ	The interfacial tension between strip or feed solution and SLM phase

References

[1] K.W. Lawson, D.R. Lloyd, Review: membrane distillation, J. Membr. Sci. 124 (1997) 1–25.

[2] D.L. Huang, J.Q. Wang, G.H. He, X.J. Zhang, B.G. Yang, Research progress of membrane contactor, Membr. Sci. Technol. 25 (2005) 63–68.

[3] L. Zeng, C.J. Gao, The application prospects of membrane distillation in metallurgy industry, Membr. Sci. Technol. 28 (2008) 87–92.

[4] M. Qtaishat, T. Matsuura, B. Kruczek, M. Khayet, Heat and mass transfer analysis in direct contact membrane distillation, Desalination 219 (2008) 272–292.

[5] X.L. Lv, Discussion on membrane distillation processes, Membr. Sci. Technol. 30 (2010) 1–9.

[6] E. Curcio, E. Drioli, Membrane distillation and related operations—a review, Sep. Purif. Rev. 34 (1) (2005) 35–86.

[7] J. Zhang, Z.Y. Song, B.A. Li, Q. Wang, S.C. Wang, Fabrication and characterization of superhydrophobic PVDF for DCMD, Desalination 324 (2013) 1–9.

[8] J. Phattaranawika, R. Jiraratananon, A.G. Fane, Effect of pore size distribution and air flux on mass transport in direct contact membrane distillation, J. Membr. Sci. 215 (2003) 75–85.

[9] M. Khayet, T. Matsuura, Application of surface modifying macromolecules for the preparation of membrane for membrane distillation, Desalination 158 (2003) 51–56.

[10] M. Khayet, J.I. Mengual, T. Matsuura, Porous hydrophobic/hydrophilic composite membranes: application in desalination using direct contact membrane distillation, J. Membr. Sci. 252 (2005) 101–113.

[11] P. Wang, M.M. Teoh, T.S. Chung, Morphological architecture of dual-layer hollow fiber for membrane distillation with higher desalination performance, Water Res. 45 (2011) 5489–5500.

[12] J.M. Wang, X.L. Lv, C.R. Wu, Q.L. Zhang, J. Liu, Study on the preparation and property of PVDF hollow fiber membrane with wavy structure diameter, Membr. Sci. Technol. 32 (2012) 39–44.
[13] T. Honjo, K. Hoe, S. Tabayashi, T. Tanaka, J. Shimada, M. Goto, H. Matsuyama, T. Maruyama, Preparation of affinity membrane using thermally induced phase separation for one-step purification of recombinant proteins, Anal. Biochem. 43 (2) (2013) 269–274.
[14] G.Q. Xu, Preparation of polyvinylidene fluoride porous membrane by thermal induced phase separation method, Master Thesis, Tianjin Polytechnic University, 2007.
[15] X.F. Li, X.L. Lv, C.F. Xiao, Research on polymer separation membrane filled by heterogeneous particles, J. Polym. Mater. Eng. 22 (2006) 24–27.
[16] X.F. Li, X.L. Lu, Morphology of polyvinylidene fluoride and its blend in thermally induced phase separation process, J. Appl. Polym. Sci. 101 (2006) 2944–2952.
[17] G.Q. Xu, X.F. Li, X.L. Lv, Effect of calcium carbonate on PVDF membrane pore structure by thermal induced phase separation method, J. Polym. Mater. Eng. 23 (2007) 234–237.
[18] N. Wang, K. Burugapalli, W. Song, J. Halls, F. Moussy, Y.D. Zheng, Y.X. Ma, Z.T. Wu, K. Li, Tailored fibro-porous structure of electrospun polyurethane membranes, their size-dependent properties and trans-membrane glucose diffusion, J. Membr. Sci. 427 (2013) 207–217.
[19] B.S. Lalia, E. Guillen, H.A. Arafat, R. Hashaikeh, Nanocrystalline cellulose reinforced PVDF-HFP membranes for membrane distillation application, Desalination 332 (2014) 134–141.
[20] T.Y. Cath, V.D. Adams, A.E. Childress, Experimental study of desalination using direct contact membrane distillation: a new approach to flux enhancement, J. Membr. Sci. 228 (2004) 5–16.
[21] X.Z. Tian, B.K. Zhu, Y.Y. Xu, P(VDF-co-HFP) membrane for recovery of aroma compounds from aqueous solutions by pervaporation I. Ethyl acetate/water system, J. Membr. Sci. 248 (2005) 109–117.
[22] L. García-Fernández, M.C. García-Payo, M. Khayet, Effects of mixed solvents on the structural morphology and membrane distillation performance of PVDF-HFP hollow fiber membranes, J. Membr. Sci. 468 (2014) 324–338.
[23] C.S. Feng, R. Wang, B. Shi, G.M. Li, Y.L. Wu, Factors affecting pore structure and performance of poly(vinylidene fluoride-co-hexafluoro propylene) asymmetric porous membrane, J. Membr. Sci. 277 (2006) 55–64.
[24] L. Shi, R. Wang, Y.M. Cao, C.S. Feng, D.T. Liang, J.H. Tay, Fabrication of poly(vinylidene fluoride-co-hexafluropropylene) (PVDF-HFP) asymmetric microporous hollow fiber membranes, J. Membr. Sci. 305 (2007) 215–225.
[25] H.Y. Lin, C.R. Wu, X.L. Lv, Modification of the superhydrophobic polyvinylidene fluoride membrane, Membr. Sci. Technol. 30 (2010) 39–44.
[26] W.B. Zhang, Z. Shi, F. Zhang, X. Liu, J. Jin, L. Jiang, Superhydrophobic and superoleophilic PVDF membranes for effective separation of water-in-oil emulsions with high flux, Adv. Mater. 25 (2013) 2071–2076.
[27] M. Khayet, T. Matsuura, Preparation and characterization of polyvinylidene fluoride membranes for membrane distillation, Ind. Eng. Chem. Res. 40 (2001) 5710–5718.
[28] T.L. Sun, L. Feng, X.F. Gao, L. Jiang, Bioinspired surfaces with special wettability, Acc. Chem. Res. 38 (2005) 644–652.
[29] Y. Wu, Y. Kong, X. Lin, W. Liu, J. Xu, Surface-modified hydrophilic membranes in membrane distillation, J. Membr. Sci. 72 (2) (1992) 189–196.
[30] L. Lin, S. Nejati, C. Boo, Y.X. Hu, C.O. Osuji, M. Elimelech, Omniphobic membrane for robust membrane distillation, Environ. Sci. Technol. Lett. 1 (2014) 443–447.
[31] T.X. Wan, Morphological and hydrophobic modifications of PVDF flat membrane with silane coupling agent grafting via plasma flow for VMD of ethanol–water mixture, J. Membr. Sci. 491 (2015) 110–120.
[32] W.T. Xu, Z.P. Zhao, M. Liu, K.C. Chen, Morphological and hydrophobic modifications of PVDF flat membrane with silane coupling agent grafting via plasma flow for VMD of ethanol–water mixture, J. Membr. Sci. 491 (2015) 110–120.
[33] A. Gabelman, S.T. Hwang, Hollow fiber membrane contactors, J. Membr. Sci. 159 (1999) 61–106.
[34] J.H. Cui, J.Q. Qi, Research and application development of membrane absorption, Chem. Equip. Technol. 26 (1) (2005) 13–17.

[35] X.S. Wang, Membrane separation technology and its application, Science Press, Beijing, (1994).
[36] L. Yuan, Z. Wang, S.C. Wang, Membrane absorption technology and its application in removing acid gas research, Membr. Sci. Technol. 22 (4) (2002) 55–59.
[37] R.H. Yang, Research of flue gas desulfurization by solvent absorption of Hollow fiber membrane Master Thesis, Nanjing University of Science and Technology, 2004.
[38] Z. Qi, E.L. Cussler, Microporous hollow fibers for gas absorption. I. Mass transfer in the liquid, J. Membr. Sci. 23 (1985) 321–332.
[39] Z. Qi, E.L. Cussler, Microporous hollow fibers for gas absorption. II. Mass transfer across the membrane, J. Membr. Sci. 23 (1985) 333–345.
[40] C.G. Sun, Y.M. Cao, L. Zuo, B. Lin, X.M. Jie, Mass transfer process of CO2 absorption through non-porous hollow fiber contactor, J. Chem. Chinese Univ. 26 (2005) 2097–2102.
[41] A.P. Luis, A. Garea, A. Irabien, Environmental and economic evaluation of SO2 recovery in a ceramic hollow fibre membrane contactor, Chem. Eng. Process. 52 (2012) 151–154.
[42] S. Zhao, P.H.M. Feron, L. Deng, E. Favre, E. Chabanon, S. Yan, J. Hou, V. Chen, H. Qi, Status and progress of membrane contactors in post-combustion carbon capture: a state-of-the-art review of new developments, J. Membr. Sci. 511 (2016) 180–206.
[43] A.C. Ghosh, S. Borthakur, N.N. Dutta, Absorption of carbon monoxide in hollow fiber membranes, J. Membr. Sci. 96 (1994) 183–192.
[44] K. Li, D.L. Wang, C.C. Koe, W.K. Teo, Use of asymmetric hollow fiber modules for elimination of H2S from gas streams via a membrane absorption method, Chem. Eng. Sci. 53 (1998) 1111–1119.
[45] X. Tan, S.P. Tan, W.K. Teo, K. Li, Polyvinylidene fluoride (PVDF) hollow fibre membranes for ammonia removal from water, J. Membr. Sci. 271 (2006) 59–68.
[46] H.H. Park, B.R. Deshwal, I.W. Kim, H.K. Lee, Absorption of SO2 from flue gas using PVDF hollow fiber membranes in a gas–liquid contactor, J. Membr. Sci. 319 (2008) 29–37.
[47] R.W. Baker, Membrane separation systems: recent developments and future directions (Noyes Dats Corporation), Chem. Eng. Sci. 18 (1991) 503–524.
[48] H.H. Park, B.R. Deshwal, H.D. Jo, W.K. Choi, I.W. Kim, H.K. Lee, Absorption of nitrogen dioxide by PVDF hollow fiber membranes in a G–L contactor, Desalination 243 (2009) 52–64.
[49] J.G. Lu, M.D. Cheng, Y. Ji, Membrane-based CO_2 absorption into blended amine solutions, J. Membr. Sci. 6 (2009) 740–746.
[50] A. Esquiroz-Molina, S. Georgaki, R. Stuetz, Influence of pH on gas phase controlled mass transfer in a membrane contactor for hydrogen sulphide (H_2S) absorption, J. Membr. Sci. 427 (2013) 276–282.
[51] J.F. Brennecke, E.J. Maginn, Ionic liquids: innovative fluids for chemical processing, AIChE J. 47 (2001) 2384–2389.
[52] J. Chau, X.M. Jie, K.K. Sirkar, Polyamidoamine-facilitated poly(ethylene glycol)/ionic liquid based pressure swing membrane absorption process for CO_2 removal from shifted syngas, Chem. Eng. J. 305 (2015) 212–220.
[53] J.-G. Lu, Z.-Y. Lu, Y. Chen, J.-T. Wang, L. Gao, X. Gao, Y.-Q. Tang, D.-G. Liu, CO_2 absorption into aqueous blends of ionic liquid and amine in a membrane contactor, Sep. Purif. Technol. 150 (2015) 278–285.
[54] L. Gómez-Coma, A. Garea, A. Irabien, Non-dispersive absorption of CO_2 in [emim][EtSO4] and [emim][Ac]: temperature influence, Sep. Purif. Technol. 132 (2014) 120–125.
[55] Z. Dai, R.D. Noble, D.L. Gin, X. Zhang, L. Deng, Combination of ionic liquids with membrane technology: a new approach for CO2 separation, J. Membr. Sci. 497 (2016) 1–20.
[56] A. Mansourizadeh, A.F. Ismail, Hollow fiber gas-liquid membrane contactors for acid gas capture: a review, J. Hazard. Mater. 171 (2009) 38–53.
[57] P. Luis, A. Garea, A. Irabien, Zero solvent emission process for sulfur dioxide recovery using a membrane contactor and ionic liquids, J. Membr. Sci. 330 (2009) 80–89.
[58] J. Balej, Water vapour partial pressures and water activities in potassium and sodium hydroxide solutions over wide concentration and temperature ranges, Int. J. Hydrogen Energ. 10 (1985) 233–243.
[59] A.K. Pabby, A.M. Sastre, State-of-the-art review on hollow fibre contactor technology and membrane-based extraction processes, J. Membr. Sci. 430 (2013) 263–303.

[60] J.A. Jönsson, L. Mathiasson, Liquid membrane extraction in analytical sample preparation: II. applications, TrAC 18 (1999) 325–334.
[61] V.V. Belova, A.E. Kostanyan, Y.A. Zakhodyaeva, A.I. Kholkin, O.A. Logutenko, On the application of bulk-supported liquid membrane techniques, Hydrometallurgy 150 (2014) 144–152.
[62] N.M. Kocherginsky, Q. Yang, L. Seelam, Recent advances in supported liquid membrane technology, Sep. Purif. Technol. 53 (2007) 171–177.
[63] L.J. Lozano, C. Godinez, A.P. de los Rios, F.J. Hernandez-Fernandez, S. Sanchez-Segado, F.J. Alguacil, Review-recent advances in supported ionic liquid membrane technology, J. Membr. Sci. 376 (2011) 1–14.
[64] R.D. Noble, Perspective on ionic liquids and ionic liquid membranes, J. Membr. Sci. 369 (2011) 1–4.
[65] A.P. de los Ríos, F.J. Hernández-Fernández, L.J. Lozano, S. Sánchez-Segado, A. Ginestá-Anzol, C. Godínez, F. Tomás-Alonso, J. Quesada-Medina, On the selective separation of metal ions from hydrochloride aqueous solution by pertraction through supported ionic liquid membranes, J. Membr. Sci. 444 (2013) 469–481.
[66] E.R. de San Miguel, X. Vital, J. de Gyves, Chromium (VI) transport via a supported ionic liquid membrane containing CYPHOS IL101 as carrier: system analysis and optimization through experimental design strategies, J. Hazard. Mater. 273 (2014) 253–262.
[67] P. Scovazzo, J. Kieft, D.A. Finan, C. Koval, D. DuBois, R. Noble, Gas separations using non-hexafluorophosphate anion supported ionic liquid membranes, J. Membr. Sci. 238 (2004) 57–63.
[68] F.J. Hernandez-Fernandez, A.P. de los Rios, F. Tomas-Alonso, D. Gomez, G. Villora, Preparation of supported ionic liquid membranes: influence of the ionic liquid immobilization method on their operational stability, J. Membr. Sci. 341 (2009) 172–177.
[69] R. Fortunato, M.J. Gonzalez-Munoz, M. Kubasiewicz, S. Luque, J.R. Alvarez, C.A.M. Afonso, I.M. Coelhoso, J.G. Crespo, Liquid membranes using ionic liquids: the influence of water on solute transport, J. Membr. Sci. 249 (2005) 153–162.
[70] F.F. Zha, A.G. Fane, C.J.D. Fell, R.W. Schofield, Critical displacement pressure of a supported membrane, J. Membr. Sci. 75 (1992) 69–80.
[71] P.R. Danesi, Separation of metal species by supported liquid membranes, Sep. Sci. Technol. 85 (1984) 857–894.
[72] H. Takeuchi, M. Nakano, Progressive wetting of supported liquid membranes by aqueous solutions, J. Membr. Sci. 42 (1989) 183–188.
[73] A.M. Neplenbroek, D. Bargeman, C.A. Smolders, Mechanism of supported liquid membrane degradation: emulsion formation, J. Membr. Sci. 67 (1992) 133–148.
[74] F.F. Zha, A.G. Fane, C.J.D. Fell, Instability mechanisms of supported liquid membranes in phenol transport process, J. Membr. Sci. 107 (1995) 59–74.
[75] N. Makoto, K. Takahashi, H. Takeuchi, A method for continuous operation of supported liquid membranes, J. Chem. Eng. Jpn. 20 (1987) 326–328.
[76] H. Zheng, B.Y. Wang, Y.X. Wu, Q.L. Ren, Instability mechanisms of supported liquid membrane for phenol transport, Chin. J. Chem. Eng. 17 (2009) 750–755.
[77] W. Rd, M.J. Semmens, E.L. Cussler, Membrane contactors, in: R.D. Noble, S.A. Stern (Eds.), Membrane Separations Technology. Principles and Applications, Elsevier, Amsterdam, 1995, pp. 474.
[78] Z. Shen, B. Han, S.R. Wickramasinghe, Cyanide removal from industrial wastewaters using gas membranes, J. Membr. Sci. 257 (2005) 171–181.
[79] G.D. Profio, E. Curcio, A. Cassetta, D. Lamba, E. Drioli, Membrane crystallization of lysozyme: kinetic aspects, J. Cryst. Growth 257 (2003) 359.
[80] B.S. Kim, P. Harriott, Critical entry pressure for liquids in hydrophobic membranes, J. Colloid Interface Sci. 115 (1987) 1–8.
[81] K. Schneider, W. Holz, R. Wollbeck, Membranes and modules for transmembrane distillation, J. Membr. Sci. 39 (1988) 25–42.
[82] R. Chiarizia, E.P. Horwitz, P.G. Rickert, K.M. Hodgson, Application of supported liquid membranes for removal of uranium from groundwater, Sep. Sci. Technol. 25 (1990) 1571–1586.
[83] J. de Gyves, E.R. de San Miguel, Metal ion separations by supported liquid membranes, Ind. Eng. Chem. Res. 38 (1999) 2182–2202.

[84] Z.F. Yang, A.K. Guha, K.K. Sirkar, Simultaneous and synergistic extraction of cationic and anionic heavy metallic species by a mixed solvent extraction system and a novel contained liquid membrane device, Ind. Eng. Chem. Res. 35 (1996) 4214–4220.
[85] W.S. Ho, T.K. Poddar, New membrane technology for removal and recovery of chromium from waste waters, Environ. Prog. 20 (2001) 44.
[86] J. Chau, G. Obuskovic, X.M. Jie, K.K. Sirkar, Pressure swing membrane absorption process for shifted syngas separation: modeling vs. experiments for pure ionic liquid, J. Membr. Sci. 453 (2014) 61–70.
[87] T. Mulukutla, G. Obuskovic, K.K. Sirkar, Novel scrubbing system for post-combustion CO_2 capture and recovery: experimental studies, J. Membr. Sci. 471 (2014) 16–26.
[88] T. Mulukutla, J. Chau, D. Singh, G. Obuskovic, K.K. Sirkar, Novel membrane contactor for CO_2 removal from flue gas by temperature swing absorption, J. Membr. Sci. 493 (2015) 321–328.
[89] K.K. Sirkar, Membrane separations: newer concepts and applications for the food industry, in: R.K. Singh, S.S.H. Rizvi (Eds.), Bioseparation Processes in Foods, Marcel Dekker, New York, 1995, pp. 353–356.

CHAPTER 14

Ceramic Membranes

W.-H. Xing

National Engineering Research Center for Special Separation Membranes, Nanjing Tech University, Nanjing, Jiangsu Province, China

14.1 Introduction

Ceramic membranes are also known as inorganic ceramic membranes, which is formed by developing a layer or multilayer of ceramic materials upon the prefabricated base matrix. The major materials include alumina (Al_2O_3), zirconia (ZrO_2), titanium oxide (TiO_2), silicon oxide (SiO_2), etc. The configurations of the ceramic membrane include plate, tube, and multichannel ones. Figs. 14.1 and 14.2 show the multichannel ceramics schematics and pictures [1].

Ceramic membranes are usually characterized by a sandwich structure composed of support layer (also called vector layers), transition layer (also known as the middle tier), and function layer (also known as selective layer). The support layer aperture ranges from 1 to 20 μm with porosity of 30–65% [2] and its major role is to ensure the mechanical strength of the overall membrane. The aperture size of the transition layer is smaller relative to the support, and the role of this layer is to prevent the permeation of particles into porous support layer during selective layer formation. The thickness for transition layer is generally in the range of 20–60 μm and porosity 30–40%. The function layer has the thickness about tens to hundreds of nanometers (NM) and porosity around 40–55%.

According to the pore size of functional layer, ceramic membranes can be classified into microfiltration (MF) (pore size greater than 50 nm), ultrafiltration (UF) (pore size 2–50 nm), nanofiltration (NF) (pore size less than 2 nm), and other species (mainly gas membrane, pervaporation membrane) [2]. The advantages of with ceramic membranes as compared with polymeric membranes include sound resistance toward stimuli such as chemicals, acid, alkali, solvents and high temperature, good mechanical strength, and hence performance stability and easy regeneration. Possessing all these features, ceramic membranes have been successfully applied in metallurgical process usually involving harsh environments [3–11].

Membrane-Based Separations in Metallurgy
http://dx.doi.org/10.1016/B978-0-12-803410-1.00014-1

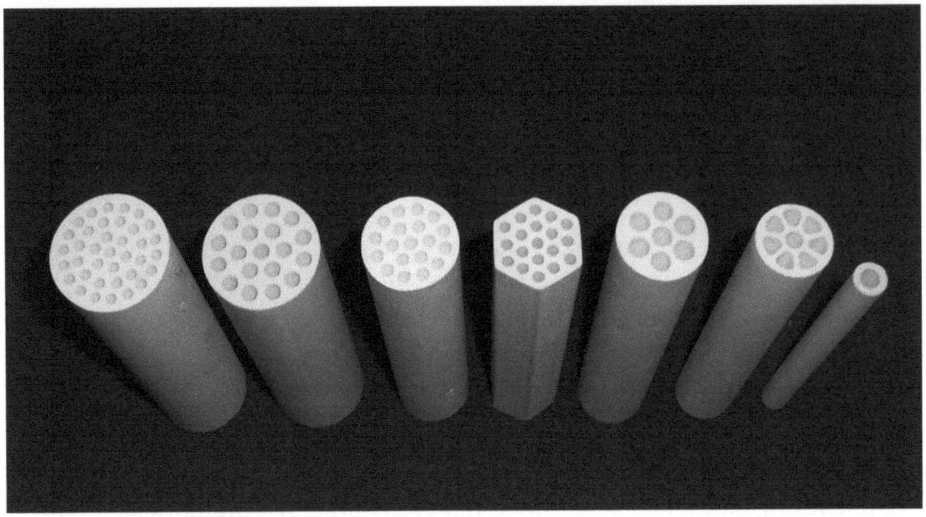

Figure 14.1: Pictures of ceramic membrane with single- and multi-channel design.

14.2 General Progress in Membrane Fabrication and Modification

14.2.1 Porous Ceramic Membranes

Preparation technologies of porous inorganic ceramic membrane have made significant progress in the past decades. So far, the methods involved in preparing porous ceramic membranes are focused on achieving desired porosity and pore size/distribution. As for the former target, the pore-forming agent, template are widely relied on, and ceramic fiber construction is a relatively new technology [12–16]. As for the latter target determining the separation precision, controlling the conditions of, the methods [17,18] such as sol–gel method, chemical vapor deposition (CVD), supercritical fluid deposition (SCFD), and many other surface modification is the principal approach.

In addition to porosity and pore size/distribution, the major factors determining the separation properties of a membrane as aforementioned, such factors as surface charge, hydrophilicity

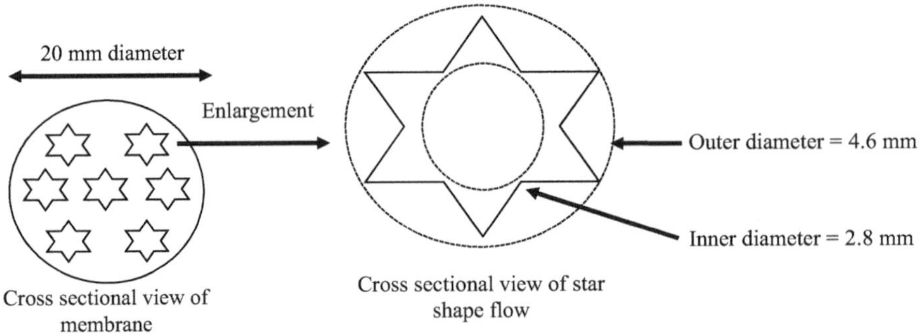

Figure 14.2: Schematic of a ceramic membrane showing arrangement of star-shaped channels.

(polarity), and hydrophobicity etc. are also known to impact membrane separations. Innovation in membrane fabrication technologies and modification of existing ceramic membrane have already successfully delivered ceramic membranes with suitable porous structure and other physiochemical properties leading to higher separation efficiency.

14.2.1.1 Sol–Gel Method

Sol–gel method is currently the principal method for preparing microporous ceramic membrane [19], which usually employs metal alkoxides as starting materials. After being dissolved in organic solvent at room temperature, the metal alkoxide hydrolyzes and a simultaneous condensation reaction occurs to form polymeric or colloidal sols. After the dealcoholization of the hydrolyzed mixture, generally, acidic condition (preferably with hydrochloric acid (HCl), and $pH < 1.1$) is made of the sol, followed with its coating on surface of porous support. Finally, inorganic ceramic membrane is formed through high temperature sintering. The simplified preparation procedure of sol–gel method is illustrated in Fig. 14.3. The mechanism for adjusting separation accuracy of ceramic membrane formed by sol–gel method is based on controlling collosol size. Generally, the size of the collosol that can be generated in sol–gel method ranges from a few nm to hundreds of nm. The membrane layer thus formed has relatively uniform pore size distribution and is suitable for UF and NF.

In recent years, improvement has been achieved with traditional sol–gel method. In Qi et al.'s work, a novel sol–gel method for preparing defect-free TiO_2 NF membranes were reported. TiO_2 sol with a mean particle size of 1.2 nm was synthesized with titanium isopropoxide as a precursor and the particle size showed a stability for more than 21 days [20]. Disk α-Al_2O_3 supported mesoporous γ-Al_2O_3 membrane with mean pore size of 3 nm was used as substrate. TiO_2 *sol was coated on the support* for just one time. The molecular weight cut-off (MWCO), mean pore size and pure water flux of TiO_2 NF membrane calcined at 350°C were 890, 1.5 nm and 5.3 L/(m^2·h)(at 8×10^5 Pa, 20°C), respectively. Higher temperature led to larger pore size, higher pure water flux, and higher MWCO for PEG.

14.2.1.2 Surface Modification

Some earlier works considered research about attaching a thin layer or even monolayer of organic or inorganic molecules on the ceramic surface was initiated in 1990s as a modification approach. This thin layer was expected, in specific cases, to improve the performances of the membrane by, on one hand reducing the pore size, and on the other hand,

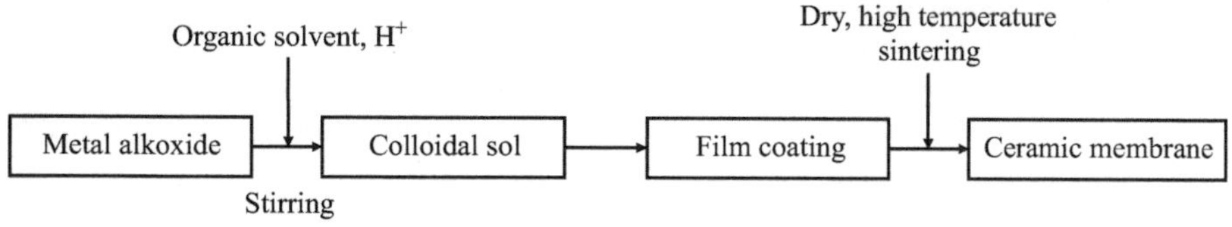

Figure 14.3: Ceramic membrane coatings prepared by sol–gel/flowchart.

promoting an eventual specific interaction between the membrane surface and the permeating molecules. The compounds that are used in the earlier years to achieve modification include phosphate derivatives [21], chlorosilane derivatives [22], or triethoxysilicon compounds [23]. Several new technologies from then on have been developed for surface design. One of the important and valuable research directions is grafting. For example, Khemakhem and Amar grafted organosilane onto mesoporous membrane surface to form a layer of organic molecules [24]. Adjusting the chain length and the radicals could actualize the controlling of pore size and special surface properties that are required by specific applications.

Another technology adopted widely nowadays for tuning ceramic membrane is CVD, which is also known as chemical evaporation. The gas-phase chemical reactions on the surface of the ceramic membrane lead to the formation of a layer of thin film. Labropoulos et al. [25] adopted circulation CVD at 573 K to successfully reduce the pore size of SiO^2 membrane from 1 to 0.56 nm.

Atomic layer deposition (ALD) is a thin film deposition method in which a film is grown on a substrate by exposing its surface to alternate gaseous state precursors [26]. In contrast to CVD, nevertheless, the precursors are not in continuous contact with the support in the reactor, but introduced as a series of nonoverlapping pulses. In each of these pulses the precursor molecules react with the surface in a self-limiting way, so that the reaction terminates once all the reactive sites on the original surface are consumed. Fig. 14.4 exhibited

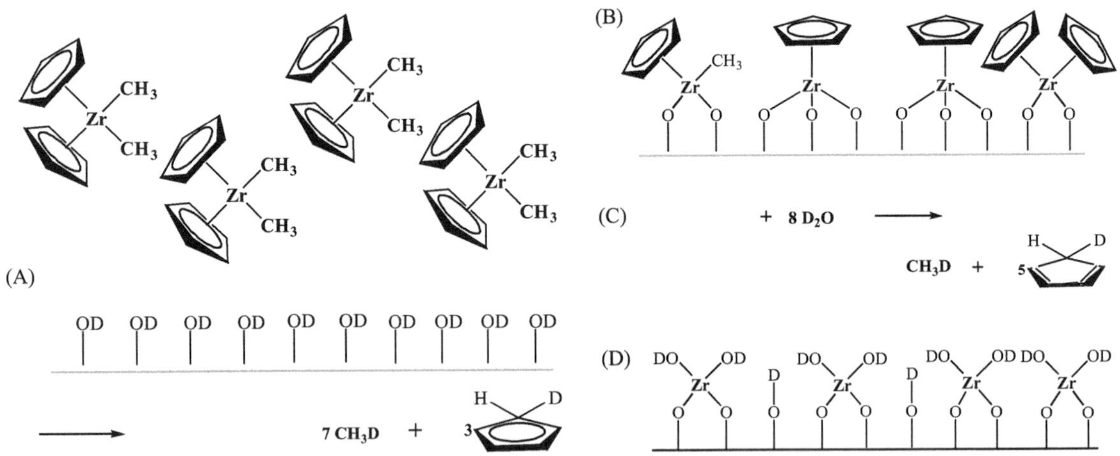

Figure 14.4: Proposed surface reactions of $Cp_2Zr(CH_3)_2/H_2O$ ALD process at 350°C. The ALD cycle consists of the (A) metal precursor pulse on the OD-terminated surface, (B) metal precursor adsorbed and CH_3D and CpD released, (C) inert gas purge and D_2O pulse, and (D) CH_3D and CpD released, the surface converted back to OD-terminated, and the inert gas purge, leaving the surface ready for the next ALD cycle.

the formation of ZrO_2 layer via ALD using bis(cyclopentadienyl)dimethylzirconium ($Cp_2Zr(CH_3)_2$). Li et al. used the ADL technology to deposit a layer of Al_2O_3 on the ceramic support with an average pore size of 50 nm. The final pore size may be adjusted by repeated deposition [27]. The 600 times of depositions can generate a rejection of 97.1%, in contrast with 2.9% obtained with only one cycle of deposition.

Supercritical fluid deposition (SCFD) is another choice for modifying membrane surface structure for more accurate discrimination. In principle, SCFD utilizes supercritical fluid (SCF) as a solvent to carry ceramic material precursor to the surface of porous support. Release of the pressure decreases the solubility of precursor in the SCF, which initiates the solid precipitation inside the pores to reduce the pore size [28,29].

A novel "On-Stream Supercritical Fluid Deposition" (OS-SFD) process has been investigated coupling the sol–gel chemistry and a filtration/compression operation in supercritical CO_2 (SC-CO_2) in Durand et al.'s work [30].

Conventionally, Sol-Gel in SCF method has following advantages: possible facilitation in efficient anchoring of the deposited material; limited cracks, and defects formation; ecofriendliness due to reduced quantity of organic solvents and wastes. However, a challenge still remains in a control of the deposit location, homogeneity, reproducibility, design, composition, and microstructure directly formed on commercial supports. In Durand et al.'s work the OS-SCFD involves the filtration of a formulated sol through a macroporous tubular support. Tetraethoxysilane (TEOS) as SiO_2 precursor and yttrium acetate [$(C_2H_3O_2)_3Y$] as a gelation retarding agent. The use of the latter chemical can also provide more flexibility to study the Si-alkoxide deposition process. The combination of the thermodynamic control associated with SCFD and the kinetic control due to sol–gel reactions may provide an additional degree for adjusting the membrane microstructure, in comparison with the classical Sol-gel process.

The schematic of the set-up used for solubilization, delivery, and deposition of the SiO_2-based sol on the ceramic tubular support (155-mm long α-Al_2O_3, with 200 nm pore sizes) is illustrated in Fig. 14.5. Using the system, membrane formation can be monitored by real-time registering the ΔP evolution pattern as a function of experimental conditions. This new design was finally demonstrated to yield uniform membranes on/in porous ceramic tubular supports. The possible membrane formation mechanisms on α-Al_2O_3 substrate were given in Fig. 14.6. For deposition at temperature lower than T_{gel}, only supersaturation and precipitation takes place. The clusters are disconnected, forming a weakly compacted layer on the support. At temperatures close to T_{gel}, the cocondensation of clusters occurs on the membrane surface, generating a mesoporous gel layer. When temperatures higher than T_{gel} is used, an intracondensation of the clusters occurs before deposition on substrate, leading to the yield of interparticles macropores.

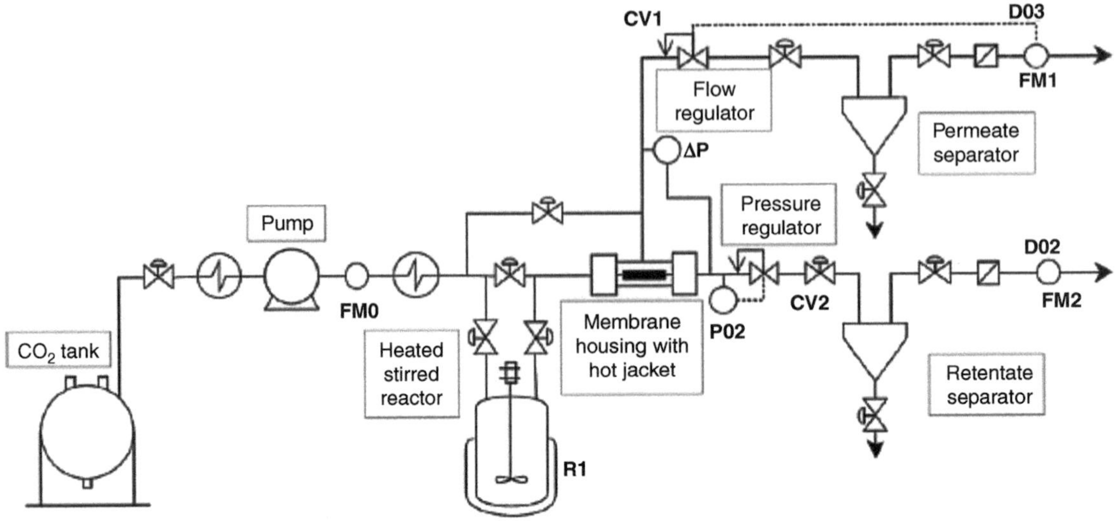

Figure 14.5: Scheme of the experimental set-up developed for on-stream deposition in SC-CO_2 media.
FM0, inlet CO_2 flow meter; FM1, permeate flow meter; FM2, retentate flowmeter; P, transmembrane pressure; D03, permeate flow; D02, retentate flow; P02, retentate pressure; CV1, control valve for permeate flow regulation; CV2, control valve for retentate pressure regulation.

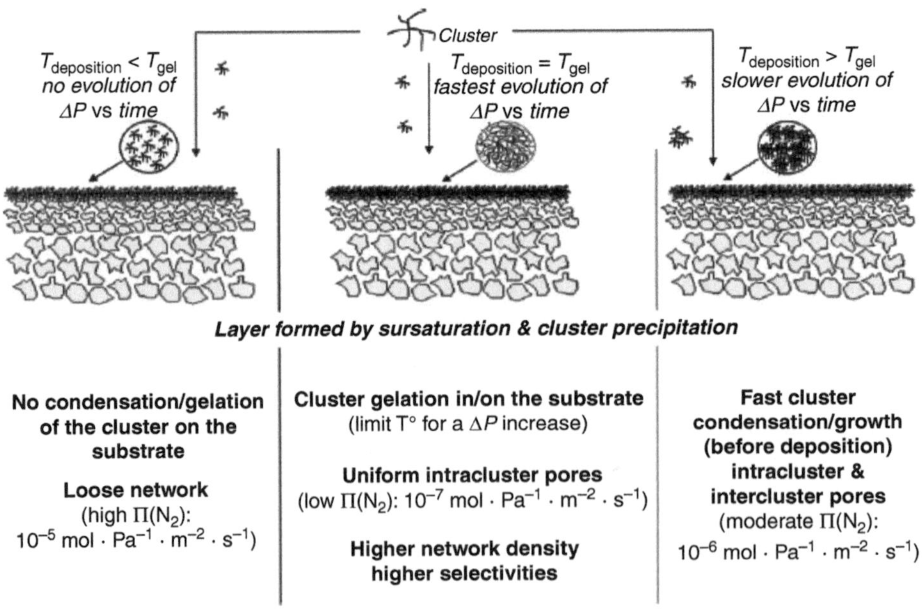

Figure 14.6: Schematic representation of the possible membrane formation mechanisms on-alumina macroporous substrate as a function of the deposition temperature selected for the OS-SFD process.

14.2.1.3 Other Methods

There are many other methods for fabricating porous ceramic membranes [31–36]. They are vacuum sputtering, physical vapor deposition (PVD), in situ granulating, solid state sintering, anodic oxidation, arc oxidation, in situ synthesis, and electrophoretic deposition. Porous Support membranes are generally prepared by direct solid state sintering method, which originated from the traditional pottery craft technology in China [37]. A type of oxide membrane with stable and high degree in pore size uniformity was produced by anodic oxidation method [33]. In this method, one face of a highly purified metal foil is oxidized at the anode inside the acidic electrolyte to form porous oxide membrane, while the other surface is dissolved by the acid. In situ synthesis method is also called in situ crystallization method [34], which has been mainly adopted preparing zeolite membranes [31,38]. Wang and Yan investigated the synthesis of oriented continuous zeolite MFI thin films on stainless steel substrate by direct in situ crystallization [38]. It was found that highly *b*-oriented MFI layer is favored by medium OH^-/Si (e.g., $OH^-/Si = 0.32$) and low Na^+/tetrapropylammonium (TPA^+) ratios (e.g., $Na^+/TPA^+ < 0.31$). What has also attracted attention during the past decade is the combination of different surface engineering technologies in developing ceramic membranes. It is proved that the innovated fashion could generate new types of function layer, which constitutes new research directions regarding surface engineering. In Giacomo et al.'s work [39], plasma was employed for assisting the laser vapor deposition in fabrication of TiO_2 membrane, of which the morphology, chemical composition, and sedimentation ratio were superior over those of membrane via mono-technique.

14.2.2 Gas Separation Membranes

Yet another type of ceramic membranes worthy of some elaboration is the gas separation membranes. For membrane with metals as function layer they will be introduced in the chapter about metal membranes. The aforementioned methods such as CVD, sol–gel can also produce membranes suitable for gas separation. Liu et al. modified an Al_2O_3 ceramic membrane with average pore size of 4 nm by CVD method to prepare a SiO_2 layer with the thickness of about 1.5μm and the pore size of 0.44 nm [37]. The resultant ceramic membrane presented N_2 permeation of 0.13 m m^{-2} h^{-1} and the separation factor of N_2/alkyl of around 400.

Currently, there exist many researches about single-component ceramic membrane for gas separation, but few are reported about mixed-component or composite ceramic membranes [35]. Composite membranes refer to those composed of two or more oxides that are "homogeneously" mixed together. The surface properties of different materials (eg, adsorption, ion exchange) have a considerable influence on separation properties. In general, the permeation rate of composite membranes is low due to the smaller pore diameter and lower porosity compared with single-component membranes. Nonetheless, they can achieve significantly higher selectivity by combining the properties such as adsorption of different component. In addition, adjustment of the ratio of different components is a convenient way leading to balance between the trans-membrane permeability and the separation factor.

Membrane technologies have the potential to separate CO_2 at lower costs and with lower energy consumption. In recent years, dense dual-phase inorganic membranes that exhibit an infinite selectivity for CO_2 over N_2, or H_2 at temperatures above 500°C have been developed [40]. Much effort has been made on inorganic membranes for CO_2 separation and some of them exhibit high CO_2 perm-selectivity and permeance [41–43]. But, at higher temperatures, the CO_2 selectivities of these inorganic membranes decrease. An amine-modified mesoporous silica membrane exhibits a CO_2/N_2 selectivity of 800 at 100°C, but it decreases to 10 at 150°C. The major challenges of present CO_2 separation membranes are therefore the poor compatibility with high-temperature process stream and the limited CO_2 selectivity.

The dense dual-phase membranes consist of a porous inorganic phase and a molten carbonate phase. The inorganic phase is usually made of solid oxide electrolytes such as yttria doped zirconia (YSZ), gadolinia doped ceria (CGO), or samaria doped ceria (SDC). They serve as a porous support to improve the stability of the membrane in a CO_2 atmosphere [44] and also as an electron and/or oxygen ion conductor.

The Li/Na/K molten carbonate phase is introduced into the porous support by a direct infiltration method. At high temperature the CO_2 can transport through the dual-phase membrane as a carbonate-ion (CO_3^{2-}) under the driving force of the CO_2 partial pressure gradient. The mechanisms for selective electrochemical CO_2 separation membranes using the dual-phase membranes are illustrated in Fig. 14.7 [44].

For the successful preparation of the asymmetric ceramic-carbonate dual-phase membrane four key factors that need to be taken into account are (1) matching thermal performance

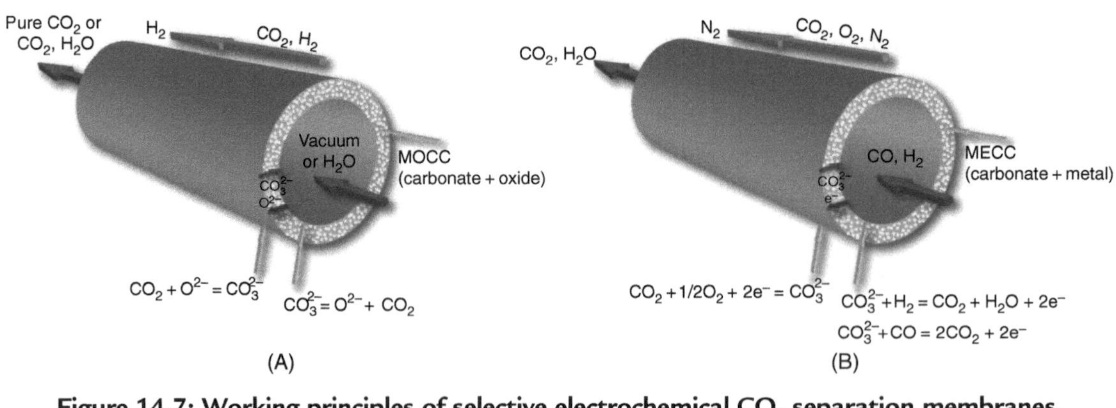

Figure 14.7: Working principles of selective electrochemical CO_2 separation membranes. (A) MOCC (mixed carbonate-ion and oxide-ion conductor) and (B) MECC (mixed carbonate-ion and electron conductor).

Figure 14.8: SEM images.
(A) Cross-section of the porous SDC/SDC-BYS substrate and (B) the porous SDC layer without infiltration; (C) cross-section of the SDC/SDC-BYS asymmetric membrane and (D) the SDC-carbonate membrane layer after infiltration.

and chemical compatibility between the membrane and the support; (2) good interfacial bonding; (3) a defect-free membrane layer; and (4) a carbonate nonwettable support. In Dong et al.'s work [40], SDC was used for preparing a thin porous inner layer, whereas an SDC and $Bi_{1.5}Y_{0.3}Sm_{0.2}O_3$–d (BYS) mixture was used for the fabrication of a porous outer support layer. As shown in Fig. 14.8, the SDC inner layer with the thickness of about 150 mm is well bound with the SDC-BYS outer layer. The SDC inner layer has high porosity and uniform pore structure (Fig. 14.8B). After infiltration with molten carbonate, the inner membrane layer becomes dense, whereas the SDC-BYS outer layer remains porous because of the carbonate nonwettable property of BYS (Fig. 14.8C). As shown in Fig. 14.8D, SDC and carbonate phases can be distinguished easily and are well mixed. The off-white phase with a clear grain boundary is SDC, whereas the dark gray phase is carbonate.

14.3 Application-Orientated Consideration for Porous Ceramic Membrane Fabrication

As obviously demonstrated in previous chapters, membrane-based separations have been successfully applied in metallurgical industry. As for ceramic membranes, an point that improve its competiviteness is their chemical and physical stability. Nevertheless, the industrial adoption of ceramic membrane technology is still rather limited [45]. The first reason behind this situation is that the cost of the membranes is so high that many applications become economically unfavorable, despite the excellent technical feasibility in terms of both membrane and system. Further analysis raises the concern about the matching between the limited membrane categories and the complexity for the characteristics of the myriads of processes. Problems such as serious membrane fouling, finite processing capacity per unit membrane area are the consequences of the mismatching. In comparision with polymeric membranes, these problems may be more severe for ceramic membranes, as the latter are almost unaminously relied on as solutions for applications with extremely complicated environment.

In Section 14.2, the general progress for ceramic membrane development has been introduced. Notwithstanding to overcome aforementioned challenges, it is highly demanded

Table 14.1: The major considerations for application-orientated ceramic membrane development

First Level	Second Level
Quantified preparation methods	Modeling and quantified control of thickness of porous ceramic membranes Modeling and quantified control of pore size of porous ceramic membranes Directional preparation of highly porous ceramic membranes Adjustment of membrane surface properties and its influence on separation efficiency
Membrane design and selection for specific application	Modeling of membrane function/microstructure relation for nanometer catalyst Modeling of membrane function/microstructure relation for colloid system
Fouling and structural variation during operation	Membrane fouling control in industrial process Evolution of membrane microstructure and function during application

to construct an integral system incorporating material design and membrane development on the basis of fully considering industrial in-service conditions (Fig. 11.1). The three aspects that have been viewed as the basic building blocks for constructing the system specifically for ceramic membranes are as follows: (1) quantified preparation methods, (2) membrane design and selection for specific application, (3) fouling and structural variation during operation. Hereinafter is Table 14.1 summarizing the factors critical for each aspect. Some achievements are introduced in this section. For a better comprehension of R&D trend and experimental/theoretical achievements, the review by Xing et al. should be referred to as the major guidance [45].

14.4 Industrialization

Research activities about ceramic membrane have been rigorous with generating membranes with a wide variety and excellent properties. Based on the strong R&D environment, some companies have already been established possessing the technologies and capacity to deliver high quality ceramic membrane products. The major overseas producers for ceramic membranes include Inopor, NGK, CeraMem, TNO, NOVEM, HOME, Smart, PervaTech, Pall, TENAMT, BWF, etc. [46–53].

R&D of ceramic membranes in China was initiated in the 1990s. With the support from the national ministries, around 20 types of products with different specifications and models have been developed using Al_2O_3, ZrO_2, and TiO_2 as materials [54]. The estimation released by Chinese Membrane Industry Association shows that at present the ceramic membrane market size in China accounts for 3% of the entire domestic membrane market size. This percentage is far below the worldwide level of around 10–20%. Nevertheless, it is growing quickly. China's membrane market occupies about 15% of the international one currently, and its growth is faster than that of the latter.

From the point of view regarding the level of the industrialization, the number of the domestic companies in China is more than 100 that are engaged in the production of porous ceramic membrane and materials, with the entire industrial scale being around hundreds of millions of Chinese CNY per annum. Except for Jiangsu Jiu-Wu Hi-Tech Co. Ltd., Jiangsu Kaimi Membrane Technology Co. Ltd., and Sinoma Advanced Material Co. Ltd, however, the other companies are not competitive due to lack of large scale production capacity. They are generally characterized by small production scale, lagging technologies, simple equipment, and inability in product quality control. Among those membranes commercialized, the MF membranes are comparable with those produced by foreign companies. As UF ceramic membrane, Jiu-Wu High Tech Company adopted template method to control the pore size with the range of 3–10 nm and succeeded in pilot-scale production. The Jiu-Wu High Tech was incubated from Nanjing Tech University. Notwithstanding, the overall industrialization scale and technology level of Chinese companies are still lagging behind those of foreign companies such as Pall, TENAMT, BWF. For example, China-based companies have no capacity to produce high-end NF membranes at present stage.

14.5 Perspectives

The emphases of future development regarding ceramic membrane fabrication and application will cover following aspects: (1) improvement in separation accuracy and stability, making possible the continuous and efficient operation of NF scale separation for liquid solution; (2) development of ceramic membrane with large pore size and high porosity, facilitating long-term stability in high temperature solid/gas separations involved in environmental protection and efficient utilization of resources; (3) adjustment of ceramic membrane surface, expanding application fields by modifying surface hydrophilicity/hydrophobicity, bio-compatibility, charge property, etc.; (4) actualization of low cost production of ceramic membrane and construction of application-orientated system incorporating membrane material design and fabrication, revolving the bottlenecks constraining the application extension; (5) development of materials with excellent resistance toward strong acid and base and improvement of stability for separation efficiency, widening the application scope in engineering industry.

Glossary and Symbols

ALD	Atomic layer deposition
Al_2O_3	Alumina
CGO	Gadolinia-doped Ceria
$(C_2H_3O_2)_3Y$	Yttrium acetate
CVD	Chemical vapor deposition
HCl	Hydrochloric acid
NF	Nanofiltration
nm	Nanometer
OS-SCFD	On-stream supercritical fluid deposition
PVD	Physical vapor deposition

SCFD	Supercritical fluid deposition
SDC	Samaria-doped Ceria
SiO_2	Silicon oxide; silica
TEOS	Tetraethoxysilane
TiO_2	Titanium oxide
UF	Ultrafiltration
YSZ	Yttria-doped Zirconia
ZrO_2	Zirconia

References

[1] T.Y. Chiu, A.E. James, Effects of axial baffles in non-circular multi-channel ceramic membranes using organic feed, Sep. Purif. Technol. 51 (2006) 233–239.

[2] T. Tsuru, Inorganic porous membranes for liquid phase separation, Sep. Purif. Rev. 30 (2001) 191–220.

[3] Z.B. Fang, G.J. Xiao, G.Y. Yin, Application in oil emulsion treatment of inorganic membrane technology, The Third National Symposium Proceedings of Application of Membrane Separation Technique in Metallurgical Industry, 2009, pp. 74–75.

[4] W. Xu, H. Tian, G.Y. Yin, The inorganic ceramic membrane filtration in the application of cutting fluid recycle technology, The Third National Symposium Proceedings of Application of Membrane Separation Technique in Metallurgical Industry, 2009, pp. 81–83.

[5] N. Yin, H. Liu, W.H. Xing, Membrane method treatment waste liquid of coking desulfurizing technology research, J. Membr. Sci. Technol. 30 (2011) 75–78.

[6] F. Zhang, F. Han, Z.X. Zhong, The research progress of ceramic membrane used for high temperature exhaust gas purification, The Fourth National Symposium Proceedings of Application of Membrane Separation Technique in Metallurgical Industry, 2014, pp. 42–47.

[7] T. Z. Li, L.H. Lin, X.S. Lin, Ceramic membrane removal technology and its application in nonferrous metallurgy flue gas dust removal, The Fourth National Symposium Proceedings of Application of Membrane Separation Technique in Metallurgical Industry, 2014, pp. 62–64.

[8] Z.G. Wang, Membrane separation technology in the application in hydrometallurgy, The Fourth National Symposium Proceedings of Application of Membrane Separation Technique in Metallurgical Industry, 2014, pp. 54–61.

[9] X.J. Zhang, Z.H. Wang, W.X. Zeng, YT membrane separation technology in the application of closed -arc furnace high temperature gas purification, The Fourth National Symposium Proceedings of Application of Membrane Separation Technique in Metallurgical Industry, 2014, pp. 25–29.

[10] X.J. Zhang, Z.H. Wang, W.X. Zeng, The application of YT membrane separation technology of removing arsenic in nonferrous smelting industry, The Fourth National Symposium Proceedings of Application of Membrane Separation Technique in Metallurgical Industry, 2014, pp. 34–37.

[11] D.S. Liu, X.J. Zhang, Q.L. Zheng, YT membrane application in wet metallurgy industry, The Fourth National Symposium Proceedings of Application of Membrane Separation in Metallurgical Industry, 2014, pp. 38–41.

[12] G.X. Dong, H. Qi, N.P. Xu, Effect of active carbon doping on structure and property of porous alumina support, J. Chin. Ceram. Soc. 40 (6) (2012) 844–850.

[13] P. Kumar, J. Ida, S. Kim, V.V. Guliants, J.Y.S. Lin, Ordered mesoporous membranes: effects of support and surfactant removal conditions on membrane quality, J. Membr. Sci. 279 (1/2) (2006) 539–547.

[14] K. Zhao, Y.Q. Fan, N.P. Xu, Preparation of three dimensionally ordered macroporous SiO_2 membranes with controllable pore size, Chem. Lett. 36 (2007) 464–465.

[15] X.B. Ke, H.Y. Zhu, X.P. Gao, J.W. Liu, Z.F. Zheng, High performance ceramic membranes with a separation layer of metal oxide nanofibers, Adv. Mater. 19 (6) (2007) 785–790.

[16] M.H. Qiu, S. Fan, Y.Y. Cai, Y.Q. Fan, N.P. Xu, Co-sintering synthesis of bi-layer titania ultrafiltration membranes with intermediate layer of sol-coated nanofibers, J. Membr. Sci. 365 (1/2) (2010) 225–231.

[17] Y.Q. Fan, H. Qi, N.P. Xu, Advance in preparation techniques of ceramic porous membrane, CIESC J. 64 (2013) 108–115.

[18] R.H. Chen, Z.D. Fang, The present and future condition of applications in water treatment of inorganic ceramic membrane, J. Chongqing College Ind. 19 (2004) 1–4.

[19] R.S.A. De Lange, J.H.A. Hekkink, K. Keizer, Formation and characterization of supported microporous ceramic membranes prepared by sol-gel modification techniques, J. Membr. Sci. 99 (1995) 57–75.

[20] H. Qi, S.-D. Li, X.-L. Jiang, J. Han, TiO_2 preparation and ion rejection performance of nanofiltration membranes, J. Inorg. Mater. 26(3) (2011) 3–6.

[21] J. Randon, P. Blanc, R. Paterson, Modification of ceramic membrane surfaces using phosphoric acid and alkyl phosphonic acids and its effect on ultrafiltration of BSA protein, J. Membr. Sci. 98 (1995) 119–129.

[22] J.J. Pesek, Plenary lecture-conversion of oxide surfaces to hydride surfaces, Spec. Publ. Roy. Soc. Chem 139 (1994) 1–23.

[23] R.E. Johnson, D.I. Ford, Stability and reactivity of dimethylethoxysilane, NASA Conference Publication, 1993, p. 811.

[24] S. Knemakhem, R.B. Amar, Grafting of fluoroalkysilanes on microfiltration Tunisian clay membrane, Ceram. Int. 37 (2011) 3323–3328.

[25] A.I. Labropoulos, G.E. Romano, G.N. Karanikolos, F.K. Katsaros, N.K. Kanellopoulos, Comparative study of the rate and locality of silica deposition during the CVD treatment of porous membranes with TEOS and TMOS, Micro. Meso. Mater. 120 (2009) 177–185.

[26] J. Niinistö, A. Rahtu, M. Putkonen, In situ quadrupole mass spectrometry study of atomic-layer deposition of ZrO_2 using $Cp_2Zr\ (CH_3)_2$ and water, Langmuir 21 (2005) 7321–7325.

[27] F.B. Li, Y.Q. Fan, W.H. Xing, Y. Wang, Modification of ceramic membranes for pore structure tailoring: the atomic layer deposition route, J. Membr. Sci. 397 (2012) 17–23.

[28] S. Sarrade, C. Guizard, G.M. Rios, New application of supercritical fluids and supercritical fluids processes in separation, Sep. Purif. Technol. 32 (2003) 57–63.

[29] Z.H. Wang, J.H. Dong, N.P. Xu, J. Shi, Process analysis and simulation on supercritical solution infiltration in ceramic membrane, J. Nanjing Univ. Chem. Technol. 20 (1988) 18–24.

[30] V. Durand, M. Drobek, A. Hertz, J.C. Ruiz, S. Sarrade, C. Guizard, A. Julbe, Evaluation of a new on-stream supercritical fluid deposition process for sol–gel preparation of silica-based membranes on tubular supports, J. Supercrit. Fluids 77 (2013) 17–24.

[31] Y.S. Tian, C.Z. Chen, D.Y. Wang, L.X. Chen, T.Q. Lei, Research progress in fabrication of ceramic coatings, Mater. Technol. 13 (2005) 427–434.

[32] T. Pavlovic, A. Ignatiev, Optical and microstructural properties of anodically oxidized aluminum, Thin Solid Films 138 (1986) 97–109.

[33] L. Zhang, The application of inorganic ceramic membrane in wastewater treatment, Ind. Water Treatm. 21 (2001) 1–4.

[34] J. Dong, Y.S. Lin, In situ synthesis of P-type zeolite membranes on porous α-alumina supports, Ind. Eng. Chem. Res. 37 (1998) 2404–2409.

[35] J.L. Zhao, X.H. Wang, J.J. Hao, R.Z. Chen, T.L. Li, Electrophoretic deposition and its application in new type ceramic technology, Funct. Mater. 36 (2) (2005) 165–169.

[36] L.H. Li, Y.M. Kong, H.W. Kim, Y.W. Kim, H.E. Kim, S.J. Heo, J.Y. Koak, Improved biological performance of Ti implants due to surface modification by micro-arc oxidation, Biomaterials 25 (14) (2004) 2867–2875.

[37] X.D. Liu, J.S. Li, A.J. Zhang, The preparation and characterization of alumina microfiltration membranes, J. Nanjing Univ. Sci. Technol. 26 (2002) 406–409.

[38] Z. Wang, Y. Yan, Controlling crystal orientation in zeolite MFI thin films by direct in situ crystallization, Chem. Mater. 13 (2001) 1101–1107.

[39] A. De Giacomo, V.A. Shakhatov, G.S. Senesi, O.D. Pascale, F. Prudenzano, Plasma-assisted pulsed laser deposition for the improvement of the film growth process, Appl. Surf. Sci. 186 (1) (2002) 533–537.

[40] X. Dong, J.O. Landeros, Y.S. Lin, An asymmetric tubular ceramic-carbonate dual phase membrane for high temperature CO_2 separation, Chem. Commun. 49 (2013) 9654–9656.

[41] M. Anderson, H.B. Wang, Y.S. Lin, Inorganic membranes for carbon dioxide and nitrogen separation, Rev. Chem. Eng. 28 (2012) 101–121.

[42] T.L. Chew, A.L. Ahmad, S. Bhatia, Ordered mesoporous silica (OMS) as an adsorbent and membrane for separation of carbon dioxide (CO_2), Adv. Colloid Interface Sci. 153 (2010) 43–57.

[43] J. Caro, M. Noack, Zeolite membranes–recent developments and progress, Micro. Meso. Mater. 115 (2008) 215–233.

[44] L. Zhang, N. Xu, X. Li, S. Wang, K. Huang, W.H. Harris, W.K.S. Chiu, High CO_2 permeation flux enabled by highly interconnected three-dimensional ionic channels in selective CO_2 separation membranes, Energy Environ. Sci. 5 (2012) 8310–8317.

[45] W.H. Xing, Y.Q. Fan, Z.X. Zhong, N.P. Xu, Recent advances in process-engineering oriented preparation and application of ceramic membranes, CIESC J. 60 (2009) 2679-2668.

[46] R.W. Van Gemert, F. Petrus-Cuperus, Newly developed ceramic membranes for dehydration and separation of organic mixtures by pervaporation, J. Membr. Sci. 105 (1995) 287–291.

[47] A.W. Verkerk, P. Van Male, M.A.G. Vorstman, J.T.F. Keurentjes, Properties of high flux ceramic pervaporation membranes for dehydration of alcohol/water mixtures, Sep. Purif. Technol. 22 and 23 (2001) 689–695.

[48] H.M. Van Veen, Y.C. Van Delft, C.W.R. Engelen, P.P.A.C. Pex, Dewatering of organics by pervaporation with silica membranes, Sep. Purif. Technol. 22 and 23 (2001) 361–366.

[49] J.E. Ten Elshof, C.R. Abadal, J. Sekulic, S.R. Chowdhury, D.H.A. Blank, Transport mechanisms of water and organic solvents through microporous silica in the pervaporation of binary liquids, Micro. Meso. Mater. 65 (2003) 197–208.

[50] T. Gallego-Lizon, E. Edwards, G. Lobiundo, L. Freitas dos Santos, Dehydration of water/*t*-butanol mixtures by pervaporation: comparative study of commercially available polymeric, microporous silica and zeolite membranes, J. Membr. Sci. 197 (2002) 309–319.

[51] T. Gallego-Lizon, Y.S. Ho, L. Freitas dos Santos, Comparative study of commercially available polymeric and microporous silica membranes for the dehydration of IPA/water mixtures by pervaporation/vapour permeation, Desalination 149 (2002) 3–8.

[52] S. Sommer, B. Klinkhammer, T. Melin, Integrated system design for dewatering of solvents with microporous silica membranes, Desalination 149 (2002) 15–21.

[53] F. Petrus-Cuperus, R.W. Van Gemert, Dehydration using ceramic silica pervaporation membranes—the influence of hydrodynamic conditions, Sep. Purif. Technol. 27 (2002) 225–229.

[54] Y.M. Cao, H.Y. Xu, J.Q. Wang, Development status and prospect of inorganic ceramic membrane in China, Membr. Sci. Technol. 33 (2) (2013) 1–11.

CHAPTER 15

Metal Membranes

Y.-H. He

School of Powder Metallurgy, Central South University, Changsha, Hunan Province, China

15.1 Introduction

Researches about metal membranes are being intensified at present. Relative to the ceramic and polymeric membranes, metal membranes are advantageous in their mechanical strength, resistance toward high temperature, ease of sealing, and integrated processing [1]. Most metal membranes are characterized of gradient composite structure comprising principally a base or primary and a separation layer. Metal membranes are broadly classified into dense membranes and porous membranes, in terms of its separation layer structure. The structure that actually performs the separation task is the surface-active layer with thickness being $1 \sim 100$ μm. With this characteristic, the flux is usually higher than that of symmetric membrane by $50 \sim 100$ times [2]. The base materials for metal membranes are metal, while the selective materials are metals, metal oxides, or metal alloys.

The pore size of dense metal membranes is at the same order of the free path for gas molecules. Its invention is traced back to 1960s and the major application is gas separation and purification. Membranes of palladium (Pd) for permeating hydrogen (H_2) are the first inorganic membranes for generating highly purified H_2 [3]. Their selectivity toward H_2 is generally thought to be infinite when perfectly dense structure is realized.

Elementary Pd is expensive; fortunately its alloys with other metals can not only reduce the cost, but also significantly improve the membrane mechanical strength. The binary alloys with Pd mostly include Pd–silver (Ag) [4–6], Pd–copper (Cu) [7–9], Pd–nickel (Ni) [10,11], etc.

One of the disadvantages of Pd is that the surface of its membranes is liable to influence of impurities such as hydrogen sulfide (H_2S), carbon monoxide (CO), leading to performance deterioration [12,13]. This situation motivates the researches about tantalum (Ta), niobium (Nb), vanadium (V), Ni, etc. In effect these new choices are cheaper than platinum (Pt) and Pd. More importantly, they possess high permeation rate and diffusion coefficient. In the past decade, great achievement has been made in the development of no-Pd membranes [14,15]. Non-Pd metal membranes are categorized into pure metal membranes, crystalline alloy membrane, and noncrystalline alloy membranes. In the past decade, great achievement

Membrane-Based Separations in Metallurgy
http://dx.doi.org/10.1016/B978-0-12-803410-1.00015-3

has been made in non-Pd dense membranes development [13,14]. Nevertheless, their immediate applications are challenged by hydrogen embrittlement, thermal stability, and catalytic stability. Optimization of alloy components, improvement of the interface between catalytic layer and alloy membrane to enhance the catalytic stability, and modification of the fabrication technology are the emphases of future effort.

Porous metal membranes utilize metal material for primary structure and metal, metal oxide, or metal alloy as membrane or selective material [16,17]. Microporous membranes are the major products among all the porous metal membrane categories; nanofiltration (NF) metal membrane has not been produced at large scale yet. Therefore, porous metal membranes are still mainly employed in liquid–solid and gas–solid separations, which fall in the domain of microfiltration (MF) separation [2,18–20].

The fabrication of porous metal membrane mainly draws the experience from the methods for preparation of ceramic and dense metal membranes. The approaches that have been used are mainly suspended particle sintering, sol–gel method, electrodeposition, chemical vapor deposition, spray coating, track etching, and anodic oxidation [16]. Suspended particle sintering is derived from traditional powder metallurgy technology. With this method the inorganic powder selected is dispersed homogeneously into organic solvent. The other components inside the solution are binding agent, additive, and dispersant. Upon stabilization the slurry containing the ingredients mentioned earlier will be made into slug via special shaping or molding, followed by desiccation and sintering. The structure thus delivered in terms of porosity, pore size, pore size distribution is dependent on the shape, size, morphology, and size distribution of particles, the sintering technologies and conditions. The elementary metals that can be used for producing porous membrane via sintering include gold (Au), Ag, Cu, Ni, aluminum (Al), magnesium (Mg), titanium (Ti), chromium (Cr), tungsten (W), molybdenum (Mo), and stainless steel (SS). SS is the most widely used.

The first metal membranes emerging in the 1940s was made of nickel (Ni) and used for separating the uranium (U) isotopes. Its thermal stability was poor, which hindered its application in industry. In the following decades, development of porous metal membranes was very slow. It was in 1990s that Graver Company of United States successfully delivered a new type of *Scepter* metal membranes using SS for primary structure and TiO_2 as a selective material. This work opened up a new era for porous metal membrane development.

15.2 Stainless Steel-Based Porous Metal Membrane

15.2.1 Neat Stainless Steel Membranes

The preparation of the composite SS membrane generally involves fabrication of substrate, coating and sintering; for achieving desired performance in the final membrane, the SS substrate should satisfy the requirements on its various physiochemical properties, for example, the pore size, pore size distribution, surface roughness, porosity, mechanical strength, and chemical stability. The research targeting the SS-supported membrane formation is very active in the past decade. Compressed molding and phase inversion are the two methods reported most frequently in recent years for producing precursors of SS metal membranes.

Graver Company adopts Metal Injection Molding (MIM) technology to fabricate microporous SS tube [21]. MIM is a rapidly developmental new technology belonging to powder metallurgy for manufacturing machine replacement parts. Its basic process involves formation of viscoelastic body by mixing metal or alloy powder with organic binder according to a certain proportion, molding by injection mold, removal of binder, and sintering.

In Huang et al.'s work using compressed molding, the first step for precursor was fixing the SS powder on the surface of the rigid tubular mold [22]. Thereafter, two types of SS powders with different particle size were mixed by certain ratio and filled into the space between the rigid and elastic molds that already contained the aforementioned layer of powder. Isostatic cold pressing was applied to the mold after the space was sealed by an elastic cap. Thereafter, the semifinished product was converted into highly asymmetric SS membrane via high-temperature cementation. In Ye's work [23], the preparation and performance of porous SS membranes by compressed molding was investigated. It was found that the major factor influencing porosity was molding pressure, whereas that for the maximum pore size was the powder size.

The research targeting the SS membrane formation is very active in the past decade.

Porous SS hollow fibers were prepared via the dry–wet spinning of a particle loaded polymer solution, followed by thermal treatment at high temperature [24–25]. The morphology of the green fibers or precursors comprises sponge-like structures and macrovoids, which is due to the dynamics of phase inversion process. The morphology can be tuned by adjusting the spinning conditions and the composition of the dope solutions. As shown by Fig. 15.1A, the cross section of a fiber prepared from a spinning mixture containing 48 wt.% SS particles has macrovoids (finger-like voids) originating from both the inner and outer surface of the fiber. With a loading of 80 wt.% (E) a sponge-like structure at the outer surface of the fiber was identified. During the follow-up sintering, the morphology in terms of macrovoids and sponge structures has been largely preserved, whereas fiber dimension shrinkage due to densification was obvious. Fibers obtained by sintering at 1100°C have a pore size distribution in the ranges 0.2–0.6 μm and 1–10 μm. The fiber sintered at 1200°C contains predominantly small pores and large pores (≥1 μm) disappeared. The fibers sintered at 1050–1100°C possessed a high strength, 750–1150 MPa with a large nitrogen (N_2) permeance (1×10^{-4} mol m^{-2} Pa^{-1} s^{-1} at 21°C) (Fig. 15.2).

In Luiten-Olieman et al.'s work [25], exceptionally small radial dimensions (shown in Fig. 15.2) have been obtained by surface energy driven viscous flow of the green fiber. The extent of viscous deformation is related to the rheology of the particle-loaded green fiber

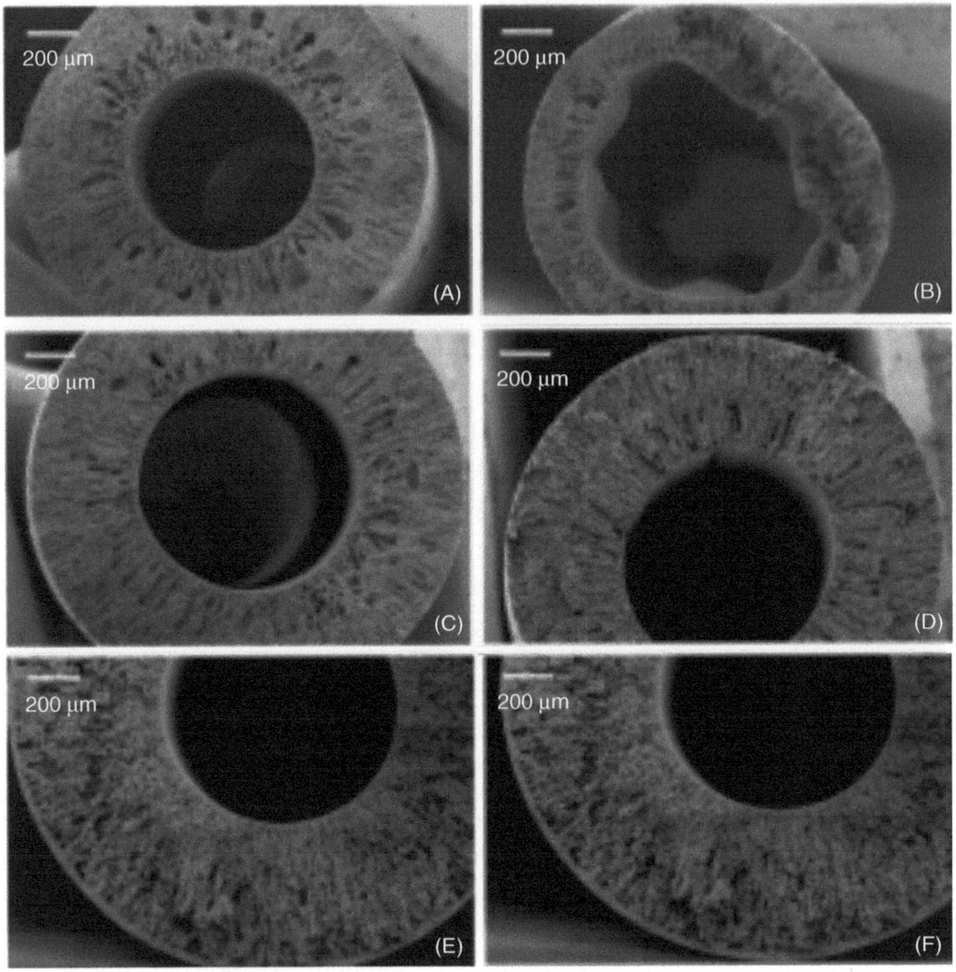

Figure 15.1: SEM images of fibers with 48–70–80 wt.% of stainless steel.
(A) 48 wt.% green compact, (B) 48 wt.% sintered 1100°C, (C) 70 wt.% green compact, (D) 70 wt.%, sintered 1100°C, (E) 80 wt.% green compact, and (F) 80 wt.% sintered 1100°C.

above the glass transition temperature of the polymer. A limited concentration range specific to particle type was identified in which viscous deformation is possible. Above a critical particle volume fraction the viscosity of the particle–polymer material increases and the time scale of viscous deformation becomes very long. On the other hand, it is not possible to sinter the particles together below a minimum concentration of particles. The method reported in this work provides the possibility to fabricate thin hollow fibers by various materials, as is demonstrated in Fig. 15.2. For small particles of Al, silicon carbide (SiC), and yttria stabilized zirconia (YSZ), nevertheless, the concentration range allowing viscous deformation is narrower as compared to that of larger particles.

In Michielsen et al.'s work, SS hollow fiber membranes were prepared using phase-inversion robotic fiber deposition (or robocasting, printing, writing) [26]. The flowchart for

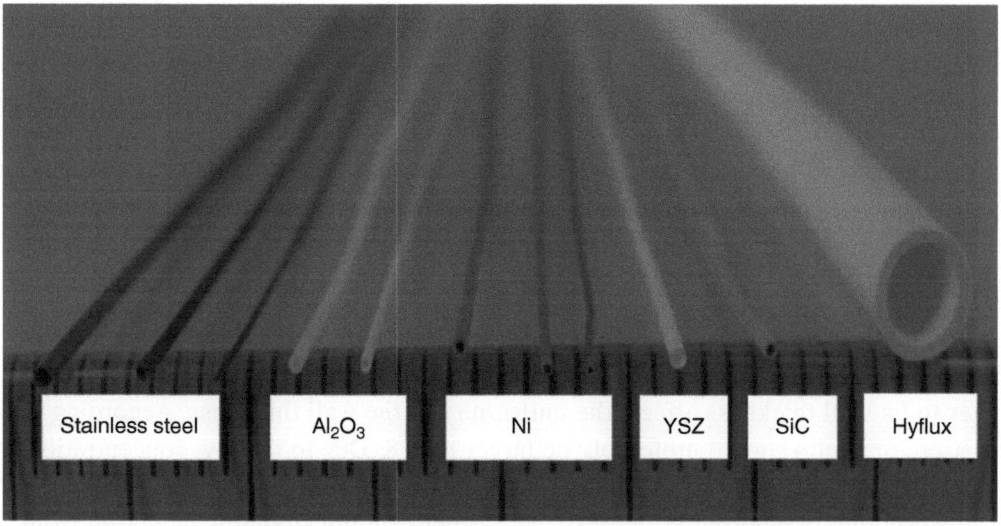

Figure 15.2: From left to right three stainless steel.
Two Al_2O_3, three Ni hollow fibers, a YSZ, a SiC fiber, and an Al_2O_3 capillary (M20, Hyflux).

the preparation procedure is shown in Fig. 15.3. The hollow fibers extruded through the nozzles were deposited on a platform of which the *xyz* direction was controlled by a Hermle UWF 600H CNC-machine. In this manner, the hollow fibers can be deposited in either simple linear or continuous patterns. The hollow fibers extruded through the nozzles were deposited on a platform of which the *xyz* direction was controlled by a Hermle UWF 600H CNC-machine. In this manner, the hollow fibers can be deposited in either simple linear or continuous patterns.

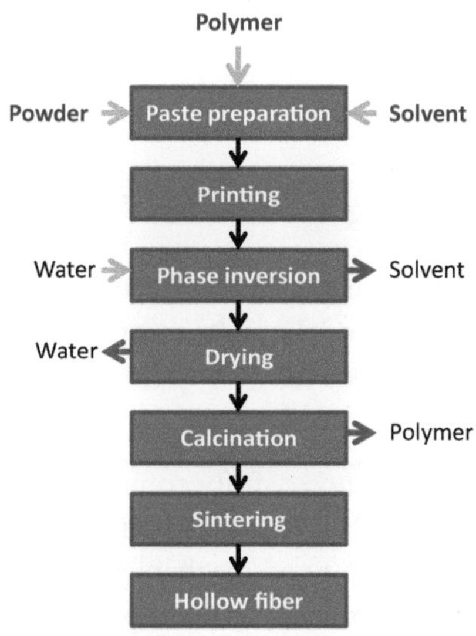

Figure 15.3: Process diagram for producing inorganic fibers using phase-inversion robotic fiber deposition induced by solvent exchange [26].

For applying the robocasting, the paste to form the fiber precursor has to fulfill a certain set of requirements: (1) extrudable under moderate pressure and at ambient temperature or in other words, shear thinning behavior; (2) low viscosity at high shear rates at which it is exposed during the extrusion; (3) high viscosity upon exiting the nozzle to retain the shape mold by the nozzle; (4) flexible enough to make bends for obtaining fiber patterns while at the same time maintaining a constant fiber diameter and wall thickness. Therefore, the resulting green SS hollow fibers are stiff and the obtained patterns can be easily relocated without being deformed.

Therefore, the resulting green SS hollow fibers are stiff and the obtained green patterns can be easily transferred without any deformation. It was found that the ratio of the diameter to its wall thickness affects the uniformity of the wall thickness. According to the observation, this ratio should preferably be larger than 8. Due to the low concentration of organic additives (2 wt.%) only a limited shrinkage occurs after sintering. The resultant SS fibers have sponge-like porous structure. The 3-point bending strengths increased from 50 to 330 MPa and N_2 permeances larger than 1.7×10^{-5} mol m^{-2} Pa^{-1} s^{-1} were found at all sintering temperatures.

Phase-inversion method has also been investigated in several other researches [27–29]. Porous 316L SS hollow fibers were successfully prepared via sintering the precursors fabricated by dry–wet spinning [29]. In Bai's work [27], the microstructure of the obtained fibers could be tuned by adjusting polyacrylnitrile (PAN) concentration. With increasing PAN concentration in the casting solution for spinning, the viscosity was increased dramatically, resulting in more compact structures in sintered SS hollow fiber, while at the same time high pure water permeability ($>3 \times 10^5$ L $m^{-2} \cdot h^{-1} \cdot Pa^{-1}$). The results showed that casting solution with 8 wt.% PAN (~10,000 mPa·s) was most suitable for forming fibers with good morphologies and microstructures. A more dense structure of SS hollow fiber would be obtained by adding additive polyvinylpyrrolidone (PVP).

15.2.2 Composite Membrane Using Stainless Steel Substrate

For fabricating composite membrane using SS base, usually a thin layer of ultrafine titania (TiO_2), alumina (Al_2O_3), or zirconia (ZrO_2) will be coated on the SS substrate followed by sintering. The smaller the particle of metal oxide is, the smaller the pore size of coating layer will be.

The metal membranes that have been successfully commercialized by German GKN Company uses porous SS as support [21]. A dense layer of TiO_2 is formed on the inner surface form composite membrane with good mechanical strength, smooth surface, and antifouling property. The TiO_2 layer dominating the separation is characterized of small pore size and thin thickness. The pore size of TiO_2 layer is around 0.1 μm, determining the filtration accuracy of the composite membrane. The porous 316 SS structure is the substrate that is responsible for the anticompression capacity of the membrane.

As aforementioned, sol–gel method is one of the conventional approaches for preparing composite metal membranes with SS porous substrate. Li et al. applied sol–gel method in coating silica (SiO_2), TiO_2, and SiO_2–TiO_2 on SS substrate [21]. One approach of sol–gel method is carried out under anhydrous condition [2]. Nitric acid (HNO_3) solution with appropriate volume is added slowly into titanium isopropylate and iso-propanol (IPA) solution; incomplete hydrolysis took place to form stable colloidal suspension. The solution is then injected into the lumen or shell channel of the membrane module for direct contact with the either surface for a specific duration. Thereafter, N_2 gas is flushed through the membrane module to facilitate thorough hydrolysis leading to sol–gel conversion. Finally, desiccation and sintering under inert gas protection are employed to obtain the composite membrane suitable for ultrafiltration (UF).

In Xiao's work [30], SS porous membranes were used as support to prepare composite metal membrane with layered double hydroxide (LDH) (i.e. hydrotalcite) as selective layer material by in situ synthesis. Nickel nitrate ($NiNO_3$) and urea of required amount and ratio were dissolved in water to form a homogeneous solution, which was stored in polytetrafluoroethylene (PTFE) reactor. The SS plate was immersed into the solution and the whole reactor was then placed in the oven set at 80°C for 24 h. The SS plate was taken out and rinsed with deionized (DI) water after the reactor cooled down to room temperature. A thin layer of NiFe–LDH composite material was deposited on the SS surface. The porosity of the composite membrane reduced by hydrogen (H_2) had a porosity of 45.79%, water flux of 4820 L m^{-2} min^{-1} MPa^{-1} and filtration accuracy of 0.324 μm.

15.3 Membranes Based on Metal Alloys

Generally, there are two types of metal alloys, namely, solid solution and intermetallic. If the crystal structure of the alloy deviates from that of any of the component inside the alloy, this phase is called intermetalllic. An intermetallic, also called an intermetallic compound, intermetallic alloy, ordered intermetallic alloy, and a long-range-ordered alloy, is a solid-state compound exhibiting metallic bonding, defined stoichiometry, and ordered crystal structure. In some situations, intermetallic compounds are often simply referred to as alloys.

Porous intermetallics have attracted considerable attention as potential functional materials for ion exchange, catalysis, energy storage, and molecular separation at high temperature or in corrosion environment, for their unique properties, such as low density, good thermal conductivity, good mechanical strength at elevated temperature, and environmental corrosion resistance [31,32].

15.3.1 Membrane Based on Titanium-Aluminum Alloys

Ti-aluminum (Al) intermetallic has a series of superior chemical [33–35], physical, and mechanical properties [36,37]. In order to capitalize on the advantages of this new type of material for separation, the concept of Ti-Al porous material has been proposed. Several methods have been developed to develop Ti–Al material, such as powder metallurgy [38] for porous Ti–Al alloy, and continuous zone melting technique [39] for porous Ni_3Al alloy with oriented pores. The other methods include unidirectional solidification [40] and high temperature synthesis with incorporating blowing agents [41,42].

However, these fabrication methods can only prepare pore size larger than 10 mm, which makes the membrane inappropriate for application in liquid–solid separation MF field which needs requiring the pore size in the range of 0.1–10 mm [35]. In addition, in effect, the metallic MF membranes fabricated via the above manner cannot be applied in practice, owing to the complication in the synthesis [36].

A new method involving differential diffusion of elementary powder mixture, reactive synthesis, and sintering procedures was put forward [43]. The schematic for this reactive synthesis of Ti-Al intermetallic is demonstrated in Fig. 15.4. During the synthesis, the volume expansion percentage of the press compact can be as high as 30%. The mechanism for the phenomenon is given later [44,45]. During the low temperature thermal treatment, the differential diffusion of Al causes Kirkendall effect, leading to the formation of many Kirkendall pores; in the further heating the residual Al melts breaks the $TiAl_3$ phase to form honeycomb structure due to the capillary force and surface tension; formation of $TiAl_3$ leads to the shift of the centric position of Ti particles; the large quantity of gas absorbed on the particles' surface and inside the press compact was released.

The most prominent advantage of reactive synthesis is its maneuverability; in other words, the pore structure and the membrane performance can be adjusted by molding technology, molding pressure, sintering approach, and operation parameters of sintering process. Associated with this manufacturing technology are the low cost, simple process, fine and homogeneous material structure. Belonging to near-net shape technology, the new approach involving differential diffusion, metal reaction, and sintering is very promising.

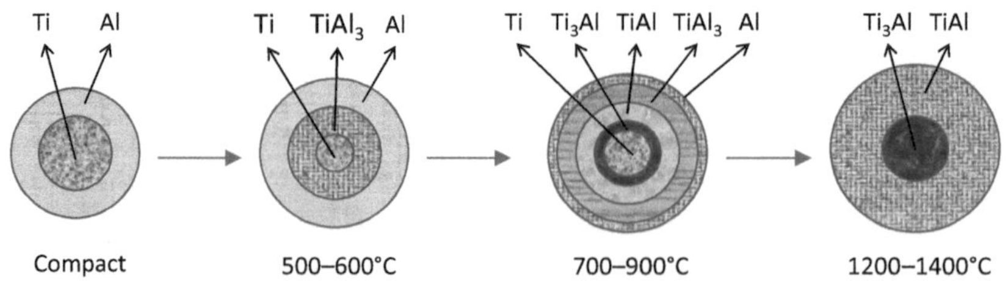

Figure 15.4: Schematic of reactive synthesis procedure of Ti/Al elemental compacts [43].

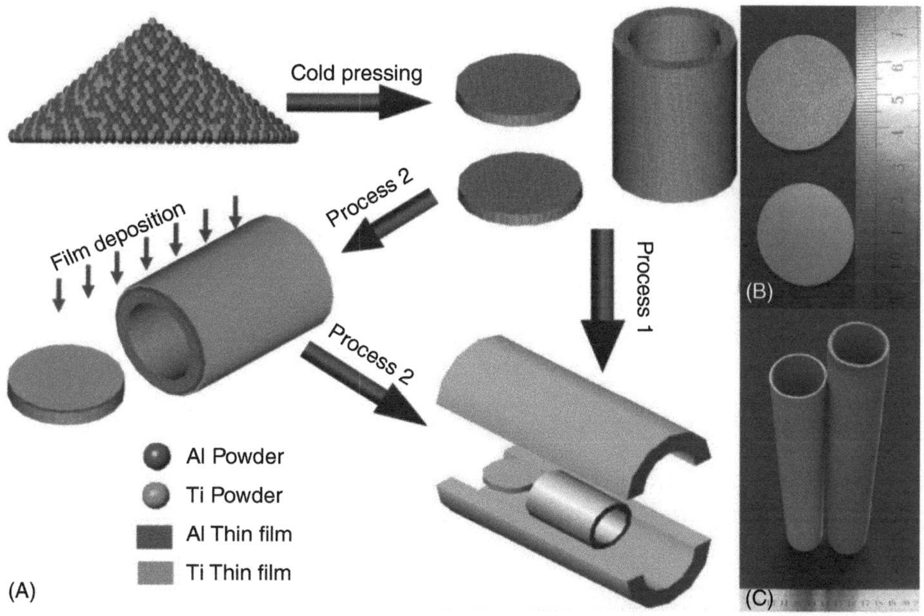

Figure 15.5: Production of porous Ti-Al alloys and characterization of porous Ti-Al structures. (A) Schematic illustration of the production of porous Ti-Al alloys. (B, C) Comparison of disc/cylinder compacts before and after the sintering process, respectively [38,46].

Two procedures for fabricating porous Ti–Al alloys via the Kirkendall effect are illustrated in Fig. 15.5A [38,46]. In the first procedure, commercial Ti and Al elemental powders were mixed and cold pressed into both disc and tubular compacts, followed by the solid sintering process (in which the Kirkendall effect occurs). Considerable expansion of these compacts has been observed and their examples are shown in Fig. 15.5B–C. Through the precise control of the sintering step and the prevention of the formation of the liquid phase, the original shapes of the compacts has been preserved (as shown in Fig. 15.5B and C) even though a volume expansion occurred. Under an identical sintering condition (initially sintered at 600°C for 60 min and then sintered at 1300°C for 30 min), the porous materials with different Ti–Al alloys (ie, α_2-Ti_3Al, c-TiAl, and $TiAl_3$) could be fabricated. These porous Ti–Al alloy membranes were used in an industrial filtration for removing suspended substances from liquid titanium tetrachloride ($TiCl_4$).

The SEM of the alloy with the composition of Ti/34 wt.% Al is shown in Fig. 15.6A, in which holes with different sizes can be clearly seen [46]. As shown in Fig. 15.6B, the crystal structures before the sintering were determined to be elemental Al (a = 0.4050 nm, Joint Committee on Powder Diffraction Standards (JCPDS) 65-2869) and Ti (a = 0.2951 nm and c = 0.4683 nm, JCPDS 44-1294); while the crystal structure after the sintering was pure Y-TiAl phase (a = 0.4001 nm and c = 0.4071 nm, JCPDS 65-0428). These results suggest

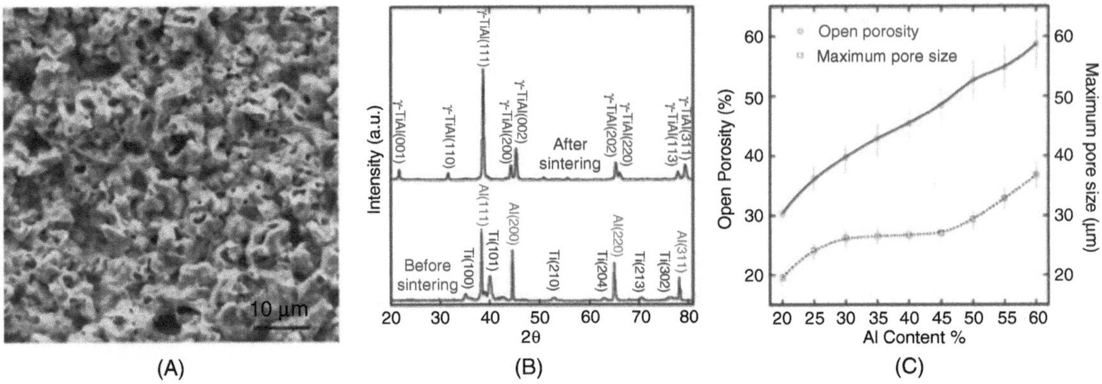

Figure 15.6: (A) A SEM image of a Ti-Al porous alloy with the nominal compositions of Ti-34 wt.% Al. (B) XRD patterns of disc compacts before and after the sintering process. (C) Plots of the open porosity and maximum pore size as a function of the Al content [38,53].

that a complete phase transformation has taken place during the sintering Fig. 15.6C shows the results measured from a series of samples with varying Al contents; the open porosity and the maximum pore-size increase with an increase of Al content. It is suggested that the pores formed during the sintering process are not caused by the density fluctuation of materials during the sintering process, based on the following facts: no weight loss during the sintering process, the similar densities (3.7–3.8 g cm^{-3}) of both the mixtures of Ti-Al elemental powders (Ti–34 wt.%Al) before the sintering process and the sintered Ti–Al alloy, and the porosity of the compact before the sintering process being less than 10%.

Niobium (Nb)-containing Ti–Al intermetallics are potentially high temperature structural materials [47]. Gradient porous membranes of high Nb-containing Ti–Al were synthesized by elemental powder method (EPM) [48–49]. The effects of Nb addition manner (Nb elemental powder and Nb–Al alloying powder) and content of Nb on pore structures and tensile properties of Ti–48Al alloys were discussed [50]. Porous structure of Ti–48Al–2Nb alloy is quite similar to that of Ti–48Al alloy. When Nb content is raised to around 4%, the surface of the solid structure shows many particles and the overall structure becomes looser. Addition of Nb causes the weakening of tensile strength. High Nb-containing Ti–Al porous and gradient porous alloys were synthesized by EPM [49]. For Ti–48Al–6Nb porous alloys, the pores are interconnected and the skeletons consist of typical γ-TiAl/α_2-Ti_3Al lamellar microstructure. Obvious decrease in size of skeletons and pores was observed when the particle size of Ti, Al, and Nb elemental powders changes from above 75 μm to 25–48 μm, 48–75 μm to 1–3 μm and 48–75 μm to 25–48 μm, respectively. Further reduction in powder size to nanoscale can generate nanostructured TiAl-based porous alloys. The gas permeability improves when the porosity and the pore size increase μA Ti–48Al–6Nb porous MF composite membrane was developed by cold gas spraying (CGS) coating followed by reactive sintering [50]. Pores in the range of 0.5–3 μm formed following the mechanism of diffusion reactions and skeletons growth during the sintering. The bending strength of the composite membrane with a thickness of 604.4 μm was 89.33 MPa and gas permeability 5.39×10^{-13} m^2.

The properties of Ti–Al intermetallics in terms of resistance toward oxidation at high temperature [34,46,51,52] and corrosion from molten salt [53] were reported.

Wu et al. investigated the corrosion resistance of Ti-Al alloy in hot (90°C) hydrochloric acid (HCl) with pH value being 2 and 3 intermetallic with 35% Al was produced by powder metallurgy [46]. Corrosion resistance of Ti, Ni, and SS was also reported. It was found that TiAl alloy exhibited the best performance; this was due to the special bonding characteristics between Ti and Al, and the strong deactivation ability of Ti and Al elements. The Ti–Al alloy membranes are suitable for MF applications in extreme environments.

In addition to be used directly as selective material, Ti–Al intermetallics have also great potential in forming substrate for composite membrane. MF membrane consisting of a Ni layer on a porous Ti-AI intermetallic compound substrate was prepared using sedimentation method and dip-coating in Wang's work [54].

As for preparing Ni membrane by sedimentation, the membrane became thicker and the pore became smaller with increasing Ni content. For dip-coating method, the appropriate coating time was 60 s. It was found that the optimized sintering temperature was 500°C. The membrane prepared by dip coating has an average pore size of 0.83 μm and a thickness of about 30 μm. The pure water flux of the membrane was about 6782 L/m^2-h and stable filtration flux for $CaCO_3$ suspension was about 380 L/m^2-h. The testing regarding the alkalescence resistance showed that the pore size and pure water flux were maintained at the original level after 41 days.

Ti–Al alloy supported Pd membranes were also fabricated in Wang's work [54]. The preparation method was electroless plating. Ni powder coating and abrading of Ti–Al substrate led to the formation of defect-free Pd membrane. Reduction of nickel oxide (NiO) will cause the failure of the Pd membrane and preparation with reduced support may solve the problem.

A TiO_2 membrane supported on a planar porous Ti–Al alloy was prepared by combination of electrophoretic deposition (Fig. 15.7) and dip coating [55]. The TiO_2 powders were dispersed into pure water with different amounts of polyethyleneimine (PEI) as the dispersant. The isoelectric (IE) point of TiO_2 shifted to pH 8.0 with 0.64 dwb% (on a dry weight of TiO_2 basis, dwb) PEI addition, while 3.0 dwb% PEI gave rise to a more noticeable shift to pH 10.3. The change in IE point is attributable to the adsorption of the positively charged PEI on the surface of the TiO_2 particles. The amine groups in PEI molecules will protonize and the resultant polyions will expand due to mutual charge repulsion. This is the mechanism for TiO_2 surface charge. The conductivities of the suspensions increased with the increasing amount of PEI additions.

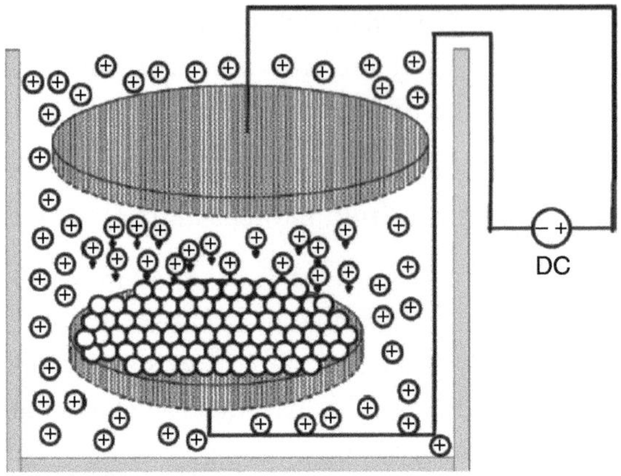

Figure 15.7: Schematic of the electrophoretic deposition (EPD) cell for the fabrication of the supported TiO_2/Ti–Al composite membrane [55].

The viscosity of suspension remained almost a constant within 24 h, and no appreciable phase separation occurred after 24 h sedimentation by visual examination. These observations indicated that the PEI addition improved the dispersion of the suspension.

In the deposition process, the membrane was thicker with longer deposition time and lower suspension viscosity. The high quality TiO_2/Ti–Al composite membrane was obtained by further dip-coating modification. The coated membrane was dried at room temperature, 70 and 110°C for 12 h, respectively, then sintered in an argon (Ar) atmosphere at 1050°C for 3 h in an electric furnace with a heating and cooling rate of 1°C min^{-1}. The surface of the membrane was defect-free. Rutile TiO_2 remained as a main phase in the membrane bulk, as suggested by XRD characterization. A new phase of titanium oxides with the form of Ti_xO_y was also revealed. The TiO_2/Ti–Al composite membrane produced in this research had an average pore size of 0.28 μm, a thickness of 40 μm or so and a pure water flux of 3037 L m^{-2} h^{-1} bar^{-1}.

15.3.2 Membrane Based on Nickel–Aluminum Alloys

Most of the Ni–Al alloys have outstanding engineering properties, such as lightweight, high melting point, excellent acid/alkali corrosion resistance, good oxidation resistance at elevated temperatures, as well as good thermal conductivity [56,57]. The major barrier for applying Ni–Al alloys in the structural applications is its low ductility at the ambient temperature and the low strength and creep resistance at elevated temperatures [58].

Porous Ni–Al intermetallics have been produced from Ni and Al powder compacts by combustion synthesis [59]. The porous NiAl intermetallic compounds have excellent oxidation and corrosion resistances owing to the formation of α-Al_2O_3 around their external and internal surfaces at high temperatures [60,61]. The porous Ni–Al alloys also presented good corrosion resistance to the acid/alkaline solution [62].

Porous NiAl alloy membranes with different ratios of Ni/Al were fabricated through reactive synthesis using Ni and Al elemental powders [63,64]. It had been found that the maximum pore size, volume expansion, and permeability of porous Ni–Al alloys increased with increasing the Al content. This observation indicated that the nature of the pores can be manipulated via adjusting the Al content. The porous Ni–Al alloys can have three crystalline phases (ie, Ni_3Al, NiAl, and Ni_2Al_3), in good agreement with the Ni–Al phase diagram. The bending strength and tensile strength of porous Ni–Al alloys decreased with increasing the Al content and were in the range of 5–80 MPa.

15.3.3 Membrane Based on Iron–Aluminum Alloys

Porous iron (Fe)-40 wt.% Al alloy membranes were fabricated by reactive synthesis of Fe and Al elemental mixture [65]. Programmed heating process (initially heated to 550°C at a heating rate of 10°C min^{-1}, and kept at 550°C for 1 h; further heated to 640°C at a heating rate of 1°C min^{-1}, and kept at 640°C for 1 h; heated to 1050°C with a heating rate of 5°C min^{-1} and held at this temperature for 1 h) has been adopted . Elemental powders of Fe (99.8% purity carbonyl powders with the average particle diameter of 3~5 μm) and Al (99.5% purity gas atomized powders with the average particle diameter of 3~5 μm) were blended with an appropriate ratio (40 wt.% Al) and homogenized. The powders without binder and lubricant were cold pressed uniaxially to form a flat compact under a pressure of 250 MPa. Fe and Al mixed powders exhibited good processibility. The compacts were sintered according the program mentioned earlier in a vacuum furnace with a pressure of 10^{-3} Pa. The discs were held at 550°C and 610°C for 1 h, respectively, to generate sufficient pores through the Kirkendall effect, to eliminate cracks and to preserve the original shapes.

The porous Fe–Al alloy exhibits much better oxidation resistance than other porous materials, such as Ti, Ni, and 316L SS. The oxidation of the porous Fe–40 wt.% Al alloy fitted a power law and no spallation was observed after cyclic oxidation for 202 h in air. This is because the Al in Fe–Al is oxidized to generate a surface layer of almost pure Al_2O_3, upon exposure to oxygen-bearing gases, to prevent further oxidation [66]. The special long-range ordered structure of Fe–Al also enhanced the resistance toward corrosion. In another work, it was confirmed that the Fe–Al intermetallic membrane has excellent sulfur resistance at higher temperature Fig. 15.8 [67].

A new graded-porosity FeAl alloy was fabricated by Shen et al. [68]. FeAl alloy with large connecting open pores and permeability were used as porous supports. The support of FeAl porous alloy was formed by powder metallurgy techniques. The novelty of the work is associated with coating which was obtained by spraying slurries consisting of mixtures

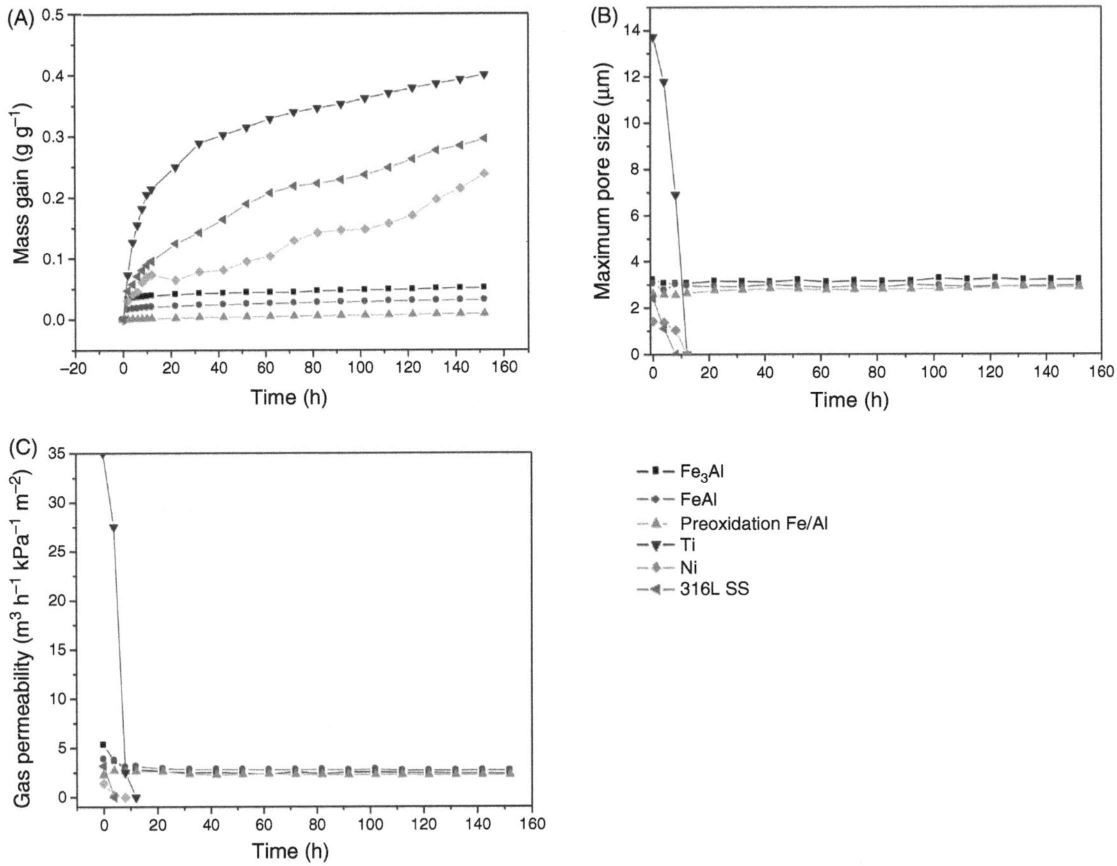

Figure 15.8: Kinetics curves of anti-vulcanization for Fe–Al intermetallics as a function of exposure time under 800°C.
(A) Mass gain; (B) maximum pore size evolution; and (C) gas permeation variation. [67]

of Fe powder and Al powder with 3–5 μm diameter onto porous FeAl support and then sintered at 1100°C. The SEM characterizations as shown in Fig. 15.9 suggest that better quality of the porous FeAl alloys with graded pores can be achieved by coating the mixed powder spray on the sintered FeAl supporting discs.

Differential flow rate distributions for the coating layer with a thickness around 80 μm are depicted in Fig. 15.8. The mean pore size of coating is 2.5 μm, maximum pore size 6 μm, and minimum pore 1.7 μm. Increase in the coating layer thickness (up to 200 μm) correlates with reduction of maximum pore progressively from 23.6 to 5.9 μm; correspondingly the permeability decreased from 184.2 to 76.2 m^3 m^{-2} kPa^{-1} h^{-1}, respectively, for a identical sintering temperature of 1100°C. The composite membranes have potential application for filtration in severe environments.

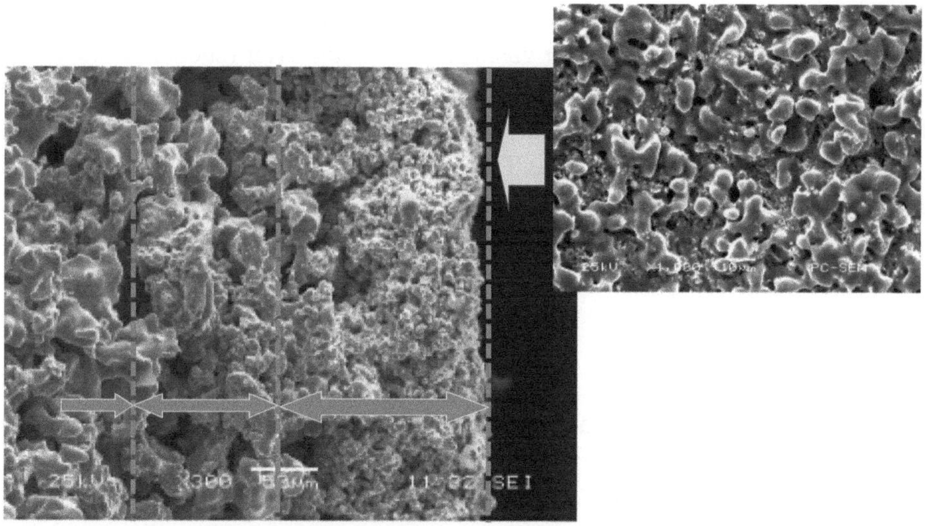

Figure 15.9: SEM micrographs of cross-section and surface of a two-layer FeAl porous alloy with a coating layer of 120 μm [68].

15.4 Industrialization

So far there are totally more than 400 types of MF membranes that have been commercialized worldwide [69,70]. Among these, most are organic membranes and metal membranes that are limited to Ni, Pd, Ag, and stainless membranes. The number of companies producing inorganic filtration membranes is 144 with 94 of them for ceramic membranes and 36 for metal membranes. These companies are concentrated in United States, Europe, and Japan, including Graver, Pall, GKN Ceramem, Osmonics, Hydrarautils, Motton, etc. In 1999, the sale of metal membranes was US$25 million with annual growth rate reaching 12%.

The SS membrane produced by GKN uses porous SS as support with a denser layer of TiO_2 sintered on its lumen surface. The SS MF membrane developed by Bekaert of Belgium exhibits a filtration accuracy of submicron size. The gas flux of the membrane is higher than 3×10^{-3}L min cm^{-2}, which is 20 times as high as that of the porous ceramic filtration material with the same filtration accuracy. A type of SS membrane of sintered power with filtration accuracy being 0.003 μm was invented by MOTT of United States. Nichadia Company developed a high-accuracy SS porous membrane with filtration rate approaching 0.2 μm.

Chengdu Intermet Technology Co. Ltd., founded in August 2007, is a China-based high-tech enterprise devoted to R&D of Al intermetallic compound asymmetric membrane filter materials and their application. The product series of the company mainly include iron Al, Ti–Al, and Ni–Al intermetallic compound membrane filter and filter systems. The membrane products have been applied in purification of high-temperature sulfurous gases, purification and recovery of high-temperature burner gas, recycle of converter (such as Blast Furnace Top Gas Recovery Turbine Unit (abbr. TRT)) gas, coal chemical industry (such as brown coal hydro gasification, coal to liquid, coal to gas, Integrated Gasification Combined Cycle (IGCC)), chloridizing metallurgy (such as $TiCl_4$), hydrometallurgy (such as $CoSO_4$ and $ZnSO_4$ purification), calcium carbide (Ca-C) smelting, new energy economy (oil shale, coal gangue, activated carbon, biomass gas production, etc.), gas–solid and liquid–solid separation and purification of ecoprotecting field (such as burner gas recovery for acid making), etc.

15.5 Conclusions

Due to their excellent resistance toward corrosion and temperature, porous metal membrane has been applied in some metallurgical processes with extreme in-service condition and exhibited reliable performance. The future work is supposed to focus on more mature fabrication technology with cheap materials, improvement of membrane integrity, uniformity, and resistance toward cracking, and decrease in thickness. In addition, smaller and more uniform pore size leading to higher filtration precision is also an important direction.

Glossary and Symbols

Ag	Silver
Al	Aluminum
Al_2O_3	Alumina
Ar	Argon
Au	Gold
CaC	Calcium carbide
CGS	Cold gas spraying
CO	Carbon monoxide
Cr	Chromium
Cu	Copper
EPM	Elemental powder method
Fe	Iron
H_2	Hydrogen
H_2S	Hydrogen sulfide
HCl	Hydrochloric acid
HNO_3	Nitric acid
IE	Isoelectric
IGCC	Integrated Gasification Combined Cycle
IPA	Iso-propanol
LDH	Layered double hydroxide
MF	Microfiltration
Mg	Magnesium
MIM	Metal injection molding
Mo	Molybdenum
Nb	Niobium
NF	Nanofiltration

Ni	Nickel
$NiNO_3$	Nickel nitrate
NiO	Nickel oxide
PAN	Polyacrylonitrile
Pd	Palladium
PEI	Polyethyleneimine
Pt	Platinum
PTFE	Polytetrafluoroethylene
PVP	Polyvinylpyrrolidone
SiC	Silicon carbide
SiO_2	Silica
SS	Stainless steel
Ta	Tantalum
Ti	Titanium
$TiCl_4$	Titanium tetrachloride
TiO_2	Titania
TRT	Blast Furnace Top Gas Recovery Turbine Unit
U	Uranium
UF	Ultrafiltration
V	Vanadium
W	Tungsten
YSZ	Yttria-stabilized zirconia
ZrO	Zirconia

References

[1] H.Y. Gao, Y.S. Lin, Y.D. Li, New advancement of the development and application of metallic membranes, Membr. Sci. Technol. 27 (2007) 76–79.

[2] G.T. Huang, X.Q. Zuo, Y.L. Sun, Y. Zhou, J.S. Lu, H. Wang, Progress in porous metal filtration materials, Mater. Rev. 24 (16) (2010) 448–456.

[3] Y. Huang, X. Li, Y.Q. Fan, N.P. Xu, Principles of hydrogen permeating pd composite membranes, Chem. Prog. 18 (2006) 230–238.

[4] W.H. Lin, Y.C. Liu, H.F. Chang, Hydrogen production from oxidative steam reforming of ethanol in a palladium silver alloy composite membrane reactor, J. Chin. Inst. Chem. Eng. 39 (2008) 435–440.

[5] J. Okazaki, T. Ikeda, D.A.P. Tanaka, K. Sato, T.M. Suzuki, F. Mizukami, An investigation of thermal stability of thin palladium—silver alloy membranes for high temperature hydrogen separation, J. Membr. Sci. 366 (2011) 212–219.

[6] T.A. Peters, M. Stange, R. Bredesen, On the high pressure performance of thin supported Pd-23%Ag membranes: Evidence of ultrahigh hydrogen flux after air treatment, J. Membr. Sci. 378 (2011) 28–34.

[7] X.L. Zhang, W.P. Wang, J. Liu, S.S. Sheng, G.X. Xiong, W.S. Yang, Hydrogen transport through thin palladium—copper alloy composite membranes at low temperatures, Thin Solid Films 516 (2008) 1849–1856.

[8] O. Iyoha, R. Enick, R. Killmeyer, B. Morreale, The influence of hydrogen sulfide-to-hydrogen partial pressure ratio on the sulfidization of Pd and 70mol% Pd-Cu membranes, J. Membr. Sci. 305 (2007) 77–92.

[9] W.H. Chen, P.C. Hsu, Hydrogen permeation measurements of Pd and Pd-Cu membranes using dynamic pressure difference method, Int. J. Hydrogen Energy 6 (2011) 9355–9366.

[10] T. Shobha, C.L. Aravinda, P. Bera, L. Gomathi Devi, S.M. Mayanna, Characterization of Ni-Pd alloy as anode for methanol oxidative fuel cell, Mater. Chem. Phys. 80 (3) (2003) 656–661.

[11] S.K. Ryi, J.S. Park, S.H. Kim, S.-C. Hong, D.-W. Kim, The study of a new porous nickel support for palladium-based alloy membrane, Desalination 200 (2006) 216–218.

[12] A. Criscuoli, A. Basile, E. Drioli, O Loiacono, An economic feasibility study for water gas shift membrane reactor, J. Membr. Sci. 181 (2001) 21–27.
[13] K.J. Bryden, J.Y. Ying, Nanostructured palladium-iron membranes for hydrogen separation and membrane hydrogenation reactions, J. Membr. Sci. 203 (2002) 29–42.
[14] M.D. Dolan, Non-Pd BCC alloy membranes for industrial hydrogen separation, J. Membr. Sci. 362 (2010) 12–28.
[15] S. Hara, K. Sakaki, N. Itoh, H.-M. Kimura, K. Asami, A. Inoue, An amorphous alloy membrane without noble metals for gaseous hydrogen separation, J. Membr. Sci. 164 (2000) 289–294.
[16] Q.B. Wang, H.P. Tang, Z.P. Xi, Research progress of porous metal membranes, Mater. Rev. 18 (2004) 26–28.
[17] R.D. Hu, Z.D. Zhou, W.Y. Zhang, L.X. Liu, W.Z. Zhuang, Y.C. Wang, Investigation and application of microporous metal membrane and membrane separation technology, Powder Metall. Ind. 7 (1997) 16–21.
[18] Y.-B. Kim, K. Lee, J.-H. Chung, Optimum cleaning-in-place conditions for stainless steel microfiltration membrane fouled by terephthalic acid solids, J. Membr. Sci. 209 (2002) 233–240.
[19] J.J. Porter, R.S. Porter, Filtration studies of selected anionic dyes using asymmetric titanium dioxide membranes on porous stainless-steel tubes, J. Membr. Sci. 101 (1995) 67–81.
[20] W. Lai, G. Li, X.Q. Liu, G.Y. Meng, Study on the preparation of fine porous membranes on macroporous alumina supports with wide pore distribution, Membr. Sci. Technol. 4 (2001) 29–33.
[21] Z.H. Li, Z.P. Yang, N.G. Qiu, G.M. Yang, A sol-gel-derived α-Al_2O_3 crystal interlayer modified 316 porous stainless steel to support TiO_2, SiO_2, and TiO_2-SiO_2 hybrid membranes, J. Mater. Sci. 46 (2011) 3127–3235.
[22] Y. Huang, J. Yu, X.H. Zhu, The fabrication technology of a stainless steel membrane, Chinese Patent, CN1013691330A, 2014.
[23] X.Y. Ye, Controllable preparation and performance research of asymmetric porous metal material, Master Thesis, East China University of Science and Technology, 2013.
[24] M.W.J. Luiten-Olieman, L. Winnubst, A. Nijmeijer, M. Wessling, N.E. Benes, Porous stainless steel hollow fiber membranes via dry–wet spinning, J. Membr. Sci. 370 (2011) 124–130.
[25] M.W.J. Luiten-Olieman, M.J.T. Raaijmakers, L. Winnubst, Ton C. Bor, M. Wessling, A. Nijmeijer, N.E. Benes, Towards a generic method for inorganic porous hollow fibers preparation with shrinkage-controlled small radial dimensions, applied to Al_2O_3, Ni, SiC, stainless steel, and YSZ, J. Membr. Sci. 407-408 (2012) 155–163.
[26] B. Michielsen, H. Chen, M. Jacobs, V. Middelkoop, S. Mullens, I. Thijs, A. Buekenhoudt, F. Snijkers, Preparation of porous stainless steel hollow fibers by robotic fiber deposition, J. Membr. Sci. 437 (2013) 17–24.
[27] Y. Bai, Fabrication and characterization of stainless steel hollow fiber membrane, Master's Thesis, East China University of Science and Technology, 2012.
[28] Z. L. Xu, X.H. Ma, Y. Bai, The fabrication technology of a hollow fiber stainless steel 316L membrane, Chinese Patent, Application No. 201310017920.X, 2013.
[29] X.H. Ma, Y. Bai, Y. Cao, Z.L. Xu, Effect of polymer and additive on the structure and property of porous stainless steel hollow fiber, Korean J. Chem. Eng. 31 (2014) 1438–1443.
[30] T. Xiao, Preparation of porous metal/LDHs composite membrane using stainless steel as substrate and its properties of membrane filtration, Master Thesis, Peking University of Chemical Engineering, 2014.
[31] W.J. Zhang, S.C. Deevi, G.L. Chen, On the origin of superior high strength of Ti-45Al-10Nb alloys, Intermetallics 10 (2002) 403–406.
[32] G.L. Chen, L.C. Zhang, Deformation mechanism at large strains in a high-Nb-containing TiAl at room temperature, Mater. Sci. Eng. A 329-331 (2002) 163–170.
[33] H. Kamide, H. Kashima, Hot corrosion behaviour of TiAl with saltin artificial sea-water, Corros. Eng. 46 (1997) 83–89.
[34] Z. Tang Z, F. Wang, W. Wu, Hot-corrosion behavior of TiAl-base intermetallics in molten salts, Oxid. Met. 51 (1999) 235–250.
[35] W.J. Wang, J.P. Lin, Y.L. Wang, Y. Zhang, G.L. Chen, Isothermal corrosion TiAl-Nb alloy in liquid zinc, Mater. Sci. Eng. A 452-453 (2007) 194–201.
[36] Y.W. Kim, Advances in fundamental understanding for designing engineering gamma TiAl alloys, Trans. Chin. Inst. Eng. 22 (1999) 13–25.

[37] Z. Zhong, D. Zou, S. Li, Advance in Ti_3Al and TiAl intermetallic materials, Acta Metall. Sin. A 8 (1995) 531–541.
[38] Y.H. He, Y. Jiang, N.P. Xu, J. Zou, B.Y. Huang, C.T. Liu, P.K. Liaw, Fabrication of Ti–Al micro/nanometer-sized porous alloys through the Kirkendall effect, Adv. Mater. 19 (2007) 2102–2106.
[39] H. Nakajima, Fabrication properties and application of porous metals with directional pores, Prog. Mater. Sci. 52 (2007) 1091–1173.
[40] T. Ide, M. Tane, H. Nakajima, Compressive deformation behavior of porous γ-TiAl with directional pores, Mater. Sci. Eng. A 508 (2009) 220–225.
[41] B.Y. Li, L.J. Rong, Y.Y. Li, The influence of addition of TiH_2 in elemental powder sintering porous Ni–Ti alloys, Mater. Sci. Eng. A 281 (2000) 169–175.
[42] Y.P. Zhang, D.S. Li, X.P. Zhang, Gradient porosity and large pore size NiTi shape memory alloys, Scr. Mater. 57 (2007) 1020–1023.
[43] K. Shibue, Production of TiAl(γ)/$Ti_3Al(\alpha_2)$ alloys, Mater. Jap. 35 (1996) 616–622.
[44] D.L. Lin, The latest development of research about high temperature ordered intermetallics, Acad. J. Shanghai Jiao Tong Univ. 32 (1998) 95–108.
[45] T. Noda, Application of cast gamma TiAl for automobiles, Intermetallics 6 (1998) 709–713.
[46] Z.F. Wu, Y.H. He, Y. Jiang, L.M. Tang, B.Y. Huang, N.P. Xu, Corrosion resistance of porous TiAl alloy in hot hydrochloric acid solution, Mater. Sci. Eng. Powder Metall. 12 (2007) 310–315.
[47] W.J. Zhang, F. Appel, Effect of Al content and Nb addition on the strength and fault energy of TiAl alloys, Mater. Sci. Eng. A 649-652 (2002) 329–331.
[48] Y.H. Wang, J.P. Lin, Y.H. He, Y.L. Wang, G.L. Chen, Effect of Nb on pore structure and tensile property of Ti–48Al cellular alloy, J. Alloy. Compd. 456 (2008) 297–303.
[49] F. Yang, L.Q. Zhang, J.P. Lin, Y.F. Liang, Y.H. He, S.L. Shang, Z.-K. Liu, Pore structure and gas permeability of high Nb-containing TiAl porous alloys by elemental powder metallurgy for microfiltration application, Intermetallics 33 (2013) 2–7.
[50] Y. Fan, J.P. Lin, Y.H. He, D. Hao, G.L. Chen, Innovative fabrication of Ti–48Al–6Nb porous coating by cold gas spraying and reactive sintering, Mater. Lett. 76 (2012) 190–193.
[51] F.H. Wang, Z.L. Tang, Oxidation and protection of TiAl intermetallics, Chin. J. Mater. Res. 12 (1998) 337–344.
[52] G.X. Jin, L.J. Qiao, K.W. Gao, et al. Effect of Mn and V on hot corrosion of Ti/Al alloy, Acta Metall. Sin. 40 (2004) 179–184.
[53] J.R. Nicholls, Hot salt corrosion of titanium aluminides, Mater. Corros. 48 (1997) 56–64.
[54] J. Wang, Study on the preparation of Ni membrane and Pd membrane supported on TiAl intermetallic alloy, Master Thesis, Nanjing University of Technology, 2006.
[55] S.Y. Zhou, Y.Q. Fan, Y.H. He, N.P. Xu, Preparation of titania microfiltration membranes supported on porous TiAl alloys, J. Membr. Sci. 325 (2008) 546–552.
[56] T. Czeppe, S. Wierzbinski, Structure and mechanical properties of NiAl and Ni_3Al-based alloys, Int. J. Mech. Sci. 42 (2000) 1499–1518.
[57] H.J. Grabke, Oxidation of NiAl and FeAl, Intermetallics 7 (1999) 1153–1158.
[58] A. Biswas, S.K. Roy, K.R. Gurumurthy, N. Prabhu, S. Banerjee, A study of self-propagating high-temperature synthesis of NiAl in thermal explosion mode, Acta Mater. 50 (2002) 757–773.
[59] A. Paul, M.J.H. van Dal, A.A. Kodentsov, F.J.J. van Loo, The Kirkendall effect in multiphase diffusion, Acta Mater. 52 (2004) 623–630.
[60] H.P. Hsieh, R.R. Bhave, H.L. Fleming, Microporous alumina membranes, J. Membr. Sci. 39 (1988) 221–241.
[61] F. Shojai, T. Mantyla, Monoclinic zirconia microfiltration membranes: preparation and characterization, J. Porous Mater. 8 (2001) 129–142.
[62] D.W. Lee, Y.G. Lee, S.E. Nam, B. Sea, K.H. Lee, Preparation and characterization of SiO_2 composite membrane for purification of hydrogen from methanol steam reforming as an energy carrier system for PEMFC, Sep. Purif. Technol. 32 (2003) 45–50.

[63] H.X. Dong, Y.H. He, Y. Jiang, L. Wu, J. Zou, N.P. Xu, B.Y. Huang, C.T. Liu, Rapid communication-effect of Al content on porous Ni–Al alloys, Mater. Sci. Eng. A 528 (2011) 4849–4855.
[64] T. Sleboda, J. Kane, R.N. Wright, N.S. Stoloff, D.J. Duquette, The effect of thermomechanical processing on the properties of Fe–40 at.%Al alloy, Mater. Sci. Eng. A 368 (2004) 332–336.
[65] P.Z. Shen, M. Song, H.Y. Gao, Y.H. He, J. Zou, N.P. Xu, B.Y. Huang, C.T. Liu, Structural characteristics and high-temperature oxidation behavior of porous Fe–40 at.%Al alloy, J. Mater. Sci. 44 (2009) 4413–4421.
[66] P.F. Tortorelli, K. Natesan, Critical factors affecting the high-temperature corrosion performance of iron aluminides, Mater. Sci. Eng. A 258 (1998) 115–125.
[67] P.Z. Shen, L. Gao, H.Y. Gao, Y.H. He, High temperature sulfidation behavior of porous FeAl intermetallics and its application in industrial SO_2-containing gas cleanup, the Third Conference of Membrane Application in Metallurgical Industry, pp. 130–135.
[68] P.Z. Shen, Y.H. He, H.Y. Gao, J. Zou, N.P. Xu, Y. Jiang, B.Y. Huang, C.T. Liu, Development of a new graded-porosity FeAl alloy by elemental reactive synthesis, Desalination 249 (2009) 29–33.
[69] Y.Z. Guo, Suggest on development of membrane industry in China, New Chem. Mater. 30 (2002) 4–8.
[70] H. Chai, X.M. Zhang, Information resource of membrane on the Internet, Membr. Sci. Technol. 21 (2001) 57–63.

Index

J

K

L

M

N

O

P

R

Kolibri 360 GmbH
Ferdinand-Jühlke-Straße 7
99095 Erfurt, Deutschland
produktsicherheit@kolibri360.de